WORLD HEALTH ORGANIZATION

INTERNATIONAL AGENCY FOR RESEARCH ON CANCER

IARC MONOGRAPHS
ON THE
EVALUATION OF CARCINOGENIC RISKS TO HUMANS

Wood Dust and Formaldehyde

VOLUME 62

This publication represents the views and expert opinions
of an IARC Working Group on the
Evaluation of Carcinogenic Risks to Humans,
which met in Lyon,

11–18 October 1994

1995

IARC MONOGRAPHS

In 1969, the International Agency for Research on Cancer (IARC) initiated a programme on the evaluation of the carcinogenic risk of chemicals to humans involving the production of critically evaluated monographs on individual chemicals. In 1980 and 1986, the programme was expanded to include evaluations of carcinogenic risks associated with exposures to complex mixtures and other agents.

The objective of the programme is to elaborate and publish in the form of monographs critical reviews of data on carcinogenicity for agents to which humans are known to be exposed and on specific exposure situations; to evaluate these data in terms of human risk with the help of international working groups of experts in chemical carcinogenesis and related fields; and to indicate where additional research efforts are needed.

This project is supported by PHS Grant No. 5-UO1 CA33193-13 awarded by the United States National Cancer Institute, Department of Health and Human Services. Additional support has been provided since 1986 by the European Commission.

©International Agency for Research on Cancer 1995

IARC Library Cataloguing in Publication Data

IARC Working Group on the Evaluation of Carcinogenic Risks to
   Humans (1994 : Lyon, France)
   Wood dust and formaldehyde : views and expert opinions of an IARC
   Working Group on the Evaluation of Carcinogenic Risks to Humans
   which met in Lyon, 11–18 October 1994.

(IARC monographs on the evaluation of carcinogenic risks to humans ; 62)

1. Dust – adverse effects – congresses 2. Formaldehyde – adverse effects – congresses
3. Neoplasms – chemically induced – congresses 4. Wood – congresses I. Series

ISBN 92 832 1262 2   (NLM Classification: W1)

ISSN 0250-9555

Publications of the World Health Organization enjoy copyright protection in accordance with the provisions of Protocol 2 of the Universal Copyright Convention.

All rights reserved. Application for rights of reproduction or translation, in part or in toto, should be made to the International Agency for Research on Cancer.

Distributed for the International Agency for Research on Cancer
by the Secretariat of the World Health Organization, Geneva

PRINTED IN THE UNITED KINGDOM

CONTENTS

NOTE TO THE READER 1

LIST OF PARTICIPANTS 3

PREAMBLE 9
   Background 9
   Objective and Scope 9
   Selection of Topics for Monographs 10
   Data for Monographs 11
   The Working Group 11
   Working Procedures 11
   Exposure Data 12
   Studies of Cancer in Humans 14
   Studies of Cancer in Experimental Animals 17
   Other Data Relevant to an Evaluation of Carcinogenicity and Its Mechanisms 20
   Summary of Data Reported 21
   Evaluation 23
   References 27

GENERAL REMARKS 31

THE MONOGRAPHS

Wood dust 35
   1. Exposure data 35
      1.1 Composition of wood 35
          1.1.1 Classification and nomenclature 35
          1.1.2 Anatomical features 38
          1.1.3 Cell wall structure and distribution of components of wood 41
          1.1.4 Chemical components 41
               (a) Macromolecular components 48
               (b) Low-relative-molecular-mass components 48
      1.2 Wood-related industries and occupations 51
          1.2.1 Major woodworking processes 53
               (a) Debarking 53
               (b) Sawing 54
               (c) Sanding 55

		(d) Planing, jointing, moulding and shaping	55
		(e) Turning (lathing)	55
		(f) Boring (drilling), routing and carving	55
		(g) Mortising and tenoning	56
		(h) Veneer cutting	56
		(i) Chipping, flaking, hogging and grinding	56
		(j) Mechanical defibrating	57
	1.2.2	Sawmilling	57
	1.2.3	Manufacture of plywood and other boards	59
		(a) Plywood manufacture	59
		(b) Manufacture of particle-board and related boards	60
		(c) Fibre-board manufacture	60
	1.2.4	Wooden furniture manufacture and cabinet-making	61
	1.2.5	Manufacture of other wood products	63
	1.2.6	Construction, carpentry and other wood-related occupations	63
		(a) The construction industry	63
		(b) Maintenance and repair	64
		(c) Pattern and model making	65
		(d) Wood shop teachers and artists	65
1.3	Analytical methods		65
	1.3.1	Characterization and measurement of wood dust	65
		(a) Type of wood	66
		(b) Airborne dust concentrations	66
		(c) Particle size distribution	67
		(d) Other characteristics of wood dust	68
	1.3.2	Chemical analysis of wood constituents	68
		(a) Extractives	68
		(b) Inorganic compounds in ash	69
		(c) Polysaccharides	69
		(d) Lignin	70
1.4	Exposure to wood dust and other agents in the workplace		70
	1.4.1	General influences on occupational exposure to wood dust	70
	1.4.2	Extent of exposure to wood dust	71
	1.4.3	Exposure during sawmilling	72
	1.4.4	Exposure during the manufacture of plywood and other boards	74
	1.4.5	Exposure during wooden furniture manufacture and cabinet-making	78
	1.4.6	Exposure during the manufacture of other wood products	82
	1.4.7	Exposure in other wood-related occupations	86
	1.4.8	Particle size distribution of wood dust in workroom air	88
1.5	Regulations and guidelines		93

2. Studies of cancer in humans 94
  2.1 Case reports 94
  2.2 Descriptive studies 95
  2.3 Cohort studies.. 101
  2.4 Case–control studies... 102
    2.4.1 Cancer of the nasal cavity and paranasal sinuses 102
      (a) Exposure to wood dust 102
      (b) Occupational group... 111
    2.4.2 Cancers of other parts of the respiratory system 127
      (a) Nasopharyngeal cancer.. 127
      (b) Pharyngeal cancer other than cancer of the nasopharynx..... 129
      (c) Laryngeal cancer.. 132
      (d) Lung cancer 136
    2.4.3 Cancers of the lymphatic and haematopoeitic system......................... 146
      (a) Non-Hodgkin's lymphoma 146
      (b) Hodgkin's disease.. 152
      (c) Multiple myeloma 154
      (d) Leukaemia 156
    2.4.4 Cancers of the digestive tract.. 158
      (a) Exposure to wood dust 158
      (b) Occupational group... 164
3. Studies of cancer in experimental animals 165
  3.1 Inhalation.. 165
    3.1.1 Rat... 165
    3.1.2 Hamster.. 169
  3.2 Intraperitoneal injection... 169
  3.3 Administration with known carcinogens or other modifying factors......... 169
    3.3.1 Rat... 169
    3.3.2 Hamster.. 170
  3.4 Skin application of wood dust extracts... 171
  3.5 Experimental data on wood shavings 173
4. Other data relevant to an evaluation of carcinogenicity and its mechanisms..... 173
  4.1 Deposition and clearance... 173
    4.1.1 Humans.. 173
    4.1.2 Experimental systems 173
  4.2 Toxic effects... 173
    4.2.1 Humans.. 173
      (a) Effects on the nose... 173
      (b) Effects on the lung... 177
      (c) Other effects 183
    4.2.2 Experimental systems 184

    4.3 Genetic and related effects 185
        4.3.1 Humans 185
        4.3.2 Experimental systems 185
5. Summary of data reported and evaluation 191
    5.1 Exposure data 191
    5.2 Human carcinogenicity data 192
    5.3 Animal carcinogenicity data 193
    5.4 Other relevant data 193
    5.5 Evaluation 194
6. References 194

Formaldehyde 217
    1. Exposure data 217
        1.1 Chemical and physical data 217
            1.1.1 Nomenclature 217
            1.1.2 Structural and molecular formulae and relative molecular mass 217
            1.1.3 Chemical and physical properties of the pure substance 217
            1.1.4 Technical products and impurities 218
            1.1.5 Analysis 218
        1.2 Production and use 219
            1.2.1 Production 219
            1.2.2 Use 221
        1.3 Occurrence 223
            1.3.1 Natural occurrence 223
            1.3.2 Occupational exposure 224
                (a) Extent of exposure 224
                (b) Manufacture of formaldehyde, formaldehyde-based resins and
                    other chemical products 224
                (c) Manufacture of wood products and paper 225
                (d) Manufacture of textiles and garments 231
                (e) Manufacture of metal products, mineral wool and other products .. 232
                (f) Mortuaries, hospitals and laboratories 237
                (g) Building, agriculture, forestry and other activities 240
            1.3.3 Ambient air 242
            1.3.4 Residential indoor air 242
            1.3.5 Other exposures 243
        1.4 Regulations and guidelines 243
    2. Studies of cancer in humans 246
        2.1 Case reports 246
        2.2 Descriptive studies 246
        2.3 Cohort studies 246
            2.3.1 Professional groups 247
            2.3.2 Industrial groups 251

2.4	Case–control studies		262
	2.4.1	Cancers of the nasal cavity, paranasal sinuses, nasopharynx, oropharynx and pharynx (unclassified)	262
	2.4.2	Cancers of the lung and larynx	272
	2.4.3	Cancers at other sites	275
2.5	Meta-analyses		282

3. Studies of cancer in experimental animals 282
   3.1 Inhalation 282
       3.1.1 Mouse 282
       3.1.2 Rat 284
       3.1.3 Hamster 289
   3.2 Oral administration 289
   3.3 Skin application 292
   3.4 Subcutaneous injection 292
   3.5 Administration with known carcinogens and other modifying factors 292
       3.5.1 Mouse 292
       3.5.2 Rat 293
       3.5.3 Hamster 293

4. Other data relevant to an evaluation of carcinogenicity and its mechanisms 294
   4.1 Absorption, distribution, metabolism and excretion 294
       4.1.1 Humans 294
       4.1.2 Experimental systems 295
   4.2 Toxic effects 301
       4.2.1 Humans 302
           (a) Acute effects 302
           (b) Chronic effects 304
           (c) Allergy 308
       4.2.2 Experimental systems 308
           (a) Acute effects 309
           (b) Chronic effects 315
           (c) Immunotoxicity 319
   4.3 Reproductive and developmental effects 319
       4.3.1 Humans 319
       4.3.2 Experimental systems 320
   4.4 Genetic and related effects 321
       4.4.1 Humans 321
           (a) DNA–protein cross-links 321
           (b) Mutation and allied effects 321
           (c) Sperm abnormalities 322
           (d) Urinary mutagenicity 323

         4.4.2 Experimental systems 323
                (a) DNA–protein cross-links 323
                (b) Mutation and allied effects 323
                (c) Mutational spectra 330
   5. Summary of data reported and evaluation 332
      5.1 Exposure data 332
      5.2 Human carcinogenicity data 333
      5.3 Animal carcinogenicity data 334
      5.4 Other relevant data 334
      5.5 Evaluation 335
   6. References 335

APPENDIX 1. SUMMARY TABLE OF GENETIC AND RELATED EFFECTS 365

APPENDIX 2. ACTIVITY PROFILE FOR GENETIC AND RELATED EFFECTS 369

CUMULATIVE INDEX TO THE *MONOGRAPHS* SERIES 377

NOTE TO THE READER

The term 'carcinogenic risk' in the *IARC Monographs* series is taken to mean the probability that exposure to an agent will lead to cancer in humans.

Inclusion of an agent in the *Monographs* does not imply that it is a carcinogen, only that the published data have been examined. Equally, the fact that an agent has not yet been evaluated in a monograph does not mean that it is not carcinogenic.

The evaluations of carcinogenic risk are made by international working groups of independent scientists and are qualitative in nature. No recommendation is given for regulation or legislation.

Anyone who is aware of published data that may alter the evaluation of the carcinogenic risk of an agent to humans is encouraged to make this information available to the Unit of Carcinogen Identification and Evaluation, International Agency for Research on Cancer, 150 cours Albert Thomas, 69372 Lyon Cedex 08, France, in order that the agent may be considered for re-evaluation by a future Working Group.

Although every effort is made to prepare the monographs as accurately as possible, mistakes may occur. Readers are requested to communicate any errors to the Unit of Carcinogen Identification and Evaluation, so that corrections can be reported in future volumes.

IARC WORKING GROUP ON THE EVALUATION OF CARCINOGENIC RISKS TO HUMANS: WOOD DUST AND FORMALDEHYDE

Lyon, 11–18 October 1994

LIST OF PARTICIPANTS

Members[1]

A. Blair, Occupational Studies Section, Environmental Epidemiology Branch, National Cancer Institute, Executive Plaza North, Room 418, Bethesda, MD 20892-7364, United States (*Vice-Chairman; representative of the National Cancer Institute*)

M. Casanova, Chemical Industry Institute of Toxicology, PO Box 12137, Research Triangle Park, NC 27709, United States

P. Comba, Istituto Superiore di Sanita, viale Regina Elena 299, 00161 Rome, Italy

P. Demers, Occupational Hygiene Programme, University of British Columbia, 3rd floor, Library Processing Center, 2206 E Mall, Vancouver, BC V6T 1Z3, Canada

B.D. Goldstein, Environmental and Occupational Health Sciences Institute, PO Box 1179, Piscataway, NJ 08855-1179, United States (*Chairman*)

R.C. Grafström, Division of Toxicology, Institute of Environmental Medicine, Karolinska Institute, Box 210, 171 77 Stockholm, Sweden

B. Järvholm, Occupational Medicine, Department of Internal Medicine, Sahlgrenska Hospital, St Sigfridsgatan 85, 412 66 Göteborg, Sweden

D.G. Kaufman, Department of Pathology, University of North Carolina at Chapel Hill, School of Medicine, 515 Brinkhous-Bullitt Building, CB 7525, Chapel Hill, NC 27599-7525, United States

T. Kauppinen, Institute of Occupational Health, Topeliuksenkatu 41 a A, 00250 Helsinki, Finland

A. Leclerc, INSERM U 88, 91 boulevard de l'Hôpital, 75634 Paris Cédex 13, France

K. Norpoth, Institute of Hygiene and Occupational Medicine, University of Essen, Hufelandstrasse 55, 45122 Essen, Germany

---

[1] Unable to attend: V.J. Feron, Division of Toxicology, TNO Nutrition and Food Research Institute, PO Box 360, 3700 AJ Zeist, Netherlands

S.S. Olin, Risk Science Institute, International Life Sciences Institute, 1126 Sixteenth Street NW, Washington DC 20036, United States

J.H. Olsen, Danish Cancer Society, Division of Cancer Epidemiology, Strandboulevarden 49, Box 839, 2100 Copenhagen Ø, Denmark

T. Partanen, Institute of Occupational Health, Topeliuksenkatu 41 a A, 00250 Helsinki, Finland

B. Schmezer, German Cancer Research Centre, Im Neuenheimer Feld 280, 69120 Heidelberg, Germany

C.M. Shy, University of North Carolina at Chapel Hill, School of Public Health, Department of Epidemiology, CB-7400, Chapel Hill, NC 27599-7400, United States

S. Takahashi, First Department of Pathology, Nagoya City University Medical School, 1 Kawasumi, Mizuho-cho, Mizuho-ku, Nagoya 467, Japan

S. Venitt, Institute of Cancer Research, 15 Cotswold Road, Belmont, Sutton, Surrey SM2 5NG, United Kingdom

E. Windeisen, Institute for Wood Research of Munich University, Winzererstrasse 45, 80797 Munich, Germany

Representatives/observers

American Forest and Paper Association and American Furniture Manufacturers Association

H. Pastides, Department of Epidemiology/Biostatistics, University of Massachusetts, School of Public Health, Amherst, MA 01003, United States

American Industrial Health Council/European Centre for Ecotoxicology and Toxicology of Chemicals

J.F. Acquavella, Monsanto Company, A2SL, 800 North Lindbergh Boulevard, Saint Louis, MO 63167, United States

European Confederation of Woodworking Industries

R. Schiele, Ordinarius for Occupational Medicine, Friedrich Schiller University, Institute for Occupational Medicine, Jahnstrasse 3, 07743 Jena, Germany

European Commission

P. Papadopoulos, Directorate General V, European Commission, Bâtiment Jean Monnet C4/81, Plateau du Kirchberg, 2920 Luxembourg, Grand Duchy of Luxembourg

IARC Secretariat

P. Boffetta, Unit of Analytical Epidemiology
M. Friesen, Unit of Environmental Carcinogenesis
M.-J. Ghess, Unit of Carcinogen Identification and Evaluation
E. Heseltine, 24290 St Léon-sur-Vézère, France
P. Kleihues, Director
V. Krutovskikh, Unit of Multistage Carcinogenesis
D. McGregor, Unit of Carcinogen Identification and Evaluation

D. Mietton, Unit of Carcinogen Identification and Evaluation
H. Møller, Unit of Carcinogen Identification and Evaluation
A. Mylvaganam, Unit of Analytical Epidemiology
C. Partensky, Unit of Carcinogen Identification and Evaluation
I. Peterschmitt, Unit of Carcinogen Identification and Evaluation
B. Stewart, Director's Office[1]
P. Webb, Unit of Carcinogen Identification and Evaluation
J. Wilbourn, Unit of Carcinogen Identification and Evaluation
H. Yamasaki, Unit of Multistage Carcinogenesis

Secretarial assistance

M. Lézère
J. Mitchell
S. Reynaud

---

[1] Present address: Children's Leukaemia & Cancer Research Centre, High Street, Randwick, Sydney NSW 2031, Australia

PREAMBLE

IARC MONOGRAPHS PROGRAMME ON THE EVALUATION OF CARCINOGENIC RISKS TO HUMANS[1]

PREAMBLE

1. BACKGROUND

In 1969, the International Agency for Research on Cancer (IARC) initiated a programme to evaluate the carcinogenic risk of chemicals to humans and to produce monographs on individual chemicals. The *Monographs* programme has since been expanded to include consideration of exposures to complex mixtures of chemicals (which occur, for example, in some occupations and as a result of human habits) and of exposures to other agents, such as radiation and viruses. With Supplement 6 (IARC, 1987a), the title of the series was modified from *IARC Monographs on the Evaluation of the Carcinogenic Risk of Chemicals to Humans* to *IARC Monographs on the Evaluation of Carcinogenic Risks to Humans*, in order to reflect the widened scope of the programme.

The criteria established in 1971 to evaluate carcinogenic risk to humans were adopted by the working groups whose deliberations resulted in the first 16 volumes of the *IARC Monographs series*. Those criteria were subsequently updated by further ad-hoc working groups (IARC, 1977, 1978, 1979, 1982, 1983, 1987b, 1988, 1991a; Vainio *et al.*, 1992).

2. OBJECTIVE AND SCOPE

The objective of the programme is to prepare, with the help of international working groups of experts, and to publish in the form of monographs, critical reviews and evaluations of evidence on the carcinogenicity of a wide range of human exposures. The *Monographs* may also indicate where additional research efforts are needed.

The *Monographs* represent the first step in carcinogenic risk assessment, which involves examination of all relevant information in order to assess the strength of the available evidence that certain exposures could alter the incidence of cancer in humans. The second step is quantitative risk estimation. Detailed, quantitative evaluations of epidemiological data may be

---

[1] This project is supported by PHS Grant No. 5-UO1 CA33193-13 awarded by the United States National Cancer Institute, Department of Health and Human Services. Since 1986, the programme has also been supported by the European Commission.

made in the *Monographs*, but without extrapolation beyond the range of the data available. Quantitative extrapolation from experimental data to the human situation is not undertaken.

The term 'carcinogen' is used in these monographs to denote an exposure that is capable of increasing the incidence of malignant neoplasms; the induction of benign neoplasms may in some circumstances (see p. 18) contribute to the judgement that the exposure is carcinogenic. The terms 'neoplasm' and 'tumour' are used interchangeably.

Some epidemiological and experimental studies indicate that different agents may act at different stages in the carcinogenic process, and several different mechanisms may be involved. The aim of the *Monographs* has been, from their inception, to evaluate evidence of carcinogenicity at any stage in the carcinogenesis process, independently of the underlying mechanisms. Information on mechanisms may, however, be used in making the overall evaluation (IARC, 1991a; Vainio *et al.*, 1992; see also pp. 25–27).

The *Monographs* may assist national and international authorities in making risk assessments and in formulating decisions concerning any necessary preventive measures. The evaluations of IARC working groups are scientific, qualitative judgements about the evidence for or against carcinogenicity provided by the available data. These evaluations represent only one part of the body of information on which regulatory measures may be based. Other components of regulatory decisions may vary from one situation to another and from country to country, responding to different socioeconomic and national priorities. **Therefore, no recommendation is given with regard to regulation or legislation, which are the responsibility of individual governments and/or other international organizations.**

The *IARC Monographs* are recognized as an authoritative source of information on the carcinogenicity of a wide range of human exposures. A users' survey, made in 1988, indicated that the *Monographs* are consulted by various agencies in 57 countries. Each volume is generally printed in 4000 copies for distribution to governments, regulatory bodies and interested scientists. The Monographs are also available via the Distribution and Sales Service of the World Health Organization.

3. SELECTION OF TOPICS FOR MONOGRAPHS

Topics are selected on the basis of two main criteria: (a) there is evidence of human exposure, and (b) there is some evidence or suspicion of carcinogenicity. The term 'agent' is used to include individual chemical compounds, groups of related chemical compounds, physical agents (such as radiation) and biological factors (such as viruses). Exposures to mixtures of agents may occur in occupational exposures and as a result of personal and cultural habits (like smoking and dietary practices). Chemical analogues and compounds with biological or physical characteristics similar to those of suspected carcinogens may also be considered, even in the absence of data on a possible carcinogenic effect in humans or experimental animals.

The scientific literature is surveyed for published data relevant to an assessment of carcinogenicity. The IARC directories of agents being tested for carcinogenicity (IARC, 1973–1994) and directories of on-going research in cancer epidemiology (IARC, 1976–1994) often indicate those exposures that may be scheduled for future meetings. Ad-hoc working groups convened by IARC in 1984, 1989, 1991 and 1993 gave recommendations as to which agents should be evaluated in the IARC Monographs series (IARC, 1984, 1989, 1991b, 1993).

As significant new data on subjects on which monographs have already been prepared become available, re-evaluations are made at subsequent meetings, and revised monographs are published.

4. DATA FOR MONOGRAPHS

The *Monographs* do not necessarily cite all the literature concerning the subject of an evaluation. Only those data considered by the Working Group to be relevant to making the evaluation are included.

With regard to biological and epidemiological data, only reports that have been published or accepted for publication in the openly available scientific literature are reviewed by the working groups. In certain instances, government agency reports that have undergone peer review and are widely available are considered. Exceptions may be made on an ad-hoc basis to include unpublished reports that are in their final form and publicly available, if their inclusion is considered pertinent to making a final evaluation (see pp. 25–27). In the sections on chemical and physical properties, on analysis, on production and use and on occurrence, unpublished sources of information may be used.

5. THE WORKING GROUP

Reviews and evaluations are formulated by a working group of experts. The tasks of the group are: (i) to ascertain that all appropriate data have been collected; (ii) to select the data relevant for the evaluation on the basis of scientific merit; (iii) to prepare accurate summaries of the data to enable the reader to follow the reasoning of the Working Group; (iv) to evaluate the results of epidemiological and experimental studies on cancer; (v) to evaluate data relevant to the understanding of mechanism of action; and (vi) to make an overall evaluation of the carcinogenicity of the exposure to humans.

Working Group participants who contributed to the considerations and evaluations within a particular volume are listed, with their addresses, at the beginning of each publication. Each participant who is a member of a working group serves as an individual scientist and not as a representative of any organization, government or industry. In addition, nominees of national and international agencies and industrial associations may be invited as observers.

6. WORKING PROCEDURES

Approximately one year in advance of a meeting of a working group, the topics of the monographs are announced and participants are selected by IARC staff in consultation with other experts. Subsequently, relevant biological and epidemiological data are collected by IARC from recognized sources of information on carcinogenesis, including data storage and retrieval systems such as MEDLINE and TOXLINE, and EMIC and ETIC for data on genetic and related effects and reproductive and developmental effects, respectively.

For chemicals and some complex mixtures, the major collection of data and the preparation of first drafts of the sections on chemical and physical properties, on analysis, on production and use and on occurrence are carried out under a separate contract funded by the United States National Cancer Institute. Representatives from industrial associations may assist in the preparation of sections on production and use. Information on production and trade is obtained from governmental and trade publications and, in some cases, by direct contact with industries. Separate production data on some agents may not be available because their publication could disclose confidential information. Information on uses may be obtained from published sources but is often complemented by direct contact with manufacturers. Efforts are made to supplement this information with data from other national and international sources.

Six months before the meeting, the material obtained is sent to meeting participants, or is used by IARC staff, to prepare sections for the first drafts of monographs. The first drafts are compiled by IARC staff and sent, prior to the meeting, to all participants of the Working Group for review.

The Working Group meets in Lyon for seven to eight days to discuss and finalize the texts of the monographs and to formulate the evaluations. After the meeting, the master copy of each monograph is verified by consulting the original literature, edited and prepared for publication. The aim is to publish monographs within six months of the Working Group meeting.

The available studies are summarized by the Working Group, with particular regard to the qualitative aspects discussed below. In general, numerical findings are indicated as they appear in the original report; units are converted when necessary for easier comparison. The Working Group may conduct additional analyses of the published data and use them in their assessment of the evidence; the results of such supplementary analyses are given in square brackets. When an important aspect of a study, directly impinging on its interpretation, should be brought to the attention of the reader, a comment is given in square brackets.

7. EXPOSURE DATA

Sections that indicate the extent of past and present human exposure, the sources of exposure, the people most likely to be exposed and the factors that contribute to the exposure are included at the beginning of each monograph.

Most monographs on individual chemicals, groups of chemicals or complex mixtures include sections on chemical and physical data, on analysis, on production and use and on occurrence. In monographs on, for example, physical agents, occupational exposures and cultural habits, other sections may be included, such as: historical perspectives, description of an industry or habit, chemistry of the complex mixture or taxonomy. Monographs on biological agents have sections on structure and biology, methods of detection, epidemiology of infection and clinical disease other than cancer.

For chemical exposures, the Chemical Abstracts Services Registry Number, the latest Chemical Abstracts Primary Name and the IUPAC Systematic Name are recorded; other synonyms are given, but the list is not necessarily comprehensive. For biological agents, taxonomy and structure are described, and the degree of variability is given, when applicable.

Information on chemical and physical properties and, in particular, data relevant to identification, occurrence and biological activity are included. For biological agents, mode of replication, life cycle, target cells, persistence and latency and host response are given. A description of technical products of chemicals includes trades names, relevant specifications and available information on composition and impurities. Some of the trade names given may be those of mixtures in which the agent being evaluated is only one of the ingredients.

The purpose of the section on analysis or detection is to give the reader an overview of current methods, with emphasis on those widely used for regulatory purposes. Methods for monitoring human exposure are also given, when available. No critical evaluation or recommendation of any of the methods is meant or implied. The IARC publishes a series of volumes, *Environmental Carcinogens: Methods of Analysis and Exposure Measurement* (IARC, 1978–93), that describe validated methods for analysing a wide variety of chemicals and mixtures. For biological agents, methods of detection and exposure assessment are described, including their sensitivity, specificity and reproducibility.

The dates of first synthesis and of first commercial production of a chemical or mixture are provided; for agents which do not occur naturally, this information may allow a reasonable estimate to be made of the date before which no human exposure to the agent could have occurred. The dates of first reported occurrence of an exposure are also provided. In addition, methods of synthesis used in past and present commercial production and different methods of production which may give rise to different impurities are described.

Data on production, international trade and uses are obtained for representative regions, which usually include Europe, Japan and the United States of America. It should not, however, be inferred that those areas or nations are necessarily the sole or major sources or users of the agent. Some identified uses may not be current or major applications, and the coverage is not necessarily comprehensive. In the case of drugs, mention of their therapeutic uses does not necessarily represent current practice nor does it imply judgement as to their therapeutic efficacy.

Information on the occurrence of an agent or mixture in the environment is obtained from data derived from the monitoring and surveillance of levels in occupational environments, air, water, soil, foods and animal and human tissues. When available, data on the generation, persistence and bioaccumulation of the agent are also included. In the case of mixtures, industries, occupations or processes, information is given about all agents present. For processes, industries and occupations, a historical description is also given, noting variations in chemical composition, physical properties and levels of occupational exposure with time and place. For biological agents, the epidemiology of infection is described.

Statements concerning regulations and guidelines (e.g. pesticide registrations, maximal levels permitted in foods, occupational exposure limits) are included for some countries as indications of potential exposures, but they may not reflect the most recent situation, since such limits are continuously reviewed and modified. The absence of information on regulatory status for a country should not be taken to imply that that country does not have regulations with regard to the exposure. For biological agents, legislation and control, including vaccines and therapy, are described.

8. STUDIES OF CANCER IN HUMANS

(a) Types of studies considered

Three types of epidemiological studies of cancer contribute to the assessment of carcinogenicity in humans—cohort studies, case–control studies and correlation (or ecological) studies. Rarely, results from randomized trials may be available. Case series and case reports of cancer in humans may also be reviewed.

Cohort and case–control studies relate individual exposures under study to the occurrence of cancer in individuals and provide an estimate of relative risk (ratio of incidence or mortality in those exposed to incidence or mortality in those not exposed) as the main measure of association.

In correlation studies, the units of investigation are usually whole populations (e.g. in particular geographical areas or at particular times), and cancer frequency is related to a summary measure of the exposure of the population to the agent, mixture or exposure circumstance under study. Because individual exposure is not documented, however, a causal relationship is less easy to infer from correlation studies than from cohort and case–control studies. Case reports generally arise from a suspicion, based on clinical experience, that the concurrence of two events—that is, a particular exposure and occurrence of a cancer—has happened rather more frequently than would be expected by chance. Case reports usually lack complete ascertainment of cases in any population, definition or enumeration of the population at risk and estimation of the expected number of cases in the absence of exposure. The uncertainties surrounding interpretation of case reports and correlation studies make them inadequate, except in rare instances, to form the sole basis for inferring a causal relationship. When taken together with case–control and cohort studies, however, relevant case reports or correlation studies may add materially to the judgement that a causal relationship is present.

Epidemiological studies of benign neoplasms, presumed preneoplastic lesions and other end-points thought to be relevant to cancer are also reviewed by working groups. They may, in some instances, strengthen inferences drawn from studies of cancer itself.

(b) Quality of studies considered

The Monographs are not intended to summarize all published studies. Those that are judged to be inadequate or irrelevant to the evaluation are generally omitted. They may be mentioned briefly, particularly when the information is considered to be a useful supplement to that in other reports or when they provide the only data available. Their inclusion does not imply acceptance of the adequacy of the study design or of the analysis and interpretation of the results, and limitations are clearly outlined in square brackets at the end of the study description.

It is necessary to take into account the possible roles of bias, confounding and chance in the interpretation of epidemiological studies. By 'bias' is meant the operation of factors in study design or execution that lead erroneously to a stronger or weaker association than in fact exists between disease and an agent, mixture or exposure circumstance. By 'confounding' is meant a situation in which the relationship with disease is made to appear stronger or to appear weaker than it truly is as a result of an association between the apparent causal factor and another factor that is associated with either an increase or decrease in the incidence of the disease. In evaluating the extent to which these factors have been minimized in an individual study, working groups consider a number of aspects of design and analysis as described in the report of the study. Most of these considerations apply equally to case–control, cohort and correlation studies. Lack of clarity of any of these aspects in the reporting of a study can decrease its credibility and the weight given to it in the final evaluation of the exposure.

Firstly, the study population, disease (or diseases) and exposure should have been well defined by the authors. Cases of disease in the study population should have been identified in a way that was independent of the exposure of interest, and exposure should have been assessed in a way that was not related to disease status.

Secondly, the authors should have taken account in the study design and analysis of other variables that can influence the risk of disease and may have been related to the exposure of interest. Potential confounding by such variables should have been dealt with either in the design of the study, such as by matching, or in the analysis, by statistical adjustment. In cohort studies, comparisons with local rates of disease may be more appropriate than those with national rates. Internal comparisons of disease frequency among individuals at different levels of exposure should also have been made in the study.

Thirdly, the authors should have reported the basic data on which the conclusions are founded, even if sophisticated statistical analyses were employed. At the very least, they should have given the numbers of exposed and unexposed cases and controls in a case–control study and the numbers of cases observed and expected in a cohort study. Further tabulations by time since exposure began and other temporal factors are also important. In a cohort study, data on all cancer sites and all causes of death should have been given, to reveal the possibility of reporting bias. In a case–control study, the effects of investigated factors other than the exposure of interest should have been reported.

Finally, the statistical methods used to obtain estimates of relative risk, absolute rates of cancer, confidence intervals and significance tests, and to adjust for confounding should have been clearly stated by the authors. The methods used should preferably have been the generally accepted techniques that have been refined since the mid-1970s. These methods have been reviewed for case–control studies (Breslow & Day, 1980) and for cohort studies (Breslow & Day, 1987).

(c) Inferences about mechanism of action

Detailed analyses of both relative and absolute risks in relation to temporal variables, such as age at first exposure, time since first exposure, duration of exposure, cumulative exposure and time since exposure ceased, are reviewed and summarized when available. The analysis of temporal relationships can be useful in formulating models of carcinogenesis. In particular, such analyses may suggest whether a carcinogen acts early or late in the process of carcinogenesis, although at best they allow only indirect inferences about the mechanism of action. Special attention is given to measurements of biological markers of carcinogen exposure or action, such as DNA or protein adducts, as well as markers of early steps in the carcinogenic process, such as proto-oncogene mutation, when these are incorporated into epidemiological studies focused on cancer incidence or mortality. Such measurements may allow inferences to be made about putative mechanisms of action (IARC, 1991a; Vainio et al., 1992).

(d) *Criteria for causality*

After the quality of individual epidemiological studies of cancer has been summarized and assessed, a judgement is made concerning the strength of evidence that the agent, mixture or exposure circumstance in question is carcinogenic for humans. In making their judgement, the Working Group considers several criteria for causality. A strong association (i.e. a large relative risk) is more likely to indicate causality than a weak association, although it is recognized that relative risks of small magnitude do not imply lack of causality and may be important if the disease is common. Associations that are replicated in several studies of the same design or using different epidemiological approaches or under different circumstances of exposure are more likely to represent a causal relationship than isolated observations from single studies. If there are inconsistent results among investigations, possible reasons are sought (such as differences in amount of exposure), and results of studies judged to be of high quality are given more weight than those of studies judged to be methodologically less sound. When suspicion of carcinogenicity arises largely from a single study, these data are not combined with those from later studies in any subsequent reassessment of the strength of the evidence.

If the risk of the disease in question increases with the amount of exposure, this is considered to be a strong indication of causality, although absence of a graded response is not necessarily evidence against a causal relationship. Demonstration of a decline in risk after cessation of or reduction in exposure in individuals or in whole populations also supports a causal interpretation of the findings.

Although a carcinogen may act upon more than one target, the specificity of an association (i.e. an increased occurrence of cancer at one anatomical site or of one morphological type) adds plausibility to a causal relationship, particularly when excess cancer occurrence is limited to one morphological type within the same organ.

Although rarely available, results from randomized trials showing different rates among exposed and unexposed individuals provide particularly strong evidence for causality.

When several epidemiological studies show little or no indication of an association between an exposure and cancer, the judgement may be made that, in the aggregate, they show evidence of lack of carcinogenicity. Such a judgement requires first of all that the studies giving rise to it meet, to a sufficient degree, the standards of design and analysis described above. Specifically, the possibility that bias, confounding or misclassification of exposure or outcome could explain the observed results should be considered and excluded with reasonable certainty. In addition, all studies that are judged to be methodologically sound should be consistent with a relative risk of unity for any observed level of exposure and, when considered together, should provide a pooled estimate of relative risk which is at or near unity and has a narrow confidence interval, due to sufficient population size. Moreover, no individual study nor the pooled results of all the studies should show any consistent tendency for relative risk of cancer to increase with increasing level of exposure. It is important to note that evidence of lack of carcinogenicity obtained in this way from several epidemiological studies can apply only to the type(s) of cancer studied and to dose levels and intervals between first exposure and observation of disease that are the same as or less than those observed in all the studies. Experience with human cancer indicates that, in some cases, the period from first exposure to the development of clinical cancer is seldom less than 20 years; latent periods substantially shorter than 30 years cannot provide evidence for lack of carcinogenicity.

9. STUDIES OF CANCER IN EXPERIMENTAL ANIMALS

All known human carcinogens that have been studied adequately in experimental animals have produced positive results in one or more animal species (Wilbourn *et al.*, 1986; Tomatis *et al.*, 1989). For several agents (aflatoxins, 4-aminobiphenyl, azathioprine, betel quid with tobacco, BCME and CMME (technical grade), chlorambucil, chlornaphazine, ciclosporin, coal-tar pitches, coal-tars, combined oral contraceptives, cyclophosphamide, diethylstilboestrol, melphalan, 8-methoxypsoralen plus UVA, mustard gas, myleran, 2-naphthylamine, nonsteroidal oestrogens, oestrogen replacement therapy/steroidal oestrogens, solar radiation, thiotepa and vinyl chloride), carcinogenicity in experimental animals was established or highly suspected before epidemiological studies confirmed the carcinogenicity in humans (Vainio *et al.*, 1995). Although this association cannot establish that all agents and mixtures that cause cancer in experimental animals also cause cancer in humans, nevertheless, **in the absence of adequate data on humans, it is biologically plausible and prudent to regard agents and mixtures for which there is sufficient evidence (see p. 24) of carcinogenicity in experimental animals as if they presented a carcinogenic risk to humans**. The possibility that a given agent may cause cancer through a species-specific mechanism which does not operate in humans (see p. 25) should also be taken into consideration.

The nature and extent of impurities or contaminants present in the chemical or mixture being evaluated are given when available. Animal strain, sex, numbers per group, age at start of treatment and survival are reported.

Other types of studies summarized include: experiments in which the agent or mixture was administered in conjunction with known carcinogens or factors that modify carcinogenic effects; studies in which the end-point was not cancer but a defined precancerous lesion; and experiments on the carcinogenicity of known metabolites and derivatives.

For experimental studies of mixtures, consideration is given to the possibility of changes in the physicochemical properties of the test substance during collection, storage, extraction, concentration and delivery. Chemical and toxicological interactions of the components of mixtures may result in nonlinear dose–response relationships.

An assessment is made as to the relevance to human exposure of samples tested in experimental animals, which may involve consideration of: (i) physical and chemical characteristics, (ii) constituent substances that indicate the presence of a class of substances, (iii) the results of tests for genetic and related effects, including genetic activity profiles, DNA adduct profiles, proto-oncogene mutation and expression and suppressor gene inactivation. The relevance of results obtained, for example, with animal viruses analogous to the virus being evaluated in the monograph must also be considered. They may provide biological and mechanistic information relevant to the understanding of the process of carcinogenesis in humans and may strengthen the plausibility of a conclusion that the biological agent that is being evaluated is carcinogenic in humans.

(a) Qualitative aspects

An assessment of carcinogenicity involves several considerations of qualitative importance, including (i) the experimental conditions under which the test was performed, including route and schedule of exposure, species, strain, sex, age, duration of follow-up; (ii) the consistency of the results, for example, across species and target organ(s); (iii) the spectrum of neoplastic response, from preneoplastic lesions and benign tumours to malignant neoplasms; and (iv) the possible role of modifying factors.

As mentioned earlier (p. 11), the *Monographs* are not intended to summarize all published studies. Those studies in experimental animals that are inadequate (e.g. too short a duration, too few animals, poor survival; see below) or are judged irrelevant to the evaluation are generally omitted. Guidelines for conducting adequate long-term carcinogenicity experiments have been outlined (e.g. Montesano *et al.*, 1986).

Considerations of importance to the Working Group in the interpretation and evaluation of a particular study include: (i) how clearly the agent was defined and, in the case of mixtures, how adequately the sample characterization was reported; (ii) whether the dose was adequately monitored, particularly in inhalation experiments; (iii) whether the doses and duration of treatment were appropriate and whether the survival of treated animals was similar to that of controls; (iv) whether there were adequate numbers of animals per group; (v) whether animals of both sexes were used; (vi) whether animals were allocated randomly to groups; (vii) whether the duration of observation was adequate; and (viii) whether the data were adequately reported. If available, recent data on the incidence of specific tumours in historical controls, as well as in concurrent controls, should be taken into account in the evaluation of tumour response.

When benign tumours occur together with and originate from the same cell type in an organ or tissue as malignant tumours in a particular study and appear to represent a stage in the progression to malignancy, it may be valid to combine them in assessing tumour incidence (Huff *et al.*, 1989). The occurrence of lesions presumed to be preneoplastic may in certain instances aid in assessing the biological plausibility of any neoplastic response observed. If an agent or mixture induces only benign neoplasms that appear to be end-points that do not readily undergo transition to malignancy, it should nevertheless be suspected of being a carcinogen and requires further investigation.

(b) Quantitative aspects

The probability that tumours will occur may depend on the species, sex, strain and age of the animal, the dose of the carcinogen and the route and length of exposure. Evidence of an increased incidence of neoplasms with increased level of exposure strengthens the inference of a causal association between the exposure and the development of neoplasms.

The form of the dose–response relationship can vary widely, depending on the particular agent under study and the target organ. Both DNA damage and increased cell division are important aspects of carcinogenesis, and cell proliferation is a strong determinant of dose–response relationships for some carcinogens (Cohen & Ellwein, 1990). Since many chemicals require metabolic activation before being converted into their reactive intermediates, both metabolic and pharmacokinetic aspects are important in determining the dose–response pattern. Saturation of steps such as absorption, activation, inactivation and elimination may produce nonlinearity in the dose–response relationship, as could saturation of processes such as DNA repair (Hoel *et al.*, 1983; Gart *et al.*, 1986).

(c) Statistical analysis of long-term experiments in animals

Factors considered by the Working Group include the adequacy of the information given for each treatment group: (i) the number of animals studied and the number examined histologically, (ii) the number of animals with a given tumour type and (iii) length of survival. The statistical methods used should be clearly stated and should be the generally accepted techniques refined for this purpose (Peto *et al.*, 1980; Gart *et al.*, 1986). When there is no difference in survival between control and treatment groups, the Working Group usually compares the proportions of animals developing each tumour type in each of the groups. Otherwise, consideration is given as to whether or not appropriate adjustments have been made for differences in survival. These adjustments can include: comparisons of the proportions of tumour-bearing animals among the effective number of animals (alive at the time the first tumour is discovered), in the case where most differences in survival occur before tumours appear; life-table methods, when tumours are visible or when they may be considered 'fatal' because mortality rapidly follows tumour development; and the Mantel-Haenszel test or logistic regression, when occult tumours do not affect the animals' risk of dying but are 'incidental' findings at autopsy.

In practice, classifying tumours as fatal or incidental may be difficult. Several survival-adjusted methods have been developed that do not require this distinction (Gart *et al.*, 1986), although they have not been fully evaluated.

10. OTHER DATA RELEVANT TO AN EVALUATION OF CARCINOGENICITY AND ITS MECHANISMS

In coming to an overall evaluation of carcinogenicity in humans (see p. 25), the Working Group also considers related data. The nature of the information selected for the summary depends on the agent being considered.

For chemicals and complex mixtures of chemicals such as those in some occupational situations and involving cultural habits (e.g. tobacco smoking), the other data considered to be relevant are divided into those on absorption, distribution, metabolism and excretion; toxic effects; reproductive and developmental effects; and genetic and related effects.

Concise information is given on absorption, distribution (including placental transfer) and excretion in both humans and experimental animals. Kinetic factors that may affect the dose–response relationship, such as saturation of uptake, protein binding, metabolic activation, detoxification and DNA repair processes, are mentioned. Studies that indicate the metabolic fate of the agent in humans and in experimental animals are summarized briefly, and comparisons of data from humans and animals are made when possible. Comparative information on the relationship between exposure and the dose that reaches the target site may be of particular importance for extrapolation between species. Data are given on acute and chronic toxic effects (other than cancer), such as organ toxicity, increased cell proliferation, immunotoxicity and endocrine effects. The presence and toxicological significance of cellular receptors is described. Effects on reproduction, teratogenicity, fetotoxicity and embryotoxicity are also summarized briefly.

Tests of genetic and related effects are described in view of the relevance of gene mutation and chromosomal damage to carcinogenesis (Vainio et al., 1992). The adequacy of the reporting of sample characterization is considered and, where necessary, commented upon; with regard to complex mixtures, such comments are similar to those described for animal carcinogenicity tests on p. 17. The available data are interpreted critically by phylogenetic group according to the end-points detected, which may include DNA damage, gene mutation, sister chromatid exchange, micronucleus formation, chromosomal aberrations, aneuploidy and cell transformation. The concentrations employed are given, and mention is made of whether use of an exogenous metabolic system *in vitro* affected the test result. These data are given as listings of test systems, data and references; bar graphs (activity profiles) and corresponding summary tables with detailed information on the preparation of the profiles (Waters et al., 1987) are given in appendices.

Positive results in tests using prokaryotes, lower eukaryotes, plants, insects and cultured mammalian cells suggest that genetic and related effects could occur in mammals. Results from such tests may also give information about the types of genetic effect produced and about the involvement of metabolic activation. Some end-points described are clearly genetic in nature (e.g. gene mutations and chromosomal aberrations), while others are to a greater or lesser degree associated with genetic effects (e.g. unscheduled DNA synthesis). In-vitro tests for tumour-promoting activity and for cell transformation may be sensitive to changes that are not necessarily the result of genetic alterations but that may have specific relevance to the process of carcinogenesis. A critical appraisal of these tests has been published (Montesano et al., 1986).

Genetic or other activity manifest in experimental mammals and humans is regarded as being of greater relevance than that in other organisms. The demonstration that an agent or mixture can induce gene and chromosomal mutations in whole mammals indicates that it may have carcinogenic activity, although this activity may not be detectably expressed in any or all species. Relative potency in tests for mutagenicity and related effects is not a reliable indicator of carcinogenic potency. Negative results in tests for mutagenicity in selected tissues from animals treated *in vivo* provide less weight, partly because they do not exclude the possibility of an effect in tissues other than those examined. Moreover, negative results in short-term tests with genetic end-points cannot be considered to provide evidence to rule out carcinogenicity of agents or mixtures that act through other mechanisms (e.g. receptor-mediated effects, cellular toxicity with regenerative proliferation, peroxisome proliferation) (Vainio et al., 1992). Factors that may lead to misleading results in short-term tests have been discussed in detail elsewhere (Montesano et al., 1986).

When available, data relevant to mechanisms of carcinogenesis that do not involve structural changes at the level of the gene are also described.

The adequacy of epidemiological studies of reproductive outcome and genetic and related effects in humans is evaluated by the same criteria as are applied to epidemiological studies of cancer.

Structure–activity relationships that may be relevant to an evaluation of the carcinogenicity of an agent are also described.

For biological agents—viruses, bacteria and parasites—other data relevant to carcinogenicity include descriptions of the pathology of infection, molecular biology (integration and expression of viruses, and any genetic alterations seen in human tumours) and other observations, which might include cellular and tissue responses to infection, immune response and the presence of tumour markers.

11. SUMMARY OF DATA REPORTED

In this section, the relevant epidemiological and experimental data are summarized. Only reports, other than in abstract form, that meet the criteria outlined on p. 11 are considered for evaluating carcinogenicity. Inadequate studies are generally not summarized: such studies are usually identified by a square-bracketed comment in the preceding text.

(a) Exposures

Human exposure to chemicals and complex mixtures is summarized on the basis of elements such as production, use, occurrence in the environment and determinations in human tissues and body fluids. Quantitative data are given when available. Exposure to biological agents is described in terms of transmission, and prevalence of infection.

(b) Carcinogenicity in humans

Results of epidemiological studies that are considered to be pertinent to an assessment of human carcinogenicity are summarized. When relevant, case reports and correlation studies are also summarized.

(c) Carcinogenicity in experimental animals

Data relevant to an evaluation of carcinogenicity in animals are summarized. For each animal species and route of administration, it is stated whether an increased incidence of neoplasms or preneoplastic lesions was observed, and the tumour sites are indicated. If the agent or mixture produced tumours after prenatal exposure or in single-dose experiments, this is also indicated. Negative findings are also summarized. Dose–response and other quantitative data may be given when available.

(d) Other data relevant to an evaluation of carcinogenicity and its mechanisms

Data on biological effects in humans that are of particular relevance are summarized. These may include toxicological, kinetic and metabolic considerations and evidence of DNA binding, persistence of DNA lesions or genetic damage in exposed humans. Toxicological information, such as that on cytotoxicity and regeneration, receptor binding and hormonal and immunological effects, and data on kinetics and metabolism in experimental animals are given when considered relevant to the possible mechanism of the carcinogenic action of the agent. The results of tests for genetic and related effects are summarized for whole mammals, cultured mammalian cells and nonmammalian systems.

When available, comparisons of such data for humans and for animals, and particularly animals that have developed cancer, are described.

Structure–activity relationships are mentioned when relevant.

For the agent, mixture or exposure circumstance being evaluated, the available data on endpoints or other phenomena relevant to mechanisms of carcinogenesis from studies in humans, experimental animals and tissue and cell test systems are summarized within one or more of the following descriptive dimensions:

(i) Evidence of genotoxicity (i.e. structural changes at the level of the gene): for example, structure–activity considerations, adduct formation, mutagenicity (effect on specific genes), chromosomal mutation/aneuploidy

(ii) Evidence of effects on the expression of relevant genes (i.e. functional changes at the intracellular level): for example, alterations to the structure or quantity of the product of a proto-oncogene or tumour suppressor gene, alterations to metabolic activation/inactivation/DNA repair

(iii) Evidence of relevant effects on cell behaviour (i.e. morphological or behavioural changes at the cellular or tissue level): for example, induction of mitogenesis, compensatory cell proliferation, preneoplasia and hyperplasia, survival of premalignant or malignant cells (immortalization, immunosuppression), effects on metastatic potential

(iv) Evidence from dose and time relationships of carcinogenic effects and interactions between agents: for example, early/late stage, as inferred from epidemiological studies; initiation/promotion/progression/malignant conversion, as defined in animal carcinogenicity experiments; toxicokinetics

These dimensions are not mutually exclusive, and an agent may fall within more than one of them. Thus, for example, the action of an agent on the expression of relevant genes could be summarized under both the first and second dimension, even if it were known with reasonable certainty that those effects resulted from genotoxicity.

12. EVALUATION

Evaluations of the strength of the evidence for carcinogenicity arising from human and experimental animal data are made, using standard terms.

It is recognized that the criteria for these evaluations, described below, cannot encompass all of the factors that may be relevant to an evaluation of carcinogenicity. In considering all of the relevant scientific data, the Working Group may assign the agent, mixture or exposure circumstance to a higher or lower category than a strict interpretation of these criteria would indicate.

(a) Degrees of evidence for carcinogenicity in humans and in experimental animals and supporting evidence

These categories refer only to the strength of the evidence that an exposure is carcinogenic and not to the extent of its carcinogenic activity (potency) nor to the mechanisms involved. A classification may change as new information becomes available.

An evaluation of degree of evidence, whether for a single agent or a mixture, is limited to the materials tested, as defined physically, chemically or biologically. When the agents evaluated are considered by the Working Group to be sufficiently closely related, they may be grouped together for the purpose of a single evaluation of degree of evidence.

(i) Carcinogenicity in humans

The applicability of an evaluation of the carcinogenicity of a mixture, process, occupation or industry on the basis of evidence from epidemiological studies depends on the variability over time and place of the mixtures, processes, occupations and industries. The Working Group seeks to identify the specific exposure, process or activity which is considered most likely to be responsible for any excess risk. The evaluation is focused as narrowly as the available data on exposure and other aspects permit.

The evidence relevant to carcinogenicity from studies in humans is classified into one of the following categories:

Sufficient evidence of carcinogenicity: The Working Group considers that a causal relationship has been established between exposure to the agent, mixture or exposure circumstance and human cancer. That is, a positive relationship has been observed between the exposure and cancer in studies in which chance, bias and confounding could be ruled out with reasonable confidence.

Limited evidence of carcinogenicity: A positive association has been observed between exposure to the agent, mixture or exposure circumstance and cancer for which a causal interpretation is considered by the Working Group to be credible, but chance, bias or confounding could not be ruled out with reasonable confidence.

Inadequate evidence of carcinogenicity: The available studies are of insufficient quality, consistency or statistical power to permit a conclusion regarding the presence or absence of a causal association, or no data on cancer in humans are available.

Evidence suggesting lack of carcinogenicity: There are several adequate studies covering the full range of levels of exposure that human beings are known to encounter, which are mutually consistent in not showing a positive association between exposure to the agent, mixture or exposure circumstance and any studied cancer at any observed level of exposure. A conclusion of 'evidence suggesting lack of carcinogenicity' is inevitably limited to the cancer sites, conditions and levels of exposure and length of observation covered by the available studies. In addition, the possibility of a very small risk at the levels of exposure studied can never be excluded.

In some instances, the above categories may be used to classify the degree of evidence related to carcinogenicity in specific organs or tissues.

(ii) Carcinogenicity in experimental animals

The evidence relevant to carcinogenicity in experimental animals is classified into one of the following categories:

Sufficient evidence of carcinogenicity: The Working Group considers that a causal relationship has been established between the agent or mixture and an increased incidence of malignant neoplasms or of an appropriate combination of benign and malignant neoplasms in (a) two or more species of animals or (b) in two or more independent studies in one species carried out at different times or in different laboratories or under different protocols.

Exceptionally, a single study in one species might be considered to provide sufficient evidence of carcinogenicity when malignant neoplasms occur to an unusual degree with regard to incidence, site, type of tumour or age at onset.

Limited evidence of carcinogenicity: The data suggest a carcinogenic effect but are limited for making a definitive evaluation because, e.g. (a) the evidence of carcinogenicity is restricted to a single experiment; or (b) there are unresolved questions regarding the adequacy of the design, conduct or interpretation of the study; or (c) the agent or mixture increases the incidence only of benign neoplasms or lesions of uncertain neoplastic potential, or of certain neoplasms which may occur spontaneously in high incidences in certain strains.

Inadequate evidence of carcinogenicity: The studies cannot be interpreted as showing either the presence or absence of a carcinogenic effect because of major qualitative or quantitative limitations, or no data on cancer in experimental animals are available.

Evidence suggesting lack of carcinogenicity: Adequate studies involving at least two species are available which show that, within the limits of the tests used, the agent or mixture is not carcinogenic. A conclusion of evidence suggesting lack of carcinogenicity is inevitably limited to the species, tumour sites and levels of exposure studied.

(b) Other data relevant to the evaluation of carcinogenicity and its mechanisms

Other evidence judged to be relevant to an evaluation of carcinogenicity and of sufficient importance to affect the overall evaluation is then described. This may include data on preneoplastic lesions, tumour pathology, genetic and related effects, structure–activity relationships, metabolism and pharmacokinetics, physicochemical parameters and analogous biological agents.

Data relevant to mechanisms of the carcinogenic action are also evaluated. The strength of the evidence that any carcinogenic effect observed is due to a particular mechanism is assessed, using terms such as weak, moderate or strong. Then, the Working Group assesses if that particular mechanism is likely to be operative in humans. The strongest indications that a particular mechanism operates in humans come from data on humans or biological specimens obtained from exposed humans. The data may be considered to be especially relevant if they show that the agent in question has caused changes in exposed humans that are on the causal pathway to carcinogenesis. Such data may, however, never become available, because it is at least conceivable that certain compounds may be kept from human use solely on the basis of evidence of their toxicity and/or carcinogenicity in experimental systems.

For complex exposures, including occupational and industrial exposures, the chemical composition and the potential contribution of carcinogens known to be present are considered by the Working Group in its overall evaluation of human carcinogenicity. The Working Group also determines the extent to which the materials tested in experimental systems are related to those to which humans are exposed.

(c) Overall evaluation

Finally, the body of evidence is considered as a whole, in order to reach an overall evaluation of the carcinogenicity to humans of an agent, mixture or circumstance of exposure.

An evaluation may be made for a group of chemical compounds that have been evaluated by the Working Group. In addition, when supporting data indicate that other, related compounds for which there is no direct evidence of capacity to induce cancer in humans or in animals may also be carcinogenic, a statement describing the rationale for this conclusion is added to the evaluation narrative; an additional evaluation may be made for this broader group of compounds if the strength of the evidence warrants it.

The agent, mixture or exposure circumstance is described according to the wording of one of the following categories, and the designated group is given. The categorization of an agent, mixture or exposure circumstance is a matter of scientific judgement, reflecting the strength of the evidence derived from studies in humans and in experimental animals and from other relevant data.

Group 1—The agent (mixture) is carcinogenic to humans.
The exposure circumstance entails exposures that are carcinogenic to humans.

This category is used when there is *sufficient evidence* of carcinogenicity in humans. Exceptionally, an agent (mixture) may be placed in this category when evidence in humans is less than sufficient but there is *sufficient evidence* of carcinogenicity in experimental animals and strong evidence in exposed humans that the agent (mixture) acts through a relevant mechanism of carcinogenicity.

Group 2

This category includes agents, mixtures and exposure circumstances for which, at one extreme, the degree of evidence of carcinogenicity in humans is almost sufficient, as well as those for which, at the other extreme, there are no human data but for which there is evidence of carcinogenicity in experimental animals. Agents, mixtures and exposure circumstances are assigned to either group 2A (probably carcinogenic to humans) or group 2B (possibly carcinogenic to humans) on the basis of epidemiological and experimental evidence of carcinogenicity and other relevant data.

Group 2A—The agent (mixture) is probably carcinogenic to humans.
The exposure circumstance entails exposures that are probably carcinogenic to humans.

This category is used when there is *limited evidence* of carcinogenicity in humans and sufficient evidence of carcinogenicity in experimental animals. In some cases, an agent (mixture) may be classified in this category when there is inadequate evidence of carcinogenicity in humans and *sufficient evidence* of carcinogenicity in experimental animals and strong evidence that the carcinogenesis is mediated by a mechanism that also operates in humans. Exceptionally, an agent, mixture or exposure circumstance may be classified in this category solely on the basis of limited evidence of carcinogenicity in humans.

Group 2B—The agent (mixture) is possibly carcinogenic to humans.
The exposure circumstance entails exposures that are possibly carcinogenic to humans.

This category is used for agents, mixtures and exposure circumstances for which there is *limited evidence* of carcinogenicity in humans and less than *sufficient evidence* of carcinogenicity in experimental animals. It may also be used when there is *inadequate evidence* of carcinogenicity in humans but there is *sufficient evidence* of carcinogenicity in experimental animals. In some instances, an agent, mixture or exposure circumstance for which there is *inadequate evidence* of carcinogenicity in humans but *limited evidence* of carcinogenicity in experimental animals together with supporting evidence from other relevant data may be placed in this group.

Group 3—The agent (mixture or exposure circumstance) is not classifiable as to its carcinogenicity to humans.

This category is used most commonly for agents, mixtures and exposure circumstances for which the evidence of carcinogenicity is inadequate in humans and inadequate or limited in experimental animals.

Exceptionally, agents (mixtures) for which the evidence of carcinogenicity is inadequate in humans but sufficient in experimental animals may be placed in this category when there is strong evidence that the mechanism of carcinogenicity in experimental animals does not operate in humans.

Agents, mixtures and exposure circumstances that do not fall into any other group are also placed in this category.

Group 4—The agent (mixture) is probably not carcinogenic to humans.

This category is used for agents or mixtures for which there is *evidence suggesting lack of carcinogenicity* in humans and in experimental animals. In some instances, agents or mixtures for which there is *inadequate evidence* of carcinogenicity in humans but *evidence suggesting lack of carcinogenicity* in experimental animals, consistently and strongly supported by a broad range of other relevant data, may be classified in this group.

References

Breslow, N.E. & Day, N.E. (1980) *Statistical Methods in Cancer Research*, Vol. 1, *The Analysis of Case–Control Studies* (IARC Scientific Publications No. 32), Lyon, IARC

Breslow, N.E. & Day, N.E. (1987) *Statistical Methods in Cancer Research*, Vol. 2, *The Design and Analysis of Cohort Studies* (IARC Scientific Publications No. 82), Lyon, IARC

Cohen, S.M. & Ellwein, L.B. (1990) Cell proliferation in carcinogenesis. *Science*, **249**, 1007–1011

Gart, J.J., Krewski, D., Lee, P.N., Tarone, R.E. & Wahrendorf, J. (1986) *Statistical Methods in Cancer Research*, Vol. 3, *The Design and Analysis of Long-term Animal Experiments* (IARC Scientific Publications No. 79), Lyon, IARC

Hoel, D.G., Kaplan, N.L. & Anderson, M.W. (1983) Implication of nonlinear kinetics on risk estimation in carcinogenesis. *Science*, **219**, 1032–1037

Huff, J.E., Eustis, S.L. & Haseman, J.K. (1989) Occurrence and relevance of chemically induced benign neoplasms in long-term carcinogenicity studies. *Cancer Metastasis Rev.*, **8**, 1–21

IARC (1973–1994) *Information Bulletin on the Survey of Chemicals Being Tested for Carcinogenicity/Directory of Agents Being Tested for Carcinogenicity*, Numbers 1–16, Lyon

Number 1 (1973)	52 pages
Number 2 (1973)	77 pages
Number 3 (1974)	67 pages
Number 4 (1974)	97 pages
Number 5 (1975)	8 pages
Number 6 (1976)	360 pages

Number 7 (1978)   460 pages
Number 8 (1979)   604 pages
Number 9 (1981)   294 pages
Number 10 (1983) 326 pages
Number 11 (1984) 370 pages
Number 12 (1986) 385 pages
Number 13 (1988) 404 pages
Number 14 (1990) 369 pages
Number 15 (1992) 317 pages
Number 16 (1994) 293 pages

IARC (1976–1994)

Directory of On-going Research in Cancer Epidemiology 1976. Edited by C.S. Muir & G. Wagner, Lyon

Directory of On-going Research in Cancer Epidemiology 1977 (IARC Scientific Publications No. 17). Edited by C.S. Muir & G. Wagner, Lyon

Directory of On-going Research in Cancer Epidemiology 1978 (IARC Scientific Publications No. 26). Edited by C.S. Muir & G. Wagner, Lyon

Directory of On-going Research in Cancer Epidemiology 1979 (IARC Scientific Publications No. 28). Edited by C.S. Muir & G. Wagner, Lyon

Directory of On-going Research in Cancer Epidemiology 1980 (IARC Scientific Publications No. 35). Edited by C.S. Muir & G. Wagner, Lyon

Directory of On-going Research in Cancer Epidemiology 1981 (IARC Scientific Publications No. 38). Edited by C.S. Muir & G. Wagner, Lyon

Directory of On-going Research in Cancer Epidemiology 1982 (IARC Scientific Publications No. 46). Edited by C.S. Muir & G. Wagner, Lyon

Directory of On-going Research in Cancer Epidemiology 1983 (IARC Scientific Publications No. 50). Edited by C.S. Muir & G. Wagner, Lyon

Directory of On-going Research in Cancer Epidemiology 1984 (IARC Scientific Publications No. 62). Edited by C.S. Muir & G. Wagner, Lyon

Directory of On-going Research in Cancer Epidemiology 1985 (IARC Scientific Publications No. 69). Edited by C.S. Muir & G. Wagner, Lyon

Directory of On-going Research in Cancer Epidemiology 1986 (IARC Scientific Publications No. 80). Edited by C.S. Muir & G. Wagner, Lyon

Directory of On-going Research in Cancer Epidemiology 1987 (IARC Scientific Publications No. 86). Edited by D.M. Parkin & J. Wahrendorf, Lyon

Directory of On-going Research in Cancer Epidemiology 1988 (IARC Scientific Publications No. 93). Edited by M. Coleman & J. Wahrendorf, Lyon

Directory of On-going Research in Cancer Epidemiology 1989/90 (IARC Scientific Publications No. 101). Edited by M. Coleman & J. Wahrendorf, Lyon

Directory of On-going Research in Cancer Epidemiology 1991 (IARC Scientific Publications No.110). Edited by M. Coleman & J. Wahrendorf, Lyon

Directory of On-going Research in Cancer Epidemiology 1992 (IARC Scientific Publications No. 117). Edited by M. Coleman, J. Wahrendorf & E. Démaret, Lyon

Directory of On-going Research in Cancer Epidemiology 1994 (IARC Scientific Publications No. 130). Edited by R. Sankaranarayanan, J. Wahrendorf & E. Démaret, Lyon

IARC (1977) *IARC Monographs Programme on the Evaluation of the Carcinogenic Risk of Chemicals to Humans*. Preamble (IARC intern. tech. Rep. No. 77/002), Lyon

IARC (1978) *Chemicals with* Sufficient Evidence *of Carcinogenicity in Experimental Animals*—IARC Monographs *Volumes 1–17* (IARC intern. tech. Rep. No. 78/003), Lyon

IARC (1978–1993) *Environmental Carcinogens. Methods of Analysis and Exposure Measurement*:
- Vol. 1. *Analysis of Volatile Nitrosamines in Food* (IARC Scientific Publications No. 18). Edited by R. Preussmann, M. Castegnaro, E.A. Walker & A.E. Wasserman (1978)
- Vol. 2. *Methods for the Measurement of Vinyl Chloride in Poly(vinyl chloride), Air, Water and Foodstuffs* (IARC Scientific Publications No. 22). Edited by D.C.M. Squirrell & W. Thain (1978)
- Vol. 3. Analysis of Polycyclic Aromatic Hydrocarbons in Environmental Samples (IARC Scientific Publications No. 29). Edited by M. Castegnaro, P. Bogovski, H. Kunte & E.A. Walker (1979)
- Vol. 4. *Some Aromatic Amines and Azo Dyes in the General and Industrial Environment* (IARC Scientific Publications No. 40). Edited by L. Fishbein, M. Castegnaro, I.K. O'Neill & H. Bartsch (1981)
- Vol. 5. *Some Mycotoxins* (IARC Scientific Publications No. 44). Edited by L. Stoloff, M. Castegnaro, P. Scott, I.K. O'Neill & H. Bartsch (1983)
- Vol. 6. N-*Nitroso Compounds* (IARC Scientific Publications No. 45). Edited by R. Preussmann, I.K. O'Neill, G. Eisenbrand, B. Spiegelhalder & H. Bartsch (1983)
- Vol. 7. *Some Volatile Halogenated Hydrocarbons* (IARC Scientific Publications No. 68). Edited by L. Fishbein & I.K. O'Neill (1985)
- Vol. 8. *Some Metals: As, Be, Cd, Cr, Ni, Pb, Se, Zn* (IARC Scientific Publications No. 71). Edited by I.K. O'Neill, P. Schuller & L. Fishbein (1986)
- Vol. 9. *Passive Smoking* (IARC Scientific Publications No. 81). Edited by I.K. O'Neill, K.D. Brunnemann, B. Dodet & D. Hoffmann (1987)
- Vol. 10. *Benzene and Alkylated Benzenes* (IARC Scientific Publications No. 85). Edited by L. Fishbein & I.K. O'Neill (1988)
- Vol. 11. *Polychlorinated Dioxins and Dibenzofurans)* (IARC Scientific Publications No. 108). Edited by C. Rappe, H.R. Buser, B. Dodet & I.K. O'Neill (1991)
- Vol. 12. *Indoor Air* (IARC Scientific Publications No. 109). Edited by B. Seifert, H. van de Wiel, B. Dodet & I.K. O'Neill (1993)

IARC (1979) *Criteria to Select Chemicals for* IARC Monographs (IARC intern. tech. Rep. No. 79/003), Lyon

IARC (1982) *IARC Monographs on the Evaluation of the Carcinogenic Risk of Chemicals to Humans*, Supplement 4, *Chemicals, Industrial Processes and Industries Associated with Cancer in Humans* (IARC Monographs, Volumes 1 to 29), Lyon

IARC (1983) *Approaches to Classifying Chemical Carcinogens According to Mechanism of Action* (IARC intern. tech. Rep. No. 83/001), Lyon

IARC (1984) *Chemicals and Exposures to Complex Mixtures Recommended for Evaluation in IARC Monographs and Chemicals and Complex Mixtures Recommended for Long-term Carcinogenicity Testing* (IARC intern. tech. Rep. No. 84/002), Lyon

IARC (1987a) *IARC Monographs on the Evaluation of Carcinogenic Risks to Humans*, Supplement 6, *Genetic and Related Effects: An Updating of Selected* IARC Monographs *from Volumes 1 to 42*, Lyon

IARC (1987b) *IARC Monographs on the Evaluation of Carcinogenic Risks to Humans*, Supplement 7, *Overall Evaluations of Carcinogenicity: An Updating of* IARC Monographs *Volumes 1 to 42*, Lyon

IARC (1988) *Report of an IARC Working Group to Review the Approaches and Processes Used to Evaluate the Carcinogenicity of Mixtures and Groups of Chemicals* (IARC intern. tech. Rep.No. 88/002), Lyon

IARC (1989) *Chemicals, Groups of Chemicals, Mixtures and Exposure Circumstances to be Evaluated in Future IARC Monographs, Report of an ad hoc Working Group* (IARC intern. tech. Rep. No. 89/004), Lyon

IARC (1991a) *A Consensus Report of an IARC Monographs Working Group on the Use of Mechanisms of Carcinogenesis in Risk Identification* (IARC intern. tech. Rep. No. 91/002), Lyon

IARC (1991b) *Report of an Ad-hoc* IARC Monographs *Advisory Group on Viruses and Other Biological Agents Such as Parasites* (IARC intern. tech. Rep. No. 91/001), Lyon

IARC (1993) *Chemicals, Groups of Chemicals, Complex Mixtures, Physical and Biological Agents and Exposure Circumstances to be Evaluated in Future* IARC Monographs, *Report of an ad-hoc Working Group* (IARC intern. Rep. No. 93/005), Lyon

Montesano, R., Bartsch, H., Vainio, H., Wilbourn, J. & Yamasaki, H., eds (1986) *Long-term and Short-term Assays for Carcinogenesis—A Critical Appraisal* (IARC Scientific Publications No. 83), Lyon, IARC

Peto, R., Pike, M.C., Day, N.E., Gray, R.G., Lee, P.N., Parish, S., Peto, J., Richards, S. & Wahrendorf, J. (1980) Guidelines for simple, sensitive significance tests for carcinogenic effects in long-term animal experiments. In: *IARC Monographs on the Evaluation of the Carcinogenic Risk of Chemicals to Humans*, Supplement 2, *Long-term and Short-term Screening Assays for Carcinogens: A Critical Appraisal*, Lyon, pp. 311–426

Tomatis, L., Aitio, A., Wilbourn, J. & Shuker, L. (1989) Human carcinogens so far identified. *Jpn. J. Cancer Res.*, **80**, 795–807

Vainio, H., Magee, P., McGregor, D. & McMichael, A., eds (1992) *Mechanisms of Carcinogenesis in Risk Identification* (IARC Scientific Publications No. 116), Lyon, IARC

Vainio, H., Wilbourn, J. & Tomatis, L. (1995) Identification of environmental carcinogens: the first step in risk assessment. In: Mehlman, M.A. & Upton, A., eds, *The Identification and Control of Environmental and Occupational Diseases*, Princeton, Princeton Scientific Publishing Company (in press)

Waters, M.D., Stack, H.F., Brady, A.L., Lohman, P.H.M., Haroun, L. & Vainio, H. (1987) Appendix 1. Activity profiles for genetic and related tests. In: *IARC Monographs on the Evaluation of Carcinogenic Risks to Humans*, Suppl. 6, *Genetic and Related Effects: An Updating of Selected IARC Monographs from Volumes 1 to 42*, Lyon, IARC, pp. 687–696

Wilbourn, J., Haroun, L., Heseltine, E., Kaldor, J., Partensky, C. & Vainio, H. (1986) Response of experimental animals to human carcinogens: an analysis based upon the IARC Monographs Programme. *Carcinogenesis*, 7, 1853–1863

GENERAL REMARKS

This sixty-second volume of *IARC Monographs* addresses wood dust and formaldehyde. Wood dust has not been evaluated before in the *Monographs*; however, a previous working group, which met in June 1980, prepared a series of monographs (IARC, 1981) on industries and occupations in which exposure to wood dust can occur: the lumber and sawmill industry (including logging), the furniture and cabinet-making industry, carpentry and joinery and the pulp and paper industry. The information was updated during the preparation of Supplement 7 to the *Monographs* (IARC, 1987), and it was concluded that furniture and cabinet-making entail exposures that are carcinogenic to humans (Group 1), and carpentry and joinery entail exposures that are possibly carcinogenic to humans (Group 2B); exposures in the lumber and sawmill industries and the pulp and paper industries could not be classified as to their carcinogenicity to humans (Group 3). Formaldehyde has been considered previously in the *IARC Monographs* series; a monograph was first prepared in 1981 (IARC, 1982), and the epidemiological and experimental data were updated during preparation of Supplement 7 (IARC, 1987), when it was concluded that formaldehyde is probably carcinogenic to humans (Group 2A).

The present volume of *Monographs* addresses wood dust rather than wood industries. Few epidemiological studies are available to characterize the carcinogenic risks of specific concentrations or types of wood dust. Most of the available evidence comes from case series and case–control studies of cancer of the nasal cavities and paranasal sinuses among workers in various types of employment. It was difficult to obtain direct measurements of exposure or to characterize the concentration or type of wood dust or types of additives; in addition, no account could be taken of changes in the work environment over time. No studies were found on nonoccupational exposure to wood dust. Some industries and occupations not previously evaluated that are described in the present monograph on wood dust include plywood and particle-board manufacture and pattern and model making. Exposures in the pulp and paper industry were not included.

Progress in the understanding of the carcinogenesis of wood dust would be greatly facilitated by additional studies in experimental animals; in particular, an experimental model should be developed. Although the Working Group was aware of on-going experiments on the carcinogenicity of oak dust, with and without various additives, sustained support of such studies would be valuable, ideally at two or more research laboratories. These studies are needed in order to test various hypotheses about the mechanism of carcinogenicity of wood dust. The results of such studies could be used to generate hypotheses for further epidemiological studies of human carcinogenesis.

A number of occupational situations that involve exposure to wood dust also entail exposure to formaldehyde, for instance in plywood and particle-board manufacture, during furniture and cabinet-making and during parquet floor sanding and varnishing. Formaldehyde is a constituent of living tissue and in this and other contexts is widely distributed in the environment, often at low concentrations. As described in the monograph, it also has specific uses in the chemical industry and is widely used as a preservative for biological materials. Humans are therefore occupationally exposed to formaldehyde, and studies have been conducted to investigate a possible relationship between exposure to formaldehyde and cancer. Cancer has not been associated with exposure to formaldehyde in an environmental context, nor with the occurrence of formaldehyde as a constituent of living tissue.

Many studies are available of occupational exposure to formaldehyde (sometimes in known association with other chemicals or types of exposure). Evaluation of the epidemiological data on formaldehyde is particularly difficult, from the point of view of classification of tumour type and establishment of causation in relation to rare tumours.

Evidence that formaldehyde causes cancer in experimental animals provides an inference of carcinogenic hazard for humans. Extrapolation of the available experimental findings to humans should take into consideration all other relevant data (including possible levels of human exposure and data indicating mechanism of action). Although human tissue may be inherently susceptible to the carcinogenicity of formaldehyde, any such effect may require exposure to concentrations of the chemical that are so high that they would not be tolerated because of its pungent odour. The Working Group gave detailed consideration to this and all other matters impinging on an evaluation of the carcinogenicity of formaldehyde to humans.

References

IARC (1981) *IARC Monographs on the Evaluation of the Carcinogenic Risk of Chemicals to Humans*, Vol. 25, *Wood, Leather and Some Associated Industries*, Lyon, pp. 49–157

IARC (1982) *IARC Monographs on the Evaluation of the Carcinogenic Risk of Chemicals to Humans*, Vol. 29, *Some Industrial Chemicals and Dyestuffs*, Lyon, pp. 345–389

IARC (1987) *IARC Monographs on the Evaluation of Carcinogenic Risks to Humans*, Suppl. 7, *Overall Evaluations of Carcinogenicity: An Updating of* IARC Monographs *Volumes 1 to 42*, Lyon, pp. 211–216, 378–387

THE MONOGRAPHS

WOOD DUST

1. Exposure Data

1.1 Composition of wood

Wood is one of the world's most important renewable resources and grows in forests all over the world. Forests cover about one-third of the globe's total land area, about 3.4 million km^2. This area represents more than 1.0 trillion m^3 of total tree biomass; of this biomass, about 3.5 thousand million m^3 per year are harvested, about half of which is used as fuel, predominantly in developing countries. 'Industrial roundwood' (1.7 thousand million m^3/year) is the term applied to all sawn wood (54%), pulpwood (21%), poles, pit props (14%) and wood used for other purposes such as particle-board and fibre-board (11%) (FAO, 1992; Schulz, 1993).

The species of trees that grow and are harvested in different countries vary considerably. Hardwoods dominate, for instance, in Italy, where oak, chestnut and beech are important species (Haden-Guest *et al.*, 1956). In other, primarily colder regions, conifers dominate: for example, pine and spruce in the Nordic countries and pine, spruce, hemlock, cedar and fir in Canada. Table 1 shows trees harvested for industrial use by broad category (conifer versus non-conifer) in some of the countries in which epidemiological studies have been conducted on wood dust. Even within countries, however, there is considerable variation: in western United States of America, conifers, such as Douglas fir and Ponderosa pine, were of primary economic importance in the 1950s, while various non-conifers were important in the mid-west and northeast, and southern yellow pine was the single most important species in the south-east (Haden-Guest *et al.*, 1956).

Many countries with little domestic production of lumber, such as the Netherlands, import wood, and even countries with much domestic production import wood for specific uses: for example, Finland, which produces pine, spruce and birch, imports some tropical woods, such as mahogany and teak, for furniture production (Welling & Kallas, 1991). The data in Table 1 probably represent the species used in the logging, sawmill and pulp and paper industries, which usually consume wood from nearby regions. The species of trees used in different branches of the wood industry are described in sections 1.4.2–1.4.6. About two-thirds of all wood used in the world for industrial purposes is from softwood species (FAO, 1993).

1.1.1 *Classification and nomenclature*

The Earth has an estimated 12 000 species of tree, each producing a characteristic type of wood. Spermatophytes are subdivided into two classes on the basis of seed type: gymnosperms, which have exposed seeds, and angiosperms, with encapsulated seeds. These classes are further separated into orders, families, genera and species. As an illustration of the main divisions, the full classification of Scots pine (*Pinus sylvestris* L.) is given in Figure 1.

Table 1. Industrial roundwood[a] production in 1980 (thousands of cubic metres) by country

Country	Conifers	Non-conifers
Europe		
Denmark	1 185 (64%)	665 (36%)
Finland	38 010 (88%)	5 010 (12%)
France	14 069 (49%)	14 897 (51%)
Germany (western)	21 670 (74%)	7 657 (26%)
Italy	1 404 (28%)	3 705 (72%)
Netherlands	577 (72%)	228 (28%)
Norway	8 158 (96%)	316 (4%)
Sweden	40 000 (89%)	4 795 (11%)
United Kingdom	2 609 (68%)	1 232 (32%)
Former USSR	246 400 (89%)	31 300 (11%)
Asia		
China[b]	50 016 (63%)	29 186 (37%)
Japan	21 427 (63%)	12 624 (37%)
North America		
Canada	144 100 (93%)	10 124 (7%)
United States	246 525 (75%)	80 570 (25%)
Australia	4 009 (26%)	11 642 (74%)
New Zealand	9 698 (98%)	247 (2%)

From FAO (1993)
[a] All wood used for sawn wood and veneer logs, pulpwood, chips, particles, poles and pit props
[b] Estimates

Most species are deciduous trees or hardwoods, and only about 800 species are coniferous trees or softwoods (Bauch, 1975). Wood-producing tree-like plants, such as bamboo (*Graminaceae*) and palm (*Palmae*), differ from trees in that they lack secondary thickness growth.

The terms 'hardwood' and 'softwood' refer to the species of tree and not necessarily to the hardness of the wood. While hardwoods are generally more dense than softwoods, the density varies considerably within each family and the hardness of the two groups overlaps somewhat (Fengel & Wegener, 1989; see also Table 6). Gymnosperms comprise all trees that yield softwood lumber. Only one order, Coniferales, is important from the point of view of industrial use. The angiosperms are separated into two classes on the basis of initial seed leaf: monocots (e.g. bamboo and palms) have one initial seed leaf, and dicots (e.g. oak and birch) have two. Dicots comprise all tree-sized plants that yield hardwood lumber and occur mostly in temperate zones.

Figure 1. Classification of Scots pine

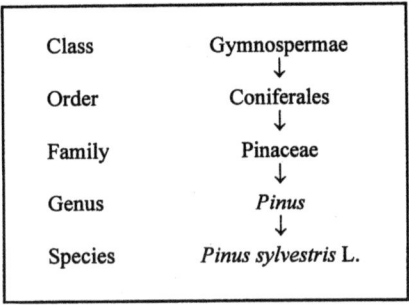

Class	Gymnospermae
Order	Coniferales
Family	Pinaceae
Genus	*Pinus*
Species	*Pinus sylvestris* L.

From Jane (1970)

The scientific and common names of some softwoods and hardwoods are given in Table 2.

Table 2. Nomenclature of some softwoods and hardwoods

Genus and species	Common name
Softwood	
Abies	Fir
Chamaecyparis	Cedar
Cupressus	Cypress
Larix	Larch
Picea	Spruce
Pinus	Pine
Pseudotsuga menziesii	Douglas fir
Sequoia sempervirens	Redwood
Thuja	Thuja, arbor vitae
Tsuga	Hemlock
Hardwood	
Acer	Maple
Alnus	Alder
Betula	Birch
Carya	Hickory
Carpinus	Hornbeam, white beech
Castanea	Chestnut
Fagus	Beech

Table 2 (contd)

Genus and species	Common name
Hardwood (contd)	
Fraxinus	Ash
Juglans	Walnut
Platanus	Sycamore
Populus	Aspen, poplar
Prunus	Cherry
Salix	Willow
Quercus	Oak
Tilia	Lime, basswood
Ulmus	Elm
Tropical hardwood	
Agathis australis	Kauri pine
Chlorophora excelsa	Iroko
Dacrydium cupressinum	Rimu, red pine
Dalbergia	Palisander
Dalbergia nigra	Brazilian rosewood
Diospyros	Ebony
Khaya	African mahogany
Mansonia	Mansonia, bete
Ochroma	Balsa
Palaquium hexandrum	Nyatoh
Pericopsis elata	Afrormosia
Shorea	Meranti
Tectona grandis	Teak
Terminalia superba	Limba, afara
Triplochiton scleroxylon	Obeche

From Vaucher (1986)

1.1.2 Anatomical features

Detailed information on wood anatomy is given by Jane (1970), Panshin and de Zeeuw (1970), Wagenführ and Scheiber (1974), Grosser (1977), Barefoot and Hankins (1982) and Hoadley (1990).

A look at the cross-section of a stem or a stem segment (Fig. 2) reveals a differentiation between bark and wood and, in many species, inner and outer areas with different coloration. Whereas some cells in the outer part (sapwood) are still alive (parenchyma), all cells in the inner part (heartwood) are dead. There is strong biosynthetic activity at the sapwood–heartwood boundary, where stored materials such as starch and other carbohydrates are transformed into low- and medium-relative-molecular-mass substances (extractives) and deposited in the heartwood (Streit & Fengel, 1994). It is assumed that these substances contribute to the conservation and protection of wood. The cells of wood tissue are produced in the cambium, a cell monolayer between the phloem (inner bark) and xylem (wood) (Fig. 2), where growth in length and thickness of a tree occurs.

The morphology of softwood tissue is simpler than that of hardwood tissue. Softwood consists, in its bulk, of only one cell type, tracheids, which are elongated, fibre-like cells with a square or polygonal cross-view. Less than 10% of the wood consists of short, brick-like parenchymal cells arranged radially. Moreover, some softwoods contain epithelial cells that secrete resin into canals which run horizontally and radially through the wood.

Figure 2. Segment of a stem showing the various macroscopically visible areas of wood

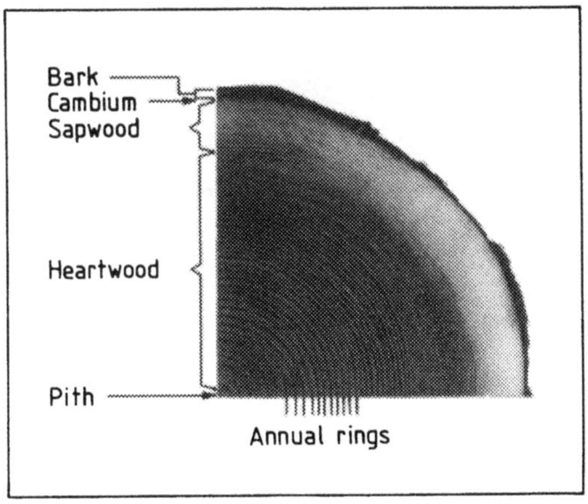

Adapted from Hoadley (1990)

Tissue formed in springtime (in temperate zones) is called 'earlywood' and consists of tracheids with wide lumina and thin walls (Fig. 3), which transport water from the roots to the top of the tree. Rows of valve-like openings (pits) at the ends of the tracheids allow exchange of water between adjacent cells. Tracheids produced in summertime ('latewood') have thick walls and small lumina; they serve predominantly to stabilize the tree.

In hardwoods, there is more detailed differentiation between stabilizing, conducting and storage tissue. Stabilizing tissues contain libriform fibres and fibre tracheids, which are elongated cells with thick polygonal walls and small lumina (Fig. 4). The conducting system is composed of vessel elements fitted together to form long tubes of up to several metres. The vessels have thin walls and large diameters. At the junction of two elements, there are stiffening rings or plates with perforations characteristic of different wood species. Storage tissues consist of longitudinally and radially arranged parenchymal cells. Hardwoods that contain resin canals also have a secretory system of epithelial cells.

Figure 3. Border between latewood (left) and earlywood (right) in a softwood (spruce); light micrograph

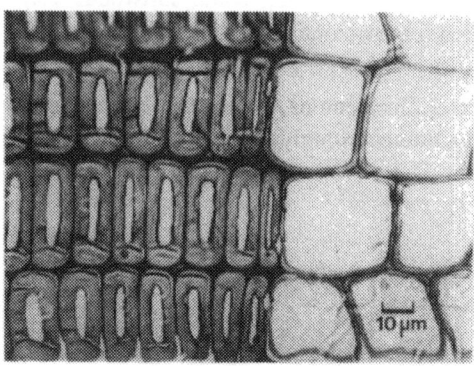

From Fengel & Wegener (1989)

Figure 4. Hardwood tissue (beech) with vessels (V) and parenchymal cells (P) surrounded by libriform fibres (F); scanning electron micrograph

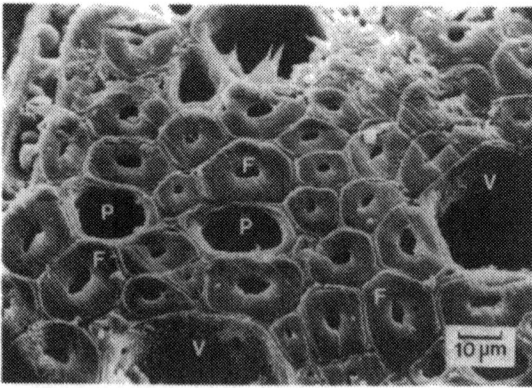

Adapted from Fengel & Wegener (1989)

1.1.3 Cell wall structure and distribution of components of wood

The walls of wood cells consist of various layers, which differ in structure and chemical composition (Fig. 5; Fengel & Wegener, 1989). The individual cells of wood tissue are glued together in the middle lamella, which consists mainly of lignin, polyoses and pectins. Often there is no exact, visible border between the pure middle lamella and the outer cell wall layer, which is called the primary wall and is formed by a net-like arrangement of cellulose fibrils (Fig. 5a) embedded in a matrix of lignin and polyoses. The middle lamella and primary wall are often called the 'compound middle lamella' (Fig. 5b).

The next layer is the secondary wall, which is subdivided into secondary walls 1 (S1) and 2 (S2). S1 and S2 contain densely packed cellulose fibrils arranged in parallel, which differ in the angle at which the fibrils run: in the S1, the fibrils run at a wide angle in relation to the fibre axis and in the S2, at a small angle. The S2 is the thickest wall layer and accounts for 50% (vessels, parenchymal cells) to 90% (tracheids, libriform fibres) of the whole cell wall. Parenchymal cells are equipped with a third secondary wall (S3) which has a fibril arrangement that is more open than that in the S2.

At the inner border of the cell wall, there is a final thin layer called the tertiary wall (T), in which the cellulose fibrils run at an angle similar to that of the S1. In some species, the tertiary walls of tracheids, fibres and vessels are covered with a wart-bearing amorphous layer (Fig. 5a,b).

The lignin content decreases from the compound middle lamella through the S2, while the cellulose content increases in the same direction (Table 3). The polyose content is highest in the S1, but because of the thickness of the S2, most lignin and polyoses are deposited in this layer.

The 'extractives', the organic matter of wood, are found in the resin canals and parenchymal cells. In heartwood, extractives are also deposited in the compound middle lamella and the secondary walls. Trees with a high content also have extractives in the lumina (Fengel, 1989).

In areas where mechanical deformation of stems and branches has occurred, special tissues are found. In such 'compression areas' in softwood tracheids, the S2 is provided with helical cavities and contains a high percentage of lignin (compression wood). In hardwoods, fibres with a thick additional gelatinous layer consisting of relatively pure cellulose are formed in tension areas (tension wood). Compression and tension woods are together referred to as 'reaction wood' and may influence the processing of wood.

1.1.4 Chemical components

In this section, the main constituents of woods are identified and differences between hardwood and softwood are indicated. The formulae of a number of chemical constituents are shown in Figure 6. Chemicals found only in fruit or flowers or in fungi growing on trees are not described. Extensive reviews, with detailed accounts of numerous extractives, have been published (Beecher et al., 1989; Fengel & Wegener, 1989; Swan, 1989). Hillis (1987) specifically reviewed heartwood.

Figure 5. Cell wall structure

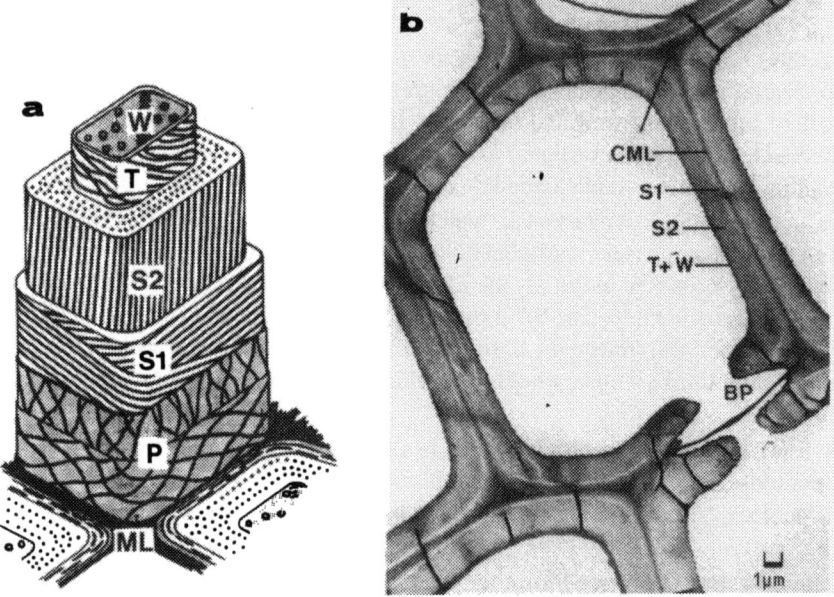

From Fengel & Wegener (1989)

(a) Model of a vascular cell (softwood tracheids, hardwood libriform fibres); (b) cross-section of a softwood tracheid (umbrella fir) with a bordered pit (BP); transmission electron micrograph. ML, middle lamella; P, primary wall; S1, secondary wall 1; S2, secondary wall 2; T, tertiary wall; W, warty layer; CML, compound middle lamella (ML + P)

Table 3. Approximate content of lignin, cellulose and polyoses in the cell wall layers of softwood tracheids

Wall layer	Lignin (%)	Cellulose (%)	Polyoses (%)	Main polyoses
Compound middle lamella	60	15	25	Pectins, galactan, xylan
Secondary wall 1	30	35	35	Xylan, mannan
Secondary wall 2 + tertiary wall	25	60	15	Mannan

From Fengel & Wegener (1989)

Figure 6. Chemical formulae of certain chemical components of wood

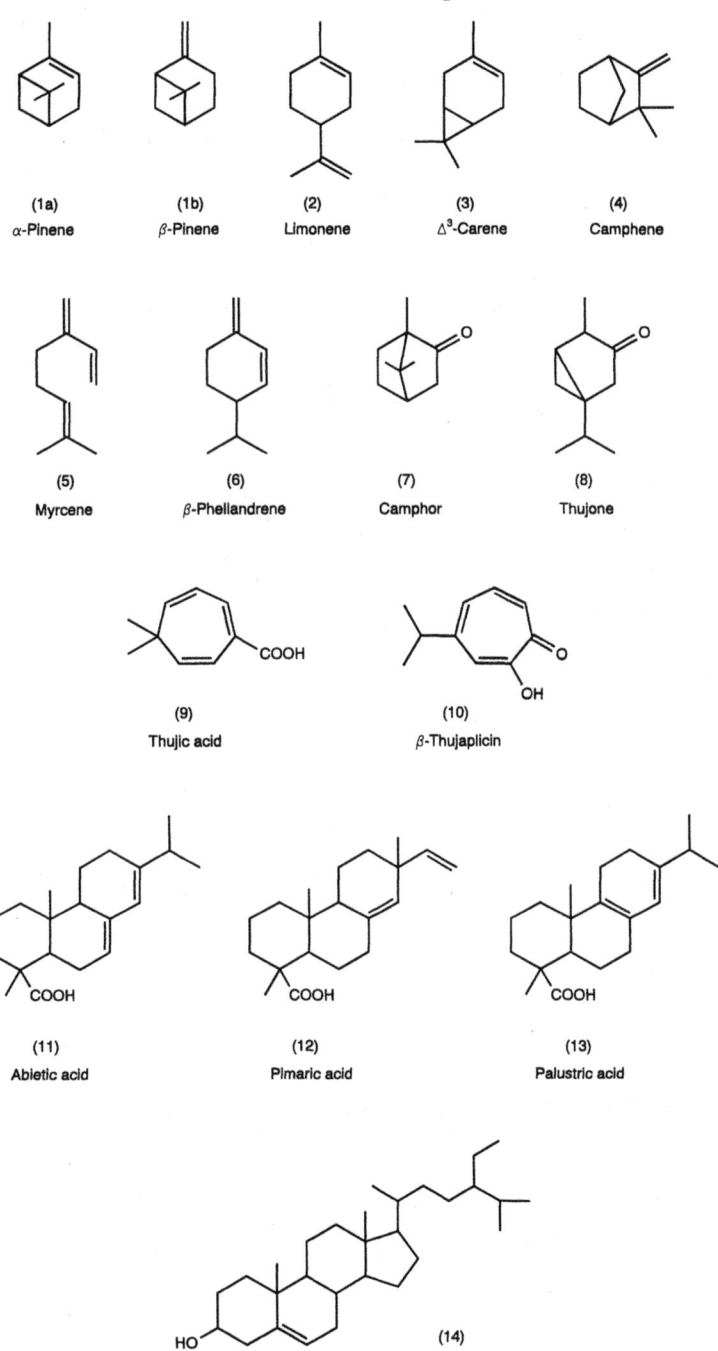

Figure 6 (contd)

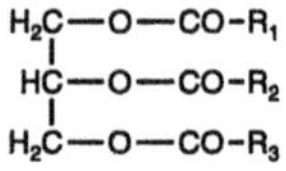

(15)
Fats and Oils

$H_3C—(CH_2)_n—CO—O—(CH_2)_m—CH_3$

(16)
Waxes

$H_3C—(CH_2)_n—COOH$

(17)
Fatty acids

$HO—(CH_2)_m—CH_3$

(18)
Fatty alcohols

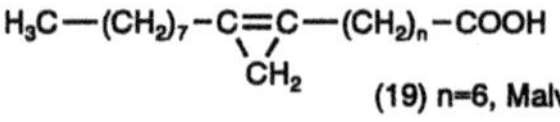

(19) n=6, Malvalic acid
(20) n=7, Sterculic acid

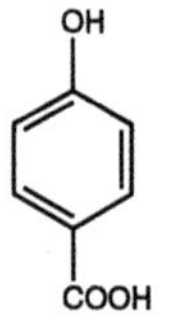

(21)
para-Hydroxy-
benzoic acid

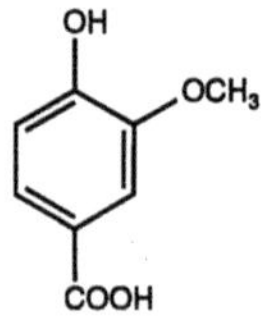

(22)
Vanillic acid

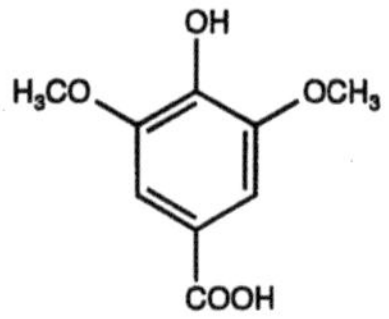

(23)
Syringic acid

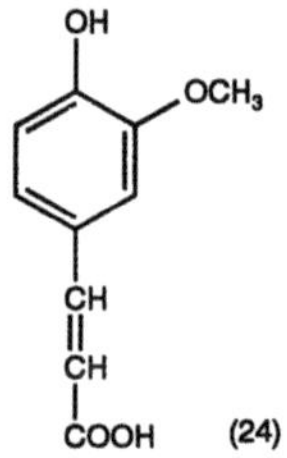

(24)
Ferulic acid

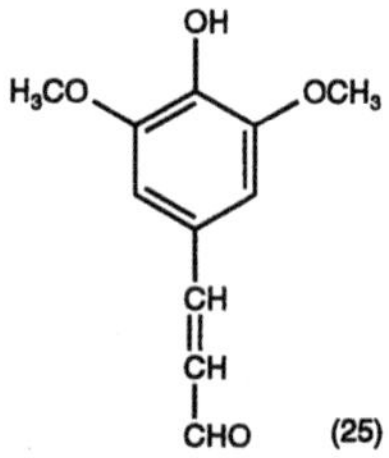

(25)
Sinapaldehyde

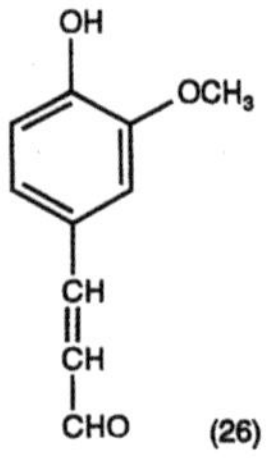

(26)
Coniferyl aldehyde

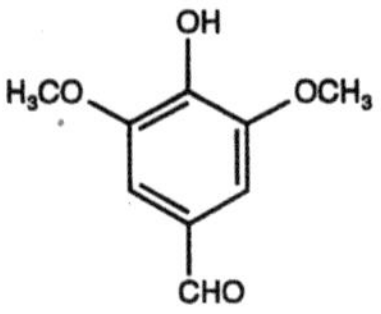

(27)
Syringaldehyde

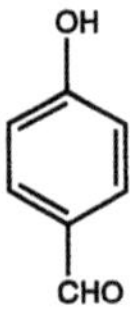

(28)
para-Hydroxy-
benzaldehyde

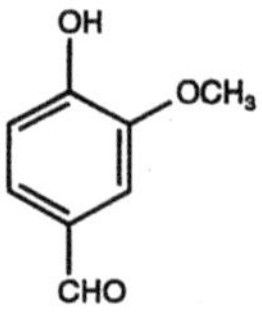

(29)
Vanillin

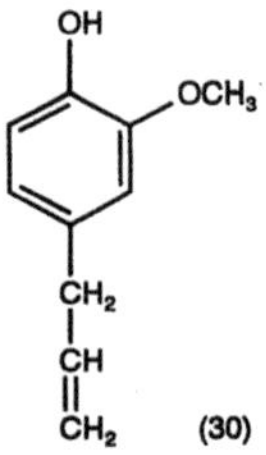

(30)
Eugenol

Figure 6 (contd)

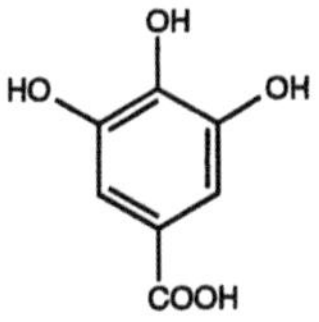

(31) Gallic acid

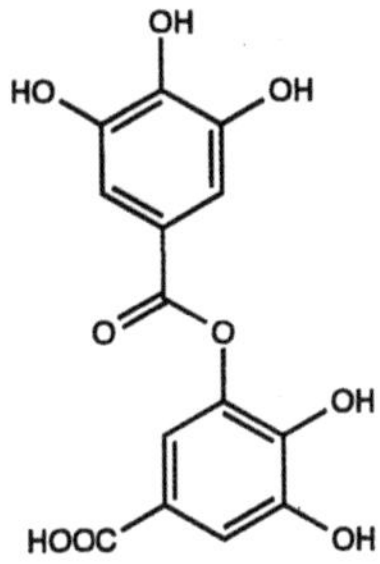

(32) Digallic acid

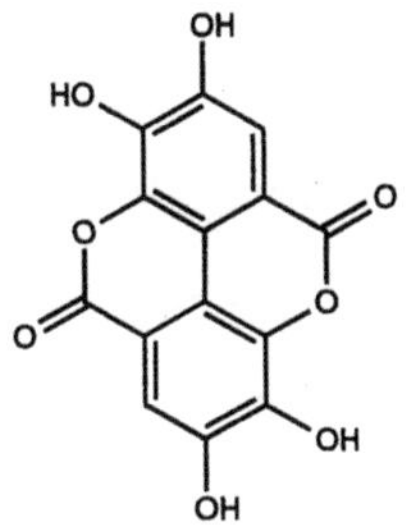

(33) Ellagic acid

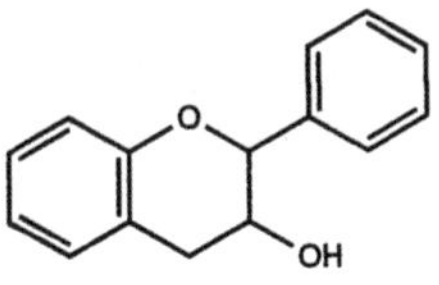

(34) Flavan-3-ol

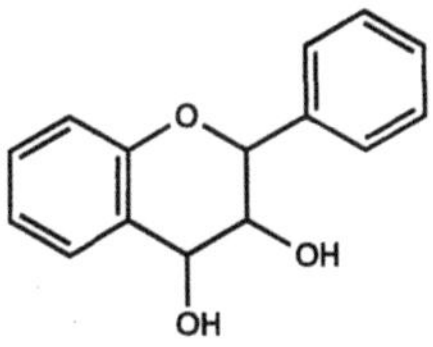

(35) Flavan-3,4-diol

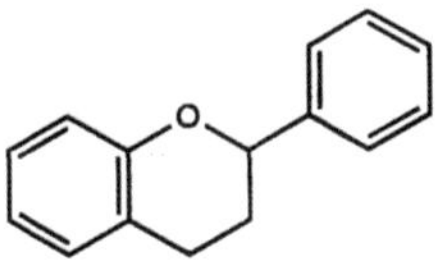

(36) Flavanes

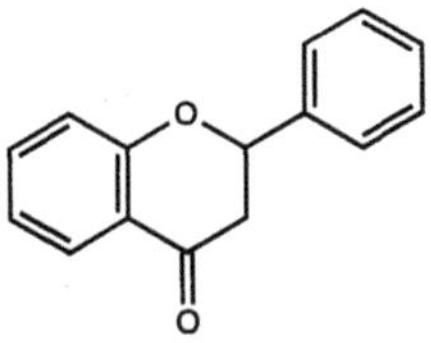

(37) Flavanones

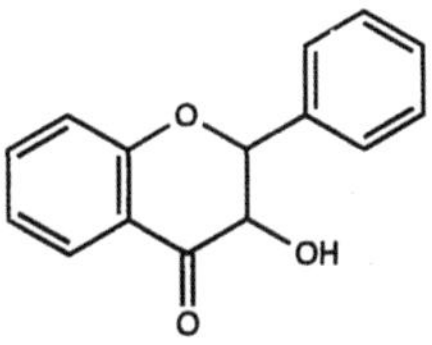

(38) Flavanonols

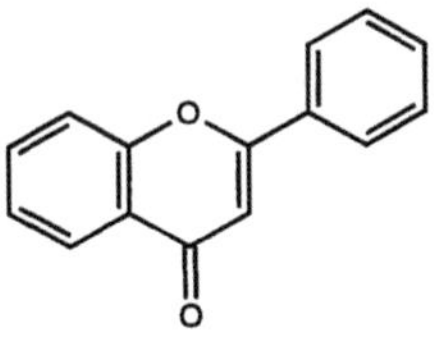

(39) Flavones

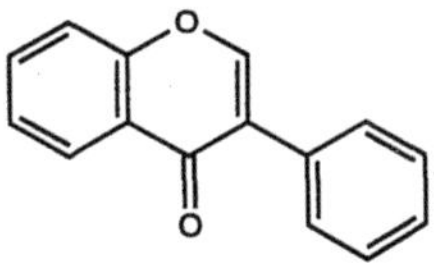

(40) Isoflavones

Figure 6 (contd)

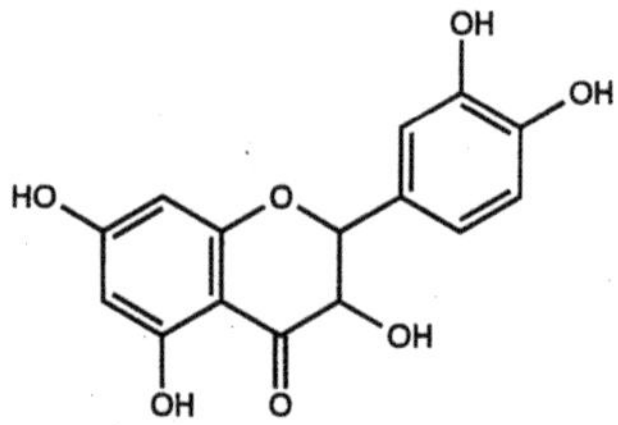

(41) Taxifolin

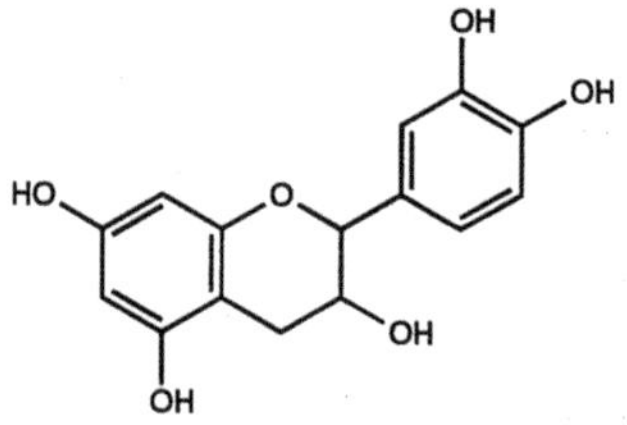

(42) Catechin

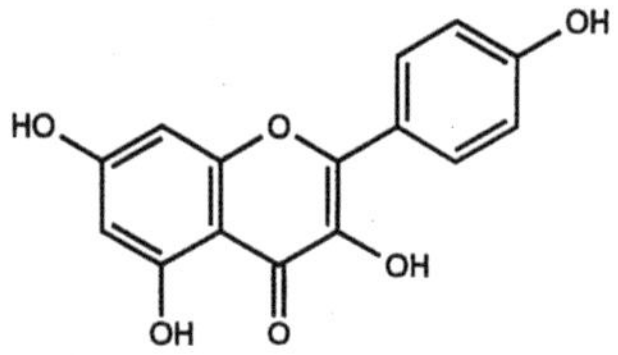

(43) Kaempherol

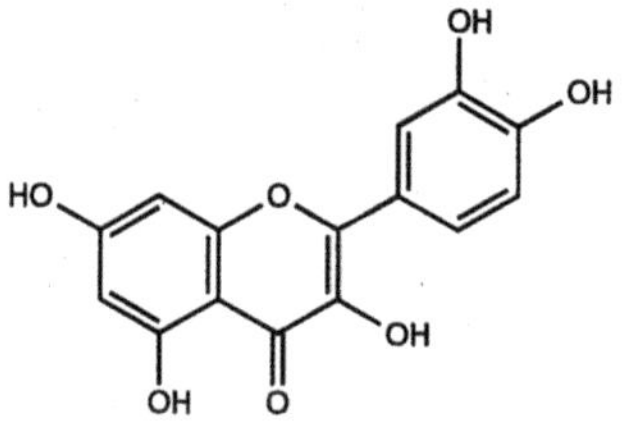

(44) Quercetin

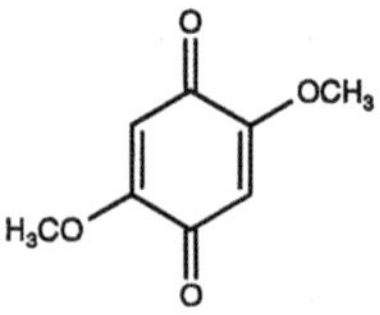

(45a) 2,5-Dimethoxy-benzoquinone

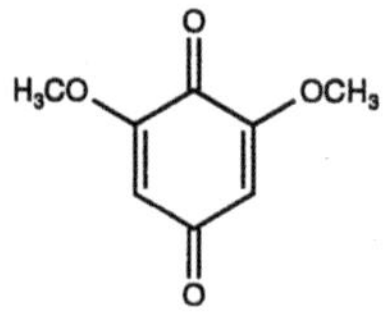

(45b) 2,6-Dimethoxy-benzoquinone

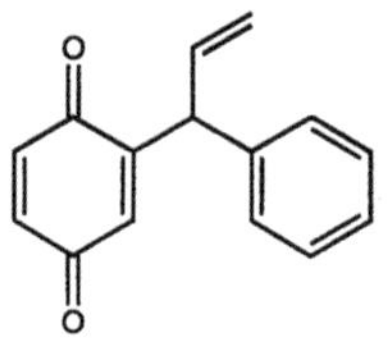

(46) Dalbergione

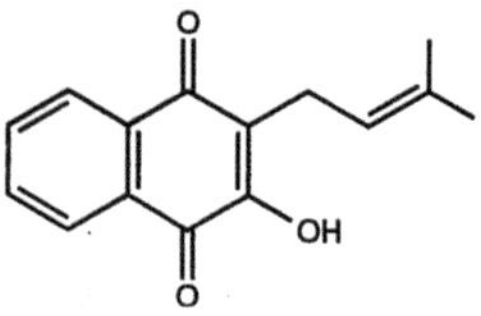

(47) Lapachol

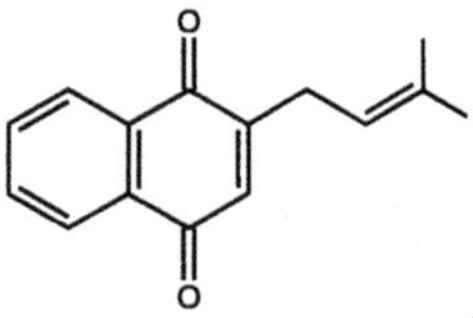

(48) Desoxylapachol

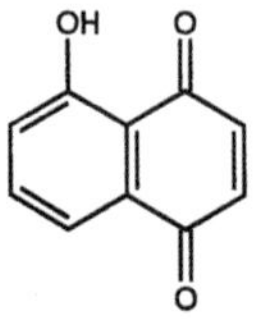

(49) Juglone

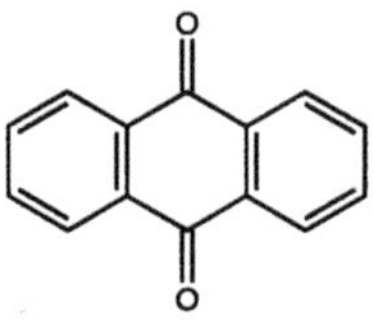

(50) Anthraquinones

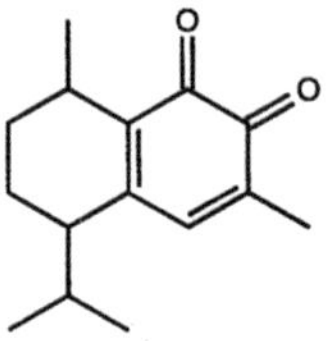

(51) Mansanone A

Figure 6 (contd)

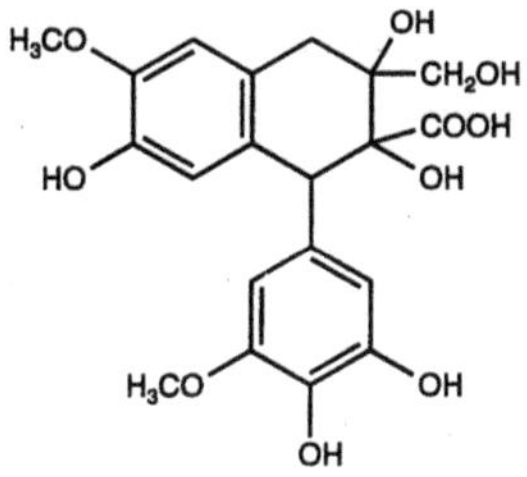

(52) Plicatic acid

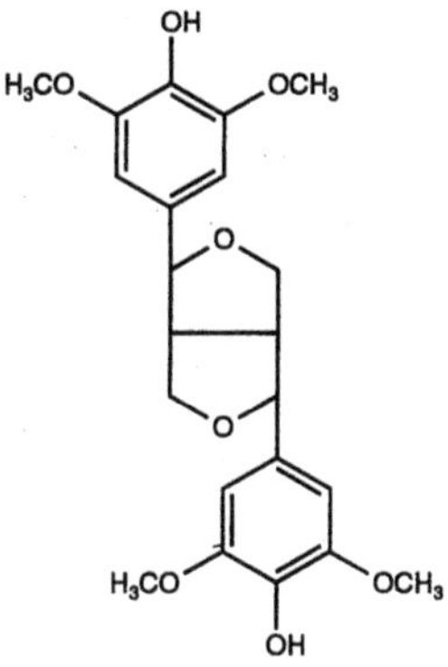

(53) Syringaresinol

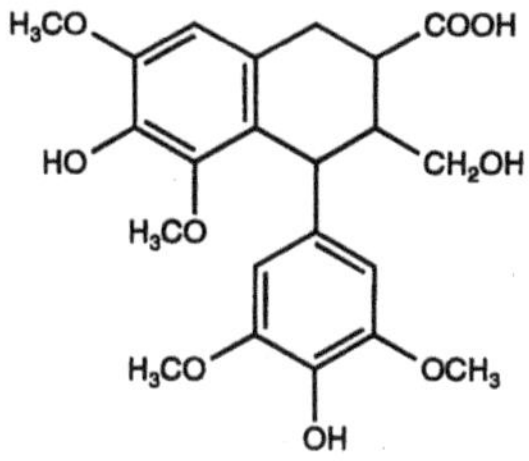

(54) Thomasic acid

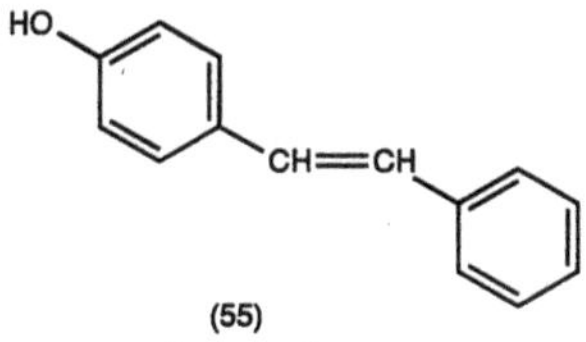

(55) 4-Hydroxystilbene

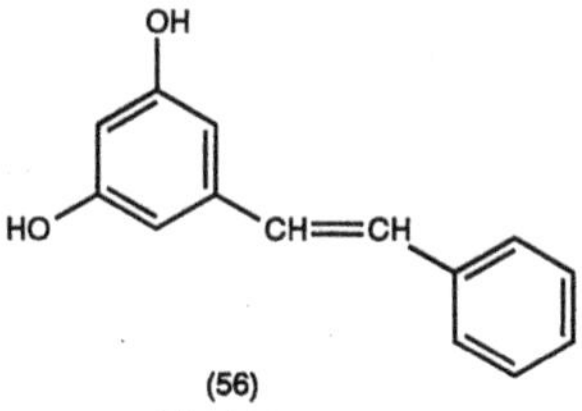

(56) Pinosylvin

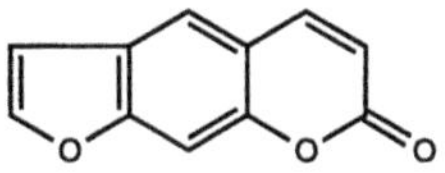

(57) Psoralen

(a) Macromolecular components

The essential chemical constituents of wood are cellulose, polyoses (hemicelluloses) and lignin, which has a macromolecular structure. Although cellulose is the uniform structural element of all woods, the proportions and chemical composition of lignin and polyoses differ in softwood and hardwood. Wood generally also contains small amounts of polymeric compounds, such as starch, pectic substances and proteins.

Cellulose is the major component (40–50%) of both softwood and hardwood. It can be briefly characterized as a linear high-relative-molecular-mass polysaccharide built up exclusively of D-glucose units joined by $\beta(1\rightarrow4)$glycosidic linkages.

Polyoses (hemicelluloses) are present in larger amounts in hardwood than in softwood and differ in their sugar composition. They are composed mainly of five neutral sugar units—the hexoses, glucose, mannose and galactose, and the pentoses, xylose and arabinose. Some polyoses also contain uronic acid units. The molecular chains of polyoses are much shorter than those of cellulose and are branched and/or contain side groups. Softwood has a higher proportion of mannose units and more galactose units than hardwood; hardwood has a higher proportion of xylose units (Table 4).

Table 4. Non-glucosic sugars in polyoses of some woods (%)

Species	Mannose	Xylose	Galactose	Arabinose	Uronic acid	Rhamnose
European spruce	13.6	5.6	2.8	1.2	1.8[a]	0.3
Scots pine	12.4	7.6	1.9	1.5	5.0	NR
European beech	0.9	19.0	1.4	0.7	4.8[a]	0.5

From Fengel & Wegener (1989); NR, not reported

[a]4-*O*-Methylglucuronic acid

Lignin, the third macromolecular component of wood, is quite different from the polysaccharides. The monomers of lignin are phenylpropane units joined by various linkages, resulting in complicated three-dimensional macromolecules. The lignin content of softwood is higher than that of hardwood, and softwood and hardwood lignins differ structurally and with regard to their contents of guaiacyl, syringyl and *para*-hydroxyphenyl units. Most softwood lignins are typical guaiacyl lignins containing minor amounts of syringyl and *para*-hydroxyphenyl units. The composition of lignin varies much more in hardwood than in softwood, and hardwood lignins have higher proportions of syringyl units. The syringyl content of typical hardwood lignins varies between 20 and 60%.

(b) Low-relative-molecular-mass components

A heterogeneous mixture of organic and inorganic compounds occurs in different species of wood in various amounts. The organic matter that can be extracted from wood with nonpolar or polar solvents is commonly called 'extractives'; the inorganic part is reduced mainly to ash in the analysis of wood. The 'extractives' represent 0.1–1% of the wood mass in trees of temperate zones and 15% or more in tropical wood. As some of these compounds protect against injury or attack from fungi, insects and bacteria, they may have toxic, irritant or sensitizing properties.

Organic extractives can have aliphatic, alicyclic or aromatic structures. Non-polar extractives comprise mainly terpenes, fatty acids, resin acids, waxes, alcohols, sterols, steryl esters and glycerides. The polar extractives of wood generally consist of aromatic (phenolic) compounds, i.e. tannins, flavonoids, quinones and lignans. The common water-soluble extractives of wood are carbohydrates and their derivatives, alkaloids, proteins and inorganic material. Hardwood tends to have a higher percentage of polar extractives than softwood (see section 1.3.2 (*a*) and Table 7).

(i) Terpenes and terpenoids

Terpenes are a ubiquitous group of natural compounds, of which over 4000 have been isolated and identified. They can be derived from isoprene (2-methyl-1,3-butadiene) units, which are usually connected to form one or more rings. Two or more isoprene units build up the mono-, sesqui-, di-, tri-, tetra- and polyterpenes.

The extractives of softwood include all classes of terpenes, whereas hardwoods contain mainly higher terpenes. Monoterpenes are found only in some hardwood tropical species, while the volatile oil (turpentine) of softwoods consists mainly of these compounds. The commonest are α- and β-pinene (Fig. 6, **1a, 1b**) and limonene (**2**) (see IARC, 1993), which are present in all softwoods; but Δ^3-carene (**3**), camphene (**4**), myrcene (**5**) and β-phellandrene (**6**) are also widespread. α-Pinene has been suggested to be a sensitizing agent, and terpenes with a keto group (e.g. camphor (**7**) and thujone (**8**)) appear to be more toxic than related compounds.

The tropolones, structures with seven-membered rings, are considered to be derivatives of monoterpenes. Compounds such as thujic acid (**9**) and β-thujaplicin (**10**) are found only in species of the *Cypressaceae* family. Sesquiterpenes are found in many tropical woods but are quite rare in hardwood in temperate zones and in softwood.

The diterpenes seem to be restricted to softwood species and occur mainly in the form of resin acids. They are mostly tricyclic compounds, such as abietic (**11**), pimaric (**12**) and palustric acids (**13**). The neutral diterpenes consist of hydrocarbons, oxides, alcohols and aldehydes.

A great variety of triterpenes are present in many hardwoods in tropical and temperate zones and also in softwoods. Most have a steran substructure and must therefore be assigned to the steroids. The main component of the steroid group in both softwood and hardwood is β-sitosterol (**14**). Some saponins, which are glycosides of triterpenes and steroids, cause dermatitis and other diseases.

(ii) Fats, waxes and their components

Although saturated and unsaturated higher fatty acids are found in wood mostly as esters with glycerol (fats and oils (**15**)) or with higher alcohols (waxes (**16**)), they also occur in free form (**17**). Free and combined fatty acids typically include linoleic, palmitic (**17**: $n=14$) and stearic acid (**17**: $n=16$) and some cyclopropenic acids, such as malvalic (**19**) and sterculic (**20**) acid. The corresponding fatty alcohols (**18**) are also found. Triglycerides dominate the glycerides (fats), as compared with mono- and diglycerides, especially in hardwood species.

(iii) Phenolic compounds

Extracts of woods contain low-relative-molecular-mass phenols. Some are probably degradation products of compounds such as lignin, which may be hydrolysed during several extractions or during steam distillation. These simple phenols are represented in hardwood predominantly by acids (including *para*-hydroxybenzoic (**21**), vanillic (**22**), syringic (**23**) and ferulic acids (**24**)) and in both softwood and hardwood by aldehydes (including sinapaldehyde (**25**), coniferyl aldehyde (**26**), syringaldehyde (**27**), *para*-hydroxybenzaldehyde (**28**), vanillin (**29**) and eugenol (**30**) [see IARC, 1985a]).

The extractives of wood also contain various compounds with phenolic substructures, including tannins, flavonoids, quinones, lignans and stilbenes.

Tannins: Tannins can be separated into hydrolysable and condensed types (phlobaphenes). The hydrolysable tannins are esters of gallic acid (**31**) and its dimer (digallic acid (**32**)) and of ellagic acid (**33**) with monosaccharides, mainly glucose. They may be subdivided into gallotannins, which yield gallic acid, and ellagitannins, which yield ellagic acid upon hydrolysis. The main components of condensed tannins are catechin derivatives (flavan-3-ol (**34**)) and leucoanthocyanidin derivatives (flavan-3,4-diol (**35**)). Tannins can precipitate proteins and influence cell metabolism.

In general, tannins are found mainly in hardwoods, although they may also occur in the heartwood of certain softwoods (conifers) that contain condensed tannins. Hydrolysable tannins (gallic acid type) occur less frequently than condensed tannins in hardwoods and are found predominantly in oak, chestnut and eucalyptus. All other woods, particularly tropical woods, contain mainly condensed tannins (catechin type). An overview of tannin chemistry is presented by Hemingway and Karchesy (1989).

Flavonoids: Condensed tannins belong to the large group of extractives known as flavonoids, which comprise the subgroups flavanes (**36**), flavanones (**37**), flavanonols (**38**), flavones (**39**) and isoflavones (**40**). They have various numbers of hydroxyl and methoxyl groups on the aromatic rings, either as glycosides or aglycones. The large number of flavonoids is due not only to the degree of saturation of the heterocyclic ring but also to the variation in degree of hydroxylation of the rings. The colour of some heartwoods is a result of the presence of flavonoids and related compounds. These compounds also occur more frequently in hard- than softwood, although several flavonoids, such as taxifolin (**41**) and catechin (**42**), have also been identified in softwood. Other flavonoids found in hardwood are kaempherol (**43**) (see IARC, 1983a) and quercetin (**44**) (see IARC, 1983b).

Quinones: A range of aromatic quinones is present in extracts of various woods, including substituted benzoquinones (2,5- (**45a**) and 2,6-dimethoxybenzoquinone (**45b**), dalbergione (**46**)), various naphthoquinones (lapachol (**47**), desoxylapachol (**48**), juglone (**49**)) and anthraquinones (**50**) and their derivatives, such as quinoid sesquiterpenes (mansonone A (**51**)). They are responsible for the strong colours, high durability and dermatitic properties of some woods. Dimethoxybenzoquinone has been found in extracts of about 50 wood species.

Lignans: Lignans consist of two β–β-linked phenylpropane units. Some are dimeric structures that are also present in the lignin molecule. Lignans are typical heartwood components and occur in small or negligible amounts in sapwood. Although they are often found in softwood (e.g. plicatic acid (**52**) in western red cedar), some occur in hardwood, including syringaresinol (**53**) and thomasic acid (**54**), particularly in alder, oak and elm species.

Stilbenes: Stilbenes occur especially in the heartwood of both softwood and hardwood. These compounds, e.g. 4-hydroxystilbene (**55**) and pinosylvin (**56**), are responsible for the photosensitive reactions that cause darkening of woods.

(iv) Miscellaneous organic compounds

Free amino acids and their linked structures (proteins) are also found, but in small amounts. A high protein content encourages the development of wood-destroying organisms. Some hardwood species, particularly many tropical woods, also contain alkaloids of various chemical structures, including very toxic compounds such as berberine and strychnine. Phototoxic compounds belonging to the furocoumarin group, such as psoralen (**57**) and its derivatives, occur in some tropical wood species. Table 5 includes some biologically active compounds found in wood.

(v) Inorganic compounds

The main inorganic components of wood are potassium, calcium and magnesium; silicon is found in some tropical woods (Fengel & Wegener, 1989). These components comprise 0.2–0.5% of wood in temperate zones but often much more in tropical woods. Potentially carcinogenic inorganic elements such as chromium have been found in very small amounts in some wood species (Saka & Goring, 1983).

Table 6 summarizes several characteristics of softwoods and hardwoods; these are generalizations to which there are specific exceptions. Some characteristics are the same for softwoods and hardwoods; for others, the ranges among species overlap; and for a few characteristics, there are marked distinctions.

1.2 Wood-related industries and occupations

Workers in a wide variety of industries may be exposed to wood dust. In this section, the main industries and occupations in which such exposure may occur and the steps in the processes used are described (see also Koch, 1964; Maloney, 1977; FAO, 1981; IARC, 1981; Darcy, 1984; Industrial Accident Prevention Association, 1985; McCammon *et al.*, 1985; Holliday *et al.*, 1986; Suchsland & Woodson, 1986; Clayton Environmental Consultants, 1988; Williston, 1988). Detailed descriptions are generally provided only for processes that result in wood dust. The history of woodworking processes in the sawmill, furniture and construction industries was reviewed previously (IARC, 1981) and is not repeated here; however, major changes in processes that occurred during this century and affect exposure to wood dust are discussed.

Table 5. Examples of biologically active organic compounds found in wood

Substance class	Compound	Wood type
Terpenes	α-Pinene	Softwood
	Δ^3-Carene	Softwood
	Camphor	Softwood
	Thujone	Softwood
	β-Thujaplicin	Softwood
	Sesquiterpene lactones	Softwood/hardwood
	Abietic/Neoabietic acid	Softwood/hardwood
	Saponins	Hardwood
Phenols	Coniferyl aldehyde	Softwood/hardwood
	Sinapaldehyde	Hardwood
	Eugenol	Softwood/hardwood
	3-(Pentadecyl)catechol	Hardwood
	5-(Pentadec-10-enyl)resorcinol	Hardwood
Tannins	Catechin derivatives	Hardwood
	Leucoanthocyanidin derivatives	Hardwood
Flavonoids	Kaempherol	Hardwood
	Quercetin	Hardwood
Quinones	2,5- and 2,6-Dimethoxybenzoquinone	Softwood/hardwood
	3,4-Dimethoxydalbergione	Hardwood (tropical)
	Lapachol	Hardwood
	Desoxylapachol	Hardwood
	Juglone	Hardwood
	Mansonone A	Hardwood (tropical)
Lignans	Plicatic acid	Softwood
Stilbenes	2,3′,4′,5′-Tetrahydroxystilbene	Softwood
	Chlorophorin	Softwood
	Pinosylvin	Softwood
Miscellaneous	Alkaloids (berberin)	Hardwood
	Furocoumarins (psoralen)	Hardwood (tropical)

From Hausen (1981), Henschler (1983) and Swan (1989)

Table 6. Comparison of softwoods and hardwoods

Characteristic	Gymnosperms/conifers/ softwoods	Angiosperms/deciduous wood/hardwoods
World production of industrial roundwood (1980) ($\times$ 1000 m^3)	990 000	450 000
Density (g/cm^3)	White (silver) fir: mean, 0.41 (0.32–0.71) European spruce: mean, 0.43 (0.30–0.64) Scots pine: mean, 0.49 (0.30–0.86)	European beech 0.68 (0.49–0.88) European oak 0.65 (0.39–0.93)
Fibres	Long (1.4–4.4 mm)	Short (0.2–2.4 mm)
Cell type	One (tracheids)	Various
Cellulose	~40–50%	~40–50%
Unit	β-D-Glucose	β-D-Glucose
Fibre pulp	Long	Short
Polyoses	~15–30%	~25–35%
Units	More mannose More galactose	More xylose
Lignin	~25–35%	~20–30%
Units	Mainly guaiacyl	Mainly syringyl or guaiacyl
Methoxy group content	~15%	~20%
Extractives content		
Non-polar (e.g. terpenes)	High	Low
Polar (e.g. tannins)	Low	High

From Fengel & Wegener (1989)

1.2.1 Major woodworking processes

(a) Debarking

Debarking is the mechanical removal of bark from a log and is performed in sawmills and other mills where logs are first processed. Debarking can be done in a number of ways: in wood-to-wood abrasion, the pounding and friction of logs against each other in a rotating drum removes the bark; in the 'flail' method, chains pound against the log to loosen the bark; in peripheral milling, logs are rotated against knives. Bark can also be removed by pressing tool points against the log to loosen the bark and then using a ring debarker or high-pressure water jets. In general, debarking involves little or no exposure to wood dust, because the wood is 'green' (fresh) and thus has a high moisture content. Furthermore, the main goal of the operation is to leave the wood intact.

(b) Sawing

Saws are used to cut logs or large pieces of wood into appropriate sizes for further modification and use. Sawing is performed by drawing a blade with a series of sharpened teeth through the wood. As with many woodworking machines, the amount of wood dust generated by mechanical sawing operations is influenced by the speed of the sawing action, the angle of cutting relative to the wood grain and the sharpness and width of the blade. Sawing against the wood grain (cross-cutting) is more likely to shatter wood cells than sawing lengthwise (ripping). Sharp, thin blades produce less wood dust by volume because the kerf, the cut made in the wood, is narrower, but the particle sizes are also likely to be smaller.

The simplest saws are blades with a series of teeth along one edge and a handle on one or both ends; they are powered by a human operator or operators, who move the blade back and forth. Almost all saws used for commercial purposes are mechanically operated. In recent decades, high-energy jets and lasers have been introduced, which generate less wood dust but are not in broad commercial use. The commonest types of mechanical saws are described below.

(i) Band saw

The blade of a band saw is a continuous metal strip with teeth on one or both edges which rotates around two wheels. Band saws are used in many wood industries, from sawmills to wood product manufacturing and can be powered by steam, hydraulic or electric mechanisms.

(ii) Circular saw

The blade of a circular saw is a rigid metal disk with teeth along its circumference which cuts as it rotates. Circular saws are also used in many wood industries. A table saw is a circular saw with the blade protruding through a table; a radial arm saw is a circular saw suspended above the working surface on a movable armature.

(iii) Sash gang saw

In a sash gang saw (frame saw), a series of parallel blades (a gang) fixed between two vertical members (a sash) are drawn up and down to rip boards being moved through the saw on rollers. These saws are used almost exclusively in sawmills to cut large pieces of lumber lengthwise in order to create a set of boards. Circular saws are also commonly used for gang sawing.

(iv) Jig saw

A jig saw has a short, rigid blade attached to a reciprocating mechanism to cut with an up-and-down motion. Jig saws are used in many wood products industries and can be both hand-held and stationary.

(v) Chain saw

A chain saw has a continuous articulated chain with teeth along its outer edge. It is generally powered by a gasoline engine and is used almost exclusively in logging, although it may sometimes be used as a cut-off saw in sawmills.

(c) Sanding

Sanding is smoothing the surface of wood, 'an abrasive process in which sharp edges of small, hard, crystalline particles are rapidly drawn across the surface of the wood with pressure being applied perpendicular to the surface' (Holliday *et al.*, 1986). The abrasives used include carborundum, emery, glass and pumice. The smaller the abrasive particles, the finer the dust produced; and the faster the sander, the greater the volume of dust produced. Sanding is done in many wood industries, with small, hand-held sanders or generally larger stationary machines. Sanders can range in size from hand-held to large drums or belts for smoothing a full panel. The commonest types of sander are the belt sander, a continuous strip of sandpaper rotated around two rollers; the disk sander, a circular piece of sandpaper fastened to a rotating disk; the drum sander, a continuous strip of sandpaper rotated on a drum; and the orbital sander, which operates with an elliptical, vibrating motion.

(d) Planing, jointing, moulding and shaping

Planing, jointing, moulding and shaping are milling processes. A planer is used to smooth one or more sides of a piece of wood and at the same time to reduce it to a predetermined thickness. The planer head is made up of a series of cutting blades mounted on a cylinder, which revolves at high speed. The operation is generally performed parallel to the wood grain.

A jointer is a machine for squaring and smoothing the edges of lumber or panels. Jointers are used especially in preparation for glueing and in other situations in which a smooth surface is needed. Jointers vary in size from small, hand-held devices to large, stationary machines. A jointer is similar to a planer in its operation, but its blades are generally smaller, and it is designed to smooth or true a surface rather than to change the thickness of a board.

Moulders are used to cut and shape mouldings. They generally consist of a top cutter head, followed by two sideheads and by a bottom cutterhead. The cutterheads are staggered spindles of various designs. Shapers are similar to moulders, but are used to cut and shape the outer sides of wood boards and products. Shapers generally consist of a table through which protrudes a rotating spindle with blades shaped to produce the desired contour. Shapers can be used to cut the edge designs found on furniture and picture frames and in many other applications, such as wooden model making.

(e) Turning (lathing)

Turning involves use of a lathe to produce cylindrical shapes in wooden objects. One end of the piece of wood is fixed to a clamp or plate, which is rotated. The point of operation is a tool point or a long knife, which has a cutting, scraping or shearing action (depending on the angle of contact) when applied against the wood. Some lathes rotate the piece quickly while the tool or knife remains stationary and in continuous contact until the desired shape is formed; other lathes rotate the piece slowly and have a peripheral milling cutterhead.

(f) Boring (drilling), routing and carving

Boring machines are designed to drill holes for dowel joists, screws and other purposes. They are similar to drills and drill presses used for other materials and contain rotating bits of various designs. Routers are used to shape the edges and corners of wooden objects and to cut slots of various shapes; a spindle is suspended over the piece, and the process combines aspects of boring and milling. Carving machines are rotating tools mounted on spindles, which are designed for both side and end cutting.

(g) Mortising and tenoning

A mortise is a cavity cut into a piece of wood to receive a tenon (a protrusion), which together form a mortise–tenon joint. Mortises are of various shapes and can be formed by different machines. The commonest is the hollow chisel mortiser, which forms a rectangular mortise with a hollow chisel or shell, inside of which are a rotating boring bit or bits. The bits cut the hole, which is squared by the sharp edges of the chisel. Chain mortisers, similar in design to chain saws, and oscillating bit mortisers, which are specialized routers, can also be used. Tenoning machines are also of various designs, involving both milling and sawing actions. End matchers are tenoning machines used to create both the tongue and groove for hardwood flooring.

(h) Veneer cutting

A rotary peeler is a lathe-like machine used to cut veneers, thin sheets of wood, from whole logs using a shearing action. The log is rotated against a pressure bar as it hits a cutting knife to produce a thin sheet of 0.25–5 mm in thickness. The logs used in this process may be softened before use by soaking in hot water or steaming. The edges of the sheet are usually trimmed with knives attached to the pressure bar. Veneers are used as decorative laminates or for the manufacture of plywood. Because the moisture content is high, very little wood dust is usually generated by this process.

(i) Chipping, flaking, hogging and grinding

Chippers, sometimes known as 'hackers' in Europe, are generally large rotating discs with blades embedded in the face and slots for chips to pass through. The chips are produced when logs or mill wastes are introduced to the blades by inclined gravity feed, horizontal self-feed or controlled power feeding. Generally, the cutting action of the chipper is perpendicular to the blades, and different designs are used for whole logs and for slabs and edgings. Although the term 'chipping' is sometimes used to refer to related processes, such as flaking and hogging, the end product is quite different. Chippers are used in many industries to reduce logs and wood waste to uniform-sized chips for pulp, reconstituted board and other uses.

Flaking machines convert wood into flakes for use as the raw material for particle-board, flake-board and wafer-board. They may be similar in design to a chipper, except that the wood must be fed to the flaker with the grain orientated parallel to the knives. Peripheral milling designs are also used. Wafers are generally made directly from logs that have been stored in a holding pond using a waferizer, a machine containing a series of rotating knives which peel thin wafers (Holliday *et al.*, 1986). Water-saturated wood is best for these processes, and, because the wood must be orientated, short logs are often used. Because the moisture content is high and the wood is orientated, less dust is generated than is commonly the case with chipping.

Hogs are used to reduce pieces of wood and residues into chips for use as fuel or for other purposes in which a uniform size is not required. Hogging machines are of various designs, including knife-type hogs, with rotating cylinders bearing protruding knives, rotating disks with progressively smaller teeth, and hammer mills, with rapidly revolving 'hammers' that cut the wood by impact. Hogging produces chips of maximal size but not of uniform shape or size.

Grinding is used to reduce wood chips to the consistency of flour. Hammer mills and grinding plates are used to pulverize or grind the wood, and the resulting product is sifted to control the size. Wood flour can be used for a ground cover or animal bedding or as a sweeping compound, filler or extender for composition boards and plastics, depending on the size of the particles. As the wood used for this process is generally dry, exposure to dust in uncontrolled settings is high.

(j) Mechanical defibrating

A mechanical defibrator is a grinding machine used to break wood down into fibres for wood pulp and various types of fibre-board. The wood must have a high moisture content; species such as spruce and fir are preferred because of their light colour, uniform structure and high fibre content. Short logs are forced against a grindstone, made of natural stone or of artificial stone composed of silicon carbide or aluminium oxide. The stone is showered with water to remove the pulp and cool the surface. Little or no exposure to wood dust should occur during this wet process.

1.2.2 Sawmilling

The raw materials for sawmills are supplied by the forestry and logging industries. Workers responsible for cutting down trees (felling), sawing felled trees into log lengths (bucking), trimming off branches and clearing brush are most likely to be exposed to wood dust. Sawing and cutting are usually performed with chain saws, although axes, hand saws and malls (metal wedges) may also be used. Trailer-mounted chipping or hogging machines may be used at logging sites. As wood that is sawn and chipped has a very high moisture content, the particles of dust generated are likely to be large. Trees may also be sheared with hydraulic mechanisms mounted on logging tractors; in general, little dust is produced during the shearing of fresh wood. Although other logging workers, such as those involved in yarding and loading, may be exposed to wood dust, such exposures would probably be low.

Sawmills vary greatly in size, and operations are performed outdoors or indoors depending on the size of the mill and the climate of the region. The smallest sawmills are mobile or portable units consisting of a circular saw, a simple log carriage and a two-saw edger powered by a diesel or gasoline engine and operated by two to four workers (FAO, 1981). The largest mills are permanent structures, have much more elaborate, specialized equipment and may employ more than 1000 workers. A representative production line and various phases of work at a typical Scandinavian sawmill were described previously (IARC, 1981). The equipment in sawmills varies considerably with the age and size of the mill and the type and quality of boards produced.

After transport to a sawmill, logs are stored on land, in bodies of water adjacent to the mill or in ponds constructed for the purposes. The first process is debarking; as the wood is green or has been stored in water, little wood dust is produced. This process has been described as 'messy' rather than dusty, as earth, mould and fungus particles may adhere to the surface of the bark (Holliday *et al.*, 1986).

A cut-off saw, usually a circular saw or a very large chain saw, is used to even up the ends of the trunks before primary breakdown (the first phase of sawmilling) at a headrig. The headrig is a large stationary circular or bandsaw used to cut the log longitudinally. The log is transported to the headrig on a travelling carriage, which can rotate the log 90 or 180° and carries it back and forth through the headrig. Multiple band headrigs may also be used, especially for smaller logs, so that only a single pass may be needed (Williston, 1988). The products of the headrig are a cant (the square centre of the log), a series of slabs (the rounded outer edges of the log) and, in some cases, large boards. In secondary breakdown (or 'resaw'), the cant and large boards are further processed into usable sized boards. In small mills, a circular or band saw may be used; in larger mills, the cant and large boards are generally processed with gang saws of either the sash, circular or band saw type. Boards are cut to the proper width with edgers consisting of at least two parallel saws and to the proper length with trim saws. Edging and trimming can be done with circular or band saws; in some cases, chain saws are used for trimming. Exposure to wood dust may occur in all these operations, but the concentration varies greatly, depending on the distance from the point of operation and whether or not the worker is operating the saw from an enclosed booth (Teschke *et al.*, 1994).

In many mills, the slabs and other waste wood are chipped. Chipping is generally a separate process, but in some cases a chipper may be integrated into the headrig to increase efficiency (Williston, 1988). Wood chips and sawdust may be sold for pulp or used for reconstituted board manufacture, landscaping, fuel and other uses. Exposure to wood dust may occur during chipping.

After breakdown, the boards are sorted according to dimensions and grade and then stacked by hand or machine to await drying, also referred to as seasoning. At this point, fungicides may be applied, either by dipping single boards or bundles or by various spraying procedures, to prevent the growth of fungi on the sap which stain the surface of the wood blue (Kauppinen & Lindroos, 1985).

Cut lumber is either dried in the open air or, more commonly, in various types of kilns, including serial compartment and high-temperature kilns and continuous kilns, in which stacked bundles can move in a perpendicular or parallel position and the air movement is perpendicular or parallel to the boards. Exposure to wood dust is generally low in these operations.

Either before or after drying, the wood is marketable as a green or rough lumber; however, for most industrial uses, it must be processed further. Lumber is cut to its final size and surfaced in a planing mill, usually simultaneously on two sides of the board; planers that operate on all four sides may be called matchers. Moulders are sometimes used to round the edges of the wood. Exposure to wood dust may occur during planing and moulding, because the wood is dry and the aim of the operations is to produce a relatively smooth surface. Dust control systems, such as local exhaust ventilation, may be present in these operations; however, their effectiveness in controlling exposures is not certain (Teschke *et al.*, 1994).

After processing, wood is sorted, stacked and bundled in preparation for shipping. Workers may be exposed to wood dust during these operations, especially if no measures have been taken to remove dust after surfacing operations.

1.2.3 Manufacture of plywood and other boards

Plywood, particle-board and other boards consist of wood components of varying sizes, ranging from veneers to fibres, held together by an adhesive bond. The simplest of these boards are created in two steps: generation of the components, directly from whole logs or, for some products, from woodworking waste or non-commercial or low-value tree species; and their recombination into sheets with chemical resins or, in the case of wet process fibre-board, 'natural' bonding. These steps may be carried out at different locations, especially when woodworking waste is used. The manufacture of plywood, particle-board, wafer-board, strand-board, insulation board, fibre-board and hard-board are all relatively new industries which first became commercially important during this century, especially since the 1940s. For example, although techniques for making plywood have existed for many centuries, the term 'plywood' did not enter common usage until the 1920s (Maloney, 1977).

(a) Plywood manufacture

The term 'plywood' is used for panels consisting of three or more veneers that have been glued together. Plywood can be made from either softwood or hardwood. Veneers are usually created directly from debarked whole logs by rotary peeling; decorative veneers can be created by slicing a cant with a pressure arm and blade in a manner similar to peeling. The veneers are used either for manufacturing plywood or as decorative laminates for particle-board and other reconstituted boards. After peeling or slicing, the veneers are collected on long, flat trays or rolled onto reels and are then clipped into usable lengths with a guillotine-like machine and dried by artificial heating or natural ventilation. The dried panels are inspected and, if necessary, patched with small pieces or strips of wood and formaldehyde-based resins. If the dried veneers are too small, they can be spliced together by applying a liquid formaldehyde-based adhesive to the edges, pressing the edges together and applying heat to cure the resin. As the wood used to produce veneers is wet and the peeling and clipping operations do not generally produce much dust, relatively little exposure to wood dust occurs during these operations (Holliday et al., 1986).

Plywood panels are produced by placing veneers that are roller- or spray-coated with formaldehyde-based resins between two unglued veneers. The plies are then stacked perpendicular to each other with respect to grain and transferred to a hot press, where they are subjected to pressure and heat in order to cure the resin. They are then cut to the proper dimensions with circular saws and surfaced with large drum or belt sanders. Additional machining may be done to give the plywood special characteristics.

The highest exposures to wood dust during the production of plywood occur in sanding, machining and sawing. Sanding can produce particularly large amounts of dust, since as much as 10–15% of the board may be removed during surfacing (Holliday et al., 1986). These processes are now generally enclosed or done with local exhaust ventilation.

(b) Manufacture of particle-board and related boards

Particle-board (chipboard), flake-board, strand-board and wafer-board are made from chips of wood of various sizes and shapes using similar processes. Wafer-board and strand-board are made from very large particles—wood shavings and strands, respectively—and are used primarily for structural applications. Particle-board and flake-board are made from smaller wood chips and are often used to make wood-veneered and plastic-laminated panels for the manufacture of furniture, cabinets and other wood products. Most elements are made directly from logs, branches and mill waste.

The processes used for making reconstituted panels are generally the same. The elements must be sorted by size and grade and then dried, by artificial means, to a closely controlled moisture content. The dried elements are mixed with an adhesive (a phenol–formaldehyde or urea–formaldehyde resin) and laid out in mats. The mats are cut into sections, generally with a circular cut-off saw. The panels are formed into sheets by curing the thermosetting resin in a hot press and are cooled and trimmed to size. If necessary, sanders are used to finish the surface; drum sanders were used earlier, but wide belt sanders are now generally used (Maloney, 1977). Most sanders are enclosed, and large-capacity air systems are necessary to remove the dust generated. Reconstituted boards that are to be covered with a wood veneer or plastic laminate must be sanded, and surface coatings may be applied.

Reconstituted panels are made from either hardwood or softwood. Exposure to wood dust may occur during processing but varies greatly with the moisture content of the wood and the nature of the process. The highest exposures may occur during chipping and grinding of dried wood, and high exposures occur during cutting and finishing of panels, especially in sanding operations, if engineering controls are not in place or functioning properly.

In recent decades, a new industry has emerged to produce reconstituted lumber for various structural uses, such as beams, supports and other weight-bearing elements. While the manufacturing processes used may be similar to those used for making particle-board, isocyanate-based resins are used to add strength.

(c) Fibre-board manufacture

Fibre-boards are panels consisting of bonded wood fibres. The fibres are made by reducing (pulping) short logs or wood chips in a manner similar to that used for producing pulp for the paper industry (see IARC, 1981). A mechanical (groundwood) pulping process is usually used, in which chips are soaked in hot water and then ground mechanically. A wet or a dry process may be used to bond the fibres and create the panels. The wet process, based on paper production, was developed first; the dry process, which stems from particle-board techniques, was developed later. In the wet process, a slurry of pulp and water is distributed on a screen to form a mat, which is pressed, dried, cut and surfaced. The boards created by wet processes are held together by natural adhesive-like wood components and the formation of hydrogen bonds (Suchsland & Woodson, 1986). The dry process is similar, except that the fibres are distributed on the mat after addition of a binder (a thermosetting or thermoplastic resin or a drying oil) which forms a bond between the fibres. Fibre-boards vary greatly in density. Hard-board (high-density fibre-board) and medium-density fibre-board can be produced by wet or dry processes, while insulation board (low-density fibre-board) can be produced only by the wet process.

Fibre-boards can be made from either softwoods or hardwoods. Hardwoods generally make better hard-board, while softwoods make better insulation board (Suchsland & Woodson, 1986). Exposure to wood dust in the fibre-board industry may occur during debarking, cutting of logs to size and chipping (if these are not performed elsewhere) or the handling of wood chips before pulping. The processes involved in pulping have a chemical effect on the groundwood and some of the lignin and extractive materials may be removed; the dust generated during cutting of fibre-board and finishing operations may therefore differ from unprocessed wood dust (Holliday et al., 1986).

1.2.4 Wooden furniture manufacture and cabinet-making

Traditionally, furniture is made from solid wood, and many different tree species have been used. Common species include hardwoods, such as sycamore, birch, oak, hickory, cherry, beech, ash and walnut, tropical woods such as mahogany, ebony and teak, and softwoods such as pine, fir, redwood, cedar and larch (Darcy, 1984). In this century, veneer- and plastic-covered chip-board and fibre-board panels have been used increasingly for the manufacture of cabinets, table tops and similar products. Solid hardwoods and hardwood-veneered panels are used for high-quality furniture because of the attractive patterns formed by their grains. In the furniture factories of the High Wycombe area of England, mainly beech, ash and elm are used for making tables and chairs, while elm, ash, veneered chip-board and fibre-board are used for cabinets and similar products (Jones & Smith, 1986).

The wooden furniture industry includes a wide variety of woodworking and non-woodworking operations. Various phases of the production of furniture and cabinets were described previously (IARC, 1981). In order to examine patterns of exposure to wood dust in the furniture industry in High Wycombe, Jones and Smith (1986) identified three stages of furniture production during which exposure may occur: conversion, component making and assembly. Although manufacturing processes vary by country and type of furniture produced, these three stages, sometimes under different names, generally occur in most facilities.

The 'conversion' stage is also referred to as rough milling, rough sizing or breakdown, when rough lumber is cut into the standard sizes needed for further machining. A variety of sawing and planing operations are performed during this stage (Darcy, 1984; Jones & Smith, 1986), usually with stationary machines. The wood used for furniture must generally be well seasoned (dried), and some facilities have kilns to further lower the moisture content before sawing and planing. Wood waste may be reduced with a hogger or chipper. Exposure to wood dust may occur during sawing, planing and chipping. Because of the nature of the operations, exposures should not be high, but if local exhaust ventilation is not used exposure to wood dust could occur.

The second stage, 'component making', is also referred to as machining or machine room operations. The converted pieces of lumber are cut to finished sizes and machined into the components (arms, legs, tops, sides) needed to make furniture. Some sawing and planing is done; in particular, bandsaws are used to shape pieces roughly before machining. To produce a variety of end-products, different milling machines are used, including jointers, routers, moulders, shapers, tenoners, lathes, boring machines and carving machines. The components may also be sanded, using brush, belt and drum sanders. Exposure to wood dust may occur during sawing, machining and sanding, although use of stationary machines and local exhaust ventilation may reduce exposure. If control measures are not used or are ineffective, however, exposure could be high. While most furniture is produced with machines, some workshops (in Italy and Spain, for example) produce traditional furniture or furniture that resembles antiques using many manual operations (IARC, 1981).

The final stage of production is 'assembly', when the components are put together. The potential source of wood dust at this stage is sanding, usually after assembly, which is often done with hand-held power tools. Dust control is more difficult for such operations. Although sanding was commonly performed at this stage in the English furniture industry, resulting in high exposures (Jones & Smith, 1986), assembled furniture may not require sanding.

Although the operations described above are the main sources of wood dust in the furniture industry, workers with other duties are also potentially exposed. Cleaning and maintenance workers may be exposed while removing dust from a work area or cleaning machinery or ventilation equipment. Their degree of exposure is directly related to the methods used to remove the dust; wet methods and vacuuming produce little dust, while sweeping and brushing may result in exposure, and the use of compressed air to blow off dust can result in high exposures. Assemblers, material handlers and other non-woodworkers, such as glue and upholstery workers, may be exposed if they perform their duties while the wood is still dusty. The furniture industry typically includes many operations in which there is little or no exposure to wood dust because a clean surface is needed for the operation to be effective. These operations include staining, varnishing, lacquering and painting. Generally, workers employed in these operations have little exposure to wood dust. In some situations (e.g. small factories and shops), workers may be employed in multiple phases of production.

French polish is a solution consisting of shellac dissolved in methanol. 'French polishers', however, not only apply this solution but sand down the surface after each coat has dried; the operation may be repeated tens of times. They were classified in the same category as stainers, sprayers and spray polishers in the study of Acheson *et al.* (1984) and were placed in the middle exposure category in the analysis.

Cabinet-making is a skilled trade closely related to furniture making. Cabinet-makers are highly skilled workers who are trained to operate a variety of woodworking machines and use various hand tools to fabricate and repair high-grade furniture. They may also be responsible for finishing surfaces with paints, stains and varnishes and for installing hardware, such as hinges and handles, and for other non-woodworking tasks. Cabinet-makers work in a variety of settings, from large furniture factories to small cabinet-making shops. They may also be employed in the construction industry to build, install and repair cabinets and other furniture and fixtures in both new and existing structures.

1.2.5 Manufacture of other wood products

Other products that may be manufactured from wood are musical instruments, sports equipment, kitchen utensils, wooden boxes, toys, rifle stocks, smoking pipes, coffins, doors and sashes, boats, mobile homes, wooden pallets, flooring, railroad ties, barrels and kegs, prefabricated structures, crates and fences. Exposure to wood dust varies according to the processing operations used, the type of wood used and other factors discussed in sections 1.2.1 and 1.4.1. The type of wood used is related to the use of the product; for example, products such as flooring, parquetry, baseball bats and tool handles are often made from hardwoods for aesthetic reasons and because of their durability. Doors, frames, panels and toys are often made from softwoods because of the ease with which these woods can be cut and machined.

The stages of production in which exposure to wood dust may occur can be categorized in the same way as for the wooden furniture industry in England (Jones & Smith, 1986): an initial phase mainly of machine sawing and planing to convert rough lumber into the sizes needed for further machining; an intermediate phase to cut the pieces of lumber to final size, machine and sand them; and a final stage to assemble the components, which may be sanded as part of the finishing. There is considerable variation between industries. For example, wooden pallets are made from rough lumber and require only sawing and assembly; or the final product of some industries may be a component for another industry, such as the stock for a rifle or the face of a clock. Wooden boats and other products that may be subjected to harsh environmental conditions are sometimes constructed of woods that are naturally resistant to environmental degradation, such as cypress, cedar and teak.

In some of these industries, exposures may be similar to those in the construction industry (see below). For example, the manufacture of prefabricated structures and mobile homes is very similar to construction carpentry, except that the operations are generally performed within an enclosed space. As in the construction industry, softwood is often used for framing and other structural purposes.

1.2.6 Construction, carpentry and other wood-related occupations

(a) The construction industry

Carpenters and joiners are skilled woodworkers employed extensively in the construction industry. Carpenters are responsible for the construction, erection, installation and repair of wooden structures and fixtures. Joiners are usually involved in the finer aspects of construction and the finishing of buildings. Carpenters use various saws, planers and chisels to perform their tasks, while joiners use a wider variety of tools and may perform some of their work away from the building site. The line between these two trades, which had separate guilds during the Middle Ages, is not always clear, however, and they may differ between countries. For example, in the United States, carpenters perform both types of tasks, while in France there is a greater distinction between carpenters and joiners. The term 'finish carpenter' is sometimes used to refer to workers who specialize in installing wooden trim, stairs and floors and other finishing operations.

Woodworking in the construction industry differs from that in the manufacturing industry, in that carpentry and some aspects of joinery are usually carried out at building sites, where conditions are constantly changing and hand tools are still used extensively. These two factors make it difficult to control or monitor exposures. When work is done outdoors, natural ventilation may lower the potential exposure, although operations such as sanding may still result in high exposures to wood dust. Construction-related woodwork carried out in shops with stationary equipment generates exposures to wood dust and other materials that are similar to those in other wood product industries.

Construction involves excavation, building foundations, framing, electrical installation, plumbing, roofing and finishing. Carpenters and other woodworkers may be involved in at least three of these steps: building foundations, framing and finishing. They may be responsible for constructing wooden forms for concrete foundations for both wooden and metal structures, either at the construction site or by prefabrication, from plywood and other reconstituted boards that have been specially treated with light petroleum oils (see IARC, 1989a). The framing of wooden structures involves the preparation, trimming and assembly of the various pieces of wood that comprise the weight-bearing members of a structure, including the roof timbers, beams, floor joists, wall sections, staircases and supports. This work is commonly performed on-site, although the pieces may be pre-cut and, in some situations, pre-assembled off-site. Softwoods, often pine, are most commonly used for beams, trusses, joists and studs, although hardwoods, such as oak, chestnut and elm, may also be used. Reconstituted wood products are often used for walls and underflooring. Wood that has been treated with preservatives, such as chromated copper arsenate (see IARC, 1987a) and chlorophenol derivatives (see IARC, 1986, 1987b), may be used for external walls and in other situations where the wood may be exposed to adverse conditions. A variety of saws and planers may be used, as well as simple hand tools such as chisels and hammers. Exposure to wood dust is rarely high during framing; however, insulation work performed at the same time as framing may involve exposure to insulating materials (see e.g. IARC, 1988).

Carpenters, joiners and other woodworkers, such as cabinet-makers, floor layers and parquetry workers, may be involved in finishing wooden and non-wooden structures, which involves installation of floorboards, staircases, door and window frames and sashes, moulding, cabinets and panelling of structures. Circular saws of various kinds, including table saws and radial arm saws, bandsaws, sanders of various kinds, including hand-held belt and rotary sanders, planers, routers, moulders and tenoners are commonly used. Hardwoods, softwoods, tropical woods and reconstituted panels are all used in finishing. Finishing generates the greatest potential exposure to wood dust, because sanding is done frequently, often in partly or fully enclosed spaces. Sanding, and particularly the sanding of floors, can result is high exposures to wood dust, and such operations are often carried out by workers who do not have adequate protection to avoid respiration of the dust.

(b) Maintenance and repair

Carpenters and joiners may also be responsible for the maintenance and repair of wooden structures and fixtures in industries varying from the services to manufacturing. Their exposures are similar to those of wooden construction workers, except that the work is more commonly performed indoors and may include other exposures. Although the terms 'carpenter' and 'joiner' are usually associated with the construction and maintenance trades, they are used in many industries to refer to skilled woodworkers. The term 'joiner' is also used in boat building and repair to refer to skilled workers who fabricate, assemble, install or repair wooden furnishings in ships and boats. Carpenters who work on wooden boats and ships are also referred to as 'shipwrights'.

(c) Pattern and model making

Wooden pattern makers plan, lay out and construct wooden units or sectional patterns, such as those used in forming sand moulds for casting. Wooden model makers construct precision models of products or parts used in mass production. Both pattern and model makers are highly skilled workers, who are employed by small shops or directly by mass production industries. Pattern and model making are not, however, mass production operations: each piece is made individually, starting from blueprints and ending with a finished product which must often meet very close tolerances. These workers use hand tools, measuring instruments and woodworking machinery such as bandsaws, lathes, planers, routers and shapers and a variety of hard, soft, tropical and laminated woods. The raw material used depends on the intended use; for example, in the United States automobile industry, softwoods are often used for experimental models, while harder and laminated woods are used for models that require more exact, stable dimensions (McCammon *et al.*, 1985).

(d) Wood shop teachers and artists

Exposure to wood dust may occur in a number of other occupational settings, including the teaching of woodworking, wood sculpture and design secondary schools, technical schools and universities. In such classes, a variety of woodworking machinery may be used in largely unregulated environments, under health and safety conditions that would not be acceptable in an industrial setting. For example, Lucas and Salisbury (1992) reported that the equipment used by a design materials class in a university art department included a planer, table saw, jointer, lathe, belt sander and band saw and many portable power tools. While the stationary machines had local exhaust ventilation, the fabric bag dust collector was located inside the classroom. Although students may be exposed intermittently, teachers may spend many hours per day in the same setting. Artists who create wooden sculptures and artisans making wooden objects may work under similar conditions.

1.3 Analytical methods

1.3.1 Characterization and measurement of wood dust

Wood dust and exposures to wood dust are characterized in several ways that affect the nature of exposures in woodworking industries: by type of wood, as airborne dust concentrations, by particle size distribution and by other parameters.

(a) Type of wood

Wood dust is frequently described by wood species or as hardwood or softwood (see section 1.1). Wood dust is also characterized by its moisture content: Dry wood (moisture content less than about 15–20%) is less elastic than moist (green) wood, and woodworking operations with dry wood result in a larger volume of total dust and a higher percentage of inhalable dust particles (Hinds, 1988).

(b) Airborne dust concentrations

(i) Total dust measurements

The parameter most commonly used to characterize exposures to wood dust in air is total wood dust concentration, in mass per unit volume (usually mg/m^3). Standard gravimetric methods for measuring total dust concentrations, such as NIOSH Method 0500 (Eller, 1984a), have been used routinely. In this general method, a known volume of air is drawn through a special membrane filter contained in a plastic cassette with a sampling pump. The dust concentration is calculated from the change in weight of the filter divided by the volume of air sampled, with a detection limit for personal sampling of wood dust of about 0.1 mg/m^3. Polyvinyl chloride filters are preferred for sampling wood dust because of its highly variable moisture content. Filters are environmentally equilibrated before and after sampling to avoid spurious effects from differential moisture uptake (Eller, 1984a; Holliday et al., 1986; Sass-Kortsak et al., 1989; Teschke et al., 1994).

The cassette holding the filter is either open- or closed-faced during sampling. In the closed-faced mode, a cap with a 4-mm hole is placed over the 37-mm cassette face to protect the filter. Closed-faced operation is usually recommended when total suspended particulates are being measured (see, for example, United States Occupational Safety and Health Administration, 1993). Beaulieu et al. (1980) reported, however, that the open-faced filter collected 30–60% more dust (by weight) than the closed-faced filter, and they suggested that particles larger than about 10 µm are collected very inefficiently in the closed-faced configuration, as corroborated subsequently in other studies (Clayton Environmental Consultants, 1988; Hinds, 1988). Other authors have cautioned against the collection of particles 'too large to be inhaled' (Darcy, 1984) which would contribute disproportionately to the total weight of dust (Hounam & Williams, 1974). As an alternative, the United Kingdom Health and Safety Executive recommends a sampling head with seven 4-mm holes, the sampling characteristics of which appear to approximate current definitions of inhalable dust (Jones & Smith, 1986; United Kingdom Health & Safety Executive, 1989; Hamill et al., 1991; Pisaniello et al., 1991).

(ii) Particle size-selective measurements

A number of methods have been used to measure, more or less selectively, exposures to wood dust in the respirable particle size range (Hinds, 1988). NIOSH Method 0600 (Eller, 1984b) is intended for measurement of general 'respirable dust' concentrations. The equipment used is the same as that for NIOSH Method 0500, except that air is sampled through a 10-mm nylon cyclone (centrifugal separator) designed to accept 50% of unit density spherical particles of 3.5 µm aerodynamic diameter. Another standard technique, the horizontal elutriator (gravitational separator), is designed to collect a respirable particulate fraction defined by the British Medical Research Council as 50% of particles of 5 μm aerodynamic diameter (Sass-Kortsak et al., 1993).

The performance of the nylon cyclone, the horizontal elutriator and an aluminium cyclone for measuring wood dust were compared directly in an environmentally controlled chamber at various levels of humidity. Higher levels (by about 25%) were consistently measured with the aluminium cyclone than with the elutriator, with which higher levels (by about 40%) were measured than with the nylon cyclone (Sass-Kortsak et al., 1993). It has been suggested that the nylon cyclone does not accurately separate respirable and nonrespirable wood dust particles because of static charge effects with dry wood dust. The Mine Safety Appliances (MSA) respirable dust cassette, which is similar to the standard 37-mm plastic cassette but contains an aluminium inner capsule that is weighed with the filter, is reportedly about twice as efficient for measuring respirable wood dust as the standard plastic cassette (Moore et al., 1990).

Samplers have been developed to measure exposure to the 'inspirable (or inhalable) particulate mass' fraction, which includes large particulates that may deposit and cause adverse effects on the upper airways. As defined by the American Conference of Governmental Industrial Hygienists (Phalen et al., 1986), these samplers must maintain a sampling efficiency ≥ 50% for particles up to 100 μm aerodynamic diameter. A similar definition was proposed by the International Standards Organization (Vincent & Mark, 1990). The development, evaluation and use of specific samplers for the inspirable particulate mass, including wood dust, have been reported (Mark & Vincent, 1986; Hinds, 1988; Vaughan et al., 1990; Vincent & Mark, 1990; Pisaniello et al., 1991).

An early method for characterizing exposure to wood dust is determination of the number, rather than the mass, of particles in a given volume of air. The konimeter has been used extensively to measure respirable dust of 0.5–5.0 μm by drawing random, small volume, short duration spot ('grab') samples through a small orifice where airborne particles impinge on a glass slide coated with adhesive. The particles trapped on the slide are counted electronically or visually with a microscope. The measurements in grab samples are not, however, comparable to longer duration, time-weighted average concentrations (Holliday et al., 1986).

(c) Particle size distribution

When the distribution of particle sizes in an air sample is to be assessed, other methods must be used. The commonest involves a multi-stage cascade impactor (e.g. the Anderson impactor), which separates particles by mass. The impactor consists of a series of perforated plates through which air is drawn at a constant rate. The dynamics of air flow through the holes (which are of different sizes at each stage) result in trapping of particles with a known range of aerodynamic diameters. The dust collected at each stage can be weighed, and a particle size (mass) distribution can be calculated. Results are reported in various ways, for example as percentage of total mass of dust collected at each stage or as mass median aerodynamic diameter (Whitehead et al., 1981a; Carlin et al., 1981; Holliday et al., 1986; Clayton Environmental Consultants, 1988; Pisaniello et al., 1991).

Holliday et al. (1986) reported the analysis of wood dust samples by optical microscopy and classification of particles by equivalent circular diameters calculated from their projected areas by an image analysing computer. The result is a particle size frequency distribution, rather than particle mass distribution.

(d) Other characteristics of wood dust

Other characteristics of airborne wood dust are occasionally reported. For example, the irregular shapes of wood dust particles are sometimes recorded in photomicrographs (Holliday et al., 1986) or examined by scanning electron microscopy (McCammon et al., 1985). Particle density (specific gravity) has occasionally been reported (Andersen et al., 1977).

The chemical substances that are natural components, additives or adsorbed contaminants are sometimes extracted with water or organic solvents and characterized (see below). Although there is no standard procedure for measuring the extractable fraction of wood dust, the possible role of these substances in the adverse health effects of wood dust has been the subject of considerable research and speculation.

1.3.2 Chemical analysis of wood constituents

A survey of several methods for the chemical analysis of wood constituents has been published (Fengel & Wegener, 1989). In general, organic matter (extractives), inorganic matter (ash) and the main cell wall components, polysaccharides and lignin, are determined. It is important to select a sample for analysis that is representative of the wood species. Standardized sampling procedures have been published: e.g. TAPPI Standard T 257 cm-85 (Technical Association of the Pulp and Paper Industry, 1985a). Before chemical analysis, wood must be milled (particle size, 0.05–0.40 mm) to achieve complete penetration of reagents (Fengel & Wegener, 1989).

(a) Extractives

As no modern standard method for the extraction of wood exists, every research group has its own strategy for isolating and identifying chemical constituents of wood. The differences in reported data may thus be due to differences in wood composition or the use of different analytical methods. Conventional methods for investigating compounds present in wood involve either steam distillation or extraction with organic solvents in a soxhlet extractor (Mayer et al., 1969, 1971; Nabeta et al., 1987; Christensen et al., 1988; Kubel et al., 1988; Charrier et al., 1992; Weissmann et al., 1992). Table 7 shows the results of soxhlet extraction of four common wood species with a series of organic solvents of increasing polarity (Weissmann et al., 1992). Supercritical fluid extraction has also been used to isolate these compounds (Torul & Olcay, 1984; Demirbaş, 1991), providing much higher yields of some compounds than those achieved with conventional soxhlet extraction. The thermal stability of wood components is not well characterized, and both soxhlet and supercritical fluid extraction may cause molecular changes, such as decomposition and dimerization (Fengel & Wegener, 1989).

Table 7. Yields (%) of successive extractions of four common wood species

Extraction solvents	European spruce	Scots pine		European beech	European oak	
		Sapwood	Heartwood		Sapwood	Heartwood
Non-polar fractions						
Petroleum ether	0.6	2.2	8.6	0.2	0.15	0.15
Diethyl ether	0.2	0.06	0.8	0.1	0.15	0.35
Polar fractions						
Acetone:water 9:1	0.7	0.3	0.7	1.6	3.6	5.8
Ethanol:water 8:2	0.3	0.4	0.4	1.2	0.9	1.8
Totals	1.8	3.0	10.5	3.1	4.8	8.1

From Weissmann *et al.* (1992)

Groups of non-polar and polar substances resulting from extraction with solvents of increasing polarity can be purified further by chromatographic techniques, such as normal and especially reverse-phase high-performance liquid chromatography (Zinkel, 1983; Suckling *et al.*, 1990; Charrier *et al.*, 1992) and thin-layer chromatography (Nabeta *et al.*, 1987; Kubel *et al.*, 1988). Individual substances are identified by infrared and one- or two-dimensional nuclear magnetic resonance spectroscopy and gas chromatography followed by mass spectroscopy (Mayer *et al.*, 1971; Nabeta *et al.*, 1987; Fengel & Wegener, 1989).

(b) Inorganic compounds in ash

The inorganic part of wood is analysed as ash after incineration of the organic wood material at 600–850 °C (Fengel & Wegener, 1989). Detailed methods for ash determination are described in TAPPI Standard T 211 om-85 (Technical Association of the Pulp and Paper Industry, 1985b) and ASTM Standard D 1102-56 (American Society for Testing and Materials, 1965). Particular ash constituents can be identified by methods such as energy dispersive X-ray analysis coupled with scanning or transmission electron microscopy, atomic absorption or emission spectroscopy and neutron activation analysis (Fengel & Wegener, 1989).

(c) Polysaccharides

Polyoses, the second group of cell wall polysaccharides, differ from cellulose in their solubility in alkali. Only some polyoses are soluble in water. Some can be extracted directly, and others require removal of lignin before extraction, usually by treating pre-extracted wood with an acidified solution of sodium chlorite (pH 4) at 70–80 °C for 3–4 h. A standard procedure for the isolation and determination of polyoses is successive extraction of chlorite holocellulose with 5 and 24% potassium hydroxide. The insoluble residue represents cellulose.

A general procedure for isolating and determining polysaccharides consists of hydrolysis with concentrated acids and subsequent dilution steps to achieve secondary hydrolysis. Sugars can be identified and quantified after hydrolysis by various chromatographic methods, including thin-layer and high-performance liquid chromatography, gas chromatography partly combined with mass spectroscopy and ion-exchange chromatography via sugar borate complexes.

(d) Lignin

All methods for the isolation of lignin have the disadvantage that they fundamentally change the native structure of lignin or release only parts of it relatively unchanged. All lignin samples obtained by acid hydrolysis are changed with regard to structure and properties, predominantly by condensation reactions. These preparations are therefore not suitable for investigating structures but can be used for estimating lignin content. The commonest method for obtaining relatively unchanged lignin is Björkman's procedure of vibratory milling and subsequent extraction of lignin with aqueous dioxane (Björkman, 1956). In one modification of this method, ultrasound is applied during the extraction step to reduce the isolation time (Wegener & Fengel, 1978).

1.4 Exposure to wood dust and other agents in the workplace

1.4.1 General influences on occupational exposure to wood dust

Woodworking operations such as sawing, milling and sanding both shatter lignified wood cells and break out whole cells and groups of cells (chips). The more cell shattering that occurs, the finer the dust particles that are produced. For example, sawing and milling are mixed cell shattering and chip forming processes, whereas sanding is almost exclusively cell shattering. Since wood cells usually measure about 1 mm, airborne dust concentrations depend primarily on the extent of cell shattering rather than on the size or extent of chip formation (Holliday *et al.*, 1986; Hinds, 1988).

In general, the harder the wood, the more tightly bound are the cells; therefore, more shattering occurs with hardwoods, resulting in more dust. Similarly, the cells in dry wood are less plastic and more likely to be shattered, leading to dust formation. While the moisture content of different species of trees varies, it is also influenced by the freshness (greenness) of the wood. Drying and some other processing of wood, such as sawing and machining, may change the chemical composition of wood dust. For example, some of the low-relative-molecular-mass extractives, such as monoterpenes (see section 1.1.4(*b*)(i)), may be volatilized. Terpenes evaporate from coniferous wood when it warms up during sawing of logs and edging of boards, and concentrations of 100–550 mg/m^3 have been measured in these operations in Swedish sawmills (Hedenstierna *et al.*, 1983).

The orientation of the point of operation relative to the wood surface and grain also influences the generation of dust. Woodworking operations performed parallel to the natural grain of the wood are less likely to shatter cells than processes performed perpendicular to the grain. The volume of wood dust generated also depends on how the process is carried out. For example, machine sanding normally generates more dust than manual sanding.

Woodworking machines have increased greatly in efficiency since the industrial revolution, and the increased speed of production has resulted in the generation of more dust. The increased efficiency may also result in exposure to finer wood dust particles than in the past, because smoother surfaces can be produced and because saws and bits may retain their sharpness for longer. The introduction of engineering controls in some industries in some parts of the world, especially since the 1950s, has, however, reduced the exposure of workers considerably. Various measures can be taken to control exposure. A simple, effective measure is to remove the worker from the point of operation by placing controls away from the process or by providing an enclosed booth. These measures are not infrequently taken in sawmills, primarily for safety reasons. Another option is to enclose the operation or to provide local exhaust ventilation. For example, Hampl and Johnston (1985) reported on a ventilation system for horizontal belt sanding that can significantly reduce wood dust emissions, and the American Conference of Governmental Industrial Hygienists (1992) has developed design guides for local ventilation in specific woodworking operations. Unfortunately, engineering controls, even if properly maintained, are not always effective, and the dust generated by hand-held power tools, particularly sanders, is much more difficult to control.

A number of characteristics of the workplace may also affect the level of wood dust, e.g. the age, density and types of woodworking machinery and the regulatory environment. Regulations with regard to exposure to wood dust and the enforcement of those regulations may vary between countries or even between industries in the same country. Small workplaces are difficult to regulate and, for various reasons, may have fewer engineering controls in place. The use of engineering controls and occupational health regulations have also changed over time.

The quality of and methods used for cleaning are important, because wood dust that has settled on the floor, equipment and other surfaces may become resuspended. In particular, the use of compressed air hoses to clean off surfaces results in high airborne concentrations of wood dust, while wet cleaning methods and use of vacuum systems may result in little or no exposure to dust. Respirators can be used to reduce exposure, and, because of the size of wood dust particles, even simple paper masks can be relatively effective if properly fitted and used. Woodworking operations conducted outdoors or in semi-enclosed buildings may involve lower exposures because of natural ventilation, but higher exposures could result from wind and the lack of local exhaust ventilation and other engineering controls.

1.4.2 Extent of exposure to wood dust

The number of workers exposed to wood dust worldwide has not been estimated in the literature; however, estimates are available for some western countries. The National Occupational Exposure Survey, carried out in 1981–83 in the United States, estimated that about 600 000 workers were exposed to wood dust. The largest numbers of exposed workers were employed in the building trades and the lumber/wood product industries. Forestry workers, e.g. lumberjacks using chain saws, were not considered to be exposed in this survey (United States National Institute for Occupational Safety and Health, 1990). The United States produced 24% of all sawn wood in the world in 1990 (FAO, 1992). According to a Finnish survey, about 70 000 workers were exposed to wood dust in logging or in production of sawn wood, wood products and pulp (Anttila *et al.*, 1992); of these, about 12 900 were estimated to have been exposed routinely to more than 1 mg/m^3 of wood dust and 3800 routinely to more than 5 mg/m^3 (Welling & Kallas, 1991). Some occupations, such as construction carpenters, were considered to have experienced only occasional exposure to wood dust, and they were not included in the Finnish estimates.

[Country-specific estimates and production statistics allow a crude estimate to be made of the number of the workers exposed worldwide. Assuming that the technical level (workforce demand) and the internal structure of the industries that involve exposure to wood dust are approximately the same as in the reference countries, the United States and Finland, the Working Group estimated that the number of workers occupationally exposed to wood dust worldwide is at least two million and probably much higher.]

The industrial hygienic measurements reviewed in sections 1.4.3–1.4.7 and Tables 8–12 include both personal and area sampling. The results probably represent daily average exposures reasonably well because the sampling time required is normally several hours. Many measurements are made for compliance testing, and workers who have moderate or high potential for exposure tend to be monitored more frequently than others. The results, therefore, are considered to be representative for the specific jobs and operations monitored but are not necessarily representative of exposures throughout industry or for all time periods, job categories or sites.

1.4.3 Exposure during sawmilling

Measured concentrations of wood dust in the air of sawmills and planing mills are presented in Table 8. The levels vary widely, ranging from 0.1 to over 100 mg/m^3; the mean values are more frequently below than above 1 mg/m^3. In the largest study of sawmills and planing mills in the United States, 33% of the measured concentrations exceeded 1 mg/m^3 and 8% exceeded 5 mg/m^3 (Clayton Environmental Consultants, 1988). The highest exposure often occurs in the vicinity of chippers, saws and planers, but other operations, such as cleaning, grading and maintenance, may sometimes be dusty.

The results presented in Table 8 are not directly comparable across studies and countries. For example, the low concentrations reported from Canada (Teschke *et al.*, 1994) are partially due to representative sampling. This strategy tends to provide lower results than selective sampling from sites and operations involving high exposure, which is the procedure used in many other studies. The method of measurement and the use of the geometric rather than the arithmetic mean in reporting may also affect the results (see section 1.3.1).

Most measurements reported in the literature are from the 1980s. There are no comprehensive data available to indicate any clear changes over time in the level of wood dust in sawmills or planing mills.

The species of wood processed in sawmills vary. In the Canadian study (Teschke *et al.*, 1994), coniferous (soft) wood, such as hemlock and fir, was processed. Coniferous species (pine and spruce) are also the main raw materials in Finnish sawmills (Welling & Kallas, 1991). Locally and in countries where coniferous trees are rare, deciduous trees may be the main wood used in sawmills. For example, a sawmill in West Virginia, United States, processed white oak,

Table 8. Concentrations of wood dust in sawmills and planer mills

Industry and operation (country)	No. of measurements	Mean[a] (mg/m³)	Range (mg/m³)	Year	Reference
Sawmills (Canada)	18		0.3–6.1	1985	Holliday et al. (1986)
Sawmills (Canada)	78	0.2[b]	ND–6	1982–	Vedal et al. (1986)
Sawmilling (Canada)	191[c]			1989	Teschke et al. (1994)
Sawmills		0.1[d]			
Yard		0.1[d]			
Maintenance		0.2[d]			
Powerhouse		0.2[d]			
Log boom, kiln, other		0.1[d]			
Sawmills (USA)	55	2.6	0.7–10.6	1981–82	Morey (1982)
Sawmills and planing mills (USA)	193	0.7[d]	0.10–410	1987–88	Clayton Environmental Consultants (1988)
Sawmills (Finland)				1980–85	Welling & Kallas (1991)
Sawing	18	1.6	0.1–4.9		
Stacking	3	0.2	0.1–0.3		
Trimming	33	2.8	0.1–28.0		
Packaging	14	1.4	0.23–3.3		
Chipper, hogger	7	1.8	0.6–3.0		
Sawmills (Denmark)	85	0.5[e]	0.5–0.6[f]	NR	Vinzents & Laursen (1993)
Sawmills (Germany)	6	2.7[b]	0.2–50	NR	Scheidt et al. (1989)
Planer mill (Canada)	NR	0.2[d]		1989	Teschke et al. (1994)
Planer mills (Finland)				1980–85	
Sawing	8	7.8	0.6–35.2		
Planing	11	2.0	0.1–8.4		

ND, not detectable; NR, not reported
[a] Arithmetic mean unless otherwise specified; time-weighted average personal and/or area samples
[b] Median
[c] Including planer mill
[d] Geometric mean
[e] Mean of geometric means

red oak, poplar, soft maple, basswood and cherry (Morey, P., 1982, cited in United States National Institute for Occupational Safety and Health, 1987).

Sawmill workers may be exposed to chemical agents other than wood dust. Chlorophenols have been used widely in sawmills since the 1940s to prevent staining of freshly cut timber. The chlorophenols used most commonly were pentachlorophenol, tetrachlorophenols and trichlorophenols, which were usually applied to wood as water-soluble salts by dipping or spraying. Although the levels of chlorophenols reported in the air are usually below 0.1 mg/m^3, heavy exposure may occur through the skin when boards are handled manually immediately after treatment (Kalliokoski & Kauppinen, 1990). Some impurities of chlorophenols—chlorinated phenoxyphenols and polychlorinated dibenzofurans—have also been identified in the sawdust of trimming-grading plants (Levin et al., 1976). Because of concern about the health effects of chlorophenates and their possible contamination with polychlorinated dibenzo-*para*-dioxins (IARC, 1987c), fungicides and other substitutes have been introduced. In Canada and the United States, a mixture of didecyldimethyl ammonium chloride and 3-iodo-2-propynyl butyl carbamate is used (Teschke et al., 1995). Exposure to fungicides may occur if the boards are handled while still wet during grading, sorting and other operations. Many woods, especially those that have been kiln dried, may not need to be treated with fungicides, and some species, such as red cedar, are not susceptible to sapstain fungus.

The numbers of natural fungi and bacteria in wood increase during storage and drying and become suspended in air when wood is processed or handled. For example, the average concentration measured with an Andersen sampler in sawing departments and stacking sites of Finnish sawmills was about 14 000 colony-forming units (cfu)/m^3 of mesophilic bacteria, 130 cfu/m^3 of actinomycetes, 660 cfu/m^3 of xerophilic fungi, 6500 cfu/m^3 of mesophilic fungi and 3000 cfu/m^3 of thermotolerant fungi (Kotimaa, 1990).

In specialized mills, wood may be further treated with preservatives, fire retardants or chemicals that protect the surface from mechanical wear or weathering. For example, railroad ties, pilings, fence posts, telephone poles and other wood expected to be in contact with soil or water may be treated with creosote oils (see IARC, 1985b), pentachlorophenol solutions or salts containing copper, chromium (see IARC, 1990) and arsenic (see IARC, 1987a). Stains and colourants may also be used, and paint may be applied to seal the ends of boards or to add company marks.

1.4.4 *Exposure during the manufacture of plywood and other boards*

Concentrations of wood dust in the air of plywood, particle-board and other wood-based panel mills are presented in Table 9. The mean levels in plywood mills are often close to 1 mg/m^3. In a study in the United States, 27% of the measured values exceeded 1 mg/m^3 in hardwood veneer/plywood mills and 11% in softwood veneer/plywood mills (Clayton Environmental Consultants, 1988). The heaviest exposures usually occur in finishing departments where plywood is sawn and sanded. Some operations, such as drying, assembly and hot pressing, entail hardly any exposure to wood dust.

Table 9. Concentrations of wood dust in plywood, particle-board and related industries

Industry and operation (country)	No. of measurements	Mean[a] (mg/m³)	Range (mg/m³)	Year	Reference
Plywood mills (USA)					
Edge sawing, sanding, plywood machining	12	1.7	0.7–3.2	1978	Whitehead et al. (1981a)
Veneer lathe, clipper, dryer, dry veneer handling, gluing and pressing	13	0.4	0.1–0.7		
Hardwood veneer and plywood mills (USA)	48	0.8[b]	0.1–21	1987–88	Clayton Environmental Consultants (1988)
Softwood veneer and plywood mills (USA)	56	0.6[b]	<0.1–6.4	1987–88	Clayton Environmental Consultants (1988)
Veneer and plywood mill (Canada)	7		0.1–2.6	1985	Holliday et al. (1986)
Plywood mills (Finland)					Kauppinen (1986)
Log debarking/cutting	4	0.4	0.2–0.7	1975–84	
Peeling	2	NR	0.2–0.3	1975–84	
Sawing of veneers	3	1.6	0.6–3.0	1965–74	
Sawing of veneers	4	1.3	1.1–1.5	1975–84	
Sawing of plywood	6	3.3	0.5–12	1965–74	
Sawing of plywood	11	3.7	0.3–19	1975–84	
Sanding of plywood	5	3.0	0.3–6.4	1965–74	
Sanding of plywood	21	3.8	0.8–22	1975–84	
Chipping in finishing department	11	2.6	0.7–7.1	1975–84	
Finishing department	18	0.7	0.3–2.4	1975–84	
Plywood mills (Finland)					Welling & Kallas (1991)
Sawing	24	2.1	0.4–5.0	1980–85	
Sorting, cleaning, glue mixing, hogger	4	11.1	7.1–15.0		
Particle-board mills (Finland)					Kauppinen & Niemelä (1985)
Hogging	3	11	0.1–29	1975–84	
Chipping	3	1.1	0.7–1.7	1975–84	
Drying of chips	2	NR	24–29	1965–74	
Blending	3	5.3	1.0–8.0	1965–74	
Blending	3	0.8	0.6–0.9	1975–84	
Forming	9	13	4.0–26	1965–74	

Table 9 (contd)

Industry and operation (country)	No. of measurements	Mean[a] (mg/m³)	Range (mg/m³)	Year	Reference
Particle-board mills (Finland) (contd)					
Forming	4	0.4	<0.1–0.5	1975–84	Kauppinen & Niemelä (1985)
Hot pressing	6	4.1	1.0–6.1	1965–74	
Hot pressing	5	0.8	<0.1–2.1	1975–83	
Sawing	4	14	10–20	1965–74	
Sawing	9	1.1	<0.1–2.3	1975–84	
Reconstituted-board mills (USA)	112	0.7[b]	0.1–205	1987–88	Clayton Environmental Consultants (1988)
Reconstituted-board mill (Canada)	5		1.5–5.1	1985	Holliday et al. (1986)
Process hard-board mills (USA)	116	0.6[b]	<0.1–45	1987–88	Clayton Environmental Consultants (1988)
Fibre-board mill (Finland)					Welling & Kallas (1991)
Piling of boards	2	2.6	1.8–3.3	1980–85	
Sawing of boards	2	3.2	1.8–4.6	1980–85	

NR, not reported
[a] Arithmetic mean unless otherwise specified; time-weighted average personal and/or area samples
[b] Geometric mean

The levels of wood dust in various reconstituted-board (particle-board, fibre-board, hardboard, strand-board) mills exceeded 1 mg/m^3 in 22% of measurements in a large study in the United States (Clayton Environmental Consultants, 1988); however, much higher concentrations have been reported, e.g. in forming and sawing of particle-board, especially before the 1980s (Kauppinen & Niemelä, 1985).

Some data on changes in exposure levels over time are presented in the Table. The level of exposure during sawing and sanding in Finnish plywood mills did not change significantly in two consecutive 10-year periods (Kauppinen, 1986). A substantial decrease in exposure since the mid-1970s has been seen, however, in dusty operations in particle-board mills (Kauppinen & Niemelä, 1985).

Phenol–formaldehyde resin adhesives are widely used to produce softwood plywood for use under severe service conditions, such as for construction and boat building. Urea–formaldehyde resin adhesives are used extensively in producing hardwood plywood for furniture and interior panelling and can be fortified with melamine resin to increase their strength. Before the introduction of formaldehyde-based resins in the 1940s, soya bean and blood-albumin adhesives were used, and cold pressing of panels was common. These operations are still used, but are increasingly rare.

Other agents to which some plywood workers may be exposed include formaldehyde (see monograph, p. 217) and phenol (see IARC, 1989b) emitted from glues, pesticides, heating emissions from coniferous veneers, solvents from coating materials and engine exhaust from forklift trucks. Pesticides that have been used in plywood glues include lindane (see IARC, 1987d), aldrin (see IARC, 1974), heptachlor (see IARC, 1991), chloronaphthalenes, chlorophenols and tributyltin oxide. The mean level of formaldehyde in most operations is now below 1 ppm (1.23 mg/m^3), and exposure to phenol is usually well below that concentration. Most pesticides mixed in glues are only slightly volatile and are not detectable in workroom air; the exception is chloronaphthalenes, which are more volatile. Exposure to pesticides may also occur through the skin. The levels of solvents during painting and other surface treatment of plywood are 1–50 ppm. Levels of terpenes in plywood mills are not detectable (< 1.5 ppm) in most operations, and the levels were only 1–6 ppm during debarking of pine logs and peeling, drying and sorting of pine veneers, in spite of the obvious presence of a blue haze during processing of pine (Kauppinen, 1986).

Formation of polycyclic aromatic hydrocarbons due to heating during sawing and sanding of plywood could not be detected in measurements carried out in a Finnish plywood mill; however, these compounds may occur in glueing and finishing departments of plywood mills, from exhausts of forklift trucks (Kauppinen, 1986).

Exposures in reconstituted-board mills are similar to those in plywood mills. Formaldehyde-based resins, and especially urea–formaldehyde resin (Kauppinen & Niemelä, 1985), are commonly used in glueing particle-board and other wood-based panels, and the level of formaldehyde in particle-board mills may exceed 1 ppm (1.23 mg/m^3). Urea–formaldehyde resins release formaldehyde during curing more readily than phenol–formaldehyde resins; however, improvements in resin formulation have reduced exposures (Holliday et al., 1986). Exposure may also occur to pesticides, such as heptachlor, and solvents in surface coatings

(Kauppinen & Niemelä, 1985). Workers in the area of stockpiles of untreated wood chips or conveyors used to transport the chips may be exposed to moulds, bacteria and fungi (Cohn et al., 1984).

1.4.5 Exposure during wooden furniture manufacture and cabinet-making

Table 10 summarizes the levels of wood dust in the wooden furniture industry, including cabinet-making. The reported mean levels are higher than in sawmilling and wood-based panel manufacture: a concentration of 1 mg/m^3 was exceeded in 41% of measurements in household furniture manufacture, in 22% in office furniture manufacture and in 52% in kitchen cabinet manufacture in the United States (Clayton Environmental Consultants, 1988). The mean levels shown in Table 10 are frequently between 1 and 10 mg/m^3. The highest exposures occur in wood machining jobs, such as sanding, and in cabinet-making. Wood is usually machined in separate departments of large furniture plants, but some dusty jobs, such as sanding between applications of varnish layers (e.g. French polishing), may be done in surface coating departments (Welling & Kallas, 1991).

Both hardwood and softwood are commonly used in the manufacture of furniture. The proportions of different species of wood used depends on many factors, such as the country, product, plant and period considered. For example, in the British plants surveyed in 1983 (Jones & Smith, 1986), beech, ash, elm, mahogany, walnut, veneered particle-board and medium density fibre-board were used. In the Finnish furniture industry in 1986, mainly pine and birch were used, but spruce, oak, mahogany, teak and other wood species were employed to some extent (Welling & Kallas, 1991). Case reports and epidemiological studies provide some additional information on the species of wood used in the past in furniture factories of different countries (see section 2).

The mean level of wood dust in seven British furniture factories decreased from 7.8 mg/m^3 (138 samples) in 1976–77 to 4.2 mg/m^3 (209 samples) in 1983, probably due mostly to improvements in local exhaust ventilation of machines (Jones & Smith, 1986).

Other agents than wood dust to which workers in furniture and cabinet-making may be exposed include formaldehyde and solvents from varnishes, paints and glues. In spray-varnishing and -painting and in sanding of surface-coated furniture, workers may also be exposed to nonvolatile components of surface coatings, such as pigments and resins. The level of formaldehyde in the air of surface coating departments of furniture plants varies from 0.1 to over 5 ppm (0.12–> 6.15 mg/m^3), often averaging close to 1 ppm (1.23 mg/m^3). Workers who machine wood may occasionally be exposed to formaldehyde if, for example, formaldehyde-based glues are used in veneering and the hot press is situated close to wood processing machines. In addition, formaldehyde may be released from reconstituted panels during machining (Sass-Kortsak et al., 1986) or may be bound in wood dust aerosol (Stumpf et al., 1986).

Glueing, staining and varnishing are generally done at a distance from woodworking operations, however, so that machine operators and cabinet-makers, who are usually heavily

Table 10. Concentrations of wood dust during furniture and cabinet-making

Industry and operation (country)	No. of measurements	Mean[a] (mg/m^3)	Range (mg/m^3)	Year	Reference
Furniture manufacture (United Kingdom)				1973	Hounam & Williams (1974)
Turning	2	8.6	4.6–12.5		
Band sawing	6	4.3	1.0–7.3		
Routing	6	4.1	1.8–8.6		
Assembly	9	5.5	2.1–9.8		
Planing	9	5.0	1.8–10.9		
Sanding	9	7.2	2.0–22.6		
Spindle moulding	8	5.1	1.5–8.4		
Furniture manufacture (United Kingdom)				1983	Jones & Smith (1986)
Conversion	43	2.3	1.0–4.8		
Component making	106	3.4	0.3–53		
Assembly	60	7	0.5–27		
Furniture manufacture (Denmark)				1974–75	Andersen et al. (1977)
Drilling, planing, sawing	27	5.2			
Machine and hand sanding	41	14.3			
Furniture manufacture (USA)				1978	Whitehead et al. (1981a)
Rough mill (softwood)	5	0.6	0.2–1.1		
Rough mill (hardwood)	7	0.8	0.2–2.6		
Assembly (softwood)	2	2.8	2.5–3.1		
Assembly (hardwood)	3	1.5	1.1–2.1		
Lathe, planer, router (hardwood)	9	1.8	0.2–6.3		
Lathe, planer, router (softwood)	9	1.6	0.3–4.3		
Sanding (hardwood)	12	4.5	1.4–11.4		
Sanding (softwood)	13	3.2	0.6–14.3		
Manufacture of household furniture (Canada)				1985	Holliday et al. (1986)
Processing of hardwood	11		0.3–5.2		
Processing of particle-board	6		0.5–6.8		
Processing of softwood, particle-board	5		1.7–15.6		
Manufacture of office furniture (Canada)				1985	Holliday et al. (1986)
Processing of hardwood	7		0.5–1.7		
Processing of particle-board	9		0.4–5.6		

Table 10 (contd)

Industry and operation (country)	No. of measurements	Mean[a] (mg/m^3)	Range (mg/m^3)	Year	Reference
Manufacture of household furniture (USA)	112	1.3[b]	0.2–240	1987–88	Clayton Environmental Consultants (1988)
Manufacture of office furniture (USA)	23	0.8[b]	0.2–3.8	1987–88	Clayton Environmental Consultants (1988)
Furniture factories (Germany)				NR	Scheidt et al. (1989)
Routing, planing	10	6.0[c]	3–59		
Sanding	8	2.8[c]	1.2–9.1		
Sanding (manual)	8	6.1[c]	2.7–17		
Sawing	19	2.9[c]	0.4–123		
Furniture manufacture (Australia)				1989–90	Pisaniello et al. (1991)
Wood machinists	99	3.2	0.4–24		
Cabinet-makers	57	5.2	0.4–19		
Chair framemakers	15	3.5	2.0–7		
Manufacture of furniture and fixtures (Finland)				1980–85	Welling & Kallas (1991)
Boring	3	8.7	0.9–22		
Lathing	6	14	1.8–64		
Machine sanding	47	18	0.6–320		
Planing	9	0.8	0.1–2.5		
Routing	10	6.4	0.7–15		
Sanding between varnishing operations	17	16	0.4–81		
Sawing	44	6.8	0.3–73		
Spindle moulding	5	12	0.6–45		
Trimming	2	0.8	0.3–1.3		
Hand sanding	8	20	0.5–60		
Furniture factories (Sweden)	28	2.0	0.3–5.1	NR	Wilhelmsson & Drettner (1984)
Furniture manufacture (Denmark)	396	1.1[d]	1.1–1.2[e]	NR	Vinzents & Laursen (1993)
Cabinet-making (Czechoslovakia)				1961–62	Kubiš (1963)
Belt sander	10	24	3.6–65		
Cabinet-making shop (United Kingdom)	71	8.1		NR	Al Zuhair et al. (1981)

Table 10 (contd)

Industry and operation (country)	No. of measurements	Mean[a] (mg/m^3)	Range (mg/m^3)	Year	Reference
Cabinet-making (Canada)				1984	Sass-Kortsak et al. (1986)
Assembly	19	NR			
Laminating, graphics, glueing	3	1.9			
Sanding	7	1.1			
Sawing	12	2.9			
Miscellaneous work	7	1.7			
		1.2			
Manufacture of kitchen cabinets (Canada)				1985	Holliday et al. (1986)
From hardwood	12	0.3–5.1			
From particle-board	5	0.7–3.7			
Manufacture of kitchen cabinets (USA)	42	1.6[b]	0.3–13	1987–88	Clayton Environmental Consultants (1988)

NR, not reported
[a] Arithmetic mean unless otherwise specified; time-weighted average personal and/or area sample
[b] Geometric mean
[c] Median
[d] Mean of geometric means
[e] Range of geometric means

exposed to wood dust, are not exposed regularly to other chemicals: The mean exposure of cabinet-makers to formaldehyde was usually low (< 0.1 ppm [< 0.23 mg/m^3]) in a Canadian study (Sass-Kortsak et al., 1986). The mean level of solvents in finishing departments was about 20% of the national exposure limit of a mixture in Danish measurements in the early 1990s (Vinzents & Laursen, 1993) and about 40% in Finnish measurements in 1975–84 (Priha et al., 1986). The solvents used are typically mixtures of several chemicals, such as aliphatic hydrocarbons (solvent naphtha, white spirit [see IARC, 1989c]), aromatic hydrocarbons (toluene [see IARC, 1989d], xylene [see IARC, 1989e]; less often benzene [see IARC, 1987e] and styrene [see IARC, 1994a]), esters, alcohols, ketones and glycol ethers (Partanen et al., 1993). The constituents of solvent mixtures can vary, e.g. by country, period and facility. Pigments and other agents that may be used in furniture factories have been listed elsewhere (IARC, 1981).

1.4.6 Exposure during the manufacture of other wood products

The concentrations of wood dust measured during the manufacture of wooden doors, windows, prefabricated buildings, boats and other wood products are presented in Table 11. Measurements made during unspecified woodworking, which may be related to production of furniture or other wood products, are also included in the Table. The mean concentrations are similar to those in furniture manufacture because largely similar machining operations are used. The highest exposures occur in wood machining operations, where the mean levels are usually 1–10 mg/m^3.

No studies on changes over time in the levels of wood dust in these industries were available to the Working Group, but in some countries levels may have declined during the last few decades, as in the furniture industry (see section 1.4.5). The main reason is use of local exhaust systems for woodworking machines. Comparative measurements made in woodworking shops in Germany with the exhaust on and off indicate that the concentration of wood dust is very high when the exhaust system is out of operation (see Table 11; Wolf et al., 1986).

As in the furniture industry, many species of softwoods and hardwoods are used. For example, in the study in Germany mentioned above, mainly oak and beech were used but many workers had also been exposed to pine, spruce and other species (Wolf et al., 1986). Coniferous wood species are frequently used in the manufacture of window frames, doors and prefabricated buildings, and about 90% of all wood used in this way in Finland in 1986 was pine or spruce (Welling & Kallas, 1991).

Other agents that occur in workroom air depend on the products and processing methods used; they may include surface coatings (solvents, resins, pigments), glues (formaldehyde, phenol, epoxy compounds, polyurethanes) and engine exhaust, and the levels may be comparable to those found in the furniture industry (see section 1.4.5). Wood is usually coated and treated away from dust-generating operations, because dust may interfere with the application of chemicals; however, solvents, formaldehyde and other vapours may spread to areas where wood processing operations are being performed, and exposure may also occur through dermal contact from handling treated wood, through the release of chemicals into the air when treated wood is heated during wood processing operations, or through inhalation of dust from treated wood.

Table 11. Concentrations of wood dust in other wood product industries

Industry and operation (country)	No. of measurements	Mean^a (mg/m^3)	Range (mg/m^3)	Year	Reference
Woodworking shops (Germany)				NR	Wolf et al. (1986)
Sawing with exhaust	91	[5.9]	0.2–47		
Sawing without exhaust	22	[34.4]	1.5–184		
Moulding with exhaust	64	[5.6]	0.1–60		
Moulding without exhaust	12	[17.3]	1.2–113		
Sanding with exhaust	69	[8.3]	0.3–55		
Sanding without exhaust	13	[56.7]	3.7–500		
Assembly with exhaust	6	[5.2]	1.0–11		
Assembly without exhaust	19	[9.3]	0.7–40		
Woodworking shops (Germany)				1987–88	Albracht et al. (1989)
Sanding	84	3.6^b			
Sawing	88	2.4^b			
Moulding	38	1.0^b			
Planing	27	1.1^b			
All-round woodworkers	42	2.0^b			
Woodworking shops (Germany)				NR	Scheidt et al. (1989)
Sawing, routing, sanding	6	5.1^b	2.9–6.6		
Woodworking (USA)				1987–88	Clayton Environmental Consultants (1988)
Saw operators	191	0.8^c	<0.1–240		
Sander operators	85	1.2^c	0.1–41		
Milling machine operators	111	1.2^c	0.1–250		
Woodworking machine shops (United Kingdom)				NR	Hamill et al. (1991)
Hard- and softwood processing	7		0.5–5.1		
Softwood processing	37		0.3–55		
Hard- and softwood processing	51		0.5–33		
Woodworking factories (Denmark)	153	0.9^d	0.4–1.3^e	NR	Vinzents & Laursen (1993)
Joinery workshops (France)	6	22	2.4–73	NR	IARC (1981)

Table 11 (contd)

Industry and operation (country)	No. of measurements	Mean[a] (mg/m³)	Range (mg/m³)	Year	Reference
Joinery shops (Sweden)					Nygren et al. (1992)
Circular sawing	13	0.5[c]		NR	
Sanding	15	1.2[c]			
Cutting	20	0.3[c]			
Manufacture of doors and windows (Finland)				1980–85	Welling & Kallas (1991)
Machine sanding	5	3.4	1.4–6.7		
Packaging	2	1.5	1.4–1.5		
Sawing	6	2.0	1.2–3.3		
Spindle moulding	2	1.4	1.3–1.4		
Manufacture of doors and windows (Denmark)	118	0.6[d]	0.6–0.8[e]	NR	Vinzents & Laursen (1993)
Manufacture of prefabricated buildings (Canada)	8		0.4–2.5	1985	Holliday et al. (1986)
Manufacture of prefabricated houses (Finland)				1980–85	Welling & Kallas (1991)
Sawing	5	1.8	0.6–4.6		
Spindle moulding	2	2.9	0.6–5.1		
Manufacture of signs and plaques (USA) Router, sander	18	3.2	1.0–8.1	1983	McCawley, M. (1983; cited in United States National Institute for Occupational Safety and Health, 1987)
Wood component fabrication (USA)				1975	Kominsky & Anstadt (1976)
Router/groover operator	5	21.8	1.4–51.0		
Saw operator (total particulate)	19	68.8	0.7–688		

Table 11 (contd)

Industry and operation (country)	No. of measurements	Mean[a] (mg/m^3)	Range (mg/m^3)	Year	Reference
Boat building (USA) Carpenters in assembly	27	2.4	0.3–16.2	1983	Crandall, M.S. & Hartle, R.W. (1984; cited in United States National Institute for Occupational Safety and Health, 1987)
Manufacture and repair of wooden boats (Finland)	4	1.2	0.8–1.8	1980–85	Welling & Kallas (1991)

NR, not reported
[a] Arithmetic mean unless otherwise specified; time-weighted average personal and/or area samples
[b] Median
[c] Geometric mean
[d] Mean of geometric means
[e] Range of geometric means

Potential exposure to pesticides is high in the building of wooden boats because the wood must be protected from decay and marine borers (Jagels, 1985). Manufacture of windows, garden furniture, balcony decks, railroad ties, piers and other wooden structures for outdoor use may entail exposure to wood preservatives, such as chlorophenols, creosote, chromated copper arsenate and ammoniacal copper arsenate. The concentration of arsenic around various types of joinery machines was 0.5–3.1 µg/m^3 in six Swedish joinery shops using wood impregnated with copper–chromium–arsenic salt. The concentration of chromium was 0.4–2.3 µg/m^3 and that of copper was 0.4–1.9 µg/m^3. No hexavalent chromium was found (Nygren et al., 1992). Exposure to low levels of arsenic has also been reported in factories where wood is impregnated with arsenic-containing preservatives (Rosenberg et al., 1980). Insulation materials used in the manufacture of prefabricated houses often contain man-made mineral fibre products, such as glasswool and stonewool (Rockwool©) (see IARC, 1988). A number of other chemicals may be used as additives, including inorganic salts as fire retardants and chlorophenates as preservatives (Suchsland & Woodson, 1986).

1.4.7 Exposures in other wood-related occupations

Table 12 summarizes the concentrations of wood dust in some other wood-related occupations and operations, including flooring and parquet laying, pattern and model making, wood handling in pulp mills and teaching art and vocational skills.

The level of exposure in wooden model making in the automotive industry and in metal foundries averaged about 1 mg/m^3 in a study in the United States (McCammon et al., 1985). Model makers use a wide variety of woodworking machines and hand tools in preparing models. Prototypes are made of softwoods, such as pine, bass, jelutong, plywood lavan and mahogany. Mahogany has been used for die models, but cativo wood impregnated with phenol–formaldehyde resin is now commonly used. Paints, sealers and lacquers that release various solvents are used to coat models. Model-making may also involve use of adhesive systems, such as white glues and epoxy resins, and plastics like carvable putties, fibre glass and poly-foams. Model-making also requires the use of glues that contain epoxy compounds and amines. The highest solvent concentration measured in the United States was 10% of the exposure limit of the mixture; no formaldehyde or amines were detected in air (McCammon et al., 1985).

Some building trades entail exposure to wood dust. Sanding of parquet before varnishing is a dusty operation, which is usually carried out by specialized workers. Varnishes applied to parquets and wooden floors often contain formaldehyde-based resins and organic solvents, and the level of exposure to formaldehyde during varnishing may be over 1 ppm (1.23 mg/m^3). Construction carpenters use handsaws and circular saws both indoors and outdoors; however, since no measurements of exposure to wood dust were available to the Working Group, the mean level is probably low. Construction carpenters may also be exposed occasionally to other agents in the wide variety of activities carried out at construction sites. Most tasks on a construction site are performed by specialized workers, and it is unlikely that carpenters would be involved in e.g. painting, but different trades often work side-by-side, resulting in potential cross-exposure to e.g. other dusts and insulation materials. In addition, small construction

Table 12. Concentrations of wood dust in other wood-related occupations

Industry and operation (country)	No. of measurements	Mean[a] (mg/m³)	Range (mg/m³)	Year	Reference
Wooden model making (USA)					
Research and safety model shop	10	0.9	0.2–3.4	1980	Enright, J.C. (1980; cited in United States National Institute for Occupational Safety and Health, 1987)
Wood mill	10	4.7	1.2–10.2	1985	Holliday et al. (1986)
Pattern making (Canada)	5		1.0–2.6	NR	McCammon et al. (1985)
Automotive wood model shop (USA)					
Model makers/hardwood	12	0.6	0.2–0.3		
Model makers/soft- and hardwood	23	0.8	0.2–8.3		
Model makers/softwood	4	0.3	0.2–0.5		
Multi-axis machine operators	4	0.5	0.2–1.0		
Sweepers	5	1.6	0.1–6.1		
Shapers	7	2.7	0.3–13.9		
Wood mill (general)	10	0.3	0.05–0.5		
Parquet sanding (Germany)	5	6.6[b]		1987–88	Albracht et al. (1989)
Flooring/hard- and softwood (Canada)	7		0.3–1.7	1985	Holliday et al. (1986)
Parquet sanding (Germany)	2	9.3	4.4–14	NR	Scheidt et al. (1989)
Pulp/paper mill (USA)					
Chipping, debarking, screening, loading	19	0.3[c]	<0.1–18	1987–88	Clayton Environmental Consultants (1988)
Art school (USA)					
Sawing, sanding, planing	8	6.0	0.9–24.2	1976	Levy, B.S.B. (1976; cited in United States National Institute for Occupational Safety and Health, 1987)
University art department (USA)	4	3.5	1.6–5.7	1980	Lucas & Salisbury (1992)

[a] Arithmetic mean unless otherwise specified; time-weighted average personal and/or area samples
[b] Median
[c] Geometric mean

companies and those using non-union labour may not make clear distinctions between the responsibilities of different trades.

Pulp making and some papermaking processes start with retrieval of logs from storage, debarking and then chipping, and workers handling wood are exposed to wood dust, although the level of exposure is usually below 1 mg/m^3 (Table 12). High levels of fungal spores and bacteria have been found occasionally at wood and chip handling sites of pulp and paper mills (Kotimaa, 1990). Teachers and other personnel working in vocational and art schools may also have occupational exposure to wood dust (Table 12). Exposure may be high, owing to poor ventilation, but it is generally not continuous.

Forestry workers are a large occupational group who process and handle wood regularly. Lumberjacks have cut trees for centuries, first with axes and handsaws and, since about 1950, with chain saws. No measurements were available of their level of exposure to wood dust, but it is probably lower than those usually found in wood industries. Other exposures of forestry workers include engine exhaust (see IARC, 1989f) from chain saws and forest vehicles, chain oils and gasoline (see IARC, 1989g) used as fuel for chain saws.

1.4.8 Particle size distribution of wood dust in workroom air

Exposures to wood dust can be characterized not only by the mass (or number of particles) per unit volume of air but also by the distribution of particle sizes. Wood dust particles are typically irregular in shape and have rough surfaces, as observed by scanning electron microscopy; however, no differences in morphological pattern have been noted among samples from different operations (Liu *et al.*, 1985).

Several investigators have reported particle size distributions for wood dust in workplace air in various industries. Representative studies are summarized in Tables 13 and 14. In most studies, the major portion of the wood dust mass is contributed by particles larger than 10 μm in aerodynamic diameter (Whitehead *et al.*, 1981a; Darcy, 1984; Lehmann & Fröhlich, 1987, 1988; Hinds, 1988; Pisaniello *et al.*, 1991). This is attributable, in part, to the fact that larger particles are also heavier. Holliday *et al.* (1986) used an optical microscopy method (see section 1.3.1) to count particles in various size ranges and found that 61–65% (as calculated by the United States National Institute for Occupational Safety and Health, 1987) of the particles measured 1–5 μm.

Some investigators have reported that the particle size distribution varies substantially according to woodworking operation, sanding producing more small particles and sawing producing more large particles (Hounam & Williams, 1974; Darcy, 1984; Liu *et al.*, 1985). Other investigators, however, have found no consistent differences (Holliday *et al.*, 1986; Lehmann & Fröhlich, 1988; Pisaniello *et al.*, 1991).

There is also some evidence that processing (especially sanding) of hardwoods can generate a higher percentage of small particulates than processing of softwoods, although again the evidence is by no means consistent and other studies have shown no differences. Whitehead *et al.* (1981a) suggested that processing of hardwoods may lead to higher concentrations of respirable dust than processing of softwoods, on the basis of a comparison of 15 samples taken

Table 13. Particle size distribution of hardwood dust (%)

Wood, operation	Total dust (mg/m^3)	Stage^a								Back-up filter
		0	1	2	3	4	5	6	7	
Oak, hand sanding	6.9	72.6	9.6	5.1	3.3	2.3	1.7	1.7	1.5	2.2
Oak, machine sanding	2.7	65.0	12.2	3.9	4.2	3.0	3.4	3.0	2.1	3.3
Oak, sanding (hand portable machine)	2.7	47.2	14.6	7.2	9.1	7.0	4.8	2.6	2.3	5.2
Oak and beech, sawing and machine sanding	5.4	44.4	21.9	7.0	7.2	2.9	2.4	1.1	2.4	10.7
Particle-board and beech, sawing and planing	9.4	65.1	15.9	6.1	8.2	2.8	0.9	0.5	0.5	0.0
Ash, hand sanding	1.9	49.5	16.7	14.3	10.1	4.3	2.2	1.2	0.0	1.7
Beech, sawing	4.1	62.7	12.7	9.5	3.5	2.9	2.8	2.4	1.9	1.6

From Lehmann & Fröhlich (1988)

^{a}Stage 0, > 9.0 mm; stage 1, 5.8–9.0; stage 2, 4.7–5.8; stage 3, 3.3–4.7; stage 4, 2.1–3.3; stage 5, 1.1–2.1; stage 6, 0.65–1.1; stage 7, 0.43–0.65

Table 14. Wood dust sizes measured in the workplace in various studies

Study description	Equipment/operation	Sampling device	Mass median aerodynamic diameter (mm)	Reference
Cabinet-making (Czechoslovakia), 1 plant, 1961–62; area samples/total dust	Belt sander	NR	Up to 95%, < 5 Most, 2–3	Kubiš (1963)
Furniture (England), 5 plants, 1973; personal samples/total dust	Band sawing, turning Planing Routing, moulding Sanding Assembly	Four-stage cascade centripeter	11.5 9.2 10.0 8.4 7.6 (< 25%, < 5)	Hounam & Williams (1974)
Furniture (Denmark), 8 plants, 1974–75; personal samples/total dust	Sanding, drilling, planing, sawing	NR	33% (mass), < 5 41% (mass), 6–10 11% (mass), 11–15 15% (mass), > 16	Andersen et al. (1977)
Wooden component fabrication (USA), 1 plant, 1975; personal samples	Saw operator, router/ groover operator	Six-stage cascade impactor	> 10	Kominsky & Anstadt (1976)
Wooden products (USA), 2 plants, 1976; personal samples	Shake mill (western red cedar)	Cyclone unit	39%, < 10 23%, 10–20 38%, > 20	Edwards et al. (1978)
	New planer mill (Douglas fir/hemlock)	Cyclone unit	47%, < 10 25%, 10–20 28%, > 20	
Plywood/furniture (USA), 12 plants, 1978; area samples/total dust	Veneer lathe/clipper, dryer, dry veneer handling, edge sawing/ sanding, machining, assembly, milling, sanding	Six-stage cascade impactor	[1.3] mg/m^3 < 5.5[a] [3.3] mg/m^3 < 14.1[a]	Whitehead et al. (1981a)

Table 14 contd

Study description	Equipment/operation	Sampling device	Mass median aerodynamic diameter (mm)	Reference
Furniture (England), 2 plants, 1981; personal samples/total dust	Sawmill Assembly Machine floor Cabinet shop	Seven-stage cascade impactor	17.3 18.0 9.3 12.5	Al Zuhair et al. (1981)
Wooden model making (USA), 3 shops, 1981–82; personal samples/total dust	Model maker, sweeper, shaper operator, plastic shop worker, multi-axis machine operator	Nine-stage cascade impactor	7.7 (range, 5.2–10); 18–61% respirable dust[b]	McCammon et al. (1985)
Furniture (England), 7 plants, 1983; personal samples/total dust	Machine sanding, hand sanding, sawing, other cutting	Impactor	9 (54% (mass), 4–10)	Jones & Smith (1986)
Signs/plaques (USA), 1 shop, 1983; area samples/total dust	Router, sander	Four-stage cascade impactor	46–60% (mass), < 3.5 30–35% (mass), 3.5–20 5–20% (mass), > 20	McCawley, M. (1983; cited in United States National Institute for Occupational Safety and Health, 1987)
Woodworking (Finland), 1 shop, 1983; personal samples/total dust	Unloading wood, sawing, other machines, planing	Optical microscopy	97.8%, < 5	Lindroos (1983)
Plywood (Finland), 6 mills, 1984; personal and area samples/total dust	Sawing Finishing General workroom	MSA cyclone	40% respirable dust 29% respirable dust 65% respirable dust	Kauppinen et al. (1984)
Particle-board (USA), 1 plant, 1986; area samples/total dust	Sanding	Six-stage cascade impactor	8.26	Stumpf et al. (1986)
Various industries (Canada), 23 plants, 1985; personal samples/total dust	Sawing Sanding Planing/routing/shaping	Optical microscopy	62%, 1–5 61%, 1–5 65%, 1–5	Holliday et al. (1986)

Table 14 contd)

Study description	Equipment/operation	Sampling device	Mass median aerodynamic diameter (mm)	Reference
Various industries (Germany), 17 factories, 2 training shops, 1983–85; personal and area samples	Sanding, sawing, routing/planing	Anderson impactor	44.4–72.6%, > 9 9.6–21.9%, 5.8–9 3.9–14.3%, 4.7–5.8 3.3–10.1%, 3.3–4.7 2.3–7.0%, 2.1–3.3 0.9–4.8%, 1.1–2.1 0.5–3.0%, 0.7–1.1 0–2.4%, 0.4–0.7 0–10.7%, 0–0.4	Lehmann & Fröhlich (1987)
Various industries (Hong Kong), 3 factories; personal samples	Sawmill, sanding (furniture factory), mixing (mosquito-coil factory)	Scanning electron microscopy	9–12.8%, > 10 18–26.1%, 5–10 61.7–73%, 0–5	Liu et al. (1985)
Furniture (Australia), 15 factories, 1989; personal samples	Sanding Sawing Mixed woodworking	IOM/7-hole; cascade impactor	16–19 17–22 15–23	Pisaniello et al. (1991)

NR, not reported; MSA, Mine Safety Appliances Co. (Pittsburgh, PA); IOM, Institute of Occupational Medicine

[a] Gravimetric concentrations are given, rather than percentages. Of the 15 samples reported, only two contained dust of a mass median aerodynamic diameter <5.5 μm at concentrations ≥ 1 mg/m^3, and only three contained dust < 14.1 μm at > 2 mg/m^3.

[b] Scanning electron microscopy indicated that length-to-width ratios were 2.3:1 and 1.9:1 in two air samples collected with a 0.5-in [1.3-cm] stainless-steel cyclone.

during furniture and plywood manufacture. Darcy (1982), however, found that the distribution of particle sizes from sanding pine and oak were very similar (see distribution curves reproduced by Hinds, 1988). Pisaniello *et al.* (1991) reported only a very slight difference in the average mass median aerodynamic diameter of dust from hardwood (18.7 μm; geometric standard deviation (GSD), 2.0) and from softwood/reconstituted wood (19.6 μm; GSD, 2.1).

1.5 Regulations and guidelines

Several countries have set standards or guidelines for occupational exposures to wood dust, with 8-h time-weighted average (TWA) exposure limits ranging from 1 to 10 mg/m^3 (United States National Institute for Occupational Safety and Health, 1987, 1992). In the regulations of some countries, a particular class of wood dust is named (e.g. 'hardwood' and 'softwood'). For example, in the United Kingdom, the long-term exposure limit (8-h TWA) for hardwood and softwood is 5 mg/m^3; for hardwood, there is a notation that the substance can cause respiratory sensitization, and the limit for softwood is noted for intended change (United Kingdom Health and Safety Executive, 1992). In Canada, the limits are 1 mg/m^3 for hardwood and 5 mg/m^3 for softwood (United States National Institute for Occupational Safety and Health, 1987). In Germany, dusts of oak and beech have been classified as human carcinogens (group III A1) since 1985, and other wood species are suspected of having carcinogenic potential (group III B). The technical exposure limit for total wood dust was set at 2 mg/m^3 for all industrial plants in 1993. The limit of 5 mg/m^3 set up for old industrial plants in 1987 will no longer be allowed, with a few exceptions, by 31 December 1995 (Deutsche Forschungsgemeinschaft, 1993). In Sweden also, wood dust is considered potentially carcinogenic (United States Occupational Safety and Health Administration, 1987).

In other countries, wood dust is regulated under more general categories of particulate matter. In Hungary and Poland, for example, dust of vegetable and animal origin is regulated; dust containing various percentages of free silica are regulated in Poland (United States Occupational Safety and Health Administration, 1987). Standards for organic dusts are used for wood dust in Finland (8-h threshold limit value [TLV], 5 mg/m^3; maximum for 15 min, 10 mg/m^3) (Työministeriö, 1993). Switzerland has no specific standards for wood dust but controls 'total dust' and 'fine dust' (United States National Institute for Occupational Safety and Health, 1987).

The American Conference of Governmental Industrial Hygienists (1993) recommended the following TLVs: 1 mg/m^3 for an 8-h TWA for wood dust (certain hardwoods such as beech and oak); and 5 mg/m^3 TWA and 10 mg/m^3 for the short-term exposure limit to softwoods, with the notation that the substance has been identified elsewhere as a suspected human carcinogen. Similar exposure limits have been adopted by several other countries (e.g. Australia, New Zealand and Norway) as regulations or guidelines (United States National Institute for Occupational Safety and Health, 1987). 'Particulates not otherwise regulated' are covered in the United States (United States Occupational Safety and Health Administration, 1993).

2. Studies of Cancer in Humans

2.1 Case reports

Many cases of cancer of the sinonasal cavities and paranasal sinuses (referred to below as 'sinonasal cancer') have been reported among woodworkers. The earliest reports of cases of cancer of the upper respiratory tract in association with woodworking were published in Germany (reviewed by Schroeder, 1989).

Macbeth (1965) reported 20 patients (17 men) from High Wycombe, United Kingdom, presenting with a malignant disease of the paranasal sinuses; 15 of the cases, all in men, were associated with the making of wooden chairs. Macbeth noted later that the tumours were all adenocarcinomas (Acheson, 1976). Several furniture factories are located in High Wycombe which specialize in chair-making from a variety of domestic and imported hardwoods. The cross-sectional prevalence of woodworkers in the local male population at the time of the study was 23.5%.

Following these observations, a survey was carried out in the Oxford area, including High Wycombe, of 148 cases (98 males) of nasal cancer diagnosed in 1951–65 (Acheson *et al.*, 1968; Acheson, 1976). Cases were classified according to sex and histological type of cancer. The results for men indicated a strong relationship between adenocarcinoma and present or past work in the furniture industry. Of the 33 cases of adenocarcinoma in men, 24 (73%) were in woodworkers, 22 of whom (67%) worked in furniture manufacture. Among the 65 remaining male cases of nasal tumour, the corresponding numbers were five (8%) and three (5%). For the subgroup of men who were employed at the onset of their illness, which was diagnosed in 1956 or later, a detailed occupational history was obtained and was compared with the findings of the 1961 census. The estimated rate for adenocarcinoma in cabinet- and chair makers and wood machinists in High Wycombe was similar, namely 0.7 ± 0.2 per 1000 men per annum during the decade 1956–65, which was at least 500 times the rate in adult males in southern England. The risk was extended to workers making products other than chairs. The results suggested that carpenters and joiners in High Wycombe had no increase in risk. The species of wood used by the 16 patients for whom information was available before the Second World War were oak (14/16), beech (11/16) and mahogany (13/16). Walnut was also frequently used.

Acheson *et al.* (1972) performed a survey of nasal adenocarcinoma in England excluding the Oxford area. Cases of nasal adenocarcinoma were collected from cancer registries (for the period 1961–66 for most registries) and were compared with cases of nasal cancer other than adenocarcinomas. The study comprised 107 cases of adenocarcinoma (80 male) and 110 cases of nasal cancer other than adenocarcinoma (85 male) when restricted to cases 'accepted' mainly on the basis of confirmation of the histological classification. Thirty-three men (41%) and one woman (3.7%) with nasal adenocarcinoma had at some time worked as woodworkers; of these, 24 (73%) men had worked in the furniture industry. The ratio of observed cases:expected cases was 95 for furniture workers and 5 for other woodworkers (principally, carpenters and joiners) on the basis of the distribution at the 1961 census. Among nasal cancer cases other than adenocarcinoma, a significant excess in woodworkers was also observed. The main types of wood dusts were known for some of the woodworkers (with both adenocarcinoma and other histological types): most were exposed to more than one species. The species most often indicated in the furniture industry were oak (eight adenocarcinoma patients), mahogany (six adenocarcinoma patients) and beech, birch and walnut (four adenocarcinoma patients for each of the species). Four patients had used mainly softwoods; three of those with adenocarcinomas were joiners or joiners and carpenters, and the last had worked as a packing case maker and had a squamous-cell carcinoma.

After the results from the United Kingdom were published, cases were reported from many countries, including Belgium (Debois, 1969), the Netherlands (Delemarre & Themans, 1971), Denmark (Andersen, 1975; Andersen et al., 1976, 1977), France (Trotel, 1976), Australia (Ironside & Matthews, 1975), western (Gülzow, 1975; Kleinsasser & Schroeder, 1989) and eastern Germany (Löbe & Ehrhardt, 1983; Wolf et al., 1986), Sweden (Engzell et al., 1978; Klintenberg et al., 1984), Austria (Smetana & Horak, 1983), Norway (Voss et al., 1985), Switzerland (Rüttner & Makek, 1985) and Spain (López et al., 1990). Subsequently, many more case reports were published in different countries (for an extensive review, see for instance Mohtashamipur et al., 1989a). A systematic analysis of these studies is not included, as many analytical studies were available.

The studies that presented data on occupational exposures to specific species of wood are summarized in Table 15.

2.2 Descriptive studies

The studies placed under this heading were mainly designed for generating hypotheses, especially by use of record linkage with routinely collected data.

A number of descriptive studies (Table 16) have dealt with cancer mortality or incidence among woodworkers, defined on the basis of occupational title and/or industrial branch reported on death certificates (Menck & Henderson, 1976; Milham, 1976; Gallagher et al., 1985), in hospital files (Grufferman et al., 1976; Menck & Henderson, 1976; Bross et al., 1978), in cancer registries (Acheson, 1967; Nandakumar et al., 1988) or in the records of pension funds (Olsen & Jensen, 1987; Olsen et al., 1988), in union files (Milham, 1978) or declared at censuses (Pukkala et al., 1983; Gerhardsson et al., 1985; Pearce & Howard, 1986; Linet et al., 1988, 1993). None of these studies provides quantitative or semi-quantitative information on exposure to wood dust.

Some studies based on occupational titles specifically address nasal cancer in woodworkers (Table 17), both in terms of mortality (Minder & Vader, 1987) and morbidity (Malker et al., 1986; Vetrugno & Comba, 1987; Olsen, 1988). These studies corroborate the well-established indication of an increased risk for nasal cancer in woodworkers. The same observation applies to incidence studies of nasal cancer in which information on occupation was provided by questionnaires and/or interviews with patients (Ghezzi et al., 1983; Petronio et al., 1983).

Table 15. Case reports of sinonasal cancer according to occupation and type of wood

Study, year, country	Sex	Origin	Histological type	Exposed cases/ total cases	Occupations	Main types of wood
Acheson et al. (1968); Acheson (1976), Oxfordshire, United Kingdom	M	Registry	Adenocarcinoma	24/33	22 in the furniture industry (mainly wood machinists and furniture makers)	Oak, beech, mahogany, walnut
Acheson et al. (1972), United Kingdom	M	Registries	Adenocarcinoma	33/80	24 in the furniture industry (mainly cabinet- and chair makers, wood machinists and turners) 6 not in the furniture industry (mainly joiners and carpenters)	Oak, mahogany, beech, birch, walnut Softwoods in 3 cases
Leroux-Robert (1974), France	M	Hospital	Adenocarcinoma, ethmoid sinus	26/92	Not reported	All patients had used European hardwoods, some exclusively, and mainly oak (22/26), exclusively oak in one case
Luboinski & Marandas (1975), Paris, France	M	Hospital	Adenocarcinoma, ethmoid sinus	21/43	8 joiners, 4 joiners and cabinet-makers, 7 cabinet-makers, 1 cooper, 1 coffin maker	European hardwoods (oak, chestnut, wild cherry, walnut, beech, poplar); tropical species, including mahogany
Ironside & Matthews (1975), Victoria, Australia	M	Hospital	Adenocarcinoma	10/18	3 carpenters, 2 woodturners/ wood machinists, 2 builders, 1 sawmill owner, 1 timber worker, 1 joiner	Native timber

Table 15 (contd)

Study, year, country	Sex	Origin	Histological type	Exposed cases/ total cases	Occupations	Main types of wood
Andersen et al. (1976, 1977), Aarhus, Denmark	MF	Hospital	Adenocarcinoma	12/17	10 cabinet- and chair makers, 1 turner, 1 coach builder	Several kinds of wood used by each cabinet- and chair maker; primarily beech, oak, walnut; periodically mahogany and teak
	MF	Hospital	Squamous-cell carcinoma	5/71	1 carpenter, 1 turner, 1 sawyer, 1 forestry worker, 1 brushmaker	
Engzell et al. (1978), Sweden	M	Registry	Adenocarcinoma	19/36	19 joiners or cabinet-makers (at least 12 cabinet-makers)	Hardwoods such as oak, teak, mahogany and birch; never exclusively softwood
Voss et al. (1985), Norway	M	Hospital	Adenocarcinoma	1/2	Cabinet-maker	Birch, pine, spruce, teak, mahoghany
	M	Hospital	Squamous-cell carcinoma and undifferentiated carcinoma	8/30	3 joiners/carpenters, 3 loggers, 1 sawmill/carpenter, 1 cabinet-maker	Pine and spruce (7 cases), pine and lime (1 case)
Kleinsasser & Schroeder (1989), Germany	M	National recruitment	Adenocarcinoma, intestinal type	77/85	55 joiners and cabinet-makers, 11 wheelwrights, 6 coopers, 5 parquet floor layers, 5 carpenters (non-exclusive categories)	Oak, beech No case patient had handled softwood or exotic wood exclusively.

Table 16. Studies of mortality and morbidity in woodworkers

Reference	Method	Results
Milham & Hesser (1967)	Analysis of occupations reported on death certificates (New York State, USA, 1940–53 and 1957–64). Comparison with all other causes of death	Hodgkin's disease: significant association with woodworking (χ^2, 14.59; $p<0.001$)
Acheson (1967)	Analysis of occupations of patients in the Oxford area, United Kingdom, 1956–65	Hodgkin's disease: three cases among woodworkers versus four expected
Grufferman et al. (1976)	Incidence by occupation in Boston, USA, 1959–73. Reference rates: whole Boston population	Hodgkin's disease: woodworkers: RR, 1.6 (95% CI, 0.9–2.6); 15 observed
Menck & Henderson (1976)	Mortality by occupation in Los Angeles County, USA, 1968–70. Reference rates: mortality in all occupations	Lung cancer: lumber, wood, furniture: SMR, 1.1 [0.7–1.8]; 20 observed
Milham (1976)	Proportionate analysis of mortality by occupation from death certificates, Washington State, USA, 1950–71	Plywood mill workers, stomach cancer: PMR, 1.5 [1.0–2.2]; 32 observed; leukaemia: PMR, 1.9 [1.2–2.9]; 23 observed Sawmill workers, cancer of the testis: PMR, 1.7 [0.9–2.7]; 15 observed Carpenters, stomach cancer: PMR, 1.3 [1.1–1.4]; 271 observed; Hodgkin's disease: PMR, 1.6 [1.1–2.2]; 38 observed
Bross et al. (1978)	Analysis of occupations of cancer patients attending Roswell Park Memorial Institute, USA, 1956–65. Comparison groups: noncancer patients	Oesophageal cancer: lumber workers: significantly increased risk
Milham (1978)	Analysis of mortality of woodworkers, USA, 1969–73. Reference rates from US population	Malignant neoplasm of stomach: SMR, 1.1 [1.0–1.2]; 407 observed
Pukkala et al. (1983)	Incidence by occupation declared at 1970 census in Finland, 1971–75. Reference rates from total economically active population	Lung cancer: woodworking: SIR, 1.3 [1.2–1.4]; 366 observed Joiners: SIR, 1.4 [1.3–1.6]; 264 observed
Gallagher et al. (1985)	Proportionate cancer mortality analysis of woodworkers in British Columbia, Canada, 1950–78	Loggers: nasal sinus: PCMR, 3.6 (1.2–8.5); 5 observed Woodworkers: stomach: PCMR, 1.3 (1.1–1.5); 116 observed; non-Hodgkin's lymphoma, PCMR, 1.4 (1.0–1.9); 42 observed

Table 16 (contd)

Reference	Method	Results
Pearce & Howard (1986)	Analysis of cancer mortality rates by occupation in New Zealand, 1974–78. Reference: mortality rates of all employed people	Large-bowel cancer in cabinet-makers and woodworkers: RR, 2.6 (1.3–4.6); 11 observed
Olsen & Jensen (1987)	Proportionate analysis of cancer incidence by occupation in Denmark, 1970–79	Among men: woodworking: stomach cancer: SPIR, 1.8 (1.3–2.5), 41 observed; breast cancer: SPIR, 3.9 (1.2–12) Among men: manufacture of wooden furniture: cancer of nasal cavities and sinuses: SPIR, 5.9 (2.5–14), 5 observed
Nandekumar et al. (1988)	Incidence by occupation in Western Australia, 1960–84. Reference rates: other occupations except farming	Multiple myeloma: woodworking: RR, 1.7 (0.78–3.9); 3 cases
Linet et al. (1988)	Incidence by industry through record linkage between 1960 census and National Cancer Registry in Sweden, 1961–79. Reference: national incidence rates	Acute nonlymphocytic leukaemia: wood (men): SIR, 1.3 [1.0–1.7]; 67 observed
Linet et al. (1993)	Incidence by industry through record linkage between 1960 census and National Cancer Registry in Sweden, 1961–79. Reference: national incidence rates	Non-Hodgkin's lymphoma: furniture and furnishings (men): SIR, 1.3 [1.0–1.7]; 55 observed

CI, confidence interval; RR, relative risk; SMR, standardized mortality ratio; PMR, proportionate mortality ratio; SIR, standardized incidence ratio; PCMR, proportionate cancer mortality ratio; SPIR, standardized proportionate incidence ratio

Table 17. Mortality and morbidity studies on nasal cancer in wood workers

Reference	Method	Results
Malker et al. (1986)	Incidence in Swedish subjects who reported their occupation as woodworker at 1960 census. Follow-up through 1979. Comparison with population.	Woodworkers (males) - All nasal cancers: SIR, 1.3 [0.8–1.9], 24 observed - Adenocarcinomas: SIR, 2.2 [0.8–4.8], 6 observed Furniture workers (males) - All nasal cancers: SIR, 4.1 [2.7–6.1], 25 observed - Adenocarcinoma: SIR, 17 [10–26], 19 observed
Minder & Vader (1987)	Mortality of Swiss subjects who reported their occupation as woodworker at 1980 census. Follow-up through 1985. Comparison with all workers.	SMR: 6.2 (3.6–10), 16 observed
Vetrugno & Comba (1987)	Analysis of the occupations reported by 189 cases diagnosed and/or treated in 1983–85 at 61 Italian ear-nose-and-throat clinics and hospital departments.	Among males, woodworkers account for 11% of the case series and 22% of adenocarcinoma cases.
Olsen (1988)	Analysis of employment histories reported by cases diagnosed in Denmark 1970–84.	Among males: wooden furniture production: SIR, 3.6 (1.3–8.0), 5 observed
Ghezzi et al. (1983)	Incidence among woodworkers in Brianza (Italy), 1976–80. Comparison with incidence in other occupations.	Rate ratio: 4.4 (1.8–9.1), 7 observed
Petronio et al. (1983)	Incidence in woodworkers in Trieste (Italy), 1968–80. Comparison with incidence in all occupations.	Woodworkers: incidence rate: 6.4×10^{-5} All occupations: incidence rate: 0.54×10^{-5}
Gerhardsson et al. (1985)	Record linkage between 1960 census and 1961–79 cancer registry in Sweden for morbidity from respiratory cancers in furniture workers. Reference rates: all other employed men.	Sinonasal carcinoma: SMR, 7.1; 90% CI, 4.4–11, 15 observed Sinonasal adenocarcinoma: SMR, 44; 90% CI, 27–69, 14 observed

CI, confidence interval; SMR, standardized mortality ratio; SIR, standardized incidence ratio

2.3 Cohort studies

The only cohort study that addressed the issue of exposure to wood dust was that conducted in High Wycombe, United Kingdom (Acheson *et al.*, 1984); the others assessed exposure by occupational title.

One cohort study was conducted in Finland of 1223 sawmill workers followed-up during 1945–80 (Jäppinen *et al.*, 1989). Cancer incidence was not in excess overall, and no cases of nasal cancer were found (0.3 expected). The only cancer for which an increased incidence was seen was non-melanocytic skin cancer (excluding basal-cell carcinoma), with six cases in men (standardized incidence ratio [SIR], 3.1; 95% confidence interval [CI], 1.2–6.8) and two cases in women [SIR, 1.8; 95% CI, 0.2–6.6]; however, four of the six male patients were first employed after 1945, when chlorophenols were used.

Four cohort studies of furniture workers are available, from the United Kingdom, Denmark, Germany and the United States. In the Danish study, 40 428 members of the carpenters' and cabinet-makers' union in 1971 were followed up to 1976 (Olsen & Sabroe, 1979). The overall mortality of both active and retired workers was below that expected; the only cancer for which increased mortality was seen was that of the nose and sinuses (standardized mortality ratio [SMR], 4.7; 95%CI, 2.5–6.8; four deaths).

In the study of 5108 furniture workers in High Wycombe, United Kingdom, followed through 1982, overall mortality and mortality from all cancers were below expectation; the only cancer site for which there was increased mortality was that of the nose and sinuses ([SMR, 8.2; 95%CI, 3.7–16] nine deaths) (Rang & Acheson, 1981; Acheson *et al.*, 1984). When workers were divided into three groups according to dustiness of workplace, all nasal cancer deaths were found in the group exposed to the most dust [SMR, 16; 95% CI, 7.2–30].

In a cohort of 759 cabinet-makers or joiners studied during 1973–84 in Germany, no cases of nasal cancer were found [expected number not reported]; the only cancer for which the incidence was increased was malignant melanoma (SIR, 9.5; 95%CI, 2.4–28; two cases) (Barthel & Dietrich, 1989).

A cohort of furniture makers in the United States, which included 34 801 subjects (of whom 12 158 were employed in wooden furniture facilities), was studied between 1946 and 1983 (Miller *et al.*, 1989, 1994). Overall mortality and mortality from all cancers were below expectation for the wooden furniture workers; the only neoplasm for which mortality was increased was myeloid leukaemia (SMR, 1.9; 95%CI, 1.0–3.5; 11 deaths); seven deaths were from acute leukaemia. A significant increase in mortality from chronic nephritis was also found (SMR, 2.5; 95% CI, 1.1–5.0; eight deaths). Two cohort members died from nasal cancer (2.5 expected); one case occurred in the cohort of wooden furniture workers [expected number not reported].

A cohort of 2283 plywood workers from four mills in Washington and Oregon, United States, was studied between 1945 and 1977 (Robinson *et al.*, 1990). Overall mortality and mortality from all cancers were below expectation, and no significantly increased mortality was seen. No deaths from nasal cancer were found (0.4 expected).

A total of 10 322 American Cancer Society volunteers enrolled in a large prospective study conducted between 1959 and 1972 reported wood-related occupations, and their mortality was compared with that of over 400 000 volunteers with other occupations (Stellman & Garfinkel, 1984). Overall mortality and mortality from all cancers were close to expectation; a significant increase in mortality was found from cancers of the stomach (SMR, 1.5; 44 deaths) and urinary bladder (SMR, 1.4; 29 deaths); non-significant increases in mortality were found for lung cancer (SMR, 1.1; 135 deaths) and nasal cancer (SMR, 2.0; two deaths); the two deaths from nasal cancer occurred among carpenters and joiners (SMR, 3.3).

Cohort studies on woodworkers are summarized in Table 18.

During the last decade, several papers raised the possibility that the risk for colorectal cancer was increased among wooden pattern and model makers exposed to wood dust in the automobile industry (Swanson & Belle, 1982; Swanson et al., 1985; Tilley et al., 1990; Becker et al., 1992; Roscoe et al., 1992). These studies are summarized in Table 19. Although various study designs were used, leading to different risk estimates (SMRs, proportionate mortality ratios [PMRs] and SIRs), the first three reported excess risks for colorectal cancer. The suggested association has been the object of some debate in the scientific literature (Chovil, 1982; Davies, 1983; Kurt, 1986). [The studies that gave positive results had several methodological problems, namely short duration of observation, high proportion of loss to follow-up and inadequate assessment of exposure; the study that was of more adequate design with respect to ascertainment of exposure and control of confounding (Roscoe et al., 1992) did not reach positive conclusions.]

2.4 Case–control studies

Information on exposure to wood dust or employment in wood-related occupations has been reported in studies dealing with many cancer sites. The Working Group reviewed in particular case–control studies of organs in the respiratory, the digestive and the lympho-haematopoietic systems. The case–control studies are grouped according to whether exposure to wood dust was addressed specifically or whether the results are based on job titles or industrial branch. The term 'relative risk' is used to cover all estimated risk ratios, which are usually given as odds ratios.

2.4.1 *Cancer of the nasal cavity and paranasal sinuses*

(a) *Exposure to wood dust*

Hernberg et al. (1983) reported the results of a joint Danish–Finnish–Swedish case–control study of 167 patients with primary malignant tumours of the nasal cavity and paranasal sinuses diagnosed in Denmark, Finland and Sweden between July 1977 and December 1980. Ninety-five cases were epidermoid carcinoma, 18 were adenocarcinoma, 17 were anaplastic carcinoma and 37 were other histological types. Controls were 167 patients with tumours of the colon and rectum, who were matched to patients on country, sex and age at diagnosis. Cases and controls were identified through the national cancer registries of Finland and Sweden and four of the five oncological centres in Denmark. Subjects were interviewed extensively by telephone to

Table 18. Cohort studies of workers in wood-related industries

Industry	Reference	Methods	Results	Notes
Sawmill workers	Jäppinen et al. (1989)	1223 sawmill workers employed between 1945 and 1961; follow-up till 1980; cancer incidence in the cohort contrasted to local incidence rates. Lost to follow-up: 0.2%	SIR: All cancers: men, 1.1 (0.9–1.3), 90 observed women, 1.2 (0.9–1.6), 55 observed Skin cancer: men, 3.1 (1.2–6.8), 6 observed women, [1.8; 0.2–6.6], 2 observed No other significantly increased SIR. No case of nasal cancer, 0.3 expected. *Women* Lip, mouth, pharynx (1 observed/0.9 expected) Stomach, 1.1 (0.4–2.5), 5 observed Colon (2 observed/2.3 expected) Rectum, 2.3 (0.6–5.8), 4 observed Larynx (0 observed/0.1 expected) Lung, 3.3 (0.9–8.3), 4 observed Lymphoma (0 observed/0.9 expected) Hodgkin's disease (0 observed/0.2 expected) Leukaemia, 2.7 (0.6–8.0), 3 observed *Men* Lip, mouth, pharynx, 1.8 (0.6–3.8), 6 observed Stomach, 0.8 (0.4–1.5), 11 observed Colon, 1.7 (0.6–4.1), 5 observed Rectum, 1.3 (0.4–3.3), 4 observed Larynx (2 observed/2.1 expected) Lung, 1.0 (0.6–1.4), 24 observed Lymphoma, 2.0 (0.6–5.2), 4 observed Hodgkin's disease (2 observed/0.8 expected) Leukaemia, 2.2 (0.6–5.5), 4 observed	Sawn timber, mainly pine and spruce; dust levels in sawmills generally below 1 mg/m^3

Table 18 (contd)

Industry	Reference	Methods	Results	Notes
Furniture workers	Olsen & Sabroe (1979)	40 428 members of the Danish carpenters'/cabinet-makers' trade union active or retired in 1971; follow-up through 1976; mortality in the cohort contrasted with national mortality rates	SMR: All causes: Active workers, 0.82 (0.76–0.88), 692 observed Retired workers, 0.70 (0.67–0.74), 1483 observed Nasal cancer: All workers, 4.7 (2.5–6.8), 4 observed (3 cases in cabinet-makers, 1 in a carpenter); no other significantly increased SMR *Active* Intestine, 0.75 (0.29–1.2) Stomach, 1.1 (0.53–1.6) Lung, 0.96 (0.68–1.1) Leukaemia, 1.3 (0.55–2.0) *Retired* Intestine, 0.94 (0.65–1.2) Stomach, 0.84 (0.59–1.1) Lung, 1.1 (0.92–1.3) Leukaemia, 0.71 (0.29–1.1)	Type of wood not reported
Furniture workers	Rang & Acheson (1981); Acheson et al. (1984)	5108 workers born before 1940 and active before 1969 in at least one of nine furniture workshops in High Wycombe (United Kingdom) Categorization of exposures: - class I (less dusty): office workers, upholsterers and yardmen - class II (dusty): polishers, veneerers and maintenance men - class III (very dusty): cabinet- and chair makers, sanders and wood machinists Mortality studied for 1941–82; rates for England and Wales used as reference. Lost to follow-up: 1.2%	SMR: All causes: 0.68 (0.62–0.76), 1638 observed All cancers: 0.88 (0.80–0.97), 435 observed Nasal cancer: 8.1 (3.7–16), 9 observed; all cases in people with very dusty occupations (0.57 expected). No other significantly increased SMR or trend with level of exposure. Mouth, pharynx, 1.2 (0.45–2.7) Stomach, 1.2 (0.92–1.5) Colon, 0.68 (0.42–1.0) Rectum, 1.1 (0.7–1.7) Larynx, 0.58 (0.12–1.7) Lung, 0.80 (0.68–0.93) All lymphatic/haematopoietic, 0.92 (0.61–1.3)	Chairs traditionally made from beech; wide range of imported hardwood used in furniture (Acheson et al., 1968)

Table 18 (contd)

Industry	Reference	Methods	Results	Notes
Furniture workers	Barthel & Dietrich (1989)	759 cabinet-makers or joiners from 170 enterprises located in Neubrandenburg district (Germany), followed from 1973 to 1984; cancer incidence in the cohort compared with incidence rates of the population of the district	SIR: All tumours, 1.1, 40 observed Malignant melanoma, 9.5 (2.4–28), 2 observed No other significantly increased SIR; no case of nasal cancer Stomach, 1.3, 7 observed Appendix, colon, sigmoid, 1.5, 3 observed Rectum, 2.1, 6 observed Lung, 0.68, 9 observed Lymphoma, 3.9, 1 observed Myeloma, 5.2, 1 observed	Species of wood most frequently worked with: pine, beech and oak
Furniture workers	Miller et al. (1989, 1994)	12 158 members of the United Furniture Workers of America first employed between 1946 and 1962 at factories producing wooden furniture. Mortality studied from 1946 to 1983; US rates used as reference. SMRs computed for subcohort followed for at least 20 years (10 497 subjects).	SMR: All causes, 0.9 (0.8–0.9), 1427 observed All malignant neoplasms, 0.9 (0.8–1.0), 342 observed Buccal cavity and pharynx, 0.7 (0.3–1.4), 7 observed Stomach, 1.0 (0.5–1.6), 14 observed Colon and rectum, 0.8 (0.6–1.1), 36 observed Nose, 1 observed Larynx, 0.6 (0.1–1.8), 3 observed Lung, 1.0 (0.8–1.1), 116 observed Hodgkin's disease, 2.2 (0.6–5.5), 4 observed Non-Hodgkin's lymphoma, 1.0 (0.5–1.9), 11 observed Multiple myeloma,1.6 (0.7–3.1), 9 observed Leukaemia, 1.4 (0.8–2.2), 17 observed	

Table 18 (contd)

Industry	Reference	Methods	Results	Notes
Plywood workers	Robinson et al. (1990)	2283 white males who worked for at least one year between 1945 and 1955 in any of four mills located in Washington and Oregon (USA); mortality studied through March 1977; US rates used as reference; 2% lost to follow-up	SMR: All causes, 0.7 [0.7–0.8], 570 observed All malignant neoplasms, 0.7 [0.6–0.9], 100 observed Lymphatic/haematopoietic, 1.6 [0.8–2.7], 12 observed Lymphosarcoma and reticulosarcoma, 1.0 [0.3–2.6], 4 observed Hodgkin's disease, 1.1 [0.1–4.0], 2 observed Multiple myeloma, 3.3 [0.7–9.7], 3 observed Other lymphatic, 2.7 [0.6–8.0], 3 observed Leukaemia, 0.9 [0.3–2.0], 5 observed Buccal cavity and pharynx, 0.6 [0.1–1.9], 3 observed Stomach, 0.4 [0.1–1.1], 4 observed Intestine, 0.6 [0.3–1.2], 8 observed Nose, 0 observed, 0.4 expected Larynx, 0.5 [0.01–2.5], 1 observed Lung 0.8 [0.5–1.1], 33 observed	Plywood manufactured from softwood (mainly Douglas fir, but also cedar, pine, spruce, hemlock, larch, true firs and redwood)
Woodworkers	Stellman & Garfinkel (1984)	10 322 volunteers enrolled in the American Cancer Society prospective study, whose occupation was in a wood-related industry, followed 1959–72; mortality compared with that of over 400 000 non-woodworkers in the study	SMR: All causes, 0.98, 2503 observed All cancers, 1.0, 513 observed Stomach, 1.5, $p < 0.01$, 44 observed Urinary bladder, 1.4, $p < 0.05$, 29 observed Lung, 1.1, $p > 0.05$, 135 observed Nasal cavity, 2.0, $p > 0.05$, 2 observed Colon and rectum, 0.75, $p < 0.05$, 57 observed Larynx, 0.68, $p > 0.05$, 3 observed Leukaemia, 1.3, $p > 0.05$, 32 observed Hodgkin's disease, 0.67, $p > 0.05$, 3 observed Other lymphatic, 0.70, $p > 0.05$, 17 observed	Stomach: among carpenters and joiners, SMR 1.7, $p < 0.01$, 36 observed Lung: among carpenters and joiners, SMR 1.2, $p < 0.05$, 101 observed Nasal cavity: both cases occurred among carpenters and joiners (SMR, 3.3)

SIR, standardized incidence ratio; CI, confidence interval; SMR, standardized mortality ratio

Table 19. Cohort studies of wooden pattern and model makers

Reference	Methods	Results	Notes
Swanson & Belle (1982)	1070 wooden model and pattern makers active in 1970 in seven automobile manufacturing plants located in Detroit (USA); cancer incidence in the cohort studied 1970–78 and compared with that of the population of metropolitan Detroit. 24.1% lost to follow-up	SIR: All cancers, 1.5 [1.1–2.0], 40 observed Colon and rectum, 2.9 [1.4–5.1], 11 observed Salivary glands, 21 [2.4–72], 2 observed No other significantly increased SIR	
Swanson et al. (1985)	316 wooden model or pattern makers employed by one US automobile manufacturing company in 1976 and followed through 1982. Colon cancer incidence in the cohort compared with that of the population of metropolitan Detroit. 36.4% lost to follow-up	SIR: Colon cancer, 4.9 [1.3–13], 4 observed	
Tilley et al. (1990)	7062 white male pattern and model makers active or retired in 1980 in the USA and Canada followed through 1985 in order to study cause-specific mortality and incidence of colorectal cancer; expected figures derived from incidence rates of Detroit SEER registry and from US death rates; 6% lost to follow-up by 1984	SMR: All causes, 0.7 (0.6–0.8), 335 observed All malignant neoplasms, 0.9 (0.7–1.1), 108 observed Large intestine, 2.0 (1.3–3.0), 22 observed No other significantly increased SMR SIR: Colorectal cancer, 1.1 (0.8–1.5), 39 observed	Incidence study limited to subcohort involved in screening programme
Becker et al. (1992)	528 model and pattern makers employed by a German automobile company between 1960 and 1985, followed through 1985; mortality compared with that of a cohort of tool makers in the same company. 2% of model makers and 6% of tool makers lost to follow-up	RR: All causes, 1.1 (0.5–2.2), 28 observed All malignant neoplasms, 1.8 (1.3–2.7), 11 observed Stomach, 6.9 (1.4–32), 3 observed Genitourinary organs, 5.7 (1.4–23), 3 observed Brain, 9.8 (1.4–70), 2 observed	

Table 19 (contd)

Reference	Methods	Results	Notes
Roscoe et al. (1992)	2294 white male wooden model makers employed for any time between 1940 and 1980 by three US automobile companies in metropolitan Detroit; vital status ascertained through 1984. US mortality rates used as reference.. 1% lost to follow-up	SMR: All causes, 0.8 (0.7–0.8), 706 observed All cancers, 1.0 (0.8–1.2), 173 observed No significantly increased SMRs; no association between colon or stomach cancer and exposure to wood in nested case–control study	Automotive model makers in the USA worked with mahogany, maple, birch, cherry and South American hardwoods; since the mid-1950s, mainly South American hardwoods, such as cativo, for dies

SIR, standardized incidence ratio; SMR, standardized mortality ratio; RR, rate ratio establish occupational and exposure histories. Excess risks were observed for exposure to hardwood dust alone (odds ratio, 2.0; 95% CI, 0.2–21), softwood dust alone (3.3; 1.1–9.4) and mixed hard and softwood dusts (12; 2.4–59) on the basis of a matched analysis. Both of the subjects who had been exposed to hardwood dust alone, and none of the 13 subjects exposed to softwood dust alone, had adenocarcinomas. Subjects who had been exposed to mixed hardwood/softwood dusts had cancers of mixed histological types.

Olsen et al. (1984) studied the relationship between occupational factors and sinonasal and nasopharyngeal cancer in Denmark. A total of 488 cases of sinonasal carcinoma (excluding sarcomas) were diagnosed in Denmark between 1970 and 1982 and identified by the Danish Cancer Registry. The 2465 controls were patients with cancers of the colon, rectum, prostate or breast, diagnosed during the same period. Employment histories (industry only) back to 1964 were collected through linkage with national pension fund records, and current occupation was established through linkage with the Central Population Registry. A group of industrial hygienists reviewed the employment histories and assigned an exposure category (unexposed, probably exposed, certainly exposed, insufficient information) to a list of predetermined compounds, which included wood dust (relative risk, 2.5; 95% CI, 1.7–3.7 for men).

Olsen and Asnaes (1986) further evaluated this data set by histologically confirmed sub-groups (squamous-cell and adenocarcinoma). The odds ratio for adenocarcinoma among men with definite exposure to wood dust was 16 (5.2–51), which increased to 30 (8.9–104) when only exposures 10 or more years before diagnosis were considered. The odds ratio for squamous-cell carcinoma associated with definite exposure to wood dust was 1.3 (0.6–2.8) and did not change when only exposure 10 or more years before diagnosis was considered.

Hayes et al. (1986a) conducted a case–control study of sinonasal cancer in the Netherlands. Cases were histologically confirmed primary epithelial sinonasal cancers newly diagnosed in men aged 35–79 years between 1978 and 1981 who were identified by the six major institutions in the Netherlands which treat head and neck tumours. The controls were a random sample of living and dead males in the Netherlands in 1980, who were selected from municipal resident registries and the records of the Central Bureau of Genealogy, respectively. A total of 91 case patients or their survivors and 195 controls or their survivors were interviewed. Detailed occupational and exposure histories were collected. An excess risk for all sinonasal cancers (odds ratio, 2.5 [95% CI, 1.2–5.1]), especially adenocarcinoma (18 [5.5–57]), was observed for employment in wood-related occupations, after adjustment for age. Exposure to wood dust was assessed by an industrial hygienist who was unaware of the case or control status of the person. The risk of adenocarcinoma was higher when only high-exposure jobs were considered (26 [7.0–99]), but no excess of squamous-cell carcinoma was observed (0.5 [0.1–2.9]) among men in the highest exposure category. The excess of adenocarcinoma was seen among workers first employed before 1942. The authors stated that the woodworkers had been exposed to both hard- and softwood dusts.

Merler et al. (1986) conducted a case–control study of nasal cancer in Vigevano, Italy, with the primary goal of examining the risk in the leather industry. Cases were malignancies of the sinonasal cavity newly diagnosed in residents of Vigevano between 1968 and 1982 and identified through the otolaryngology departments of local hospitals, the cancer registry of the

National Cancer Institute of Milan and the mortality records of the city. Two controls per case, matched on age, sex, vital status and, if dead, on year of death, were chosen from electoral rolls (living controls) and mortality records (dead controls). Interviews, which included occupational histories, were conducted with 21 case patients and 39 controls. No case patient and two controls had been exposed to wood dust.

Bolm-Audorff *et al.* (1989, 1990) conducted a case–control study of histologically confirmed cases of nasal and nasopharyngeal cancer diagnosed in hospitals in Hesse, Germany, between 1 January 1983 and 31 December 1985. Fifty-four of the 66 cases were sinonasal cancers. A single control was matched to each case on age, sex and residence and was chosen from among patients with non-occupational bone fractures. Information on exposure was collected through interviews. The relative risk for all cases (sinonasal and nasopharyngeal cancer) was 3.8 (95% CI, 0.8–17), which increased to 7.8 (1.3–48) for an exposure of five years or more. Six of the seven exposed case patients had sinonasal cancer. The two exposed cases of sinonasal adenocarcinoma were both associated with exposure to hardwood (beech and oak), while the two other carcinomas were associated with exposure to softwoods; the two remaining exposed cases, a lymphoma and a neuroblastoma, were associated with exposure to mixed woods and hardwoods, respectively. Two controls had been exposed to wood dust. [The Working Group noted that the tree species were not identified for controls.]

Vaughan (1989) performed a case–control study of squamous-cell cancers of the sinonasal cavity in western Washington State, United States. Living patients in whom the cancers were diagnosed between 1979 and 1983 were identified from a population-based tumour registry, and 27 people with sinonasal cancer were interviewed. Random-digit dialling was used to obtain 552 controls who were similar in age and sex to the cases. Interviews were used to collect life-time work histories. The analyses were adjusted for age, sex, smoking and alcohol consumption. Excesses of squamous-cell cancers of the sinonasal cavity were reported for forestry and logging workers (odds ratio, 1.8; 95% CI, 0.4–7.2) and woodworking machine operators (7.5; 1.5–37). Vaughan and Davis (1991) later categorized these cases according to exposure to wood dust. The odds ratio for employment in any wood-related occupation was 2.4 (0.8–6.7), which increased to 7.3 (1.4–34) when only exposure for 10 or more years after an induction period of 15 years was considered. The authors stated that the cases were associated with exposure predominantly to dust from softwood. Information on exposure was obtained from surrogates for half of the cases but none of the controls; however, exclusion of cases for which information was obtained from surrogates did not greatly affect the risk estimates.

Luce *et al.* (1991, 1992, 1993) conducted a case–control study of patients with sinonasal cancer diagnosed between January 1986 and February 1988 in 27 participating hospitals in France. A total of 207 patients out of 303 were alive at the time of interview and agreed to participate in the study; 57 died before being interviewed, and 39 could not be located. Controls of similar age and sex to the patients were recruited from two sources: patients with a cancer diagnosed at another site and at the same or a nearby hospital; and neighbourhood controls selected from lists provided by the patients. Of these, 323 hospital and 86 neighbourhood controls were eligible and agreed to participate in the study. Detailed occupational and exposure histories were collected by personal interviews. The degree of occupational exposure to wood dust was assessed by an industrial hygienist who was unaware of the case or control status of the person. Among men, an elevated risk for sinonasal adenocarcinoma (odds ratio, 289; 95% CI, 136–615, based on 77 exposed cases and 29 exposed controls) was associated with probable or definite, medium–high exposure to wood dust, but no relationship was observed for squamous-cell carcinoma (1.0; 0.4–2.6). The results for specific occupational groups are presented in Tables 21 and 22.

Lerclerc et al. (1994) reported the results of further analyses of this study in respect of species of wood. Eighty of the 82 male patients with adenocarcinoma had been exposed to hardwood dusts, but only seven of these to hardwood alone. The odds ratio for adenocarcinoma among men exposed to hardwood or mixed wood dust was 168 (95% CI, 78–362). Positive trends were observed for duration and intensity of exposure. The relative risk was higher among men exposed before 1946 (254) than those first exposed afterwards (137). [The odds ratio for adenocarcinoma among men exposed exclusively to hardwood was 57.] Seventeen of the 59 male squamous-cell cases were associated with exposure to wood dust—three to hardwood only, three to softwood only and the remainder to a mixture of woods. The authors stated that because few subjects were exposed to one wood type alone, the relative risks for squamous-cell carcinoma could not be calculated for exposure to each type of wood alone. The odds ratios for squamous-cell carcinoma associated with exposure to hardwood (or hardwood plus other woods) and softwood (or softwood plus other woods) were 1.4 and 1.7 (not significant), respectively. Duration and intensity of exposure to either hard- or softwoods were not clearly associated with the risk for squamous-cell carcinoma, although some evidence for an excess risk was observed for subjects exposed before 1946. [The Working Group was concerned that the procedures for selecting non-hospital controls may have artificially biased the proportion of woodworking controls downward. For non-hospital controls, cases were asked to provide 'the names of several persons (colleagues excluded)' to serve as referents. The proportion of non-hospital controls exposed to hardwoods (18.8%) and softwoods (20.3%), however, was similar to the proportion among hospital controls (18.3% and 16.3%, respectively), indicating that bias was unlikely.]

Zheng et al. (1992a) conducted a case–control study of nasal cancer in Shanghai, China. Patients with newly diagnosed sinonasal cancer between January 1988 and February 1990 were identified in the population-based cancer registry of Shanghai. Controls were randomly selected in the general population from the records of the Shanghai Resident Registry. Personal interviews were conducted with 60 cases and 414 controls, and information was collected on occupational history and exposures. Of the cases, 25 were squamous-cell carcinomas, six were adenocarcinomas, 22 were tumours of other histological types and the remainder were not evaluated histologically. The relative risks for all sinonasal cancers combined were calculated for self-reported exposure to wood dust (odds ratio, 1.9; 95% CI, 0.7–5.0) and employment in wood-related occupations (1.7; 0.5–6.3).

(b) Occupational group

Brinton et al. (1977) conducted a case–control study among people who died in North Carolina (United States) counties in which at least 1% of the population was employed in furniture manufacture according to the 1963 census. Death certificates were used to identify 37 cases of cancer of the nasal cavity and sinuses between 1956 and 1974. Two controls for each case ($n = 73$) were randomly selected from among people who had died and were of the same sex, race, county of death, age at death and year of death. Information on occupation and industry was also obtained from the death certificates, and a matched analysis was performed. Elevated risks were observed for people employed in the furniture industry (odds ratio, 4.4; 95% CI, 1.3–15) and in other woodworking occupations (sawmill workers and carpenters) (1.5; 0.4–4.3).

Cecchi *et al.* (1980) performed a case–control study of sinonasal adenocarcinoma in Florence, Italy. Cases diagnosed between 1963 and 1977 were identified from the records of the otorhinolaryngology clinic or the radiology institute of the University of Florence. Eleven of 13 patients or their survivors [numbers of patients and survivors not given] were interviewed in order to obtain information on occupation and smoking habits. Two controls per case, matched on sex, age, place of residence, smoking and year of admission to hospital, were selected from among non-cancer internal medicine patients and received the same interview. Three case patients and two controls had been employed as woodworkers. Of the exposed patients, two had worked in small woodworking shops, and the third had worked with both wood and leather.

Roush *et al.* (1980) conducted a case–control study of sinonasal cancer in Connecticut (United States) based on the tumour register. Cases were sinonasal cancers in 216 men 35 years of age or older who had died between 1935 and 1975. The 691 controls were a random sample of men aged 35 years or older who had died in Connecticut between 1935 and 1975. Occupational information was collected from death certificates and city directories, in which information is based on interviews conducted during door-to-door surveys. Job titles were classified for exposure to wood dust on the basis of a review of the literature. The odds ratio associated with wood-related occupations was 4.0 (95% CI, 1.5–11) when information from both sources was considered. The odds ratio was somewhat lower (2.8) when only information from death certificates was considered and somewhat higher (5.9) when only information from city directories was considered.

Tola *et al.* (1980) performed a case–control study of patients with malignant tumours of the nose and paranasal sinuses reported to the Finnish Cancer Registry between 1970 and 1973. For each case a single control of similar age and sex was chosen from among cancer patients (other than respiratory) from the same geographical area. Questionnaires on occupational history and exposures were completed by 45 case subjects and 45 controls. Of the cases, 20 were squamous-cell carcinomas, 10 were transitional-cell carcinomas and two were malignancies classified as adenocarcinoma. One patient with an adenocarcinoma had been employed as a joiner and had been exposed mainly to oak dust. One control had been employed as a carpenter. No other results related to exposure to wood were reported.

Elwood (1981) reported the results of a case–control study of 121 men with primary epithelial tumours of the sinonasal cavity seen at the main cancer treatment centre in British Columbia, Canada, between 1939 and 1977. Of the cases, 61 were squamous-cell carcinomas, 20 were anaplastic carcinomas, 16 were transitional-cell carcinomas, 11 were adenocarcinomas, six were sarcomas, and seven were of unknown histological type. A control group of 120 patients with cancer that was considered to be unrelated to smoking or outdoor work, matched on age and year of diagnosis, was chosen. Information on occupation and smoking was retrieved from medical records, and relative risks were calculated using conditional logistic regression after adjustment for smoking and ethnicity. An elevated risk was observed for employment in wood-related occupations (odds ratio, 2.5; $p < 0.03$). Of the 28 exposed patients, 10 were loggers, seven were carpenters, four were forestry workers, four were construction workers, two were log scalers and one was a cabinet-maker. The authors reported that the predominant exposure of all but the cabinet-maker would have been to native softwoods.

Hardell *et al.* (1982) conducted a case–control study of nasal and nasopharyngeal cancers in northern Sweden to examine their relationship with exposure to phenoxy acids or chlorophenol, which included 44 male patients with sinonasal cancer who had been reported to the Swedish Cancer Registry between 1970 and 1979. Thirty-one of the cases were squamous-cell carcinomas, four were anaplastic carcinomas and three were adenocarcinomas; six were tumours of other histological types. The 541 controls had initially been identified and interviewed for a study of soft-tissue sarcoma and lymphoma. Information on exposure and employment history were collected using postal questionnaires and supplemental telephone interviews. A crude relative risk of 2.0 [95% CI, 1.1–3.6] was observed for previous employment as a carpenter, cabinet-maker or sawmill worker (19 exposed cases, 151 exposed controls). The authors noted that little hardwood is used for furniture making in northern Sweden.

Battista *et al.* (1983) performed a case–control study of sinonasal cancer in the province of Siena, Italy, where 4–7% of the active male population is employed in wood-related industries. They studied 36 male patients in whom sinonasal cancers were diagnosed at the Ear, Nose and Throat Clinic or the Radiotherapy Unit in Siena between 1963 and 1981. Seventeen (47%) of the cases were squamous-cell carcinomas and five (14%) were adenocarcinomas. For each case, five referents were selected from among men of the same age (within one year) who were admitted to the medical clinic of Siena for other diseases at the same time. All case patients and 164 of the 180 referents or their next-of-kin completed a postal questionnaire, and occupational histories were collected. Exposure to wood dust was defined as employment as a woodworker or cabinet-maker. An elevated risk for all sinonasal cancers was reported for exposed men (odds ratio, 4.7; 95% CI, 1.7–13), and the risk was especially increased for adenocarcinoma (90; 20–407). The seven case patients with exposure to wood dust had used a wide variety of species, the commonest being chestnut (four cases), oak (four cases), poplar (three cases) and fir (two cases).

Brinton *et al.* (1984) conducted a case–control study of patients with sinonasal cancer admitted to four hospitals in North Carolina and Virginia, United States, between 1970 and 1980. Two controls for each living case patient, matched on year of admission, age, sex, race and area of residence, were selected from living hospital patients. For deceased cases, similar matching criteria were used, but one hospital patient (not required to be living) and one patient with a death certificate were chosen. Potential controls were excluded if an upper aerodigestive cancer, oesophageal cancer, benign respiratory neoplasm, mental disorder or chronic sinonasal disease had been diagnosed. Telephone interviews were conducted with 160 case patients and 290 controls or their survivors. An elevated relative risk for sinonasal adenocarcinoma was associated with previous employment in any wood-related job (odds ratio, 3.7; $p < 0.05$), but no excess of squamous-cell carcinoma was observed (odds ratio, 0.8).

Ng (1986) conducted a case–control study of cancer of the nasal cavity and sinuses among Chinese people in Hong Kong. Two series of controls were used: people with nasopharyngeal cancer and people with other malignancies, all of which were diagnosed between 1974 and 1981 at the Institute of Radiology and Oncology in Hong Kong. There were 225 cases of nasal cancer (119 squamous-cell, 50 anaplastic, four adenocarcinomas, 37 of other histological types and 15 of unknown histological type), 224 cases of nasopharyngeal cancer (112 squamous-cell, 102 anaplastic and 10 of unknown histological type) and 226 controls with other malignancies. Occupational histories were collected from medical records. Two wood-related occupational categories were considered: furniture makers and woodworkers (comprising two cancers of the nasal cavity and sinuses, five nasopharyngeal cancers and one other malignancy) and construction carpenters (comprising two cancers of the nasal cavity and sinuses and three nasopharyngeal carcinomas). None of the four nasal cavity and sinus adenocarcinomas were in wood-related workers. No odds ratios or other estimates of relative risk were presented for wood-related occupations.

Fukuda et al. (1987) and Fukuda and Shibata (1988, 1990) reported the results of a case–control study on Hokkaido Island in Japan of cases of squamous-cell carcinoma of the maxillary sinus, newly diagnosed in 1982–86 in people between 40 and 79 years of age, at the two university and the two medical college hospitals on Hokkaido Island. Two controls per case, matched on sex, age and residence, were chosen from among a pool of potential controls selected from telephone directories. A questionnaire was posted to all potential cases and controls, who were later telephoned to obtain their permission to participate in the study and to confirm the responses to the questions. The matched analysis in the latest published results included 169 eligible cases and 338 eligible controls. The exposure category of woodworkers consisted of people employed as carpenters, joiners, furniture makers and other woodworkers. Excess risks for squamous-cell carcinoma of the maxillary sinus were observed among both men (relative risk, 2.9; 95% CI, 1.5–5.6) and women (2.0; 0.3–14). A significant trend ($p < 0.05$) of increasing risk with increasing duration of employment was also seen.

Takasaka et al. (1987) performed a case–control study of male patients with nasal or paranasal cancer who were admitted to Tohoku University Hospital, Japan, between 1971 and 1982. Three to five controls of the same sex, age and date of admission were selected from among patients with other otorhinolaryngological diseases admitted to the same hospital. Mailed questionnaires requesting occupational history and information on exposures were completed by 107 case patients and 413 controls. Eighty-five of the 98 cases for which histological information was available were squamous-cell carcinomas and six were adenocarcinomas. Excess risks were associated with longest-held occupation as a forester (odds ratio, 2.0; 95% CI, 0.5–7.3), woodworker (1.6; 0.4–7.1) or carpenter (2.1; 0.8–5.8).

Bimbi et al. (1988) conducted a hospital-based case–control study in Milan, Italy, of 53 patients with malignant epithelial cancers admitted between 1982 and 1985 to the Head and Neck Oncology Department of the National Institute for the Study and Treatment of Cancer. Controls were 217 patients admitted to the same department during the same years, mainly for cancers of the nasopharynx, thyroid and salivary gland. Information on occupational history was collected from medical records. Three cases and no control had been employed as woodworkers.

Finkelstein (1989) reported the results of a study based on information from death certificates of 124 men, 35 years of age or older, who had died of cancer of the nasal cavity or paranasal cavity in Ontario, Canada, between 1973 and 1983. One control per case, matched on age and year of death, was chosen from among people who had died of other causes. Information on usual job and industry was collected from death certificates. Nine cases and six controls had been employed in wood-related occupations (odds ratio, 1.9 [95% CI, 0.6–6.4]). Workers who had been employed in nickel refining (10 cases and no control) were excluded from the unmatched analyses.

Kawachi et al. (1989) reported the results of an exploratory study to examine the risk for cancer among woodworkers. Case patients and controls were men in the New Zealand Cancer Registry in whom cancer had been diagnosed between 1980 and 1984 and whose occupation was noted in Registry records. The case patients were 46 registrants in whom cancer of the nasal cavity and sinuses had been diagnosed, while the controls were 19 858 registrants with cancers at all other sites. The only information available on exposure was the current or most recent occupation. No excess risk was observed (odds ratio, 1.0; 95% CI, 0.2–4.0; based on two exposed cases) after adjustment for age.

Loi et al. (1989) conducted a case–control study of nasal cancer in the Pisa area of Italy, a region where there are many factories manufacturing wooden products, especially furniture. Case patients were 38 male nasal cancer patients admitted to Pisa University Hospital between October 1972 and October 1983; five controls per case, matched on sex, age, province of usual residence and admission date, were chosen from among patients admitted to the same hospital for diseases other than respiratory cancer or lymphoma. A postal questionnaire was used to obtain information on occupational history and smoking habits. Workers who had been employed in wood-related occupations for six or more months at least five years before diagnosis were considered to have been exposed. Subjects who had been employed in leather-related occupations, their matched controls and other controls who had been employed in leather-related occupations were excluded from the analyses of wood-related risks. A relative risk of 9.7 (95% CI, 3.2–29) was observed for employment in wood industries. The relative risk for adenocarcinoma alone tended to infinity, as all case subjects had been exposed. Individual information on exposure to different wood species was not available, but the authors stated that chestnut, walnut and fir were the most commonly used in the region. [The Working Group noted that some of the study subjects might have died before the study was conducted, and, consequently, next-of-kin may have been interviewed.]

Shimizu et al. (1989) conducted a case–control study of 45 men and 21 women with newly diagnosed squamous-cell carcinoma of the maxillary sinus at six university hospitals in northeastern Japan, between October 1983 and October 1985. Two controls, matched on age and sex, were selected from a random sample of residents in the same area from telephone directories. Each patient was asked to complete a questionnaire during initial hospitalization, which requested information about previous occupations and other potential risk factors; controls completed the same questionnaire by post. A matched analysis was performed. The relative risk among men for woodworking or joinery was 2.1 (95% CI, 0.8–5.3); the risk was 7.5 (1.5–39) when only jobs involving sanding or turning were considered.

Viren and Imbus (1989) conducted a study based on information on death certificates of 536 people in the United States who had died of nasal cancer in the states of Washington and Oregon between 1963 and 1977, Mississippi between 1962 and 1977 and North Carolina between 1964 and 1977. Two controls per case ($n = 1072$) were chosen from among people in the same states who had died from causes other than cancer, non-malignant respiratory disease or accidents; they were matched to the case patient on sex, age, race and year of death. Usual occupation and industry were obtained from death certificates. A matched analysis was performed only for men (332 cases, 664 controls). Relative risks of 3.3 ($p < 0.01$) for forestry and logging workers, 1.3 for woodworkers and woodworking machine operators and 1.6 for carpenters were observed. [The Working Group noted that the subjects from North Carolina may also have been included by Brinton et al., 1977.]

Haguenoer et al. (1990) conducted a case–control study to investigate occupational risk factors for cancers of the upper respiratory and digestive tracts (nose, lips, buccal cavity, pharynx and larynx) in northern France. An occupational history, which included only jobs held for at least 15 years, was established by interview; people who did not have at least one job that met this criterion (one-half the subjects) were excluded from the study. There were 14 histologically confirmed sinonasal cancers among men treated in the first semester of 1983. Two controls per case, matched for sex, age, ethnic group, area of residence and histories of smoking and alcohol drinking, were chosen from among non-cancer hospitals patients in the same region. Four patients with sinonasal cancer and no matched control reported previous employment as a woodworker.

Comba et al. (1992a) reported the results of a collaborative case–control study in north-eastern and central Italy of cases of sinonasal cancer diagnosed between 1982 and 1987 at hospitals providing services to the provinces of Verona, Vicenza and Siena. Four controls per case were selected from among patients admitted for diseases other than chronic rhino-sinonasal diseases or nasal bleeding to the same hospital at the same time, who were similar to the case patient with regard to sex, age and area of residence. Personal interviews were conducted by telephone or post with 78 of 96 case patients and 254 of 378 controls or their next-of-kin, to collect information on occupational history and exposures. An elevated relative risk associated with woodworking was observed for both men (odds ratio, 5.8 [95% CI, 1.8–18]) and women (3.2 [0.2–50]). The relative risks for sinonasal adenocarcinoma (14 [2.3–83]) and squamous-cell carcinoma (1.7 [0.3–9.2]) were presented for men only. The authors noted that the types of wood used by the case patients were both hard- and softwoods of many species, including birch, fir, poplar, beech, maple, cherry, oak, mahogany, walnut and chestnut.

Comba et al. (1992b) conducted another case–control study among the residents of Brescia Province in north-eastern Italy, with the primary aim of examining the relative risk for sinonasal cancer associated with employment in the metal industry. Cases were malignant epithelial sinonasal tumours treated at the ear, nose and throat department or the radiotherapy unit of Brescia Hospital between 1980 and 1989. Four controls per case were chosen from among patients treated at the same centres for benign and malignant tumours of the head and neck, excluding epidermoid carcinomas of the tongue, oral cavity, oro- and hypopharynx and larynx, who were of the same sex and age. A total of 34 case patients (23 men) and 102 controls

(70 men) or their survivors [numbers of cases and survivors not given] were interviewed, and detailed occupational histories were obtained. The age-adjusted odds ratio for male woodworkers was 11 [95% CI, 0.5–229).

Zheng et al. (1993) conducted a case–control study of 147 white men, 45 years of age or older, who had died from sinonasal cancer in 1985 and were identified in a national survey of mortality in the United States in 1986. Controls were 449 white men who had died during the same period from causes not related to smoking or alcohol consumption. The next-of-kin of cases and controls received a structured questionnaire by post requesting information on demographic factors, histories of smoking and alcohol consumption, occupational history and dietary habits. After adjustment for age and cigarette smoking, a relative risk of 1.7 (95% CI, 0.6–4.3) was observed for previous employment as a carpenter or other wood-related worker, relative to professional, managerial, technical and sales workers.

Magnani et al. (1993) conducted a case–control study of sinonasal cancer in the district of Biella in north-western Italy with the primary goal of examining risks in the woollen textile industry. Cases were epithelial or histologically unspecified sinonasal cancers diagnosed among residents of the local health areas of Biella and Cossato between 1976 and 1988. Four controls per case were chosen from among patients of the same sex and age, who were admitted to the same hospital in the same year with diagnoses other than respiratory cancer. Mailed questionnaires or telephone interviews, which included an occupational history, were completed by 26 cases and 111 controls or their relatives. An elevated risk for sinonasal cancer was associated with employment as a wood or furniture worker (odds ratio, 4.4; 95% CI, 1.4–13); the risk was much higher when only adenocarcinomas were considered (22; 4.4–124).

The studies on sinonasal cancer are summarized in Tables 20, 21 and 22.

Demers et al. (1995) performed a pooled analysis of case–control studies of sinonasal cancer and exposure to wood dust, in which the following criteria had been met: the histological types of the cases were identified; occupational histories had been collected from patients (or their survivors) and controls by interview or questionnaire; and data on age, sex and tobacco smoking were available. The authors of the 15 studies that met these criteria were asked to participate in the pooled analysis; 12 were both able and willing to do so. The studies were those conducted in Shanghai, China (Zheng et al., 1992a); France (Luce et al., 1991, 1992, 1993); Hessia, Germany (Bolm-Audorff et al., 1989, 1990); Siena, Verona and Vicenza, Italy (Comba et al., 1992a); Brescia, Italy (Comba et al., 1992b); Biella, Italy (Magnani et al., 1989, 1993); Vigevano, Italy (Merler et al., 1986); the Netherlands (Hayes et al., 1986a,b, 1987); northern Sweden (Hardell et al., 1982); North Carolina and Virginia, United States (Brinton et al., 1984); Los Angeles, United States (Mack & Preston-Martin, unpublished data); and Seattle, United States (Vaughan, 1989; Vaughan & Davis, 1991). The aggregated data consisted of 680 male cases (169 adenocarcinomas, 329 squamous-cell cancers, 157 of other histology and 25 of unknown histology), 2349 male controls, 250 female cases (26 adenocarcinomas, 101 squamous-cell cancers, 105 of other histology and 18 of unknown histology) and 787 female controls. Seven categories of jobs with potential exposure to wood dust were defined by combining occupation and industry title: forestry workers, loggers, pulp and paper workers, sawmill

Table 20. Results of community-based case–control studies of sinonasal cancer: all histological types and unspecified

Reference	Country	Sex	Cases/controls	Source of information on exposure	Exposure to which relative risk applies	OR/RR	95% CI or p	Comments
Exposure to wood dust								
Hemberg et al. (1983)	Denmark/Finland/Sweden	MF	167/167	Interviews	Hardwood only Softwood only; primarily pine and spruce, also birch and aspen Mixed hard- and softwood	2.0 3.3 12	0.2–21 1.1–9.4 2.4–59	
Olsen et al. (1984)	Denmark	MF	488/2465	Linkage with national pension fund records	Exposure to wood dust (men only) Probable exposure Definite exposure ≥ 10 years since first exposure	1.2 2.5 2.9	0.7–2.1 1.7–3.7 1.8–4.7	Exposure based on expert assessment of work history
Merler et al. (1986)	Italy	MF	21/39	Interview	Exposure to wood dust			0/2 exposed case/controls
Bolm-Audorff et al. (1989, 1990)	Germany	MF	66/66	Interviews	Exposure to wood dust; oak, beech and softwood Duration ≥ 5 years	3.8 7.8	0.8–17 1.3–48	Cases include 12 naso-pharyngeal cancers. Six of seven exposed cases had nasal cancer. Controls matched on sex, age and residence
Zheng et al. (1992a)	China	MF	60/414	Interviews	Exposure to wood dust Wood-related occupations	1.9 1.7	0.7–5.0 0.5–6.3	Self-reported exposure; adjusted for age
Occupational group								
Brinton et al. (1977)	USA	M	37/73	Death certificates	Furniture industry Sawmill workers, carpenters and other woodworking occupations	4.4 1.5	1.3–15 0.4–4.3	Controls matched on sex, race, age, county and year of death
Roush et al. (1980)	USA	M	216/691	Death certificates and city directories	Wood-related occupation Death certificate only City directories only Either source of information	2.8 5.9 4.0	 1.5–11	

Table 20 (contd)

Reference	Country	Sex	Cases/ controls	Source of information on exposure	Exposure to which relative risk applies	OR/ RR	95% CI or p	Comments
Occupational group (contd)								
Tola et al. (1980)	Finland	MF	45/45	Questionnaires	Wood-related occupations			1/1 exposed case/control
Elwood (1981)	Canada	M	121/120	Medical records	Wood-related occupations; primarily softwood	2.5	p < 0.03	Matched on age and year of diagnosis and adjusted for smoking and ethnicity
Hardell et al. (1982)	Sweden	M	44/541	Questionnaires	Carpenter, cabinet-maker or sawmill worker	2.0	[1.1–3.6]	Crude relative risk; 19/151 exposed cases/controls
Battista et al. (1983)	Italy	M	36/164	Questionnaires	Woodworker or cabinet-maker; exposure to chestnut, oak, poplar, fir, alder, walnut, beech and acacia	4.7	1.7–13	Cases of carcinoma only; matched on age
Brinton et al. (1984)	USA	MF	160/290	Interviews	Furniture industry Lumber industry Carpentry	0.8 1.4 1.5	0.3–2.0 0.7–2.6 0.6–3.4	Adjusted for year of admission, age, sex, race and area of residence
Hayes et al. (1986a)	Netherlands	M	91/195	Interviews	Wood-related occupations Furniture and cabinet-making Factory joinery/carpentry House carpentry Other wood-related occupations	2.5 13 2.1 0.6 1.1	[1.2–5.1] [2.7–59] [0.4–11] [0.1–4.3] [0.3–4.8]	Adjusted for age Adjusted for age
Ng (1986)	Hong Kong	MF	225/226	Medical records	Furniture makers, woodworkers Construction carpenters			2/1 exposed cases/control 2/0 exposed cases/control
Takasaka et al. (1987)	Japan	M	107/337	Questionnaires	Longest-held occupation Foresters Woodworkers Carpenters	2.0 1.6 2.1	0.5–7.3 0.4–7.1 0.8–5.8	

Table 20 (contd)

Reference	Country	Sex	Cases/ controls	Source of information on exposure	Exposure to which relative risk applies	OR/ RR	95% CI or p	Comments
Occupational group (contd)								
Bimbi et al. (1988)	Italy	MF	53/217	Medical records	Wood industry			3/0 exposed cases/control
Finkelstein (1989)	Canada	M	124/124	Death certificates	Wood-related occupations	1.9	[0.6–6.4]	
Kawachi et al. (1989)	New Zealand	M	46/19 858	Tumour registry records	Woodworkers	1.0	0.2–4.0	Adjusted for age
Loi et al. (1989)	Italy	M	38/153	Questionnaires	Wood industries. Individual exposure by wood species not known, but walnut, chestnut and fir commonly used in region	9.7	3.2–29	Calculated after excluding cases and controls who were leather workers
Viren & Imbus (1989)	USA	M	332/664	Death certificates	Forestry and logging Woodworking and woodworking machine operators Carpenters	3.3 1.3 1.6	$p < 0.01$	Matched on sex, age, race, state and year of death 5/8 exposed cases/ controls 14/18 exposed cases/ controls
Haguenoer et al. (1990)	France	M	14/28	Interviews	Woodworkers			Only jobs held for at least 15 years were evaluated; 4/0 exposed cases/control
Comba et al. (1992a)	Italy	MF	78/254	Interviews and questionnaires	Woodworkers (men) Furniture makers, joiners or carpenters Lumberjack Woodworkers (women)	5.8 6.5 4.1 3.2	[1.8–18] [1.7–25] [0.9–19] [0.2–50]	Results for men adjusted for age. Exposure to both hard- and softwoods
Comba et al. (1992b)	Italy	M	23/70	Interviews	Woodworkers	11	[0.5–229]	Adjusted for age

Table 20 (contd)

Reference	Country	Sex	Cases/ controls	Source of information on exposure	Exposure to which relative risk applies	OR/ RR	95% CI or p	Comments
Occupational group (contd)								
Zheng et al. (1993)	USA	M	147/449	Questionnaire	Carpenters and other woodworkers	1.7	0.6–4.3	Adjusted for age and smoking
Magnani et al. (1993)	Italy	MF	26/111	Questionnaires	Wood and furniture workers Duration, 1–9 years Duration, ≥ 20 years	4.4 3.5 5.8	1.3–13 0.6–19 1.4–24	

OR, odds ratio; RR, relative risk; CI, confidence interval; M, male; F, female

Table 21. Results of community based case–control studies of sinonasal cancer: adenocarcinoma

Reference	Country	Sex	Cases/ controls	Source of information on exposure	Exposure to which relative risk applies	OR/ RR	95% CI or p	Comments
Exposure to wood dust								
Hayes et al. (1986a)	Netherlands	M	23/195	Interviews	Wood-related occupations	18	[5.5–57]	Adjusted for age
					Furniture and cabinet-making	140	[18–1094]	Adjusted for age
					Factory joinery/carpentry	16	[2.1–125]	
					Housing carpentry and other wood-related occupations	0.0		0/9 exposed case/controls
					Exposure to wood dust	8.5	[2.6–28]	Exposure based on expert assessment of work history
					High exposure	26	[7.0–99]	
Olsen & Asnaes (1986)	Denmark	M	39/2465	Linkage with national pension fund records	Definite wood dust exposure	16	5.2–51	Exposure based on expert assessment of work history records; adjusted for exposure to formaldehyde
					≥ 10 years since first exposure	30	8.9–104	
Luce et al. (1991, 1992, 1993); Leclerc et al. (1994)	France	M	82/320	Interviews	Loggers	0.6	0.1–4.6	All results adjusted for age
					duration >15 years	0.0		
					Cabinet-makers	35	18–69	
					duration >15 years	33	14–76	
					Woodworking machine operators	7.4	3.5–16	
					duration >15 years	48	8.8–260	
					Carpenters, joiners	25	15–44	
					duration >15 years	45	22–50	
					Exposure to wood dust			Exposure based on expert assessment of work history
					Probable or definite, medium–high exposure	289	136–615	
					Hardwood dust (incl. mixed)	168	78–362	80 of 82 cases exposed to hardwood or mixed dusts
					> 35 years of exposure	303	64–1427	
					Highest level of exposure	530	104–2696	
					First exposed before 1946	254	55–1185	

Table 21 (contd)

Reference	Country	Sex	Cases/ controls	Source of information on exposure	Exposure to which relative risk applies	OR/ RR	95% CI or p	Comments
Occupational group								
Cecchi et al. (1980)	Italy	MF	11/22	Interviews	Wood-related occupation	[3.7]	[0.5-27]	Crude odds ratio; 3/2 exposed cases/controls
Battista et al. (1983)	Italy	M	5/NR	Questionnaires	Woodworker or cabinet-maker Chesnut, oak, poplar, alder, walnut and acacia	90	20-407	
Brinton et al. (1984)	USA	M	13/183	Interviews	Furniture industry Lumber industry Carpentry Any wood-related job	5.7 1.6 2.9 3.7	1.7-19 $p < 0.05$	4 exposed cases 3 exposed cases 10 exposed cases;
Comba et al. (1992)	Italy	M	13/184	Interviews and questionnaires	Wood workers (men)	14	[2.3-83]	
Magnani et al. (1993)	Italy	MF	14/111	Questionnaires	Wood and furniture workers	22	4.4-124	

OR, odds ratio; RR, relative risk; CI, confidence interval; M, male; F, female; NR, not reported

Table 22. Results of community-based case–control studies of sinonasal cancer: squamous-cell carcinoma

Reference	Country	Sex	Cases/ controls	Source of information on exposure	Exposure to which relative risk applies	OR/ RR	95% CI or p	Comments
Exposure to wood dust								
Hayes et al. (1986a)	Netherlands	M	50/195	Interviews	Exposure to wood dust High exposure	1.3* 0.5	[0.6–2.7] [0.1–2.9]	*Adjusted for age; exposure based on expert assessment of work history
Olsen & Asnaes (1986)	Denmark	M	215/2465	Linkage with national pension fund records	Definite exposure to wood dust ≥ 10 years since first exposure	1.3 1.3	0.6–2.8 0.5–3.6	Exposure based on expert assessment of work history; adjusted for exposure to formaldehyde
Vaughan (1989); Vaughan & Davis (1991)	USA	MF	27/552	Interviews	Forestry/logging Duration ≥ 10 years Woodworking machine operator Duration ≥ 10 years Any wood-related occupation Exposures lagged by 15 years Duration ≥ 10 years after lagging by 15 years	1.8 11 7.5 29 2.4 3.1 7.3	0.4–7.2 1.5–37 0.8–6.7 1.0–9.0 1.4–34	Primary exposure to softwood; all results adjusted for age, sex, smoking and alcohol intake
Luce et al. (1991, 1992, 1993); Leclerc et al. (1994)	France	M	59/320	Interviews	Loggers Duration >15 years Cabinet-makers Duration >15 years Woodworking machine operators Duration >15 years Carpenters, joiners Duration >15 years Carpenter, joiner employed in wood manufacture Duration >15 years Exposure to wood dust Probable or definite low exposure Probable or definite medium–high exposure	2.9 3.9 1.6 3.4 1.2 0.0 1.6 1.9 2.3 8.1 1.4 1.0	0.8–10 0.3–56 0.3–8.9 0.5–22 0.2–6.6 0.5–5.1 0.4–9.0 0.6–9.0 1.3–50 0.6–3.0 0.4–2.6	All results age-adjusted Exposure based on expert assessment of work history

Table 22 (contd)

Reference	Country	Sex	Cases/ controls	Source of information on exposure	Exposure to which relative risk applies	OR/ RR	95% CI or p	Comments
	France (contd)				Hardwood dust (incl. mixed)	1.4		14/58 exposed cases/controls
					First exposed < 1946	2.2	1.0–4.8	No trend with duration/intensity
					Softwood dust (incl. mixed)	1.7		13/54 exposed cases/controls
					First exposed < 1946	2.5	1.1–6.0	No trend with duration/intensity
Occupational group								
Brinton et al. (1984)	USA	M	53/183	Interviews	Furniture industry	0.3		1 exposed case
					Lumber industry	1.1		12 exposed cases
					Carpentry	1.0		5 exposed cases
					Any wood-related job	0.8		22 exposed cases
Fukuda et al. (1987); Fukuda & Shibata (1988, 1990)	Japan	MF	169/338	Questionnaires	Woodworkers (men)	2.9	1.5–5.6	Cases were maxillary sinus only; matched on age, sex and residence
					Woodworkers (women)	2.0	0.3–14	Significant trend ($p < 0.05$):
					Duration of woodworking (men)			
					2–11 years	1.1		8/16 exposed cases/controls
					12–29 years	2.5		10/9 exposed cases/controls
					29–55 years	4.2		11/6 exposed cases/controls
Shimizu et al. (1989)	Japan	M	45/90	Questionnaires	Woodworking or joinery	2.1	0.8–5.3	Matched on age and sex
					Sanding or lathing	7.5	1.5–39	
Comba et al. (1992a)	Italy	M	25/184	Interviews and questionnaires	Woodworkers (men)	1.7	[0.3–9.2]	
Magnani et al. (1993)	Italy	MF	11/111	Questionnaires	Wood and furniture workers	0.9	0.4–8.3	

OR, odds ratio; RR, relative risk; CI, confidence interval; M, male; F, female

workers, furniture makers, other wood product workers and carpenters. The jobs were classified with regard to the level of exposure to wood dust on the basis of an ad-hoc job–exposure matrix. Logistic regression was applied, with control for age and study. No association was seen between tobacco smoking and exposure to wood dust. A high risk for adenocarcinoma in men was associated with employment in wood-related occupations (odds ratio, 14; 95% CI, 9.0–20); no excess risk was found for squamous-cell carcinoma in men (0.8; 0.6–1.1). The corresponding odds ratios for women were 2.8 (0.8–10) and 1.2 (0.5–3.1). When subjects were categorized according to exposure to wood dust, a clear exposure–response relationship was found for adenocarcinoma in men, but not for squamous-cell carcinoma in men or for either histological type in women (Table 23). Similarly, an association with duration of employment in wood-related jobs or duration of moderate or high exposure to wood dust was found only with adenocarcinoma in men. The overall relative risks for adenocarcinoma showed some heterogeneity among the studies included in the re-analysis, with particularly high risks for adenocarcinoma in European countries other than Sweden (Table 24). [This finding may suggest variability in the type or intensity of exposure in different countries. The Working Group noted that elevated risks were found for highly exposed individuals in Sweden and the United States, however, although these were based on small numbers.]

Table 23. Results of pooled analysis of studies on exposure to wood dust: cases of adeno- and squamous-cell carcinoma

Sex	Exposure to wood dust	No. of exposed controls	Adenocarcinoma		Squamous-cell carcinoma	
			No. of exposed cases	Odds ratio	No. of exposed cases	Odds ratio
Men	Low	83	1	0.6	6	0.5
	Moderate	402	14	3.1*	42	1.0
	High	82	104	46*	11	0.8
Women	Low	11	2	7.7*	2	1.5
	Moderate	10	0	0	2	4.5
	High	6	0	0	2	1.6

From Demers et al. (1995)
*$p < 0.05$

Table 24. Odds ratios for adenocarcinoma by individual study included in a pooled re-analysis of studies of exposure to wood dust

Reference	Any exposure		High exposure	
	No. of exposed cases/controls	Odds ratio	No. of exposed cases/controls	Odds ratio
Zheng et al. (1992)	0/12	0.0	0/1	0.0
Luce et al. (1991, 1992, 1993)	79/45	161	69/15	516
Bolm-Audorff et al. (1989, 1990)	2/1	64	1/0	∞
Comba et al. (1992a)	8/23	12	7/11	23
Comba et al. (1992b)	2/2	32	2/1	50
Magnani et al. (1989, 1993)	5/8	15	5/2	55
Merler et al. (1986)	1/4	0.5	1/0	∞
Hayes et al. (1986a,b, 1987)	17/35	13	16/13	36
Hardell et al. (1982)	1/277	0.5	1/22	6.1
Brinton et al. (1984)	4/58	0.9	2/9	3.0
Mack & Preston-Martin (1995)	0/12	0.0	0/2	0.0
Vaughan (1989); Vaughan & Davis (1991)	0/98	0.0	0/6	0.0

From Demers et al. (1995); odds ratios adjusted for age and study

2.4.2 Cancers of other parts of the respiratory system

(a) Nasopharyngeal cancer

(i) Exposure to wood dust

Armstrong et al. (1983) conducted a study of 100 Chinese patients (65 men and 35 women) diagnosed with nasopharyngeal cancer between 1973 and 1980 and treated at the Institute for Radiotherapy at the General Hospital of Kuala Lampur (the only hospital offering this treatment for nasopharyngeal cancer in Malaysia). A matched neighbourhood control of the same sex and of similar age was selected for each case. Both cases and controls had lived in the study region for at least five years. Interviews were used to collect information on occupational and other exposure. A matched analysis was performed. The relative risk reported for exposure to wood and sawdust was 2.2 ($p < 0.08$, one-sided).

In the study of Olsen et al. (1984) described on p. 109, there were 266 cases of nasopharyngeal carcinoma (excluding sarcomas). A relative risk of 0.4 (95% CI, 0.2–1.0) was observed among men with definite exposure to wood dust. Olsen and Asnaes (1986) further evaluated this data set by histologically confirmed sub-group; results were not presented for nasopharyngeal cancer, but the authors stated that there was no association with exposure to wood dust.

In the study of Vaughan (1989), described on p. 110, 21 people with nasopharyngeal cancer were interviewed. An excess risk for nasopharyngeal cancer was reported for carpenters (odds ratio, 3.3; 95% CI, 0.8–13) in any employment; the risk increased to 4.5 (1.1–19) after exclusion of the last 15 years. Vaughan and Davis (1991) later categorized these cases according to exposure to wood dust. The odds ratio for nasopharyngeal cancer associated with employment in any wood-related occupation was 1.2 (0.2–4.6) and increased to 4.2 (0.4–27) when only exposure for 10 or more years after an induction period of 15 years was considered. The authors stated that the case patients were predominantly exposed to softwood dust.

In the study of Bolm-Audorff et al. (1989, 1990), described on p. 110, 12 of the 66 cases were nasopharyngeal cancers. One case of undifferentiated carcinoma of the nasopharynx was associated with exposure to wood dust (oak and beech) for 24 years, while two controls had been exposed to wood dust (species not identified), one of them for fewer than five years.

The etiology of nasopharyngeal carcinoma was studied in the Philippines, in investigations addressing both viral (Hildesheim et al., 1992) and non-viral (West et al., 1993) risk factors. There were 104 people with histologically confirmed nasopharyngeal carcinoma in the Philippines General Hospital and 104 hospital controls (matched on sex, age and private versus public ward) and 101 community controls (matched on sex, age and neighbourhood). The occupational history of each subject was collected by personal interview. The occupations of carpenter, lumberman, raftsman, woodchopper, farm manager and farmer were considered to entail exposure to wood dust on the basis of an assessment by an industrial hygienist who was unaware of the case or control status of the subject. A matched analysis was conducted. The relative risk for exposure fewer than 35 years before diagnosis was 1.3, and that for exposure 35 or more years before first diagnosis was 2.1. [The Working Group noted that the authors did not control for the presence of Epstein–Barr virus antibodies, which showed a strong association with nasopharyngeal cancer (odds ratio, 21) in the study of Hildesheim et al. (1992).]

(ii) Occupational group

In the study of Hardell et al. (1982) described on p. 113, there were 27 male patients with nasopharyngeal cancer; five were squamous-cell carcinoma, 20 were anaplastic carcinoma and two were adenoid cystic carcinoma. A crude relative risk of 1.3 [95% CI, 0.6–2.9] was reported for employment as a carpenter, cabinet-maker or sawmill worker, with nine exposed cases and 151 exposed controls. [The comparability of the ages of cases and controls was unknown.]. The authors noted that little hardwood is used for furniture making in northern Sweden.

In the study of Ng (1986), described on p. 114, there were 224 cases of nasopharyngeal cancer: 112 squamous-cell, 102 anaplastic and 10 of unknown histology; 226 controls had other malignancies. Among people in the two wood-related occupational categories, furniture makers and woodworkers had five nasopharyngeal cancers and one other malignancy, and construction carpenters had three nasopharyngeal cancers and no other malignancies. No odds ratios or other estimates of relative risk were presented.

In the study of Kawachi et al. (1989), described on p. 115, excess risks were observed for nasopharyngeal cancer among woodworkers (odds ratio, 2.5; 95% CI, 0.9–6.6), forestry and logging workers (6.0; 1.0–28) and carpenters (2.5; 0.6–8.5), after adjustment for age. No cases were observed among sawmill workers, pulp and paper workers or cabinet-makers.

Sriamporn et al. (1992) performed a case–control study in North-east Thailand of 120 patients with histologically confirmed nasopharyngeal cancer diagnosed between 1987 and 1990, who were undergoing radiation therapy at the only hospital offering such therapy in the area. Sixty-nine (57.5%) of the cases were squamous-cell carcinoma, and the remaining 51 were undifferentiated carcinomas. The 120 controls were patients admitted to the same hospital for diseases other than cancer and respiratory disease, who were matched to the cases on sex and age. Occupational histories were collected by questionnaire. The results were adjusted for age, sex, smoking, consumption of alcohol and salted fish, education and area of residence. Excess risks were reported for wood cutting, excluding agriculture (odds ratio, 4.1; 95% CI, 0.8–22) and for wood cutting and farming combined (8.0; 2.3–28).

Studies on nasopharyngeal cancer are summarized in Table 25.

(b) Pharyngeal cancer other than cancer of the nasopharynx

Elwood et al. (1984) reported the results of a case–control study of 374 patients with primary epithelial cancers of the oral cavity, pharynx (excluding nasopharynx) and larynx in British Columbia, Canada. The study included 44 oropharyngeal cancers, 38 hypopharyngeal cancers and five pharyngeal cancers at 'other' subsites. Controls were 374 patients with selected other cancers, who were matched to the cases on age and sex. Lifetime occupational histories were collected by interview in 1977–80. No quantitative results for wood-related exposures were reported, but the authors stated that 'exposure to wood dust was assessed and analysed in more detail, examining nature, intensity, and duration of exposure, but no regular or significant trends were seen'.

In the study of Vaughan (1989), described on p. 110, 183 people with cancer of the oro- or hypopharynx and 552 controls were interviewed. Excess risks were reported for carpenters, construction carpenters and machine operators in the wood industry after exclusion of exposure during the previous 15 years. Vaughan and Davis (1991) later categorized these cases according to exposure to wood dust. The odds ratio for pharyngeal cancer associated with employment in any wood-related occupation was 0.5 (95% CI, 0.2–1.2); it was 1.5 (0.4–5.5) when only exposure for 10 or more years after an induction period of 15 years was considered.

In the study of Haguenoer et al. (1990), described on p. 116, there were 114 histologically confirmed cases of oro- and hypopharyngeal cancers (no nasopharyngeal cancers were included) among men treated during the first semester of 1983. A matched analysis was conducted. A history of employment in woodworking occupations was associated with an increased risk for all cancers of the upper respiratory tract combined (odds ratio, 3.5; 95% CI, 1.2–10). Four case patients with oro- and hypopharyngeal cancers and one matched control had been employed as woodworkers, but no odds ratio was presented.

Maier et al. (1991) conducted a case–control study in the Heidelberg and Giessen areas of Germany of 200 male patients with squamous-cell carcinomas of the mouth, oropharynx, hypopharynx and larynx that had been diagnosed or treated at an eye, nose and throat clinic during 1987 and 1988. For each case, four age-matched male controls were selected from among

Table 25. Community-based case-control studies on cancer of the nasopharynx

Reference	Country	Sex	Cases/ controls	Source of information on exposure	Exposure to which relative risk applies	OR/RR	95% CI or p	Comments
Exposure to wood dust								
Armstrong et al. (1983)	Malaysia	MF	100/100	Interviews	Exposed to wood/saw dust	2.2	p (one-sided) = 0.08	Self-reported exposure; matched on age and sex; ethnic Chinese
Olsen et al. (1984)	Denmark	MF	266/2465	Linkage with national pension fund records	Definite exposure to wood dust (men)	0.4	0.2–1.0	Exposure based on expert assessment of work history
Vaughan (1989); Vaughan & Davis (1991)	USA	MF	21/552	Interviews	Carpenters Construction carpenters Any wood-related occupation Duration ≥ 10 years	4.5 6.8 1.5 4.2	1.1–19 1.6–28 0.2–6.1 0.4–27	Only squamous-cell cancers; exposure primarily to softwood. Exposures lagged by 15 years. Results adjusted for age, sex, smoking and alcohol intake
Bolm-Audorff et al. (1990)	Germany	MF	12/66	Interviews	Exposed to wood dust			1/2 exposed case/controls
West et al. (1993)	Philippines	MF	104/205	Interviews	Wood dust-exposed occupation < 35 years since first employment ≥ 35 years since first employment	1.3 2.1		Exposed occupations were carpenters, farm managers, farmers, lumbermen, raftsmen and wood-choppers. Matched on sex and age
Occupational group								
Hardell et al. (1982)	Sweden	M	27/541	Questionnaires	Carpenter, cabinet-maker or sawmill worker	1.3	[0.6–2.9]	9/151 exposed cases/controls
Ng et al. (1986)	Hong Kong	MF	224/226	Medical records	Furniture makers, woodworkers Construction carpenters	[5.0]		5/1 exposed cases/control 3/0 exposed cases/control

Table 25 (contd)

Reference	Country	Sex	Cases/ controls	Source of information on exposure	Exposure to which relative risk applies	OR/RR	95% CI or p	Comments
Occupational group (contd)								
Kawachi et al. (1989)	New Zealand	M	NR/ ~19 000	Tumour registry records	Woodworkers Foresters, loggers Sawmill, pulp and paper workers Cabinet-makers Carpenters	2.5 6.0 2.5	0.9–6.6 1.0–28 0.6–8.5	Adjusted for age 0/108 exposed case/controls 0/91 exposed case/controls
Sriamporn et al. (1992)	Thailand	MF	120/120	Interview	Wood cutting, not agriculture Wood cutting and agriculture	4.1 8.0	0.8–22 2.3–28	Adjusted for age, sex, smoking, consumption of alcohol and salted fish, education and area of residence

NR, not reported; OR, odds ratio; RR, relative risk; CI, confidence interval; M, male; F, female non-cancer patients attending the same clinic and a university clinic. Information on occupational and other exposures was collected from questionnaires. An elevated relative risk associated with self-reported exposure to wood dust was found for cancers at all upper aerodigestive tract sites combined (2.2; 95% CI, 1.0–4.9), but no site-specific results were reported.

Merletti *et al.* (1991) reported the results of a case–control study on occupation and cancer of the oral cavity and oropharynx in Turin, Italy. The cases were cancers of the oropharynx ($n = 12$) and oral cavity ($n = 74$) diagnosed among male residents of Turin between 1982 and 1984. The 385 controls were selected from a random sample of city residents interviewed as part of a study on laryngeal cancer. Occupational histories were collected by interview. Results were presented only for all cases combined (oropharynx and oral cavity). The odds ratios were adjusted for age, education, area of birth, tobacco smoking and alcohol drinking. The odds ratio for previous employment as a cabinet-maker or related woodworker was 1.2 (95% CI, 0.4–3.9); that for employment in any wood industry was 0.9 (0.3–3.0), and that for employment in the wood furniture industry was 1.4 (0.4–5.5).

Huebner *et al.* (1992) performed a case–control study of oral and pharyngeal (excluding nasopharyngeal) cancer among residents of four areas of the United States: Los Angeles, metropolitan Atlanta, two counties south of San Francisco, and New Jersey. Cases were identified from population-based tumour registries, to give 1114 cases diagnosed between 1 January 1984 and 31 March 1985. Controls (1268) were obtained through random-digit dialling (aged 18–64 years) and Health Care Financing Administration files (aged 65–79 years) and were frequency matched to controls on sex, race, age and study area. Information on occupation and exposure was collected at interviews. The results were adjusted for age, race, smoking, alcohol consumption and study location. The relative risks for pharyngeal cancers in men were 2.2 (95% CI, 1.0–4.7) for previous employment in the furniture/fixture industry and 2.3 (0.7–7.4) for work with woodworking machines.

Studies on oropharyngeal and hypopharyngeal cancers are summarized in Table 26.

(c) Laryngeal cancer

(i) Exposure to wood dust

Maier *et al.* (1992) reported on a case–control study of laryngeal cancer in Germany. The cases were histologically confirmed squamous-cell carcinomas of the larynx in 164 male patients who had attended the department of Otorhinolaryngology–Head and Neck Surgery, University of Heidelberg, for treatment or follow-up examinations during 1988–89. Controls were 656 males with no known cancer, who were selected randomly from two out-patient clinics in Heidelberg and matched to the cases 4:1 by age and residential area. All subjects were interviewed on life style, education and occupational history and exposures. The percentages of cases and controls, respectively, who were exposed at least once in a week during 10 years or more to different wood dusts were as follows: wood dust in general, 12.6% and 8.3% ($p < 0.08$); beech, oak, 5.8% and 6.1% ($p < 0.7$); pine, 12.6% and 7.5% ($p < 0.06$); 'precious wood', 0.8% and 3.5% ($p < 0.3$); and exotic wood, 0% and 2.1% ($p < 0.3$). For exposure to pinewood dust, the relative risks, p values and undefined CIs, adjusted for alcohol consumption and tobacco

Table 26. Community-based case-control studies on cancer of the oro- and hypopharynx in occupational groups

Reference	Country	Sex	Cases/controls	Source of information on exposure	Exposure to which relative risk applies	OR/RR	95% CI or p	Comments
Elwood et al. (1984)	Canada	MF	87/374	Interviews	Unspecified exposure to wood			'No regular or significant trends'
Vaughan (1989), Vaughan & Davis (1991)	USA	MF	183/552	Interviews	All carpenters Construction carpenters Wood machine operator Any wood-related occupation Duration ≥ 10 years	1.3 1.8 2.8 0.5 1.5	0.5–3.4 0.7–4.8 0.3–2.4 0.2–1.2 0.4–5.5	Only squamous-cell cancers; all exposures lagged by 15 years
Haguenoer et al. (1990)	France	M	114/228	Interviews	Woodworkers			4/1 exposed cases/control. Only jobs held for at least 15 years evaluated; matched on sex, age, ethnic group, area of residence, smoking and alcohol drinking
Huebner et al. (1992)	USA	MF	1114/1268	Interviews	Furniture/fixture industry Woodworking machines	2.2 2.8	1.0–4.7 0.7–7.4	For subset of male cases of pharyngeal cancer only

OR, odds ratio; RR, relative risk; CI, confidence interval; M, male; F, female

smoking, were as follows: all laryngeal cancers, 1.9 ($p = 0.05$; CI, 0.9–3.7); glottal cancers, 3.2 ($p = 0.03$; CI, 1.1–9.0); supraglottal cancers, 1.3 ($p = 0.6$; CI, 0.4–3.5). The mean time between the beginning of exposure to pinewood dust and the expression of laryngeal cancer was 39.7 years. [The Working Group noted the incomplete documentation of the composition of the control group, statistical methods and results.]

Zheng et al. (1992b) reported on a population-based case–control study of laryngeal cancer in Shanghai, China. A total of 263 residents of urban Shanghai, aged 20–75, in whom laryngeal cancer was newly diagnosed in 1988–90, were identified from a population-based cancer registry in Shanghai; 201 (76.4%) were interviewed. Controls were randomly selected from the urban Shanghai population and frequency matched by sex and age to all cases of oral, pharyngeal, laryngeal and nasal cancers reported to the Shanghai Cancer Registry during 1985–86. Of the 414 controls interviewed, 12% were second controls. The interview covered demographic data, tobacco smoking, alcohol drinking, dietary habits and occupational history and exposures. Adjusted odds ratios were calculated by stratification and unconditional logistic regression. Self-reported occupational exposure to wood dust among males was associated with an odds ratio of 1.4 (95% CI, 0.6–3.2), adjusted for age and smoking.

(ii) Occupational group

A case–control study of laryngeal cancer was reported by Wynder et al. (1976) in the United States which included 258 men and 56 women with a histologically confirmed laryngeal cancer (ICD 161.0,1), who had been admitted to hospitals in New York City, Los Angeles, Houston, Birmingham, Miami and New Orleans during 1970–73. Controls were 516 hospitalized men and 168 women (without a current diagnosis of tobacco- or alcohol-related disease and with no history of cirrhosis, stroke, gastric ulcers or myocardial infarct), who were individually matched to the cases by year of interview, hospital status and age at diagnosis. Occupational exposure to wood dust was reported for 22% of the male cases and 1.2% of controls ($p < 0.05$). Of the current smokers, equal proportions of cases and controls had been exposed to wood dust.

Occupational risks for laryngeal cancer were examined in the Third National Cancer Survey data (Flanders & Rothman, 1982). Cancer cases occurring during 1969–71 were identified in records from seven cities and two states in the United States, and a 10% probability sample of the subjects was interviewed. Ninety men with laryngeal cancer represented the cases, and 933 men with cancers at other sites, excluding oesophagus, stomach, small intestine, colon, pancreas, liver, bladder, kidney, lung, bronchus, oral cavity and pharynx, constituted the control group. Information on age, alcohol use, tobacco use and industrial and occupational categories (longest and second-longest held job) was obtained at interview and used in the analysis. The relative risks, adjusted for age, alcohol use, tobacco smoking and race, were 3.5 [95% CI, 0.6–19] for work in the lumber industry and 1.3 [0.3–6.6] for carpenters.

A hospital-based case–control study in New Haven, Connecticut (United States) (Zagraniski et al., 1986), addressed occupational risk factors for laryngeal cancer. The cases were histologically confirmed primary squamous-cell carcinomas of the larynx (ICD 8 US adapted: 161) diagnosed in 1975–80 among white male residents of Connecticut alive in 1980–

81 and treated in one of two New Haven hospitals. Controls were white male general surgery patients with no prior diagnosis of cancer or respiratory disease, who were individually matched to the cases by hospital, calendar year of admission, decade of birth, county of residence, smoking status and type of tobacco used. The interview covered occupational history, including specific exposures, medical history, tobacco and alcohol use and demographic and home environmental data. A total of 148 cases were identified; 22 were not invited for interview because of death, move out of the state, loss to follow-up and other reasons. No proxy data were sought. Of the remaining 126 cases, 14.3% refused to be interviewed. Thus, 12.7% completed a telephone interview and 73.0% completed a personal interview. Of the 317 controls, 57.1% were interviewed personally. Any employment as a woodworker was associated with an odds ratio of 2.5 (95% CI, 0.5–13); the odds ratio for carpenters was 1.1 (0.6–2.0). These figures were adjusted for lifetime exposure to tobacco and alcohol and were based on a conditional logistic model with 87 cases and 153 controls. [The Working Group noted the low response rate among the controls.]

Morris Brown et al. (1988) reported on a case–control study of laryngeal cancer conducted in six counties of the Gulf Coast of Texas, United States. Cases were primary laryngeal cancers (ICD9: 161.X, 231.0) diagnosed in white males aged 30–79, identified through tumour registers and the records of 56 hospitals. During 1975–80, 220 living and 83 dead case patients were identified. Controls were a sample of white males resident in the six-county area, who were frequency matched to the cases by age, vital status, ethnic group and county of residence; they were identified from Texas Department of Health mortality tapes, drivers' licence records and Medicare records. Interviews including job histories were completed for 153 living case patients (69.5%) and close relatives of 56 dead case patients (67.5%). Exclusions on histological grounds resulted in 183 case patients (136 living, 47 dead). There were 250 controls (179 alive, 71 dead); the response rates were 62.8% for the dead controls, 60.9% for those identified from drivers' licence lists and 85.7% for those identified from Medicare records. The odds ratio for any employment as a woodworker or furniture maker, adjusted for tobacco smoking and alcohol drinking in the logistic model, was 8.1 (95% CI, 0.95–69; 7 exposed cases); that for employment as a carpenter was 1.7 (0.8–3.5; 19 exposed cases).

Bravo et al. (1990) reported on a case–control study of laryngeal cancer conducted in Spain. The cases were histologically confirmed epidermoid carcinomas of the larynx diagnosed in 85 patients (83 men) at La Paz Hospital, Madrid, during 1985–87. Twenty-six eligible case subjects were not included in the study because of death, having moved out of the country, refusal or loss. Controls were a random sample of 170 patients from the same hospital, excluding those with respiratory diseases or alcoholic cirrhosis; they were 'stratified' with respect to the cases according to sex, age and admission month. Personal interviews were conducted with all subjects, except for 15 case patients for whom a relative was interviewed. Occupational exposure to wood dust was associated with a crude odds ratio of 0.70. The mean duration of exposure to wood dust was 25 years for the exposed cases and 26 years for the exposed controls. [The Working Group noted the incomplete description of the exposure assessment and the crude statistical analysis].

In their study of exposure to wood dust and cancer of the upper respiratory tract, described on p. 110, Vaughan and Davis (1991) compared 234 cases of squamous-cell cancer of the larynx with 547 controls. They observed no excess risk for ever having been employed in a wood-exposed occupation (odds ratio, 1.0; 95% CI, 0.5–1.9), after adjustment for potential confounders. An elevated risk was observed for employment for 10 or more years, when allowing for a 15-year induction period (2.5; 0.6–10).

In a hospital-based case–control study in the United States (Muscat & Wynder, 1992), the associations between laryngeal cancer and exposure to tobacco, alcohol and occupational factors were examined in 194 white men with histologically confirmed primary laryngeal cancer selected from the records of the Memorial Sloan-Kettering Cancer Center and seven other hospitals in New York, Illinois, Michigan and Pennsylvania and interviewed in 1985–90. Controls were 184 men matched by hospital, age and year of interview, who included patients with gastrointestinal cancers, prostate cancers or lymphomas and bone, spinal and other 'non-neoplastic conditions'. Eighty-nine percent of the eligible case patients and controls agreed to be interviewed; no proxies were used. Self-reported occupational or recreational exposure to wood dust for at least 8 h per week for at least one year was associated with an odds ratio of 1.7 (95% CI, 0.7–4.6), adjusted for age, education, tobacco smoking, alcohol drinking and relative body weight.

A population-based case–control study in western Washington State (United States) (Wortley et al., 1992) addressed occupational risk factors for laryngeal cancer. Cases were laryngeal cancers (ICD0: 161.0–161.9) diagnosed in 1983–87 in 291 patients who were identified through the cancer surveillance system (population-based cancer registry covering 13 counties in western Washington) of the Fred Hutchinson Cancer Research Center, Seattle; 235 (80.8%) were successfully interviewed. The closest next-of-kin was interviewed if the case was dead (17 surrogate interviews). Controls were men identified by random-digit dialling, who were frequency matched to the cases in categories of age and sex; 547 (80%) of the eligible controls were successfully interviewed. Lifetime occupational histories were coded according to the 1980 United States census codes for occupation and industry. Odds ratios were derived from a multiple logistic regression model and adjusted for smoking, alcohol drinking, age and education. The odds ratio for any employment as a woodworking machine operator, lagged by 10 years, was 0.4 (95% CI, 0.1–1.3); for fewer than 10 years, the odds ratio was 0.4, and for at least 10 years, it was 2.3 (CIs not given; trend $p = 0.36$). Five cases and 18 controls had been employed as woodworking machine operators. [The Working Group noted that although the number of controls in this study (547) was similar to that in the study of Vaughan and Davis (1991), they used separate control series.]

The studies on laryngeal cancer are summarized in Table 27.

(d) Lung cancer

(i) Exposure to wood dust

Blot et al. (1982) reported on occupational determinants of lung cancer in an area of northern Florida (United States) with exceptionally high rates of lung cancer. Interviews were conducted with 181 patients with lung cancer, 342 hospital controls (1978–79), 217 next-of-kin

Table 27. Community-based case-control studies on cancer of the larynx

Reference	Country	Sex	Cases/ Controls	Source of information on exposure	Exposure to which relative risk applies	OR/RR	95% CI or p	Comments
Exposure to wood dust								
Maier et al. (1992)	Germany	M	164/656	Interview	Wood dust Pinewood dust Pinewood dust, glottal cancers only	1.9 3.2	$p < 0.08$ $p = 0.05$ $p = 0.03$	12.6% of cases, 8.3% of controls; incomplete documentation
Zheng et al. (1992b)	China	M	177/269	Interview	Wood dust (self-reported)	1.4	0.6–3.2	Adjusted for age, smoking
Occupational group								
Wynder et al. (1976)	USA	M	258/516	Interview	Wood dust		$p < 0.05$	Crude analysis; 22% of cases, 1.2% of controls
Flanders & Rothman (1982)	USA	M	90/933	Interview	Lumber industry as major employer Carpenter	3.5 1.3	[0.6–19] [0.3–6.6]	Adjusted for age, race, smoking, alcohol consumption
Zagraniski et al. (1986)	USA	M	87/153	Interview	Woodworker (ever) Carpenter (ever)	2.5 1.1	0.5–13 0.6–2.0	Adjusted for smoking, alcohol consumption; low response in controls
Morris Brown et al. (1988)	USA	M	183/250	Interview	Woodworker or furniture maker (ever) Carpenter	8.1 1.7	1.0–69 0.8–3.5	Adjusted for smoking, alcohol consumption
Bravo et al. (1990)	Spain	MF	85/170	Interview	Wood dust	0.7		Incomplete documentation

Table 27 (contd)

Reference	Country	Sex	Cases/	Source of controls on exposure	Exposure to which relative information	OR/RR risk applies	95% CI or p	Comments
Occupational group (contd)								
Vaughan & Davis (1991)	USA	MF	234/547	Interview	Any wood dust >10 years with 15-year induction	1.0 2.5	0.5-1.9 0.6-10	Squamous-cell only; exposure predominantly to softwood
Muscat & Wynder (1992)	USA	M	194/184	Interview	Wood dust	1.7	0.7-4.6	Adjusted for age, education, smoking, alcohol consumption, relative body weight
Wortley et al. (1992)	USA	MF	235/547	Interview	Woodworking machine operator > 10 years	2.3	p (trend) = 0.36	Adjusted for age, education, smoking, alcohol consumption

OR, odds ratio; RR, relative risk; CI, confidence interval; M, male; F, female

of dead lung cancer patients and 217 dead control subjects. The controls were selected from among hospital patients with diagnoses of or deaths from diseases other than lung cancer or chronic respiratory disease. The response rates were 86% for cases and 83% for controls. The final study group consisted of 321 cases and 434 controls. An excess relative risk for lung cancer (1.7; 95% CI, 1.0–2.7) was reported among people ever employed in the lumber or wood industries, after adjustment for tobacco smoking. A gradient was suggested with duration of employment in the lumber or wood industry: the relative risk was 1.3 among those employed 1–9 years and 1.6 among those employed for more than 10 years; both relative risks were calculated in comparison with those for people never employed in the wood industry. The excess was concentrated among those exposed to wood dust mainly in sawmills (1.9); the excess associated with exposure to wood dust was higher (3.4; 10 exposed cases) for small-cell carcinoma than for other cell types.

A hospital-based case–control study was conducted in an area of Louisiana (United States) with a high rate of lung cancer (Correa et al., 1984). 'Current' primary lung cancers were identified from admission and pathology records of all the major hospitals in southern Louisiana, in one central Louisiana parish and two northern Louisiana parishes, except for the city of New Orleans, where the study was limited to four large hospitals. [The period of case ascertainment was not given.] A control subject was selected for each case from the same hospital and individually matched by race, sex and age. Patients whose main diagnosis was emphysema, chronic bronchitis, chronic obstructive pulmonary disease or cancer of the larynx, oral cavity, oesophagus or urinary bladder were excluded. Acceptable personal interviews were conducted with 1338 (76%) cases and 1393 (89%) controls. The questions covered occupation, residence, diet, smoking and drinking habits, health, water supply and other related items. The odds ratios (unconditional) were adjusted for smoking. For all occupations, only white male workers who had ever been employed in the 'forestry' category (most of them sawmill workers) had a significantly elevated odds ratio (1.7). Exposure to wood dust was associated with an odds ratio for lung cancer of 1.4 ($p < 0.05$); of 45 other suspected occupational exposures, only mineral oil mist was also significantly associated with lung cancer.

A nested case–control study of lung cancer in 19 608 male employees of a chemical plant in Texas (United States) (Bond et al., 1986), was designed to examine the associations between lung cancer and a number of occupational exposures. A total of 308 lung cancer deaths were recorded during 1944–80 on death certificates as cancer of the bronchus, lung or site unspecified within the respiratory system as the underlying cause, as a contributing cause or as 'other significant conditions'. Two control groups, one a dead and the other a living series, were individually matched to the lung cancer patients. Work histories were grouped into 50 work areas of homogeneous exposures. Chemical and physical exposure profiles were developed by an industrial hygienist for each case and each control. In comparison with pooled controls, the odds ratio for any (as opposed to no) exposure to wood dust was 1.1 (95% CI, 0.72–1.8). With a 15-year lag, it was 1.3 (0.78–2.2); for low exposure to wood dust, 3.9 (1.1–14); for moderate exposure, 0.91 (0.39–2.1); and for high exposure, 0.99 (0.56–1.8). [The Working Group noted that this was a study of a multi-product chemical plant where there were numerous exposures and where exposure to wood dust would be minimal in comparison with that in wood-related industries.]

Kjuus *et al.* (1986) used data from interviews with 176 men admitted to two hospitals in Telemark and Vestfold, Norway, for lung cancer during 1979–83 and with 176 age-matched control subjects admitted during the same period to the same hospitals. Patients with conditions that would have precluded employment in heavy industry, poor general health, obvious mental conditions or chronic obstructive lung disease were excluded as controls. One case and two potential controls refused the interview. Woodwork as the main lifetime occupation was associated with an odds ratio of 0.7 (95% CI, 0.2–2.3), adjusted for cigarette smoking. The odds ratio for exposure to wood dust, adjusted for a number of confounders, was 0.5, representing a significant deficit.

A population-based multi-site case–control study of cancer addressed associations between occupational factors and cancer (Siemiatycki *et al.*, 1986; Siemiatycki, 1991). A total of 3730 histologically confirmed cases of cancer at 19 sites, newly diagnosed among male residents, aged 35–70, of Montréal, Canada, in 19 major hospitals during 1979–85 were identified, and the patients were interviewed for detailed lifetime job histories and potential confounders. The response rate was 81.5%. A team of chemists and hygienists examined the questionnaires and translated each job into a list of potential exposures with the help of a checklist of some 300 occupational exposures. Odds ratios for each cancer site were calculated in comparison with other cancers and adjusted for smoking and other factors, including other occupational exposures, and were presented for any exposure and for substantial exposure (at least 10 years' duration after the first five years). There were 1082 patients with lung cancer, and 857 (79%) responded. The odds ratio for any exposure to wood dust was 1.2 [95% CI, 1.0–1.5]; for substantial exposure, it was 1.3 [0.9–1.8]. When the analysis was restricted to French-Canadian subjects, the odds ratios for oat-cell lung cancer were 1.3 [0.9–2.0] for any exposure and 1.6 [0.9–2.8] for substantial exposure.

A population-based case–control study was conducted in New Mexico, United States (Lerchen *et al.*, 1987) of 506 white residents of New Mexico (333 males and 173 females), aged 25–84 years, with primary lung cancer other than bronchioalveolar carcinoma, which was diagnosed in 1980–82 and registered by the New Mexico Tumor Registry. A total of 771 controls (499 males and 272 females) were identified through randomly selected residential telephone numbers and, for people aged 65 or older, from a roster of Medicare participants. The controls were frequency matched to cases for sex, ethnic group and 10-year age category. The case subjects and controls or their next-of-kin were interviewed about smoking and occupational history, and a self-reported history of exposures to 18 specific agents was obtained. The interview rate was 89% for cases and 83% for controls. The odds ratio for ever having been employed for at least one year as a woodworker was 0.8 (95% CI, 0.3–1.7), after adjustment for age and ethnic group. [The Working Group noted that the numbers of case subjects and controls exposed to wood dust were exactly the same as those of people employed one or more years as a woodworker and assumed that these were the same individuals.]

Twenty-five major industrial titles were evaluated as risk factors for lung cancer in a population-based case–control study conducted in Shanghai, China (Levin *et al.*, 1988). The case series consisted of all lung cancers newly diagnosed during 1984–85 among men aged 35–64 who resided in the urban areas of Shanghai. Population controls were randomly selected from the same areas from specified age categories in sampling fractions that produced similar age distributions for the cases and controls. Personal interviews were conducted with 733 surviving case subjects (88% of the total incident cases and 99% of survivors) and 760 controls. The interview concerned lifetime occupational history, smoking and other information; detailed data were obtained on every job the subject had held for at least one year since the age of 16. The employment data were classified by the industrial and occupational headings defined for the 1982 Chinese population census. Any employment in furniture manufacture was associated with an odds ratio of 1.3 (95% CI, 0.5–3.4), adjusted for smoking and age. The adjusted odds ratio for any employment in timber processing or as a wood, bamboo, hemp, rattan, palm or straw products maker was 1.2 (0.7–2.2). Self-reported exposure to wood dust was associated with a significantly increased odds ratio of 1.7 (1.0–2.7), in contrast to people who reported no exposure to dust, smoke or fumes in the workplace. Most of the subjects who reported exposure to wood dust worked in furniture manufacture or timber production. A total of 672 female case subjects and 735 female controls were also interviewed, but the small numbers in many industrial and occupational categories precluded a detailed analysis. An increased risk, similar to that observed among men, was associated with self-reported exposure to wood dust [no odds ratio given]. The authors found only slight differences in risk by cell type for most occupational or industrial categories but did not document this statement.

(ii) Occupational group

Harrington *et al.* (1978) analysed data from death certificates on the occupations of 858 white men who died of lung cancer in coastal Georgia, United States, during 1961–74, and of 858 controls who died of conditions other than lung cancer, chronic respiratory disease or bladder cancer during the same period. The controls were individually matched to the cases by age at death, year of death, sex, race and county of usual residence, and matched-pairs analyses were conducted. The usual occupations were coded into major occupational and industrial categories. The relative risk for work in the wood and paper industry was 1.3 ($p > 0.05$). A significant excess relative risk for work in the wood and paper industry (3.3; $p < 0.01$) was found in small rural counties but not in the largest counties. The excess was greater among sawmill, lumber and forestry workers than among pulp and paper workers and carpenters, but no relative risks were presented. No data were available on tobacco smoking and possible industrial confounders.

In a pilot study, Esping and Axelson (1980) used data from death certificates on the occupations associated with 25 deaths from respiratory cancer (ICD [1965]: 160–163) and those of 370 controls who had died from diseases other than respiratory and digestive cancers in the small town of Mjölby, Sweden, where there was a comparatively large woodworking industry. The deaths had occurred among men 50 years of age or older during the period 1963–77. The age-adjusted rate ratio for 'exposure to woodwork' was 4.1 (95% CI, 1.6–11). The crude relative risks were 6.0 for furniture makers and 2.3 for other woodworkers. The smoking habits were not known.

A case–control study in Alameda County, California, United States (Milne et al., 1983), covered 925 deaths from lung cancer (747 men) and 6420 deaths from other cancers (3130 men) that occurred among county residents over 18 years of age between 1958 and 1962. The study examined associations between lung cancer and usual occupation and industry, as recorded on the death certificate and coded by the US Bureau of Census Industrial and Occupational Classification System. The odds ratio were 0.8 for men employed in sawmills, 4.2 ($p < 0.01$) for men in furniture manufacture, 1.0 for male cabinet-makers and furniture finishers and 1.2 for carpenters.

In a hospital-based case–control study in metropolitan Florence, Italy (Buiatti et al., 1985), frequency matching on smoking status was used to compare 376 people with histologically confirmed primary lung cancer (340 men) with 892 control subjects with discharge diagnoses other than lung cancer and attempted suicide (817 men). The case patients and controls had been admitted in the period 1981–83. Occupational histories were collected for all cases and controls in person, and the response rate was 100%. The odds ratio for men ever employed in woodwork, adjusted for age, smoking and place of birth, was 0.6 (95% CI, 0.3–1.1).

Coggon et al. (1986) reported on a case–control study of cancer of the bronchus among middle-aged men in Cleveland, Humberside and Cheshire, United Kingdom, diagnosed during 1975–80. Controls were patients with other cancers. Occupational and smoking histories were obtained from a postal questionnaire, addressed either to the patients or their next-of-kin. The overall response rate was 52% (738 cases, 2204 controls). For patients who reported ever having been employed as a woodworker, the relative risk was 1.7 (95% CI, 1.0–3.0), adjusted for age, residence, source of occupational history and smoking. The risk ratio for ever having been employed in the industrial order of timber and furniture was 1.6 ($n = 17$). The authors compared the distribution of occupations among respondents and nonrespondents, using information from hospital records, and found no evidence of bias in reporting by response category.

In a population-based case–control study in northern Sweden (Damber & Larsson, 1987), data on 589 dead male cases and two series of matched control subjects drawn from population registries (582 deceased, 453 living) were used to examine associations between the risk for lung cancer and occupation. The cases represented deaths during 1972–79 among people in whom lung cancer was diagnosed. The occupations were ascertained from a postal questionnaire addressed to living controls or to close relatives of the cases and dead controls. The response rates were 98% for cases, 96% for dead controls and 97% for living controls. At least one year of employment as a carpenter was associated with an odds ratio of 0.8 (95% CI, 0.5–1.3), adjusted for lifetime tobacco consumption, when compared with dead controls, and with an odds ratio of 0.7 (0.5–1.2) when live controls were used.

The association between occupation and the risk for lung cancer was examined in a case–control study conducted in six areas of New Jersey, United States (Schoenberg et al., 1987). The cases were histologically confirmed primary cancers of the trachea, bronchus and lung diagnosed in 763 white males in 1980–81. Nine hundred white male population controls were selected from files of drivers' licences and death certificates. Interviews were completed with 429 case patients and 564 controls or their next-of-kin (334 and 336, respectively), in order to obtain demographic data and information on personal and environmental risk factors, including smoking, diet and occupation. The response rate was 70% for the case patients and 64% for the controls. Information on industry and job title was coded by the index system used in the 1970 United States census. The risk for lung cancer was analysed for 42 job title categories and 34 job titles in specified industries, after adjustment for smoking. The risk among men who had ever been employed as furniture or fixture workers was 1.5 (95% CI, 0.76–3.0). Smoking-adjusted odds ratios for carpenters (46 cases, 55 controls) and lumber and wood products workers (16 cases, 17 controls) were 0.90–0.99.

A population-based case–control study of lung cancer and occupation was conducted in two industrialized areas of northern Italy (Ronco et al., 1988) involving 126 men who had died from lung cancer between 1976 and 1980. Controls were a random sample of 384 men who had died from other causes (except chronic lung conditions and smoking-related cancers) during the same period and who were individually matched to the cases by year of death and 10-year age class. Next-of-kin were interviewed at home or by telephone with regard to the lifelong tobacco consumption and occupational histories of the cases and controls. The response rate was 77% for cases and 78% for controls. Job titles were coded by the ILO classification, and industrial activities according to the United Nations international classification of industries. While no excess risk was associated with carpentry or joinery (6 cases, 28 controls), increased odds ratios were observed for woodworkers employed in furniture and cabinet-making, with an aggregated odds ratio of 2.8 (0.93–8.4), adjusted for age, smoking and having been engaged in an occupation known or suspected to be associated with increased lung cancer risk.

Hoar Zahm et al. (1989) examined the associations between different histological types of lung cancer and occupations in 4431 white Missouri (United States) residents with histologically confirmed lung cancer diagnosed in 1980–85 and reported to the Missouri Cancer Registry. The 11 326 controls were all white Missouri residents with a diagnosis of any cancer except those of the lip, oral cavity, oesophagus, lung, bladder, ill-defined sites and unknown sites, during the same period. The occupation at the time of diagnosis of cancer was abstracted from the Registry files, which obtained this information from medical records. Occupation was coded according to the index system of the United States Bureau of Census, and codable information was available for 52% of the cases and 45% of the controls. The smoking history was unknown for 15% of the cases with known occupation and for 37% of the controls. Odds ratios were calculated for a number of occupations, with adjustment for age and smoking. For all lung cancers, the odds ratio associated with cabinet- and furniture making was 1.3 (95% CI, 0.5–3.3), and that for carpenters was 1.3 (1.0–1.7). Cabinet- and furniture makers had increased risks for adenocarcinoma of the lung (2.0; 0.4–8.1), small-cell carcinoma (1.6; 0.2–7.9) and tumours of 'other' or mixed-cell types (1.9; 0.4–7.4), but not for squamous-cell carcinoma (0.7; 0.1–3.5). In carpenters, the odds ratio for adenocarcinoma was 1.6 (1.0–2.5); that for small-cell carcinoma, 1.1 (0.6–2.0); that for squamous-cell carcinoma, 1.2 (0.8–1.8); and that for other or mixed type, 1.3 (0.8–2.2).

The studies of cancer risk in wood-related occupations, using the New Zealand Cancer Registry (Kawachi et al., 1989), described on p. 115, showed age-adjusted odds ratios of 1.3 (95% CI, 1.2–1.6) for all woodworkers, 1.3 (0.85–1.9) for foresters and loggers, 1.8 (1.2–2.5) for sawmill workers, 1.2 (0.77–1.8) for cabinet-makers 1.3 (1.1–1.5) for carpenters. Although

52% of the pulp and paper mill and sawmill workers were regular smokers at the time of the 1981 census, compared with a 38% smoking prevalence in the total labour force of New Zealand, sawmill workers did not have an excess risk of other cancers (those of the larynx, oesophagus and bladder) associated with tobacco smoking. In carpenters, the prevalence of smoking was 36%.

A case–control study was conducted in France to examine the relationship between bronchial adenocarcinoma and exposure to wood dust (Schraub et al., 1989). All histologically confirmed male cases of adenocarcinoma of the lung reported to the cancer registry of the Doubs region during 1978–85 formed the case series: 22 living and 40 dead cases were identified; nine case patients could not be located, and the remaining case patients or their next-of-kin were interviewed. A population sample (three controls randomly selected from among males within five years of the ages of the cases at the time of diagnosis) of 160 men formed the control group, representing an 86% participation rate. The controls were on average five years older than the cases. Occupational exposures and consumption of tobacco and alcoholic beverages were documented by interviews with live case patients and 160 controls and with families or physicians of the dead case patients. The crude odds ratio for exposure to wood dust was 1.1 (95% CI, 0.38–2.7). Adjustments for cigarette smoking, age and urban–rural residence resulted in 'only trivial, nonsignificant increases' in the odds ratio associated with exposure to wood dust. The mean duration of exposure was 6.8 years for the cases and 17.3 years for the controls. [The Working Group noted that, although the controls were older than the cases, the authors did not indicate whether the work histories of the controls were truncated to match the time-frame of those of the cases.]

A community-based case–referent study of occupational risk factors for lung cancer was conducted in the Detroit metropolitan area, United States (Burns & Swanson, 1991). Histories of occupation and tobacco use were obtained by telephone interview for 5935 incident lung cancer case patients and 3956 controls with colorectal cancer. The cases and controls were identified through the metropolitan Detroit cancer surveillance system, in which patients are enrolled within two to six weeks after diagnosis; the overall response rate was about 93%. The odds ratio for woodworkers was 1.1 (95% CI, 0.70–1.8), after adjustment for age at diagnosis, cigarette smoking history, race and sex. For workers in wood manufacture, which included many fewer subjects than the occupational category 'woodworkers', the odds ratio was 2.3 (0.81–6.4).

In a hospital-based case–control study conducted in five German cities (Jöckel et al., 1992), 194 patients with primary lung cancer, 194 hospital controls with an admission diagnosis unrelated to tobacco smoking and 194 population controls identified from residential registries were interviewed about smoking, occupational and residential histories. Controls were individually matched to the cases by sex and age. The response rate of the population controls was 40.7%. The smoking-adjusted odds ratio was 0.9 (95% CI, 0.46–2.0) for males in the paper, wood and printing industries, 0.7 (0.28–1.9) for wood processing workers and 0.8 (0.36–1.6) for carpenters and brick masons. [The Working Group noted the low participation rate in the controls.]

In a hospital-based case–control study in nine metropolitan areas of the United States (Morabia et al., 1992), 1793 male lung cancer cases were matched by race, age, hospital, year of interview and cigarette smoking with 2230 cancer and 998 non-cancer hospital controls, some of whom had tobacco-related diagnoses. Usual occupation, exposure to potential carcinogens and cigarette smoking were addressed during interviews conducted in 1980–89. Carpenters and cabinet-makers had a nonsignificant excess risk (odds ratio, 1.4), adjusted for age and tobacco smoking.

An updating (Kauppinen et al., 1993) of a Finnish case–control study nested in a cohort of male woodworkers (Kauppinen et al., 1986) was based on a cohort of 7307 workers from 35 industrial facilities (sawmills and furniture, construction carpentry, plywood and particle-board factories). A total of 136 incident respiratory cancers (cancers of the lung, trachea, larynx, epiglottis, tongue, pharynx, mouth, nose and nasal sinuses) was identified in the cohort during 1957–82. Three control subjects in the cohort who had not contracted respiratory cancer were matched to each case by year of birth. Plant- and time-specific job exposure matrices were constructed for 12 major agents in the wood industry. Job histories were based on plant records and complemented by responses to questionnaires from the case patients and controls or their next-of-kin; the questionnaires also provided data on tobacco smoking. The smoking-adjusted odds ratio for lung cancer and exposure to wood dust was 1.3 [95% CI, 0.8–2.3] with no exposure lagging; the ratio dropped to 0.44 [0.2–1.3] after lagging by 10 years. No trend by level or cumulative exposure was observed.

Associations with occupation were examined in a case–control study conducted in India (Notani et al., 1993) of 246 male residents of Mahrashtra State with a diagnosis of lung cancer. A total of 212 sex- and age-matched hospital controls (patients with cancers of the mouth and pharynx or non-neoplastic oral diseases) were selected, such that the community distribution was similar to that of the cases. Interviews were conducted with the case patients and controls to obtain lifetime occupational history, self-reported history of specific exposures, demographic variables, tobacco use, alcohol consumption and medical history. Each job was coded according to the International Standard Classification of Occupations. The odds ratios for ever having been employed as a woodworker were 3.0 (95% CI, 1.0–9.3; adjusted for age) and 3.2 (0.9–12; adjusted for age and smoking). [The Working Group noted that the participation rates of the case patients and controls were not documented and that the inclusion of controls with cancer of the pharynx biased the odds ratio towards the null.]

The associations between lung cancer and occupation were examined in a case–control study of 965 women aged 29–70 in whom primary lung cancer was diangosed in the cities of Shenyang and Harbin, China, and notified to the cancer registries of these cities during 1985–87 (Wu-Williams et al., 1993). They represented 92% of the eligible cases. Controls were women randomly selected from the populations of the same cities and frequency matched to the cases by age. Personal interviews were conducted with the cases and controls to obtain demographic data and information on lifetime smoking habits, sources of pollution, histories of occupation and specific exposures, and medical and dietary histories. The employment data were classified by industry and occupation according to the classification of the 1982 Chinese population census. The odds ratio for workers in the manufacture of wooden products, adjusted for smoking, study area, age and education, was 0.9 (95% CI, 0.5–1.7). For nonsmokers, the odds ratio, adjusted for study area, age and education, was 1.5 ($p > 0.05$). Timber processing was associated with an odds ratio of 1.1 (0.6–2.0), adjusted for smoking, study area, age and education. For nonsmokers, the odds ratio, adjusted for study area, age and education, was 1.5 ($p > 0.05$). Self-reported exposure to wood dust was associated with an odds ratio of 1.1 (0.8–1.7; adjusted for smoking, study area, age and education). The odds ratio was 1.3 ($p > 0.05$) among nonsmokers, after adjustment for study area, age and education.

Studies on lung cancer are summarized in Table 28.

2.4.3 Cancers of the lymphatic and haematopoietic system

(a) Non-Hodgkin's lymphoma

(i) Exposure to wood dust

In the study of Siemiatycki (1991), described on p. 140, the total number of eligible cases of non-Hodgkin's lymphoma was 258; 215 responded, for a response rate of 83%. The odds ratio for any exposure to wood dust was 0.8 [95% CI, 0.6–1.2]; for substantial exposure, it was 1.0 [0.6–1.6]).

Partanen et al. (1993) reported on a small industry-based case–control study of malignant lymphoma and exposures in the wood industry. In a retrospective cohort of male woodworkers, eight cases of non-Hodgkin's lymphoma were notified to the Finnish Cancer Registry in 1957–82. Fifty-two controls from the cohort were individually matched to the cases by year of birth and survival in 1983. Individual employment histories in woodworking facilities were abstracted from factory records, and a number of exposures were reconstructed with an ad-hoc plant- and period-specific job–exposure matrix. For the cases, this information was completed by interview of selected people at the factories and from questionnaires sent to the case subjects or their next-of-kin. [The Working Group noted that the data on exposure were more detailed for cases than for controls and that this may have induced a positive bias in the results.] The unadjusted odds ratios associated with exposure to wood dust were 2.1 (95% CI, 0.43–11) for all lymphomas and 2.1 (0.23–20) for non-Hodgkin's lymphoma.

(ii) Occupational group

Cartwright et al. (1988) reported on risk factors for non-Hodgkin's lymphoma in a case–control study in the United Kingdom. Case patients were identified in 1979–84 in hospitals, the cancer registry and the lymphoma panel in Yorkshire; additional cases during the period were sought in the area. Only cases confirmed histologically were accepted. The controls were in-patients with a variety of nonmalignant conditions. Attempts were made to match two hospital controls to each case by residential health district, sex and age. Case patients and control patients were interviewed with regard to past medical history, drug use, family medical history, hobbies, occupation, smoking and alcohol consumption. Of a total of 1407 patients with non-Hodgkin's lymphoma who had been notified, 437 (244 with low-grade tumours, 177 with high-grade tumours and 36 with unspecified subtypes) were interviewed; the commonest reasons for failure to be interviewed were insufficient information on the patient's age, sex or address, lack of histopathological confirmation or death before interview. An interview was completed with 724 controls. For woodworkers, a 'nonsignificant risk ratio under 2.0' was reported, with 28 exposed cases and 35 exposed controls. Occupational or private (more than three months) exposure to

Table 28. Case–control studies of lung cancer

Reference	Country	Sex	Cases/controls	Source of information on exposure	Exposure to which relative risk applies	OR/RR	95% CI or p	Comments
Exposure to wood dust								
Blot et al. (1982)	USA	M	321/434	Interview	Wood dust Wood dust	1.9 3.4		All lung cancer Small-cell lung cancer
Correa et al. (1984)	USA	MF	1338/1393	Interview	Wood dust	1.4	$p < 0.05$	Adjusted for smoking; white men
Bond et al. (1986)	USA	M	308/588	Company records	Any wood dust (15-year lag) High exposure	1.3 1.0	0.8–2.2 0.6–1.8	Nested study
Kjuus et al. (1986)	Norway	M	176/176	Interview	Woodworking Wood dust	0.7 0.5	0.2–2.3 $p < 0.05$	Matching by age; adjusted for cigarette smoking
Siemiatycki et al. (1986), Siemiatycki (1991)	Canada	M	857/1360	Interview	Wood dust, any exposure Wood dust, substantial exposure	1.2 1.3	[1.0–1.5] [0.9–1.8]	Adjusted for a number of confounders
Lerchen et al. (1987)	USA	M	333/499	Interview	Wood dust (ever)	0.8	0.3–1.7	Adjusted for age, ethnic group, smoking
Levin et al. (1988)	China	M	733/760	Interview	Wood dust (self-reported)	1.7	1.0–2.7	Category matching by age; adjusted for smoking and age
Occupational group								
Harrington et al. (1978)	USA	M	858/858	Death certificate	Wood and paper industry (usual job)	1.3	$p > 0.05$	Matching by age, year of death, race and residence
Esping & Axelson (1980)	Sweden	M	25/370	Death register	Woodworking Furniture maker Other woodworker	4.1 6.0 (crude) 2.3 (crude)	1.6–11	Rough adjustment for age

Table 28 (contd)

Occupational group (contd)

Reference	Country	Sex	Cases/ controls	Source of information on exposure	Exposure to which relative risk applies	OR/RR	95% CI or p	Comments
Milne et al. (1983)	USA	M	747/3130	Death certificate	Sawmills Furniture manufacture Cabinet-maker, furniture finisher (usual job)	0.8 4.2 1.0	$p > 0.05$ $p < 0.01$ $p > 0.05$	
Buiatti et al. (1985)	Italy	M	340/817	Interview	Woodworking (ever)	0.6	0.3–1.1	Adjusted for age, smoking, place of birth
Coggon et al. (1986)	United Kingdom	M	738/2204	Postal questionnaire	Woodworking (ever)	1.7	1.0–3.0	Adjusted for age, residence, source of history (patient or relative), smoking
Damber & Larsson (1987)	Sweden	M	589/1035	Postal questionnaire	Carpenter (at least 1 year)	0.8 (dead controls) 0.7 (living controls)	0.5–1.3 0.5–1.2	Matching by sex, birth year. residence; adjusted for smoking
Schoenberg et al. (1987)	USA	M	763/900	Interview	Furniture and fixture worker (ever)	1.5	0.8–3.0	Adjusted for smoking
Ronco et al. (1988)	Italy	M	126/384	Interview	Furniture or cabinet-maker (ever)	2.8	0.9–8.4	Adjusted for age, smoking, other occupations
Hoar Zahm et al. (1989)	USA	M	4431/11 326	Cancer register (medical record)	Cabinet- and furniture makers (at time of diagnosis)	1.3 (all lung cancer) 2.0 (adenocarcinoma)	0.5–3.3 0.4–8.1	Adjusted for age, cigarette smoking
Kawachi et al. (1989)	New Zealand	M	4224/15 680	Cancer registry	Any woodworking Sawmill worker Cabinet-maker	1.3 1.8 1.2	1.2–1.6 1.2–2.5 0.8–1.8	Adjusted for age

Table 28 (contd)

Reference	Country	Sex	Cases/ controls	Source of information on exposure	Exposure to which relative risk applies	OR/RR	95% CI or p	Comments
Occupational group (contd)								
Schraub et al. (1989)	France	M	53/160	Interview	Wood dust	1.1	0.4–2.7	
Burns & Swanson (1991)	USA	MF	5935/3956	Interview	Woodworker Wood manufacture	1.1 2.3 (usual job)	0.7–1.8 0.8–6.4	Adjusted for age, race, sex, tobacco smoking
Jöckel et al. (1992)	Germany	M	146/292	Interview	Wood processing worker (ever)	0.7	0.3–1.9	Matching by sex and age; adjusted for tobacco smoking
Morabia et al. (1992)	USA	M	1793/3228	Interview	Carpenter and cabinet-maker (usual job)	1.4	p > 0.05	Adjusted for age and smoking
Kauppinen et al. (1993)	Finland	M	136/408	Company records	Wood dust	0.4	[0.2–1.3]	Adjusted for smoking; 10-year lagging of exposures
Notani et al. (1993)	India	M	246/212	Interview	Woodworker (ever)	3.2	0.9–12	Adjusted for age and smoking
Wu-Williams et al. (1993)	China	F	965/959	Interview	Timber processing (> 1 year) Wood dust (self-reported)	1.5 1.1	p > 0.05 0.8–1.7	In nonsmokers Adjusted for smoking, study area, age, education

OR, odds ratio; RR, relative risk; CI, confidence interval; M, male; F, female

wood dust was associated with a significantly increased risk ratio of 1.5 (95% CI, 1.0–2.1). Exposure to wood dust was associated more strongly with the low-grade subtype of non-Hodgkin's lymphoma (odds ratio, 2.0; $p < 0.05$) than with the high-grade type (odds ratio, 1.1). [The Working Group noted that the controls were insufficiently described; it is not clear whether the 'risk ratios' reported are crude or adjusted for sex and age; a source of possible bias is the fact that a large proportion of cases could not be interviewed.]

A case–control study of non-Hodgkin's lymphoma and occupation based on information from death certificates was reported by Schumacher and Delzell (1988), involving 501 male residents of North Carolina (United States), who died at 35–75 years of age from non-Hodgkin's lymphoma (ICD 8,9 200.0–200.9, 202.0–202.9) during 1968–70, 1975–77 and 1980–82. The controls were 569 male residents of North Carolina who died from causes other than cancer in the same periods and who were frequency matched to the cases by age, year of death and race. The usual occupation and industry were abstracted from death certificates and coded according to a system that combined industry and occupation. Work in paper and wood industries was associated with odds ratios of 0.79 [95% CI, 0.5–1.3] in whites and 1.3 [0.3–4.8] in blacks. For work in the furniture industry, which was a subcategory of paper and wood, the odds ratios were 0.74 [0.4–1.4] for whites and 1.9 [0.1–30] for blacks.

Franceschi et al. (1989) reported on a hospital-based case–control study of non-Hodgkin's lymphoma conducted in the province of Pordenone, north-eastern Italy. The cases were histologically confirmed non-Hodgkin's lymphomas diagnosed in 1984–88 in men and women under the age of 80, who had been admitted as in-patients or referred for follow-up at the out-patient clinics at Aviano Cancer Centre and general hospitals in the province. Of the 232 eligible case patients with lymphosarcoma and reticulosarcoma (ICD: 200) and other non-Hodgkin's lymphoma (ICD: 202), 18 died before they could be interviewed and there was no histological confirmation for six. None of the living case patients refused interview. The case series thus comprised 208 non-Hodgkin's lymphoma patients—110 men and 98 women. Controls were 401 interviewed in-patients at the same hospitals (215 men, 186 women), who were under the age of 80. Patients with admission diagnoses of malignant disorders, conditions related to alcohol and tobacco consumption, haematological, allergic and autoimmune conditions, and diseases that might have resulted in diet modifications, such as diseases of the respiratory and digestive tracts, cardiovascular disease and diabetes, were excluded from the control series. The interview covered sociodemographic characteristics, smoking, consumption of alcohol and coffee, intake of selected food items, medical history, vaccinations, tonsillectomy, medical radiation exposure, occupational history and self-reported exposure to 20 potentially carcinogenic agents, including wood dust. The job category 'wood and furniture workers', not further specified, was associated with an odds ratio of 0.66 (95% CI, 0.37–1.2), adjusted for age and sex. The number of subjects exposed to wood dust was too small for analysis.

Persson et al. (1989) examined associations between occupational exposures and Hodgkin's disease and non-Hodgkin's lymphoma in Sweden. The 106 non-Hodgkin's lymphoma cases (66 in men) were identified at the register of the Department of Oncology, Örebro Medical Centre Hospital in people who were alive in 1986 and whose cancers were diagnosed when they were 20–80 years of age, during 1984–86. Six eligible case patients were either unwilling to participate or could not be contacted. Population controls representing the catchment area of the hospital comprised 275 people aged 20–80, after replacement of 17% who were unable or unwilling to participate. 'Exposure' was defined as that occurring 5–45 years before diagnosis of a non-Hodgkin's lymphoma and lasting at least one year. An elevated risk was suggested in carpenters and cabinet-makers (crude odds ratio, 3.1). The odds ratio for exposure to fresh wood, as defined by employment as a sawmill worker, lumberjack or paper pulp worker, was 1.3. The odds ratio for work as a carpenter or cabinet-maker, adjusted for age at diagnosis, sex and farming, was 2.8 [95% CI, 0.9–8.5]; that for exposure to fresh wood was 1.0 [0.3–3.5].

A subsequent study (Persson et al., 1993) was conducted with similar methods in an adjacent region of Sweden (Östergötland, Jönköping, Kalmar). The cases were non-Hodgkin's lymphomas (ICD8: 200.0–2) diagnosed in 1975–84 among men who were alive in 1986, who were aged 20 years or more, who were living in the catchment area and were identified at the cancer registry covering the region. After 14 refusals and lack of diagnostic confirmation, 93 cases of non-Hodgkin's lymphoma remained. The 204 controls had also been used in other case–control studies (Flodin et al., 1986, 1987, 1988) and resided in the catchment area from which the cases were drawn. A postal questionnaire requested information about occupational exposures, medical care and leisure exposures. 'Exposure' was defined as occupation 5–45 years before the diagnosis of a non-Hodgkin's lymphoma and lasting at least one year. The crude odds ratio for non-Hodgkin's lymphoma in subjects exposed to fresh wood (sawmill workers, lumberjacks and paper pulp workers) was 2.6. The odds ratio for carpenters and cabinet-makers was 0.9 [95% CI, 0.3–2.4], adjusted for age at diagnosis and occupational confounders. Among the workers exposed to fresh wood, the subcategory 'lumberjacks' was associated with an adjusted odds ratio of 6.0 [95% CI, 0.8–44].

Reif et al. (1989) conducted a series of case–control studies on a number of cancer sites in New Zealand to assess associations between forestry work and cancer. All cancer cases in men aged 20 years and over were identified in 1980–84 at the New Zealand Cancer Registry. Occupation was recorded for 19 904 men (80%); 535 eligible cases of non-Hodgkin's lymphoma (ICD: 200,202) were available. Cases of other cancers formed the control group. The age-adjusted odds ratio for non-Hodgkin's lymphoma among foresters and loggers was 1.8 (95% CI, 0.85–4.0); the odds ratio for sawmill workers was 1.2 (0.43–3.2; four exposed cases).

Whittemore et al. (1989) conducted a case–control study on mycosis fungoides, a cutaneous T-cell lymphoma. They interviewed 174 people over 20 years of age in northern California, Los Angeles County and the Seattle–Puget Sound area (United States) in whom mycosis fungoides had been diagnosed in 1981–86 and identified at tumour registries and hospitals. The controls were 294 people selected by random-digit dialling who were also interviewed about potential risk factors for mycosis fungoides, and a lifetime employment history was taken. The response rate among the case subjects was 60% (23% were dead), and that among controls was 76%. Relatively fewer case subjects than controls reported previous employment in the paper and wood industry (relative risk, 0.5; $p = 0.02$). The risk was also reduced among people exposed to chromium and its salts, mercury and its salts, halogenated and aromatic hydrocarbons and uncured plastic. [The Working Group noted that employment in the wood and paper industry was insufficiently focused to provide a reasonable proxy for exposure to wood dust; in addition, the response rate for the cases was low.]

Scherr et al. (1992) conducted a hospital-based case–control study on occupational exposures and risk for non-Hodgkin's lymphoma. Interviews were conducted with 303 residents of the Boston, Massachusetts (United States), metropolitan area (80% participation rate) with confirmed non-Hodgkin's lymphoma diagnosed in 1980–82, or with their next-of-kin, and with 303 population controls (72% participation rate) matched by age, sex, town and precinct of residence. The case patients were identified at nine hospitals and the controls from town resident lists. Interviews were completed with 202 patients, 101 proxies of patients and 303 controls, and job histories were obtained. The odds ratio for non-Hodgkin's lymphoma associated with exposure to particles (dust, sawdust and fibres) was 1.4 (95% CI, 0.9–1.8); that for employment in the paper and wood industry was 1.7 (0.7–4.2). [The Working Group noted that the categories 'particles' and 'paper and wood industry' were remote proxies for exposure to wood dust; furthermore, a high proportion of next-of-kin of patients were interviewed]

A population-based case–control study of 622 white men with non-Hodgkin's lymphoma diagnosed in 1980–83 and 1245 white male population controls without haematopoietic or lymphatic malignancies in Iowa and Minnesota, United States, was conducted to examine associations between occupation and risk for non-Hodgkin's lymphoma (Blair et al., 1993). Case patients and controls who resided in the cities of St Paul, Duluth, Minneapolis and Rochester were excluded because agricultural exposures were the primary focus of the study. Case coverage was almost complete; interviews were conducted with 87% of the case subjects or their next-of-kin and with 77% of the controls. The interview covered sociodemographic characteristics, agricultural activities and exposures, exposures to chemicals in hobbies, residential history, medical history, familial history of cancer and occupational history. A job–exposure matrix was developed for job title–industry combinations and a number of exposures. Exposure to wood dust was associated with an odds ratio of 0.9 (95% CI, 0.7–1.2), adjusted for age, state, smoking, family history of malignant lymphoproliferative disease, agricultural exposure to pesticides, use of hair dyes and direct or surrogate responder. In a category of lower intensity of exposure to wood dust, the odds ratio was 0.9 (0.7–1.2). In the category of higher intensity, there were no cases and two controls.

(b) Hodgkin's disease

(i) Exposure to wood dust

In the study of Partanen et al. (1993), described on p. 146, four cases of Hodgkin's disease were notified. In comparison with 21 controls from the cohort individually matched to the cases, the unadjusted odds ratio for Hodgkin's lymphoma and exposure to wood dust, based on three exposed cases, was 2.1 (95% CI, 0.21–22).

(ii) Occupational group

Milham and Hesser (1967) reported an association between occupational exposure to wood and Hodgkin's disease in upstate New York, United States. They analysed the occupations of 1549 white men, aged 25 years or more, who had died of Hodgkin's disease during 1940–53 and 1957–64, and 1549 dead controls individually matched to the cases by age, sex, race (white only), residence and date of death. 'Exposure to wood' was defined as notification of a wood-related occupation (the commonest were carpenter and cabinet-maker) on the death certificate. The analysis revealed 69 pairs in which the case was exposed and the control unexposed, and 30 pairs in which the case was unexposed and the control exposed, yielding an odds ratio of [2.3] ($p < 0.001$). No other occupational group showed a significant excess.

Petersen and Milham (1974) evaluated the risk for Hodgkin's disease in occupations related to woodworking in Washington State, United States. The study had three phases: (i) a case–control study of deaths from Hodgkin's disease in 1950–71, with controls consisting of all residual, nonaccidental and nonviolent deaths, individually matched to each case by year of death, age at death and county of residence (707 matched pairs), in which wood-related occupations were ascertained from death certificates; (ii) a study of dead cases of Hodgkin's disease and dead controls during 1965–70 (158 matched pairs), in which occupational histories were obtained from interviews with relatives; and (iii) a proportionate mortality study of deaths from Hodgkin's disease in 1950–71. In the case–control study based on death certificates, there were 56 discordant pairs in which the case was in a woodworker and 32 discordant pairs in which the control was a woodworker, yielding an odds ratio of [1.8] ($p < 0.05$). In the study based on interviews, the frequencies of discordant pairs were 23 and 10, respectively ([odds ratio, 2.3] $p < 0.05$). The proportionate mortality ratio in woodworkers was 1.6 (56 deaths from Hodgkin's disease observed; $p < 0.001$).

Abramson et al. (1978) conducted a case–control study of Hodgkin's disease in Israel. All cases histologically diagnosed among Jewish residents of Israel in 1960–72 were eligible, giving 527 patients (454 definite, 37 probable and 36 possible). Jewish controls were drawn from the national population register, individually matched to the cases by sex, birth year, country of birth, father's region of birth (for subjects born in Israel) and year of immigration. Interviews were conducted with patients or proxies; proxy information was obtained for 68% of cases and 28% of controls. The response rate was 96%, and suitable controls were interviewed for 473 cases. The interview yielded information on occupation. Separate comparisons were made for the main histological subtypes, nodular sclerosis and mixed cellularity. Work with wood or trees (predominantly carpentry) was associated with a relative risk of 1.1 ($p > 0.05$). The risk for nodular sclerosis subtype was 0.6, but that for mixed cellularity was 5.2 ($p = 0.0005$).

Greene et al. (1978) identified 167 deaths from Hodgkin's disease among white men in North Carolina, United States, in 1956–74, and two controls for each case, matched by sex, race, county of death and age and year of death. A risk ratio of 1.4 (95% CI, 0.8–2.3) was associated with occupational exposure to wood and paper and a risk of 4.2 (1.4–13) with carpentry and lumbering.

Fonte et al. (1982) reported on a case–control study in Italy on Hodgkin's disease diagnosed in 207 men and 180 women admitted to the university hospital of Pavia during 1972–79. The controls were 441 men and 330 women admitted to an internal medical unit in Pavia in 1977–79. The occupations appearing in medical records were classified. Nine case patients worked in the wood industry, resulting in a relative risk of 7.2 (95% CI, 2.3–22). [The Working Group noted that the methods used were not well described.]

Bernard et al. (1987) reported on risk factors for Hodgkin's disease in a case–control study in the United Kingdom. All cases identified between October 1979 and December 1984 in hospitals in Yorkshire were eligible for inclusion; only those histopathologically confirmed were accepted. The controls represented in-patients with a variety of nonmalignant conditions and were matched to the cases by health district, sex and age in a ratio of 2:1. Case patients and controls were interviewed about past medical history, drug use, family medical history, hobbies, occupation, smoking and alcohol consumption. The study comprised 297 interviewed patients, who represented 70% of all histologically confirmed cases. For both woodworkers and contact with wood dust, a nonsignificant risk ratio 'under 2.0' was reported (woodworkers: 16 cases, 35 controls; wood dust: 24 cases, 46 controls). [The Working Group noted that a large proportion of case patients could not be interviewed and the results for woodworkers and exposure to wood dust were stated only as 'under 2.0'.]

Brownson and Reif (1988) evaluated occupational risks for Hodgkin's disease, mainly in farming, by identifying cases and controls through the cancer registry in Missouri (United States). Hodgkin's disease (ICD 9: 201) was diagnosed in 475 white male Missouri residents over 20 years of age in 1984 and 1985. The 1425 controls represented other cancers, excluding those of the oral cavity, pharynx, oesophagus, larynx, lung, bladder and prostate, and were individually matched 3:1 to the cases by age. Usual occupation and industry, as obtained from the routine records of the registry, were coded by the codes of the 1980 United States census. The registry also provided data on the smoking habits of the subjects. Carpenters had an increased risk for Hodgkin's disease: odds ratio, 3.1; 95% CI, 1.0–9.8, adjusted for age and smoking. [The Working Group noted that the control group included certain cancers that are potentially etiologically related to exposure to wood dust, which would have biased the odds ratio towards the null.]

In the study of Persson et al. (1989), described on p. 150, 54 cases of Hodgkin's disease were identified. No excess risk for Hodgkin's disease was associated with carpentry or cabinet-making; the odds ratio, adjusted for age at diagnosis, sex and farming, was 0.2 [95% CI, 0.01–2.8]. The adjusted ratio for exposure to fresh wood, as defined by employment as sawmill worker, lumberjack or paper pulp worker, was 0.4 [0.1–1.5].

In the study of Persson et al. (1993), reported on p. 151, 31 cases of Hodgkin's disease were identified. The odds ratio for Hodgkin's disease associated with the job title 'carpenters and cabinet-makers' was 0.2 (one exposed case, 25 exposed controls). The odds ratio for Hodgkin's disease in subjects exposed to fresh wood, adjusted for age at diagnosis and occupational confounders, was 3.8 [95% CI, 0.9–17].

(c) Multiple myeloma

(i) Exposure to wood dust

A hospital-based case–control study of multiple myeloma was reported in the United Kingdom (Cuzik & De Stavola, 1988). The cases were identified at major referral centres in six areas of England and Wales between 1978 and 1984. Two controls were sought for each case and matched by age and sex: one from the same hospital as the case and one from the list of the general practitioner of the case patient. Interviews were conducted with 409 case subjects, 399 hospital controls and 260 general practitioner controls to obtain occupational histories and information on exposures to chemicals and radiation, diseases, immunizations, family history, chronic infections and defects in immune regulation. The results were given as percentages of employed or exposed cases and controls: 1.5% of the case patients and 2.5% of the hospital controls had been employed in the production of furniture or upholstery [crude odds ratio, 0.6; 95% CI, 0.2–1.7]; 2.8% of the case patients and 1.3% in the controls had been exposed to wood dust for 1–10 years [crude odds ratio, 2.2; 0.8–6.5]; and 3.0% of the cases and 4.3% of the controls had been exposed for more than 10 years [crude odds ratio, 0.7; 0.3–1.5].

In 1982, more than 77 000 American Cancer Society members enrolled over 1.2 million friends, neighbours and relatives in a prospective mortality study, which included the completion of an initial questionnaire on medical, occupational and lifestyle factors and exposure history. In a case–control study of multiple myeloma based on these data (Boffetta et al., 1989), 282 people who had died during the first four years and for whom multiple myeloma was reported on the death certificate as the underlying or contributing cause of death were identified, after successful tracing of 98.5% of subjects and 84% coverage of death certificates. Four randomly selected controls were matched to the cases by sex, American Cancer Society division, year of birth and ethnic group, for a total of 1128. A further subdivision was made between incident cases ($n = 128$) and prevalent cases ($n = 154$) during the case ascertainment period, since a cancer detected before this period might have affected habits and occupations and the reporting of them. Self-reported exposure to wood dust was associated with an odds ratio of 1.2 (95% CI, 0.5–3.2; logistic model, with adjustment for age, sex, ethnic group, division of the American Cancer Society, education, history of diabetes, X-ray treatment, pesticide and herbicide exposure and farming) for incident cases.

A population-based case–control study (Heineman et al., 1992) was carried out of the occupational exposures of 1098 Danish men in whom multiple myeloma was diagnosed in 1970–84 and recorded at the Danish Cancer Registry, and of 4169 male population controls matched on birth year who were alive at the time of diagnosis of the case. Histological confirmation was available for 92% of the cases. Job histories from 1964 on were abstracted from the records of the nationwide Supplementary Pension Fund. A job–exposure matrix was constructed for 47 substances. The age-adjusted odds ratios related to exposure to wood dust were 1.2 (95% CI, 0.7–2.1) for furniture maker as the most recent occupation, 1.1 (0.4–2.5) for sawmill and other woodwork, 1.0 (0.6–1.4) for wood and wood products, 0.7 (0.3–1.3) for lumber and 1.6 (0.7–4.0) for miscellaneous wood products. Employment for fewer than five years in the wood and wood product industry was associated with an age-adjusted odds ratio of 1.1 (0.6–1.9); in those employed for five years or more, it was 0.9 (0.4–1.8). The results for women were reported in another article (Pottern et al., 1992). There were 1010 cases and 4040 matched controls. The industrial category 'wood/products' was associated with an age-adjusted odds ratio of 1.1 (95% CI, 0.3–3.4); the odds ratio for probable exposure to wood dust was 1.9 (0.4–8.1).

(ii) Occupational group

The risk for multiple myeloma among furniture workers was evaluated in a population-based case–control study in 20 counties of North Carolina, United States (Tollerud et al., 1985).

Listings of deaths during 1956–80 showed that 301 men were recorded with multiple myeloma or another immunoproliferative neoplasm (ICD9: 203.0,1,8; 238.6). These men were matched with one to three male controls by race, county of usual residence, age at death and year of death, to give 858 controls. The principal industry of employment, as recorded in the death certificate, was analysed: furniture manufacture was associated with an odds ratio of 1.3 ($p = 0.25$) and other woodworking with an odds ratio of 1.1 ($p = 0.69$). For furniture workers born before 1905 and who died before the age of 65, the odds ratio was 5.4 ($p = 0.05$). The odds ratios were lower for other combinations of birth year, age at death and race; none reached statistical significance. The authors noted that no information was available on time or duration of employment or specific occupational activities or exposures, including potential confounders. Underreporting of occupation in older individuals was a further concern.

Potential risk factors for multiple myeloma (ICD1965: 203.99) were evaluated in a case–control study of 131 patients and 431 referents, all of whom were alive (Flodin et al., 1987). The cases were identified from the registers of the cytological departments of three hospitals and the medical clinics of three further hospitals, all in central and south-eastern Sweden. They were diagnosed in 1973–83 and included patients who survived until 1981–83. About one-third of all incident cases in the catchment area were estimated to have been identified. Referents were a random sample from the population registers of the catchment area. A postal questionnaire sought information on potential risk factors; the response rate was 96% for cases and 80% for controls. At least one year of exposure to fresh wood (i.e. lumberjacks, paper pulp workers and sawmill workers), lagged by five years from the diagnosis of multiple myeloma, was associated with a crude rate ratio of 3.9 (95% CI 1.9–7.6; 17 cases). Another crude rate ratio was, however, presented for the same association: 2.6 [no CI given]. The rate ratio for exposure to fresh wood, adjusted for age, was 3.2 (1.5–6.5); after adjustment for age, exposure to exhaust fumes, creosote, concrete and brickwork, sulfonylurea, γ-radiation, ex-smoking and farming, the rate ratio (men only) was 2.6 (1.1–5.7). Working with dried timber was not associated with a significant excess risk.

In the study of Reif et al. (1989), in New Zealand, described on p. 151, the number of eligible cases of multiple myeloma (ICD: 203) was 295. The age-adjusted odds ratio associated with forestry and logging was 0.53 (95% CI, 0.08–3.7).

(d) Leukaemia

(i) Exposure to wood dust

In the study of Partanen et al. (1993) reported on p. 150, 12 cases of leukaemia were diagnosed, and 79 controls from the cohort were individually matched to the cases. Exposure to wood dust was not associated with an increased risk (crude odds ratio, 0.56; 95% CI, 0.2–2.2).

(ii) Occupational group

In a short communication, Burkart (1982) reported an excess risk for leukaemia among long-term workers in sawmills. Male cases (ICD9 clinically modified: 204–208) and noncancer controls were identified in four hospitals in Oregon, United States, during 1980 and administered an occupational questionnaire. With a 90% response rate, 26 leukaemia cases and 836 controls were evaluated for exposure in sawmills. The age-adjusted summary relative risk was

1.1 for < 10 years of exposure and 3.2 for > 10 years of exposure, 'with a Mantel–Haenszel summary χ^2 for dose–response significant at $p = 0.017$.' Industrial hygiene surveys in the plants indicated use of chlorophenols.

Oleske et al. (1985) reported on a case–control study of hairy-cell leukaemia, a rare, usually chronic form of leukaemia. In 1975–81, 53 patients with this cancer who were residents of Illinois and northern Indiana, United States, were identified at the Hairy Cell Tumor Registry and Treatment Center at the University of Chicago and through inquiries to 1100 haematologists, pathologists and medical oncologists. Interview responses were obtained from 36 patients and nine proxies. Three neighbourhood controls were matched to each case by age, sex and race. In the process of identifying eligible controls, 19% of those eligible refused interviews, so that 134 controls were interviewed. Working for a minimum of 20 h per week during at least six months in woodwork was associated in men with an odds ratio of 4.0 (95% CI, 0.90–18), after control for age, sex and race.

Pearce et al. (1986) reported on leukaemia among New Zealand agricultural workers. The cases were those classified as ICD: 204-208 and registered at the New Zealand Cancer Registry in 1979–83 among 546 men who were aged 20 years or more at registration. Four controls were matched to each case by age and year of registration; those with malignant lymphoma, multiple myeloma or soft-tissue sarcoma were excluded. The occupation of carpenter, as recorded on the cancer registration form (current or most recent job title), was associated with an odds ratio of 1.5 (95% CI, 1.0–2.3).

Potential risk factors for chronic lymphatic leukaemia were evaluated in a case–control study of 111 cases (ICD[1965]: 204.15) and 431 controls, all of whom were alive (Flodin et al., 1988). Cases were identified from the registers of the cytological departments of three hospitals and the medical clinics of two further hospitals, all in central and south-eastern Sweden. Most of the cases were diagnosed in 1975–84, but some as early as 1964, and included those in which the patient survived until 1981–83. Controls were a random sample from the population registers of the catchment area. A postal questionnaire sought information on potential risk factors, with a response rate of 91% for case patients and 85% for controls (non-responders were replaced by other controls). At least one year of exposure to fresh wood (i.e. lumberjacks, paper pulp workers and sawmill workers), lagged by five years from the diagnosis of chronic lymphatic leukaemia, was associated with a crude rate ratio of 3.2 (95% CI, 1.5–6.6; 13 exposed cases). The risk ratio for exposure to fresh wood (men only), adjusted for age, solvents, farming, exhausts and contact with horses, was 2.4 (1.0–5.0). Working with dried timber was not associated with a significant excess risk.

In the study of Reif et al. (1989), reported on p. 151, there were 534 eligible cases of leukaemia (ICD: 204–208). The age-adjusted odds ratio for all leukaemias associated with forestry and logging was 0.96 (95% CI, 0.36–2.6; four exposed cases); that for sawmill workers was 0.52 (0.13–2.1; two exposed cases).

Loomis and Savitz (1991) reported on a case–control study of occupation and leukaemia (ICD9: 204–208) among 5147 men in 16 states of the United States on the basis of information from death certificates. The controls were 51 470 men who had died of other causes, excluding brain cancer, during 1985–87. The results were given for usual occupation or industry, as abstracted from the death certificates. Woodworking was associated with an age- and race-adjusted odds ratio of 0.9 (95% CI, 0.7–1.0). The odds ratio for occupations in wood products industries was 0.7 (0.5–0.9) and that for carpenters was 0.9 (0.7–1.1).

Fincham *et al.* (1993) reported on a case–control study of cancers at several sites, using data from the Alberta Cancer Registry, Canada. On the basis of undocumented numbers of cases of leukaemia and controls with all other cancers, a crude odds ratio of 1.8 (95% CI, 1.2–2.8; 23 exposed cases) was reported for exposure to wood dust. [The Working Group noted the lack of detail in the description of the study and the crudeness of the statistical analysis.]

Studies on lymphohaematopoietic cancers are summarized in Table 29.

2.4.4 Cancers of the digestive tract

(a) Exposure to wood dust

Spiegelman and Wegman (1985) examined the relationship between occupational risk factors and colon and rectal cancer in a case–control study based on the Third National Cancer Survey, in which data were collected on all incident cancers occurring in 1969–71 in seven metropolitan areas and two states of the United States (a region containing 10.3% of the national population). A 10% random sample was interviewed to collect information on primary and secondary occupation and industry and duration of time in these jobs. The cases were colon or rectal cancers in 343 men and 208 women. The controls were 626 men and 1235 women with other cancers classified by the authors as not commonly associated with occupational exposures (cancers of the soft tissue, eye, brain, endocrine glands, breast, male and female reproductive tracts and lymphomas). Occupational exposure to wood was estimated from a job–exposure matrix based on the National Occupational Hazards Survey. For colon and rectal cancer combined, the odds ratios, adjusted for age, race, marital status, income, weight and nutritional scores, were 1.1 ($p = 0.69$) for men and 1.5 ($p = 0.04$) for women. For colon cancer alone, the odds ratios were 1.3 ($p = 0.24$) for men and 1.5 ($p = 0.07$) for women.

In the study of Siemiatycki (1991), described on p. 140, 251 cases of stomach cancer were identified. The odds ratios were 1.4 [95% CI, 1.0–1.9] for any exposure and 1.1 [0.7–1.7] for substantial exposure. For the 497 cases of colon cancer, the odds ratios were 1.0 [0.8–1.3] for any exposure and 0.9 [0.7–1.2] for substantial exposure; for the 257 cases of rectal cancer, the odds ratios were 1.0 [0.7–1.4] for any exposure and 1.3 [0.8–2.0] for substantial exposure.

Peters *et al.* (1989) performed a case–control study of colorectal cancer among 147 white men in Los Angeles county (United States) in whom colorectal adenocarcinoma was first diagnosed in 1974–82 when they were 25–44 years of age. A matched series of 147 neighbourhood controls of the same sex, race, date of birth and neighbourhood of residence were selected. Occupational and exposure histories were collected by interview. The odds ratio for exposure to wood dust was 3.6 (95% CI, 1.2–11) after adjustment for age and education. The results by sub-site were 2.1 (0.5–8.5) for the right side of the colon, 1.5 (0.3–6.6) for the transverse and descending colon, 3.6 (0.6–21) for the sigmoid colon and 9.4 (2.0–45) for the rectum.

Table 29. Case–control studies on lymphatic and haematopoietic cancers

Reference	Country	Sex	Cases/controls	Source of information on exposure	Exposure to which relative risk applies	OR/RR	95% CI or p	Comments
Non-Hodgkin's lymphoma								
(i) Exposure to wood dust								
Siemiatycki et al. (1986); Siemiatycki (1991)	Canada	M	117/1563 215/2357	Interview	Wood dust, substantial, >15 years' exposure. Wood dust, substantial (update)	0.5 1.0	0.2–0.9 [0.7–1.5]	Adjusted for a number of confounders
Partanen et al. (1993)	Finland	M	8/52	Company records	Wood dust	2.1	0.2–20	Nested study; matched by age and survival; low power
(ii) Occupational group								
Cartwright et al. (1988)	United Kingdom	MF	437/724	Interview	Wood dust > 3 months	1.5	1.0–2.1	Incomplete documentation. Selection and information biases possible
Schumacher & Delzell (1988)	USA	M	501/569	Death certificate	Furniture industry (usual job): Whites Blacks	0.7 1.9	0.4–1.4 0.1–30	
Franceschi et al. (1989)	Italy	MF	208/401	Interview	Wood and furniture worker	0.7	0.4–1.2	
Persson et al. (1989)	Sweden	MF	106/275	Postal questionnaire	Carpenter, cabinet-maker > 1 year; 5–45 years' latency Fresh wood	2.8 1.0	[0.9–8.5] [0.3–3.5]	Adjusted for age, sex, farming
Reif et al. (1989)	New Zealand	M	535/19 369	Cancer register	Sawmill worker	1.2	0.4–3.2	Four exposed cases
Whittemore et al. (1989)	USA	MF (Cases: mycosis fungoides)	174/294	Interview	Employment in paper and wood industry	0.5	p = 0.02	Paper and wood a remote proxy for exposure to wood dust. Low response in cases

Table 29 (contd)

Reference	Country	Sex	Cases/controls	Source of information on exposure	Exposure to which relative risk applies	OR/RR	95% CI or p	Comments
Non-Hodgkin's lymphoma (contd)								
Occupational group (contd)								
Scherr et al. (1992)	USA	MF	303/303	Interview	Employment in paper and wood industry	1.7	0.7–4.2	Paper and wood a remote proxy for wood dust
Blair et al. (1993)	USA	M	622/1245	Interview	Wood dust	0.9	0.7–1.2	Adjusted for age, state, smoking, family cancers, pesticides, hair dyes, responder
Persson et al. (1993)	Sweden	M	93/204	Postal questionnaire	Fresh wood: lumberjack	6.0	[0.8–44]	Adjusted for age, occupational confounders
Hodgkin's disease								
(i) Exposure to wood dust								
Partanen et al. (1993)	Finland	M	4/21	Company records	Wood dust	2.1	0.2–22	Nested study; matched by age and survival; low power
(ii) Occupational group								
Milham & Hesser (1967)	USA	M	1549/1549	Death certificate	Exposure to wood		$p < 0.001$	Discordant pairs 69/30
Petersen & Milham (1974)	USA	M	707/707	Death certificate	Woodworker	[1.8]	$p < 0.05$	Discordant pairs 56/32
Petersen & Milham (1974)	USA	M	158/158	Interview	Woodworker	[2.3]	$p < 0.05$	Discordant pairs 23/10
Abramson et al. (1978)	Israel	MF	506/473	Interview	Work with wood/trees	1.1 5.2	$p > 0.05$ $p < 0.0005$	All Hodgkin's disease Mixed cellularity subtype
Greene et al. (1978)	USA	M	167/334	Death certificate	Carpentry and lumbering	4.2	1.4–13	Matched by sex, race, county, age and year of death

Table 29 (contd)

Reference	Country	Sex	Cases/controls	Source of information on exposure	Exposure to which relative risk applies	OR/RR	95% CI or p	Comments
Hodgkin's disease (contd)								
Occupational group (contd)								
Fonte et al. (1982)	Italy	MF	387/771	Clinic charts	Wood industry	7.2	2.3–22	Methods not well described
Bernard et al. (1987)	United Kingdom	MF	297/489	Interview	Wood dust	'Under 2.0'	$p < 0.05$	Large non-response of cases; incomplete documentation of results for wood dust
Brownson & Reif (1988)	USA	M	475/1425	Cancer register	Carpenter	3.1	1.0–9.8	Adjusted for age and smoking
Persson et al. (1989)	Sweden	MF	54/275	Postal questionnaire	Fresh wood	0.4	[0.1–1.5]	Adjusted for age, sex and farming
Persson et al. (1993)	Sweden	M	31/204	Postal questionnaire	Fresh wood	3.8	[0.9–17]	Adjusted for age, occupational confounders
Multiple myeloma								
(i) Exposure to wood dust								
Cuzik & De Stavola (1988)	United Kingdom	MF	399/399	Interview	Wood dust > 10 years	[0.7]	[0.3–1.5]	Crude analysis
Boffetta et al. (1989)	USA	MF	128/512	Questionnaire	Wood dust (self-reported)	1.2	0.5–3.2	Adjusted for age, sex, ethnic group, residence and farming
Heineman et al. (1992)	Denmark	M	1098/4169	Pension fund records	Probable exposure to wood dust	0.8	0.6–1.2	
Potterm et al. (1992)	Denmark	F	1010/4040	Pension fund records	Probable exposure to wood dust	1.9	0.4–8.1	

Table 29 (contd)

Reference	Country	Sex	Cases/controls	Source of information on exposure	Exposure to which relative risk applies	OR/RR	95% CI or p	Comments
Multiple myeloma (contd)								
(ii) Occupational group								
Tollerud et al. (1985)	USA	M	301/858	Death certificate	Furniture manufacture Furniture manufacture, born < 1905 and died < 65 years of age	1.3 5.4	$p = 0.25$ $p = 0.05$	
Flodin et al. (1987)	Sweden	MF	131/431	Questionnaire	Fresh wood > 1 year; lagged by 5 years (men only)	2.6	1.1–5.7	Adjusted for a number of potential confounders. Possible selection of cases
Reif et al. (1989)	New Zealand	M	295/19 609	Cancer register	Forestry worker and logger	0.5	0.1–3.7	One exposed case
Leukaemia								
(i) Exposure to wood dust								
Partanen et al. (1993)	Finland	M	12/79	Company records	Wood dust	0.6 (crude)	0.2–2.2	Nested study; matched by age and survival; low power
(ii) Occupational group								
Burkart (1982)	USA	M	26/836	Questionnaire	Sawmill exposure > 10 years	3.2	$p = 0.017$ for dose–response	Adjusted for age
Oleske et al. (1985)	USA	M	35/104	Interview	Woodworking > 20 h/week, > 6 months	4.0	0.9–18	Hairy-cell leukaemia
Pearce et al. (1986)	New Zealand	M	546/2184	Cancer register	Carpenter	1.5	1.0–2.3	

Table 29 (contd)

Reference	Country	Sex	Cases/controls	Source of information on exposure	Exposure to which relative risk applies	OR/RR	95% CI or p	Comments
Leukaemia (contd)								
Occupational group (contd)								
Flodin et al. (1988)	Sweden	M	71/200	Postal questionnaire	Fresh wood	2.4	1.0–5.0	Adjusted for age, exposure to solvents, horses, farming, exhausts. Possible selection of cases
Reif et al. (1989)	New Zealand	M	534/19 370	Cancer register	Sawmill worker	0.5	0.1–2.1	Two exposed cases
Loomis & Savitz (1991)	USA	M	5147/51 470	Death certificates	Woodworker	0.9	0.7–1.0	Adjusted for age and race
Fincham et al. (1993)	Canada	NR	NR	Cancer register	Wood dust	1.8 (crude)	1.2–2.8	Incomplete documentation

OR, odds ratio; RR, relative risk; CI, confidence interval; M, male; F, female; NR, not reported

(b) Occupational group

Brownson *et al.* (1989) conducted a cancer registry-based case–control study of colon cancer involving white males with histologically confirmed colon cancer diagnosed between 1984 and 1987 who were reported to the Missouri Cancer Registry in the United States. Five controls for each case were randomly selected from among other white male cancer patients reported to the Registry. Data on occupation and industry from registry records (originally abstracted from medical records) were available for 1993 cases and 9965 controls. The odds ratio for carpenters was 0.9 (95% CI, 0.6–1.4) after adjustment for age. No other results related to exposure to wood were presented.

Fredriksson *et al.* (1989) reported the results of a case–control study of all people living in the region of Umeå aged 30–75 in whom adenocarcinoma of the colon had been diagnosed between 1980 and 1983 and reported to the Swedish Cancer Registry, who were alive at the time of data collection. For each case, two controls of similar age, sex and residence were identified from the National Population Register. A postal questionnaire, which included an occupational history, was completed by 312 case patients (156 men) and 623 controls (306 men). Decreased risks were observed for men who were previously employed as lumberers (odds ratio, 0.7; 95% CI, 0.4–1.0), pulp workers (0.7; 0.3–1.6) and sawmill workers (0.5; 0.3–0.9) after adjustment for age and physical activity.

In the study of Kawachi *et al.* (1989), reported on p. 115, 1014 cases of stomach cancer, 2043 of colon cancer, 1376 of rectal cancer, 184 of liver cancer, 120 of gall-bladder cancer and 571 of pancreatic cancer were available for analysis. The odds ratios for employment as a woodworker were 1.2 (95% CI, 0.9–1.6) for stomach cancer, 0.7 (0.5–0.9) for colon cancer and 1.1 (0.8–1.4) for rectal cancer.

In a further analysis, Dockerty *et al.* (1991) examined the risk for stomach cancer in a cancer registry-based case–control study in New Zealand. The study base and methods were the same as those described by Kawachi *et al.* (1989). There were 1016 men with stomach cancer available for analysis. The 19 042 controls consisted of registrants with cancer at all other sites. The only information available on exposure was the current or most recent occupation in the register. After adjustment for age, socioeconomic level, ethnic group and smoking, excess risks were observed for foresters and loggers (odds ratio, 1.8; 95% CI, 1.0–3.3) and cabinet-makers (1.4; 0.7–2.8), while decreased risks were observed for wood preparation and pulp and paper workers (0.8; 0.4–1.7) and carpenters (0.8; 0.5–1.2).

González *et al.* (1991) examined the association between occupation and gastric cancer in Spain. The cases were gastric adenocarcinomas diagnosed between 1987 and 1989 at 15 hospitals in Barcelona province, Zaragoza city, Soria province, Lugo province and the north of La Coruña province. Controls, matched on age, sex and area of residence, were selected from among patients at the same hospitals, excluding those with respiratory or gastric cancer, chronic respiratory disease, diabetes or chronic diseases that require a special therapeutic diet. Occupational histories were collected for 354 cases (235 men) and 354 controls by interviewers who were unaware of the case or control status of the patients. Odds ratios were calculated by logistic regression in order to adjust for socioeconomic status and diet. Relative risks were calculated for people ever employed in forestry (odds ratio, 1.0; 95% CI, 0.3–3.6), wood and paper production (0.5; 0.2–1.7) and furniture and wood manufacture (1.8; 0.5–6.9). The odds ratio for employment in any job with exposure to wood dust was 1.0 (0.4–2.3).

Arbman *et al.* (1993) performed a case–control study of colon and rectal cancer in Sweden, among patients under the age of 75 with histologically confirmed adenocarcinoma of the colon or rectum, who were identified in hospitals in the county of Östergötland in south-eastern Sweden. Two control groups were selected: hospital patients with hernias and anal disorders and a random sample of the general population. A questionnaire, which included information on occupational history, was completed by 98 patients (51 men) with colon cancer, 79 (48 men) with rectal cancer, 371 (309 men) hospital controls and 430 (203 men) general population controls. The odds ratios for men employed as carpenters were 0.5 [95% CI, 0.1–2.7] for colon cancer and 0.9 [0.3–3.2] for rectal cancer. The odds ratios for men employed as forestry workers were 0.9 [0.4–2.0] for colon cancer and 0.5 [0.2–1.5] for rectal cancer. The odds ratios for men employed as sawmill workers were 1.2 [0.4–3.3] for colon cancer and 0.4 [0.1–1.9] for rectal cancer. The prevalence of exposure was very low among women.

Studies on cancers of the digestive tract are summarized in Table 30.

3. Studies of Cancer in Experimental Animals[1]

3.1 Inhalation

3.1.1 Rat

Sixteen female Sprague-Dawley rats, 11 weeks of age, were exposed to untreated beech wood dust (approximately 70% of the dust particles with a maximal diameter of about 10 μm and 10–20% of the particles with a diameter of about ≤ 5 μm) at 25 mg/m^3 for 6 h per day on five days per week for 104 weeks. Surviving animals were killed and autopsied; only the nasal cavities and respiratory tract were examined histologically. No respiratory tract tumour and no squamous metaplastic or dysplastic lesions were found among the 15 surviving animals. About 50% of the animals were reported to have tumours outside the respiratory tract, but the incidence was said not to differ from that in untreated controls (Holmström *et al.*, 1989a). [The Working Group noted the small number of animals and the inadequate reporting of the tumours outside the respiratory tract.]

Fifteen female Wistar rats, four weeks of age, were exposed to beech wood dust (mass median aerodynamic diameter, 7.2 μm; geometric standard deviation, 2.2 μm) at 15.3 ± 13.1 mg/m^3 for 6 h per day on five days per week for six months and were observed for up to 18 months, when survivors were killed. At autopsy, the animals were examined grossly, and

---

[1]The Working Group was aware of studies in progress in which rats are exposed to oak wood dust by inhalation (IARC, 1994b).

Table 30. Community-based case-control studies of cancer of the digestive tract

Reference	Country	Sex	Cases/controls	Source of information on exposure	Exposure to which relative risk applies	OR/RR	95% CI or p	Comments
Stomach cancer								
(i) Exposure to wood dust								
Siemiatycki et al. (1986); Siemiatycki (1991)	Canada	M	156/1524 251/2397	Interviews evaluated by panel of chemists and industrial hygienists	'Substantial' exposure < 16 years ≥ 16 years Update	1.2 1.9 1.1	[0.6–2.6] [1.0–3.7] [0.7–1.7]	
(ii) Occupational group								
Dockerty et al. (1991)	New Zealand	M	1016/19 042	Tumour register	Foresters and loggers Cabinet-makers Wood preparation, pulp and paper workers Carpenters	1.8 1.4 0.8 0.8	1.0–3.3 0.7–2.8 0.4–1.7 0.5–1.2	Current or most recent occupation; adjusted for age, socioeconomic level and smoking
Kawachi et al. (1989)	New Zealand	M	1014/18 890	Tumour register	All woodworkers	1.2	0.9–1.6	Same population and methods as Dockerty et al. (1991)
González et al. (1991)	Spain	MF	354/354	Interviews	Any wood dust-exposed job Forestry Wood and paper production Furniture/wood manufacture	1.0 1.0 0.5 1.8	0.4–2.3 0.3–3.6 0.2–1.7 0.5–6.9	Cases were adenocarcinomas; controls matched on age, sex and residence; adjusted for diet and socioeconomic status
Colon cancer								
(i) Exposure to wood dust								
Spiegelman & Wegman (1985)	USA	MF	370/1861	Interviews Job-exposure matrix	Wood (men) Wood (women)	1.3 1.5	p = 0.24 p = 0.07	Adjusted for age
Siemiatycki (1991)	Canada	M	497/2056	Interviews evaluated by panel of chemists and industrial hygienists	Any exposure to wood dust 'Substantial' exposure	1.0 0.9	[0.8–1.2] [0.7–1.2]	

Table 30 (contd)

Reference	Country	Sex	Cases/controls	Source of information on exposure	Exposure to which relative risk applies	OR/RR	95% CI or p	Comments
Colon cancer (contd)								
Exposure to wood dust (contd)								
Peters et al. (1989)	USA	M	106/106	Interviews	Wood dust	2.1 1.5 3.6	0.5–8.5 0.3–3.6 0.6–21	Right side of colon Transverse and descending colon Sigmoid colon Subjects aged 25–44; adjusted for age and education
(ii) Occupational group								
Brownson et al. (1989)	USA	M	1993/9965	Tumour register	Carpenters	0.9	0.6–1.4	Adjusted for age
Fredriksson et al. (1989)	Sweden	M	156/306	Questionnaire	Lumbermen Pulp workers Sawmill workers	0.7 0.7 0.5	0.4–1.0 0.3–1.6 0.3–0.9	Adjusted for age and physical activity. Cases were adenocarcinomas.
Kawachi et al. (1989)	New Zealand	M	2043/17 861	Tumour register	All woodworkers	0.7	0.5–0.9	Same population and methods as Dockerty et al. (1991)
Arbman et al. (1993)	Sweden	M	51/512	Questionnaire	Carpenters Forestry workers Sawmill workers	0.5 0.9 1.2	[0.1–2.7] [0.4–2.0] [0.4–3.3]	Cases were adenocarcinomas
Rectal cancer								
(i) Exposure to wood dust								
Spiegelman & Wegman (1985)	USA	MF	551/1861	Interviews and a job–exposure matrix	Wood (men) Wood (women)	1.1 1.5	$p = 0.69$ $p = 0.04$	Colon and rectum combined; adjusted for age

Table 30 (contd)

Reference	Country	Sex	Cases/controls	Source of information on exposure	Exposure to which relative risk applies	OR/RR	95% CI or p	Comments
Rectal cancer (contd)								
Exposure to wood dust (contd)								
Siemiatycki (1991)	Canada	M	257/1299	Interviews evaluated by panel of chemists and industrial hygienists	Any exposure to wood dust 'Substantial' exposure	1.0 1.3	[0.7–1.4] [0.8–2.0]	
Peters *et al.* (1989)	USA	M	41/41	Interviews	Wood dust	9.4	2.0–45	Subjects aged 25–44; adjusted for age and education
(ii) Occupational group								
Kawachi *et al.* (1989)	New Zealand	M	1376/18 528	Tumour register	All woodworkers	1.1	0.8–1.4	Same population and methods as Dockerty *et al.* (1991)
Arbman *et al.* (1993)	Sweden	M	48/512	Questionnaire	Carpenters Forestry workers Sawmill workers	0.9 0.5 0.4	[0.3–3.2] [0.2–1.5] [0.1–1.9]	Cases were adenocarcinomas

OR, odds ratio; RR, relative risk; CI, confidence interval; M, male; F, female

lungs, nasal cavities, livers, spleens and kidneys were examined histologically. No respiratory tract tumour was found, and the incidence of tumours outside the respiratory tract did not differ significantly from that in untreated controls (Tanaka et al., 1991). [The Working Group noted the small number of animals in each group and the short exposure.]

3.1.2 Hamster

One group of 12 and one group of 24 male Syrian golden hamsters, 10 weeks of age, were exposed to beech wood dust (approximately 70% of the dust particles with a maximal diameter of about 10 µm and 10–20% of the particles with a diameter of about ≤ 5mm) at 15 and 30 mg/m^3 for 6 h per day on five days per week for 36 and 40 weeks, respectively. At these times, the survivors were killed and autopsied; nasal cavities, respiratory tracts, livers and kidneys were examined histologically. No respiratory tract tumour was reported in the 12 animals exposed to 15 mg/m^3, but 1/22 hamsters exposed to 30 mg/m^3 had an unclassifiable infiltrating malignant nasal tumour [not significant], and one other animal in this group had cuboidal metaplasia with mild dysplasia of the nasal epithelium (Wilhelmsson et al., 1985a,b). [The Working Group noted the short duration of the experiment.]

3.2 Intraperitoneal injection

Rat: In a preliminary report on a study of the carcinogenic activity of various fibrous and granular dusts, one group of female Wistar rats [initial number unspecified], eight weeks of age, received three weekly intraperitoneal injections of beech wood dust [size of dust particles unspecified and total dose ambiguously reported as 250 or 300 mg/animal] suspended in 0.9% sodium chloride solution (50 mg wood dust/ml). The surviving animals were killed 140 weeks after the first treatment [survival time not clearly specified]. At post-mortem examination of the abdominal cavity, no mesothelioma or sarcoma was found in the 52 rats examined (Pott et al., 1989). [The Working Group noted the limited reporting of the experimental details and that UICC chrysotile asbestos induced mesotheliomas when similarly administered in parallel groups.]

3.3 Administration with known carcinogens or other modifying factors

3.3.1 Rat

Four groups of 16 female Sprague-Dawley rats, 11 weeks old, were exposed by inhalation in chambers to (i) air (controls); (ii) about 25 mg/m^3 untreated beech wood dust (approximately 70% of the dust particles with a maximal diameter of about 10 mm and 10–20% of the particles with a diameter of about ≤ 5 mm); (iii) 12.4 ± 1.1 ppm [14.9 ± 1.3 mg/m^3] formaldehyde; or (iv) beech wood dust (as above) plus 12.7 ± 1.0 ppm [15.2 ± 1.2 mg/m^3] formaldehyde for 6 h per day on five days per week for 104 weeks. No difference in the mortality rates was reported between the groups at any time during the study [mortality rates and statistical test unspecified]. Surviving animals were killed and autopsied; only nasal cavities and respiratory tracts were examined histologically. One respiratory tract tumour, a nasal squamous-cell carcinoma, was found in the group exposed to formaldehyde alone. Pronounced squamous metaplasia, with or without keratinization of the nasal epithelium at the level of the naso- and maxillary turbinates, was found in 9/16 rats exposed to formaldehyde and in 8/15 rats exposed to wood dust plus formaldehyde. In addition, pronounced squamous metaplasia accompanied by dysplasia of the nasal epithelium occurred in 1/16 rats exposed to formaldehyde and in 4/15 rats exposed to wood dust plus formaldehyde. No such metaplastic or dysplastic nasal cavity lesions were encountered in any of the controls or in rats exposed to wood dust alone. Tumours outside the respiratory tract were reported to affect about 50% of the animals, but this incidence was said not to differ from that in controls (Holmström et al., 1989a). [The Working Group noted the small number of animals in each group and the inadequate reporting of tumours outside the respiratory tract.]

Two groups of 20 male Wistar rats, four weeks old, were exposed by inhalation in chambers to clean air (controls) or to beech wood dust (mass median aerodynamic diameter, 7.2 µm; geometric standard deviation, 2.2 µm) at 15.3 ± 13.1 mg/m^3 for 6 h per day on five days per week for six months (total exposure, 666 h). Immediately thereafter, five rats from each group were exposed to sidestream cigarette smoke (from 10 cigarettes per day) at 10.2 mg/m^3 [standard deviation unspecified] for 2 h per day on five days per week for one month (total exposure, 40 h). After clearance periods of 12 months for rats exposed only to wood dust and 11 months for rats exposed to wood dust plus cigarette smoke (i.e. 18 months after the start of the experiment), all rats, including the controls, were killed. At autopsy, animals were examined grossly, and lungs, nasal cavities, livers, spleens and kidneys were examined histologically. No intercurrent mortality occurred, and no tumours of the nose or of other segments of the respiratory tract were observed. The incidence of tumours outside the respiratory tract did not differ significantly from that in untreated controls (Tanaka et al., 1991). [The Working Group noted the small number of animals in each group and the relatively short treatment and observation periods.]

3.3.2 Hamster

Two groups of 12 male Syrian golden hamsters, about 10 weeks old, were exposed by inhalation in chambers to air (controls) or to untreated beech wood dust (about 70% of the particles had a maximal diameter of about 10 µm, and 10–20% of the particles had a diameter of about ≤ 5 µm) at 15 mg/m^3 (range, 10–20 mg/m^3) for 6 h per day on five days per week for 36 weeks. A further two groups of hamsters were treated similarly but were also given N-nitrosodiethylamine (NDEA) at 1.5 mg/animal by subcutaneous injection, weekly for the first 12 consecutive weeks. All survivors were killed at week 36. No tumours of the nose were observed in 12 hamsters exposed to wood dust alone. Tracheal squamous-cell papillomas occurred in 1/7 controls, 0/8 hamsters exposed to wood dust alone, 3/8 hamsters treated with NDEA alone and 4/8 hamsters exposed to wood dust and NDEA (Wilhelmsson et al., 1985a,b). [The Working Group noted the short duration of the experiment, the small numbers of animals in each group, the absence of data on mortality rates and the high losses of animals and tissues due to cannibalism.]

Two groups of 24 male Syrian golden hamsters, about 10 weeks old, were exposed by inhalation in chambers to air (controls) or to untreated beech wood dust (about 70% of the particles with a maximal diameter of about 10 µm, and 10–20% of the particles with a diameter of about ≤ 5 mm) at 30 mg/m^3 (range, 25–35 mg/m^3) for 6 h per day on five days per week for 40 weeks. A further two groups of hamsters were treated similarly but were also given 3.0 mg/animal NDEA by subcutaneous injection, weekly for the first 12 consecutive weeks. The survivors were killed at week 40. The death rate was very high in all groups and significantly higher in the two NDEA-treated groups than in the two other groups ($p < 0.05$; Fisher's exact test) [death rates not further specified]. One of the 22 hamsters exposed to wood dust alone had an unclassifiable, malignant, infiltrating nasal tumour, and another hamster in this group had focal cuboidal metaplasia with mild dysplasia of the nasal epithelium. No neoplastic, dysplastic or metaplastic changes occurred in the respiratory tracts of controls. The types and incidences of respiratory tract neoplasms and dysplasia in the groups exposed to NDEA and to wood dust plus NDEA were as follows: nasal tumours (papillomas and adenocarcinomas), 10/22 (46%) and 11/21 (52%); laryngeal and/or tracheal tumours (papillomas), 10/19 (53%) and 11/18 (61%); lung tumours (adenocarcinoma), 0/19 and 1/18 (6%); and nasal dysplasia, 8/18 (44%) and 4/17 (24%). The incidences of these respiratory tract lesions did not differ significantly between these two groups [Fisher exact test] (Wilhelmsson et al., 1985a,b).

3.4 Skin application of wood dust extracts

Mouse: Four groups of 70 young female NMRI mice [age unspecified], weighing 25–30 g, received skin applications of a mutagenic fraction of a methanol extract of dust from untreated, semi-dry beech wood in 30 µl acetone on a 1–2-cm^2 shaven area of the lower back twice a week for three months. The freshly prepared, weekly doses of the fraction were equivalent to 2.5, 5, 7.5 and 10 g wood dust per mouse. Five similar groups of mice served as controls: one was treated with acetone on the shaven skin, one was shaved only and one was neither shaved nor treated with acetone; two positive control groups were treated with 5 and 10 µg benzo[*a*]pyrene, respectively. All mice were observed until they died naturally or were killed to avoid severe suffering. The survival of treated mice was not significantly different from that of untreated mice ($p = 0.571$; Mann-Whitney U test). The positive controls and mice treated with the mutagenic wood dust extract developed precancerous skin lesions (epithelial hyperplasia and hyperkeratosis) and benign and malignant tumours of the skin and mammary glands just beneath the treated skin area (see Table 31). Comparison of the mice treated with wood dust with the negative controls was reported to show a significant overall carcinogenic effect ($p < 0.01$; χ^2 test) (Mohtashamipur et al., 1989b). [The Working Group noted that a significant trend is observed for skin tumours, whether or not the analysis includes the keratoacanthoma and the papillary cystadenoma. The trend test for mammary tumours is significant when mammary gland adenocarcinomas, the adenoacanthoma and the mixed mammary tumours are grouped, and it is marginally significant when only the adenocarcinomas and the adenoacanthoma are considered.]

Table 31. Results of application to the skin of mice of mutagenic fractions of a methanol extract of dust from untreated, semi-dry beech wood, with negative and positive controls

Tumour	Negative controls			Extract (g)				Benzo[a]pyrene (μg)	
	Untreated ($n = 43$)	Shaven ($n = 44$)	Shaven, acetone-treated ($n = 42$)	2.5 ($n = 43$)	5.0 ($n = 50$)	7.5 ($n = 46$)	10.0 ($n = 49$)	5 ($n = 43$)	10 ($n = 42$)
Skin squamous-cell carcinoma	-	-	-	1	-	-	1^a	1	15
Skin squamous-cell papilloma	-	-	-	1	1	6	5^a	2	5
Skin keratoacanthoma	-	-	-	-	-	1	-	-	2
Skin papillary cystadenoma	-	-	-	-	1	-	-	-	-
Sebaceous gland adenoma	-	-	-	-	-	-	-	2	-
Mammary gland adenocarcinoma	-	-	-	-	4	3	$2^{b,c}$	1	1
Mammary gland adenoacanthoma	-	-	-	-	-	-	$1^{b,c}$	-	-
Mammary gland mixed tumours	-	-	-	-	-	-	2^b	-	-
Fibrosarcoma	-	-	-	-	-	1	-	-	-
Haemangiosarcoma	-	-	-	-	1	-	-	-	-
Neurofibrosarcoma	-	-	-	-	1	-	-	-	-
Lymphoma	-	-	-	-	-	-	1	-	-
Anaplastic carcinoma	-	-	-	-	1	-	-	-	-
Precancerous skin lesions	-	1	2	2	4	8	6	13	18

Adapted from Mohtashamipur et al. (1989b); numbers of animals given are effective numbers

[a] $p < 0.01$; Cochran-Armitage test for trend] where comparisons are made for 0 (acetone-treated controls), 2.5, 5.0, 7.5 and 10 g extract groups, including both squamous-cell carcinomas and papillomas, or papillomas alone

[b] $p < 0.02$; Cochran-Armitage test for trend if included in the analysis] where comparisons are made for 0, 2.5, 5.0, 7.5 and 10 g extract groups, including mammary gland adenocarcinoma, adenoacanthoma and mixed mammary gland tumours

[c] $p < 0.06$; Cochran-Armitage test for trend] where comparisons are made for 0, 2.5, 5.0, 7.5 and 10 g extract groups and only mammary gland adenocarcinoma and adenoacanthoma are considered

3.5 Experimental data on wood shavings

It has been suggested in several studies that cedar wood shavings, used as bedding for animals, are implicated in the prominent differences in the incidences of spontaneous liver and mammary tumours in mice, mainly of the C3H strain, maintained in different laboratories (Sabine et al., 1973; Sabine, 1975). Others (Heston, 1975) have attributed these variations in incidence to different conditions of animal maintenance, such as food consumption, infestation with ectoparasites and general condition of health, rather than to use of cedar shavings as bedding. Additional attempts to demonstrate carcinogenic properties of cedar shavings used as bedding material for mice of the C3H (Vlahakis, 1977) and SWJ/Jac (Jacobs & Dieter, 1978) strains were not successful. In none of these studies were there control groups not exposed to cedar shavings.

4. Other Data Relevant to an Evaluation of Carcinogenicity and its Mechanisms

4.1 Deposition and clearance

4.1.1 Humans

No studies of the deposition of wood dust in human airways were available to the Working Group. Particle deposition in the airways has been the object of several studies (for reviews, see Brain & Valberg, 1979; Warheit, 1989). Large particles (> 10 μm) are almost entirely deposited in the nose; the deposition of smaller particles depends on size but also on flow rates and type of breathing (mouth or nose); there is also inter-individual variation (Technical Committee of the Inhalation Specialty Section, Society of Toxicology, 1987). Particles deposited in the nasal airways are removed by mucociliary transport (for reviews, see Proctor, 1982; Warheit, 1989).

4.1.2 Experimental systems

No data on the deposition or clearance of wood dusts in animals were available to the Working Group.

4.2 Toxic effects

4.2.1 Humans

(a) Effects on the nose

In a cross-sectional study in eight furniture factories in Denmark, 68 workers were exposed to total dust at concentrations > 5 mg/m^3 in 63% of the measurements (Solgaard & Andersen, 1975; Andersen et al., 1977). The workers were exposed to a variety of hardwoods, including teak, and to pine and composites, including chipboard and Masonite. Analysis of particle size showed that 33% of the particles were < 5 μm. These workers had significantly lower nasal mucociliary transport rates than a group not exposed to dust; there was also a concentration-dependent decrease in the rates of the exposed workers: mucostasis was found in 63% of workers exposed to an average of 25.5 mg/m^3 and in 11% of those exposed to 2.2 mg/m^3 (mean concentration). Of nine subjects with mucostasis re-examined after 48 h with no exposure to wood, three still had mucostasis, while the six others had clearance rates within normal limits. Middle-ear inflammation and common colds were more frequently reported by people exposed to concentrations > 5 mg/m^3 than among those exposed to lower levels.

The nasal mucociliary transport rate was investigated in nine woodworkers, 48–66 years of age, with 6–27 years of employment in the furniture industry in England (Black et al., 1974). They had slower rates than 12 people not exposed to wood dust. Only the worker with the shortest length of employment (six years) had a clearance rate within normal limits. Three workers had complete stasis. The results of cytological examination of nasal smears were reported only for the exposed workers: squamous cells were found in four workers, cuboidal cells in one and 'less mature basal cells' in another.

Boysen and Solberg (1982) studied 103 workers in five Norwegian furniture factories in a cross-sectional study. The subjects constituted about 60% of workers who had been employed for at least 16 years. Ten retired workers and 54 people without nasal disease or an occupation associated with nasal cancer, who were not employed in woodworking industries, were examined. Nasal biopsy samples taken from the middle turbinate showed metaplastic squamous epithelium in 40% of the furniture workers and 17% of controls; the corresponding figures for dysplasia were 12 and 2%, respectively. Mechanical processing of wood was associated with histological changes of the nasal mucosa. Dysplasia occurred in four of 15 furniture workers with exposure mainly to birch, spruce and pine and in nine of 84 exposed mainly to hardwoods.

Nasal biopsy samples taken from the middle turbinate of 44 workers who had been exposed for 10–43 years to softwood dust but not to hardwood dust showed more changes than biopsy samples taken from age-matched men without nasal disease or an occupation associated with nasal cancer (mean score, 2.0 versus 1.4; $p < 0.05$) (Boysen et al., 1986). Four woodworkers and no control had dysplasia; these four men had been exposed for 20 years (one man) and more than 26 years (three men). Nasal symptoms were more frequent among the furniture workers (14% versus 4%; $p < 0.05$).

Biopsy samples were taken from the nasal mucosa of the middle turbinate, at least 5 mm behind the anterior curvature, of 45 randomly selected workers in five furniture factories and one parquet flooring factory and 17 hospital workers in Sweden (Wilhemsson & Lundh, 1984). The mean length of exposure was 15 years (range, 1–39 years). Metaplastic cuboidal epithelium was significantly more prevalent among the woodworkers (26/45 versus 4/17; $p < 0.05$), and columnar epithelium was significantly less frequent (34/45 versus 17/17; $p < 0.05$). The prevalence of metaplastic squamous epithelium was not significantly increased (9/45 versus 4/17), and that of goblet-cell hyperplasia was somewhat more frequent (10/45 versus 1/17).

Cuboidal metaplasia of the nasal mucosa was found in 19 of 22 cases of ethmoidal adenocarcinoma associated with exposure to wood dust in Sweden (Wilhemsson et al.,1985c). Histological examination of non-tumour nasal mucosa from 22 woodworkers with ethmoidal adenocarcinoma, who had been exposed to wood dust for an average of 38 years (range, 18–55 years), showed cuboidal metaplasia in 19; 16 also had dysplasia. A transitional zone with dysplastic cuboidal epithelium in continuity with the tumour was observed in 10 cases. Squamous metaplasia was also seen in five cases, but there were no cases of squamous dysplasia.

A cross-sectional study in Germany involved 149 male workers with at least 15 years' exposure to wood dust in different industries and 33 workers with no exposure to dust or chemicals (controls); people who had worked as farmers, welders or metal workers or were exposed to cement dust were excluded. Current exposure to wood dust varied between $< 1 \text{ mg/m}^3$ and $> 5 \text{ mg/m}^3$. Mucociliary clearance was not significantly different in workers exposed to unprocessed woods (oak, beech and softwood) and controls. Woodworkers with concomitant exposure to formaldehyde or chromium had decreased clearances ($p = 0.04$ and 0.01, respectively), and workers exposed to particle-board had slower mucociliary clearance. The findings in nasal biopsy samples taken from the middle turbinate were reported for various single cell types (columnar-cell hyperplasia, squamous-cell metaplasia, cuboid metaplasia) and mixed cell types (Wolf et al., 1994). [The Working Group noted the inadequate reporting of the histological classification and the high prevalence of squamous epithelial metaplasia in the control group. The Group analysed the data according to hyperplasia and metaplasia in single cell types and found no significant differences between woodworkers and controls (Table 32). The odds ratios for woodworkers exposed to softwood or hardwood, but no additives, were 2.2 (95% CI, 0.81–6.2) for cuboid metaplasia, 0.40 (0.16–1.0) for squamous-cell metaplasia and 1.3 (0.47–3.5) for columnar-cell hyperplasia. Cuboid metaplasia was commoner in workers exposed to hardwood (3.5; 1.1–12), softwood (3.1; 0.77–12) or particle-board (2.5; 0.70–8.8) without additives than in controls but was significant only for workers exposed to hardwood.]

In a cross-sectional study of workers in furniture factories in Sweden who were exposed to formaldehyde alone and to formaldehyde plus wood dust, nasal discomfort was more frequent than in clerks (Holmström & Wilhemsson, 1988). The mean combined exposure to wood dust was 1.7 mg/m^3 and that to formaldehyde was 0.25 mg/m^3; however, the prevalence of symptoms was similar in workers exposed to formaldehyde alone (mean concentration, 0.26 mg/m^3). Impaired mucociliary clearance in the nose was seen in 15% of the group exposed to wood dust plus formaldehyde, in 3% of controls and in 20% of those exposed to formaldehyde alone exposure ($p < 0.05$). Workers exposed to formaldehyde alone or to formaldehyde plus wood dust had significantly ($p < 0.01$) decreased sensitivity in an olfactory test in comparison with controls. Marked histological changes were seen in the nasal mucosa of 25% of people exposed only to formaldehyde (score, 2.2; $p < 0.05$), but the difference between those exposed to both formaldehyde and wood dust (64% ($p < 0.01$); score, 2.1) and the controls (53% ($p < 0.01$); score, 1.6) was not significant. No correlation was found between histological score and either duration or concentration of exposure (Holmström et al., 1989b).

A total of 676 workers in 50 Swedish furniture factories were classified according to exposure to wood dust as 'heavily/moderately' or 'slightly/non-exposed' (Wilhemsson & Drettner, 1984) [the details of the classification were not reported]. Nasal hypersecretion (20%

Table 32. Frequency of histological findings in nasal biopsy samples from German woodworkers according to exposure

Type of wood	Additives[a]	No.[b]	Columnar-cell hyperplasia (1)	Squamous metaplasia (2)	Cuboid metaplasia (3)	Columnar hyperplasia, squamous (4)	Columnar hyperplasia, cuboid (5)	Squamous metaplasia, cuboid (6)	Any columnar hyperplasia (1+4+5)[c]	Any squamous metaplasia (2+4+6)[c]	Any cuboid metaplasia (3+5+6)[c]	Normal	Dysplasia
Hardwood	No	31	6	1	5	3	1	6	10	10	12	9	3
Softwood	No	17	2	1	4	2	1	1	8	4	6	6	0
Particle-board	No	26	5	5	4	4	3	1	13	10	8	4	0
Softwood	Yes	19	3	9	3	2	1	0	6	11	4	1	1
Hardwood	Yes	21	7	4	2	2	1	1	10	7	4	4	1
Mixed	Yes	30	9	5	3	7	0	3	16	15	6	3	2
Controls	-	33	2	12	3	3	3	1	6	14	5	9	1

From Wolf et al. (1994)
[a] Glues, solvents, etc.
[b] The Working Group noted that the total numbers of men and findings were different, indicating that some people were not biopsied.
[c] Calculated by the Working Group

versus 12%; $p < 0.05$), obstruction (40% versus 30%; $p < 0.05$) and more than two common colds per year (21% versus 9%; $p < 0.05$) were reported more often in subjects with heavy/moderate exposure than in the other group.

A cross-sectional study was conducted of 101 woodworkers and 73 people not exposed to dust in Germany. The concentrations of dust were measured for each of the men [method of sampling was not reported]: 14 were exposed to < 5 mg/m^3, 15 to 5–9 mg/m^3, 36 to 10–19 mg/m^3 and 36 to ≥ 20 mg/m^3. An increased frequency of hyperplasia and reddening of the nasal mucosa was seen in the exposed workers (50–86% versus 7% in controls) (Ruppe, 1973). Radiographic signs of sinusitis were found in 25% of the woodworkers and 5% of controls. Cough, with or without phlegm (50% versus 11%), and conjunctivitis (15% versus 0%) were also reported more frequently among the exposed workers [significance values not reported].

In a cross-sectional study of the frequency of pulmonary and nasal symptoms in 168 woodworkers and 298 workers with no significant exposure to wood dust in furniture factories in South Australia (Pisaniello et al., 1992), the mean concentration of hardwood dust was 3.8 mg/m^3, and the mean concentration of softwood dust produced by machining particle-board and medium-density fibre-board was 3.3 mg/m^3. There was a significant association (odds ratio, 2.2; 95% CI, 1.2–4.2) between exposure to hardwood dust and two or more nasal symptoms, after adjustment for smoking and age.

In a cross-sectional study, Goldsmith and Shy (1988) examined 55 people exposed to hardwood dust in the furniture industry in the United States. The mean length of employment in this industry was 16.6 years, and the current concentration of dust was ≤ 2 mg/m^3. Frequent sneezing and eye irritation were commoner in these workers than in workers with no exposure to wood dust or finishes (prevalence odds ratios, 4.1 and 4.0; $p < 0.05$) in an analysis with adjustment for age, sex and smoking habits. Significant differences were reported for nasal obstruction (61% versus 21%), nasal discharge (41% versus 13%) and sneezing (77% versus 32%).

Symptoms in the upper and lower airways were reported more frequently among 44 randomly selected woodworkers, exposed to concentrations of 1.0–24.5 mg/m^3 dust, than among 38 office workers examined in a cross-sectional study in New Zealand (Norrish et al., 1992).

The effects of exposure to wood dust on the nose are summarized in Table 33.

(b) Effects on the lung

There are several case reports of asthma due to exposure to wood dust (for reviews, see Kadlec & Hanslian, 1983; Goldsmith & Shy, 1988). The asthmatic responses to western red cedar (Chan-Yeung, 1982, 1994) and eastern white cedar (Cartier et al., 1986) are elicited by plicatic acid.

Cough (odds ratio, 2.2; $p < 0.001$), dyspnoea (2.5; $p < 0.001$) and asthma (2.7; $p < 0.001$) were reported more frequently among 652 western red cedar mill workers than among 440 office workers in a cross-sectional study in Canada (Chan-Yeung et al., 1984). Impairment of pulmonary function, as measured by forced expiratory volume in 1 sec (FEV_1), forced vital capacity (FVC), forced mid-expiratory flow between 25 and 75% of FVC ($FEF_{25-75\%}$) and

Table 33. Effects (other than cancer) of exposure to wood dust on the nose

Study population		Geographical area	Industry	Wood type	Dust concentration in air (mg/m³)	Particle size, characteristic	Period of exposure (years)	Nasal effects		Reference
Exposed (age, years)	Controls (age, years)							Nasal region	Effect	
68 men (17–66)	66 men	Denmark, Aarhus county	Eight wood-working factories	Teak, oak, chipboard, palisander and other woods	> 5 in 63% of measurements	5–10 μm maximum	1–51; mean, 16		Mucostasis: 15% in controls, 38% in exposed, 63% in exposed with dust ≥ 10 mg/m³ ($n = 17$)	Solgaard & Andersen (1975); Andersen et al. (1977)
9 (48–66)	12 (31–69)	England, High Wycome area	One wood-working factory				6–27		Mucociliary clearance of polystyrene particles in controls, 6.8 (1.9–18.5) mm/min. Mucostasis in 7/9 exposed. In workers, nasal mucosa was normal columnar (3); normal + squamous cells (1); normal + cuboidal (1); normal + squamous metaplasia (3); normal + less mature 'basal' cells (1). Results for controls not given	Black et al. (1974)
103 active (32–69); 10 retired (68–81)	54 (35–79)	Norway, western	Five furniture factories	Birch, beech, oak, pine, mahogany, teak, chipboard (made of pine and spruce)			Active, 16–57; mean, 34; retired, 28–57; mean, 44	Anterior curvature of middle turbinate	Rhinoscopy: hyperplastic rhinitis: 5 controls, 37 workers ($p < 0.05$); mucosal polyps: 1 controls, 3 workers Histological score: controls, 1.5; all workers, 2.4 ($p < 0.05$); active workers, 2.4; retired workers, 2.9	Boysen & Solberg (1982)
44 (29–64)	37 men (35–66)	Norway	Six furniture factories	Exclusively softwood			10–43; mean, 24	Anterior curvature of middle turbinate	Histological score Controls Workers All 1.4 2.0 ($p < 0.05$) Age ≤ 44 1.2 1.6 Age 44–54 1.4 2.4 Age ≥ 55 1.6 1.9 Smokers 1.6 2.4 ($p < 0.05$) Non-smokers 1.3 1.6	Boysen et al. (1986)
45 (mean, 40)	17 (mean, 39)	Sweden, Småland county	Five furniture factories		0.3–5.1; mean, 2.0		1–39; mean, 15	Middle turbinate in widest nasal cavity	Columnar epithelium: 14/17 controls, 23/45* workers ciliated: 12/17 controls, 16/45 workers ($p < 0.05$); unciliated, 2/17 controls, 7/45 workers; Cuboidal epithelial metaplasia, 1/17 controls, 16/45 workers ($p < 0.05$) Squamous epithelial metaplasia, 2/17 controls; 6/45 workers	Wilhelmsson & Lundh (1984)

Table 33 (contd)

Study population		Geographical area	Industry	Wood type	Dust concentration in air (mg/m^3)	Particle size, characteristic	Period of exposure (years)	Nasal effects		Reference
Exposed (age, years)	Controls (age, years)							Nasal region	Effect	
22 men with ethmoidal adeno-carcinoma (57–86)		Sweden	14 furniture makers, 3 french polishers, 2 boat builders, 2 wood machinists, 1 woodwork teacher				18–55; mean, 38		19 cases of cuboidal metaplasia, 16 with dysplasia 10 cases of transitional zone with dysplastic cuboidal epithelium in continuity with the tumour 5 cases of squamous metaplasia	Wilhelmsson et al. (1985c)
100 exposed to wood dust and HCHO (mean, 40.5)	36 (mean, 39.7)	Sweden	Furniture workers	Particle-board	Wood, 1.65±1.06; HCHO, 0.25±0.05	82% < 5 μm	1–30; mean, 9		Nasal discomfort: 6 controls, 53 wood + HCHO, 45 HCHO; eye discomfort: 2 controls, 21 wood + HCHO, 17 HCHO; deep airway discomfort: 5 control, 39 wood + HCHO, 31 HCHO; frequent headache: 2 control, 17 wood + HCHO, 17 HCHO	Holmström & Wilhelmsson (1988)
70 exposed to HCHO (mean, 36.0)			Chemical plant		0.05–0.5		1–36; mean, 10.4			
89 exposed to wood dust and HCHO	32	Sweden	Furniture workers	Particle-board	Wood, 1.65±1.06; HCHO, 0.25±0.05	82% < 5 μm	1–30; mean, 9	Median or inferior aspect of middle turbinate	Nasal biopsy scores: control, 1.56; wood + HCHO, 2.07; HCHO, 2.16 ($p < 0.05$)	Holmström et al. (1989b)
62 exposed to HCHO			Chemical plant		0.05–0.5		1–36; mean, 10.4			
484 with heavy or moderate exposure	192 with light or no exposure	Sweden, Småland county	50 furniture factories		Mean, 2.0; range, 0.30–5.06		1–60; mean, 27		Nasal hypersecretion: 12% light or no exposure 20% moderate/heavy exposure Nasal obstruction: 30% light or no exposure 40% moderate/heavy exposure	Wilhelmsson & Drettner (1984)

Table 33 (contd)

Study population		Geographical area	Industry	Wood type	Dust concentration in air (mg/m^3)	Particle size, characteristic	Period of exposure (years)	Nasal effects		Reference
Exposed (age, years)	Controls (age, years)							Nasal region	Effect	
101 (18–65)	73 (18–65)	Germany			≤ 5–≥ 20				Sneezing: 0/73 controls; 7/14 < 5 mg/m^3; 11/15 5–9 mg/m^3; 32/36 10–19 mg/m^3; 30/36 ≥ 20 mg/m^3. Mucosal changes: 5/73 controls; 7/14 < 5 mg/m^3; 8/15 5–9 mg/m^3; 31/36 10–19 mg/m^3; 25/36 ≥ 20 mg/m^3	Ruppe (1973)
134 (mean, 33.6)	298 (mean, 40.1)	South Australia	15 furniture factories	Oak, teak, nyardoh, radiase pine, particle-board, fibre-board	Hardwood dust, 3.2; particle-board, fibre-board, 3.3				Two or more nasal symptoms (out of five): exposure to hardwood dust, odds ratio, 2.2 (1.2–4.2)	Pisaniello et al. (1992)
149 men (> 35)	33	Germany	Woodworkers	Oak, beech, softwood, particle-board	< 1–> 5		≥ 15	Middle turbinate	Mucociliary clearance longer in workers exposed to particle-board dust. Columnar-cell hyperplasia: all wood workers versus controls, odds ratio, 4.4 (p = 0.05). Squamous-cell metaplasia: woodworkers versus controls, odds ratio, 0.37 (p = 0.02). Cuboid metaplasia: all woodworkers versus controls, odds ratio, 2.9 (p = 0.3)	Wolf et al. (1993)

Table 33 (contd)

Study population		Geographical area	Industry	Wood type	Dust concentration in air (mg/m³)	Particle size, characteristic	Period of exposure (years)	Nasal effects		Reference
Exposed (age, years)	Controls (age, years)							Nasal region	Effect	
44 men (mean, 47.2) 11 women (mean, 41.3)	12 men (mean, 42.2) 4 women (mean, 44.5) and 7 men and 14 women in a finishing department	USA, North Carolina	One furniture factory	Hardwoods, fibre-board			Men, mean, 18.5; women, mean, 9		Frequent sneezing; odds ratio, 4.1 (1.1–15)	Goldsmith & Shy (1988)
44 men (mean, 33)	38 men (mean, 33)	New Zealand	11 furniture and joinery facilities	Rimu wood, kauri, tawa, medium-density fibre, Californian red wood	1–25.4				Nasal obstruction: 27/44 versus 8/38, $p < 0.01$ Nasal discharge: 12/44 versus 5/38, $p < 0.01$ Sneezing: 34/44 versus 12/38, $p < 0.01$	Norrish et al. (1992)

*$p < 0.05$

FEV_1/FVC, was significantly correlated ($p < 0.001$) with increasing length of employment in cedar mills. The odds ratios were adjusted for smoking, race and age.

Occupational asthma was diagnosed in 10 of 73 workers exposed to red cedar dust in a cross-sectional investigation in the United States, which also included 132 mill workers and 22 clerks and engineers not exposed to wood dust (Brooks et al., 1981). The mean concentration of total dust was 4.7 mg/m^3. Pulmonary diseases (chronic bronchitis, occupational asthma, chronic nonspecific airways disease and non-occupational asthma) were commoner among the workers than among controls (34% versus 16%) [p value not reported]. The prevalence of chronic bronchitis in workers exposed to a mixture of woods, mainly Douglas fir, West Coast hemlock and red alder, was similar to that of the workers exposed to western red cedar.

In the study of Norrish et al. (1992), described on p. 177, differences were reported for persistent cough in winter (30% versus 5%; $p < 0.01$) and work-related cough (32% versus 0%; $p < 0.01$). Five woodworkers were identified as having occupational asthma. The authors stated that adjustment for smoking did not alter the results.

In a cohort study based on census data on occupational title in Sweden, the rate of mortality from asthma was greater in woodworking machine operators (SMR, 2.3; 95% CI, 1.1–3.4), after adjustment for smoking (Torén et al., 1991).

Ávila (1972) reported on 23 Portuguese cork workers with bronchial asthma, all of whom gave positive responses in inhalation tests for immediate and late reactions to a skin prick with cork. A further 12 cork workers with diseases affecting mainly peripheral gas-exchange tissues all gave positive responses in skin tests for late (type III, arthris) reaction to cork; they showed diffuse, fine miliary mottling on chest radiographs, which disappeared within five weeks, except in a few cases where lesions attributable to fibrosis were reported. [The Working Group noted the lack of information on examination procedures].

The exposure of 334 workers to total dust was determined from job title and job location in a cross-sectional study in Canada (Vedal et al., 1986). The workers were exposed mainly to wood dust from western red cedar. In 78 samples, the total dust concentration ranged from undetectable to 6.0 mg/m^3, with a mean of 0.46 mg/m^3; 33 workers were considered to be exposed to > 1.0 mg/m^3. Spirometric measurements (FVC and FEV_1) gave lower values ($p < 0.05$) for 13 men exposed to concentrations > 2.0 mg/m^3; chronic cough, dyspnoea, persistent wheeze and asthma were not related to duration of work or dust concentration.

Al Zuhair et al. (1981) studied workers in two furniture factories in the United Kingdom. In the first factory, 53 workers in a sawmill and an assembly department were exposed to dust concentrations of 2.9 and 0.5 mg/m^3, respectively. In the second factory, 60 workers on a machine floor and in a cabinet shop were exposed to mean total dust concentrations of 1.4 and 8.3 mg/m^3, respectively. These workers had significantly decreased FEV_1 and FVC over the workshift period (0.08–0.12 L; $p < 0.001$), while there was no consistent decrease in lung function over the workshift period among workers in the first factory.

Pulmonary function (FVC, FEV_1, FEV_1/FVC and maximal mid-expiratory flow [MMEF; identical to $FEF_{25-75\%}$]) was determined in 1151 subjects exposed to maple or pine wood dust in a cross-sectional survey in the United States (Whitehead et al., 1981b). Suspended dust concentrations were measured in area samples, and a cumulative index of the dose was constructed for each person by multiplying the concentration in the job area by the working time. The workers were classified as having low (0–2 mg–years/m^3), medium (2–10 mg–years/m^3) or high (10 or more mg–years/m^3) exposure to wood dust. The authors classified the results of the spirometric tests as 'normal' or 'impaired' on the basis of external reference values and calculated the odds ratios between different categories of exposure. The ratio for FVC or FEV$_1$ was not significantly increased with increasing levels of exposure in the groups exposed to maple or pine wood, but FEV$_1$/FVC and MMEF were lower in people with high exposure. In a comparison of high and low exposure categories, the odds ratios for FEV$_1$/FVC and MMEF were 3.1 ($p = 0.01$) and 2.1 ($p = 0.02$) for workers exposed to maple dust and 4.0 ($p = 0.01$) and 2.5 ($p = 0.02$) for workers exposed to pine dust, after adjustment for smoking.

In a study of 145 nonsmoking furniture workers and 152 nonsmoking workers in a bottling firm with no exposure to dust in South Africa, cough (40.6% versus 23.7%; $p < 0.01$), phlegm (4.1% versus 10.5%; $p < 0.05$), dyspnoea (18.7% versus 5.7%; $p < 0.05$), wheezing (12.8% versus 4.8%; $p < 0.05$) and nasal symptoms (49.5% versus 18.7%; $p < 0.01$) were two to three times commoner in exposed than unexposed workers (Shamssain, 1992). Spirometric measurements were significantly lower for exposed men than for male controls (FVC: 3.64 versus 4.14 L, $p < 0.001$; FEV$_1$: 2.65 versus 3.20 L, $p < 0.001$; FEV$_1$/FVC: 73.2 versus 77.6%, $p < 0.01$; forced mid-expiratory flow between 25 and 75% of FVC [FMF$_{25-75\%}$]: 3.09 versus 3.68 L/s, $p < 0.01$; forced expiratory flow between the first 200 and 1200 ml of FVC [FEF$_{200-1200}$]: 4.94 versus 7.06 L/s, $p < 0.001$; peak expiratory flow [PEF], 6.14 versus 7.92 L/s, $p < 0.001$); there was no significant difference in these measurements between exposed and unexposed women. The frequency of an FEV$_1$/FVC below 70% was significantly higher among the woodworkers than the controls (30% versus 17%, $p < 0.01$), and the proportion was higher in men with 10–19 years of employment than in men with 1–9 years of employment (56% versus 27%, $p < 0.01$); 20% of the workers handled pine wood and 80% medium-density fibre-board. The mean total dust concentration in the factory was 3.8 mg/m^3.

In the study of Goldsmith and Shy (1988), described on p. 177, peak flow (but no other test of pulmonary function) was correlated with duration in jobs with exposure to wood dust.

The decrease in lung function over a work shift was greater in 50 carpenters and joiners than in 49 hospital workers (Holness *et al.*, 1985). The decreases in FVC were 2.4 ($p = 0.001$) and 0.15% ($p = 0.77$), respectively. The mean total dust concentration was 1.8 mg/m^3.

(c) Other effects

Exposure to wood may cause irritant dermatitis, contact urticaria and allergic contact dermatitis (for reviews, see Woods & Calnan, 1976; Hausen, 1986). The contact allergens in a number of woods have been identified, e.g. R-3,4-dimethoxydalbergione was found in a tropical hardwood, *Machaerium scleroxylum* (Beck *et al.*, 1984). Allergic conjunctivitis was reported in a worker exposed to spindle tree dust (Herold *et al.*, 1991).

Of 162 patients with a positive response in a skin prick test to one of 14 woods, 107 had no allergic symptoms (Oehling, 1963).

Inhalation fever and extrinsic allergic alveolitis have been observed in studies of workers exposed to wood contaminated with moulds (Emanuel et al., 1962; Belin, 1987; Dykewicz et al., 1988).

4.2.2 Experimental systems

A mouse hepatoma cell line, Hepa-1, was used to study cytotoxicity (effect on cell growth) and induction of enzymes (cytochrome P450IA1 and aldehyde dehydrogenase). The cells were exposed for 24 h to acetone extracts (final concentration of acetone, 0.5%) of bleached cellulose materials, softwoods (pine and a mixture of pine and spruce) and hardwoods (alder and aspen), all of which are used as bedding materials in cages for small laboratory animals. The softwood and alder extracts (final concentrations corresponding to 1.25–5 mg bedding material/ml cell culture medium) were more cytotoxic to the hepatoma cells than the aspen extract, whereas the bleached cellulose materials were found to be nontoxic at doses up to and including 20 mg/ml. Both softwood and hardwood extracts induced the activity of cytochrome P450IA1 and aldehyde dehydrogenase at concentrations which caused little toxicity (Törrönen et al., 1989).

In order to investigate the toxicity of plicatic and abietic acids, which are constituents of Western red cedar and pine woods, respectively, primary cultures of rat type II cells (isolated from Sprague-Dawley rats [sex unspecified]) and of human lung carcinoma cell line A549 were exposed to solutions of up to 1 mg/ml abietic acid and 5 mg/ml plicatic acid for 2–24 h. A time- and dose-dependent induction of cell lysis was seen with both cell types. Abietic acid was significantly more toxic (first observable effect after 2 h at 0.1 mg/ml) than plicatic acid (first effect after 4 h at 2.5 mg/ml). In studies with cultured tracheal explants from Sprague-Dawley rats [sex unspecified], both compounds produced dose-dependent desquamation of epithelial cells, abietic acid again having a higher toxic potential than plicatic acid (Ayars et al., 1989).

In order to assess the tumorigenic effect of the combination of beech wood dust and formaldehyde (see also section 3.1), groups of 16 female Sprague-Dawley rats (11 weeks old) were exposed by inhalation in whole-body exposure chambers to freshly prepared wood dust (70% with a longest dimension of about 10 μm, 10–20% ≤ 5 μm) at 25 mg/m^3 for 6 h per day on five days per week for 104 weeks, with or without formaldehyde. There was also an untreated control group. Animals were exposed in an inversed 24-h cycle, which ensured that they were as active as possible during exposure. Apart from neoplastic and preneoplastic lesions (see section 3.1), histopathological evaluation showed a greater prevalence of pulmonary emphysema in rats exposed to wood dust than in the control animals ($p < 0.05$), but no differences in mortality rates and no significant difference in the histological appearance of pulmonary epithelium were observed (Holmström, et al., 1989a).

In a study to assess the tumorigenic effects of a combination of beech wood dust and NDEA (see also section 3.1), groups of 19–23 male Syrian hamsters, weighing 90–120 g, received either wood dust or NDEA alone or the combination The animals were exposed by inhalation in whole-body exposure chambers to particles of fresh beech wood dust (30 mg/m^3) for 6 h per day on five days per week over a period of 40 weeks. In the group of hamsters exposed to wood dust alone, slight inflammatory reactions of the respiratory epithelium and submucosal stroma were detected, which were not observed in the respective control animals (Wilhemsson et al., 1985a,b).

Sixteen male guinea-pigs weighing about 300 g were given a single intratracheal instillation of 75 mg of sheesham or mango wood dust as an autoclaved suspension. Animals were killed 60 and 90 days after treatment. Treatment induced disintegration of giant cells, centrilobular emphysema and slight fibrosis in the lungs at both times (Bhattacharjee et al., 1979).

The enzyme induction activity of shavings from Eastern red cedar and oil of cedarwood was studied indirectly in groups of 6–18 C3H-A, CBA/J and Swiss albino mice [sex unspecified] as barbiturate sleeping time, the time between loss and restoration of the righting reflex after intraperitoneal injection of hexobarbital at 125 mg/kg bw). In five separate experiments in which mice were reared and/or housed with cedar bedding material for at least three weeks, a reduction in sleeping time ($p < 0.01$) was seen, which was attributed to the induction of enzymes responsible for hexobarbital oxidation (Sabine, 1975).

4.3 Genetic and related effects

4.3.1 Humans

Chromosomal aberrations in peripheral lymphocytes were studied in 13 male nonsmokers employed in three plywood factories in Finland, who were reported to be exposed to fumes emitted from heated wood. The controls were 15 male nonsmokers matched for age but not employed in wood industries. The frequency of chromatid breaks was 2.1% in the exposed group and 1.0% in the controls ($p < 0.01$) (Kurttio et al., 1993). [The Working Group noted that exposure to wood dust was not mentioned.]

4.3.2 Experimental systems

Extracts of certain woods prepared by a variety of methods (see Table 34 and section 1.3.2) gave weak positive or borderline effects for reverse mutation in *Salmonella typhimurium*. Unequivocal positive results have been obtained only with beech wood (Mohtashamipur et al., 1986); however, other woods have not been examined to the same extent. [The Working Group noted that wood contains constituents that can reduce the activity of mutagens such as benzo[*a*]-pyrene, aflatoxin B_1 and methylmethane sulfonate.] Chemically and bacterially degraded beech wood lignin significantly induced reverse mutation in *S. typhimurium* (Mohtashamipur & Norpoth, 1990), but fumes produced during the drying of birch and spruce wood were not mutagenic to *S. typhimurium* (Kurttio et al., 1990). [The Working Group noted the inappropriate correction for cell survival applied by the authors, which resulted in a different conclusion.]

Cyclohexane–ethanol extracts of beech, oak and particle-board increased the number of DNA single-strand breaks per fragile sites in rat hepatocytes *in vitro* (Schmezer et al., 1994).

Alcoholic extracts of beech wood increased the frequency of micronuclei in the crypts of the small intestine of mice treated by gavage and in the nasal epithelium of rats after topical application (Nelson et al., 1993).

Of several compounds isolated from wood, only quercetin and Δ^3-carene showed mutagenic activity (Table 35).

Table 34. Genetic and related effects of wood dusts

Test system	Result[a] Without exogenous metabolic system	Result[a] With exogenous metabolic system	Extraction medium	Reference
Ash				
SA0, *Salmonella typhimurium* TA100, reverse mutation	(+)	(+)	Methanol	McGregor (1982)
Beech				
SA0, *Salmonella typhimurium* TA100, reverse mutation	(+)	0	Methanol; cyclohexane/water	Brockmeier & Norpoth (1981)
SA0, *Salmonella typhimurium* TA100, reverse mutation	0	(+)	Methanol	Mohtashamipur et al. (1984)
SA0, *Salmonella typhimurium* TA100, reverse mutation	-	-	Acetone/water	Kubel et al. (1988)
SA0, *Salmonella typhimurium* TA100, reverse mutation	0	+	Methanol/ethyl acetate	Mohtashamipur et al. (1986)
SA0, *Salmonella typhimurium* TA100, reverse mutation	0	+	Acetone/water; lignin degradation	Mohtashamipur & Norpoth (1990)
SA9, *Salmonella typhimurium* TA98, reverse mutation	-	(+)	Methanol	McGregor (1982)
SA9, *Salmonella typhimurium* TA98, reverse mutation	-	-	Acetone/water	Kubel et al. (1988)
DIA, DNA strand breaks, rat hepatocytes *in vitro*	(+)		Cyclohexane/ethanol	Schmezer et al. (1994)
MVM, Micronucleus induction, mouse duodenal crypts *in vivo*	+		Methanol/ethyl acetate	Mohtashamipur & Norpoth (1989)
MVR, Micronucleus induction, rat nasal epithelial cells *in vivo*	+		Methanol/ethyl acetate	Nelson et al. (1993)
Birch				
SA0, *Salmonella typhimurium* TA100, reverse mutation	-	-	Drying fumes	Kurttio et al. (1990)
Chestnut				
SA0, *Salmonella typhimurium* TA100, reverse mutation	-	-	Acetone/water	Weissmann et al. (1989)
SA9, *Salmonella typhimurium* TA98, reverse mutation	-	-	Acetone/water	Weissmann et al. (1989)
Elm				
SA0, *Salmonella typhimurium* TA100, reverse mutation	-	-	Methanol	McGregor (1982)

Table 34 (contd)

Test system	Result[a]		Extraction medium	Reference
	Without exogenous metabolic system	With exogenous metabolic system		
Limba, obeche and walnut				
SA0, *Salmonella typhimurium* TA100, reverse mutation	(+)	0	Methanol; cyclohexane/water	Brockmeier & Norpoth (1981)
SA9, *Salmonella typhimurium* TA98, reverse mutation	(+)	0	Methanol; cyclohexane/water	Brockmeier & Norpoth (1981)
Mahogany				
SA0, *Salmonella typhimurium* TA100, reverse mutation	−	−	Methanol; cyclohexane/water	Brockmeier & Norpoth (1981)
SA0, *Salmonella typhimurium* TA100, reverse mutation	−	−	Methanol	McGregor (1982)
SA9, *Salmonella typhimurium* TA98, reverse mutation	−	−	Methanol; cyclohexane/water	Brockmeier & Norpoth (1981)
Oak				
SA0, *Salmonella typhimurium* TA100, reverse mutation	(+)	0	Methanol; cyclohexane/water	Brockmeier & Norpoth (1981)
SA0, *Salmonella typhimurium* TA100, reverse mutation	−	−	Methanol	McGregor (1982)
SA0, *Salmonella typhimurium* TA100, reverse mutation	−	−	Acetone/water	Weissmann et al. (1989)
SA9, *Salmonella typhimurium* TA98, reverse mutation	(+)	0	Methanol; cyclohexane/water	Brockmeier & Norpoth (1981)
SA9, *Salmonella typhimurium* TA98, reverse mutation	−	−	Acetone/water	Weissmann et al. (1989)
DIA, DNA strand breaks, rat hepatocytes *in vitro*	+		Cyclohexane/ethanol	Schmezer et al. (1994)
Spruce				
SA0, *Salmonella typhimurium* TA100, reverse mutation	−	−	Acetone/water	Kubel et al. (1988)
SA0, *Salmonella typhimurium* TA100, reverse mutation	−	−	Drying fumes	Kurttio et al. (1990)
SA2, *Salmonella typhimurium* TA102, reverse mutation	−	−	Drying fumes	Kurttio et al. (1990)
SA9, *Salmonella typhimurium* TA98, reverse mutation	−	−	Acetone/water	Kubel et al. (1988)

Table 34 (contd)

Test system	Result[a]		Extraction medium	Reference
	Without exogenous metabolic system	With exogenous metabolic system		
Spruce (contd)				
SA9, *Salmonella typhimurium* TA98, reverse mutation	–	–	Drying fumes	Kurttio *et al.* (1990)
DIA, DNA strand breaks, rat hepatocytes *in vitro*	–	–	Cyclohexane/ethanol	Schmezer *et al.* (1994)
Particle-board				
DIA, DNA strand breaks, rat hepatocytes *in vitro*	+	(+)	Cyclohexane/ethanol	Schmezer *et al.* (1994)

[a] +, considered to be positive; (+), considered to be weakly positive in an adequate study; –, considered to be negative; ?, considered to be inconclusive (variable responses in several experiments within an adequate study); 0, not tested

Table 35. Genetic and related effects of wood-related compounds

Test system	Result[a]		Reference
	Without exogenous metabolic system	With exogenous metabolic system	
Δ³-Carene			
SA0, *Salmonella typhimurium* TA100, reverse mutation	(+)	−	Kurttio *et al.* (1990)
SA2, *Salmonella typhimurium* TA102, reverse mutation	−	−	Kurttio *et al.* (1990)
SA9, *Salmonella typhimurium* TA98, reverse mutation	(+)	−	Kurttio *et al.* (1990)
Coniferyl alcohol			
SA0, *Salmonella typhimurium* TA100, reverse mutation	−	−	Mohtashamipur & Norpoth (1984)
SA9, *Salmonella typhimurium* TA98, reverse mutation	−	−	Mohtashamipur & Norpoth (1984)
Deoxypodophyllotoxin			
SA0, *Salmonella typhimurium* TA100, reverse mutation	−	−	Mohtashamipur & Norpoth (1984)
SA9, *Salmonella typhimurium* TA98, reverse mutation	−	−	Mohtashamipur & Norpoth (1984)
2,6-Dimethoxybenzoquinone			
SA0, *Salmonella typhimurium* TA100, reverse mutation	−	−	Mohtashamipur & Norpoth (1984)
SA9, *Salmonella typhimurium* TA98, reverse mutation	−	−	Mohtashamipur & Norpoth (1984)
Eugenol			
SA0, *Salmonella typhimurium* TA100, reverse mutation	−		IARC (1985a)
Quercetin			
SA0, *Salmonella typhimurium* TA100, reverse mutation	0	+	Bjeldanes & Chang (1977)
SA8, *Salmonella typhimurium* TA1538, reverse mutation	0	(+)	Bjeldanes & Chang (1977)
SA9, *Salmonella typhimurium* TA98, reverse mutation	0	+	Bjeldanes & Chang (1977)
Scopoletin			
SA0, *Salmonella typhimurium* TA100, reverse mutation	−	−	Mohtashamipur & Norpoth (1984)
SA9, *Salmonella typhimurium* TA98, reverse mutation	−	−	Mohtashamipur & Norpoth (1984)

Table 35 (contd)

Test system	Result[a]		Reference
	Without exogenous metabolic system	With exogenous metabolic system	
3,4,5-Trimethoxycinnamic acid			
SA0, *Salmonella typhimurium* TA100, reverse mutation	–	–	Mohtashamipur & Norpoth (1984)
SA9, *Salmonella typhimurium* TA98, reverse mutation	–	–	Mohtashamipur & Norpoth (1984)
Vanillic acid			
SA0, *Salmonella typhimurium* TA100, reverse mutation	–	–	Mohtashamipur & Norpoth (1984)
SA9, *Salmonella typhimurium* TA98, reverse mutation	–	–	Mohtashamipur & Norpoth (1984)

[a] +, considered to be positive; (+), considered to be weakly positive in an adequate study; –, considered to be negative; ?, considered to be inconclusive (variable responses in several experiments within an adequate study); 0, not tested

5. Summary of Data Reported and Evaluation

5.1 Exposure data

Wood is one of the world's most important renewable resources. At least 1700 million m^3 are harvested for industrial use each year. Wood dust, generated in the processing of wood for a wide range of uses, is a complex substance. Its composition varies considerably according to species of tree. Wood dust is composed mainly of cellulose, polyoses and lignin and a large and variable number of substances of lower relative molecular mass which may significantly affect the properties of the wood. These include non-polar organic extractives (fatty acids, resin acids, waxes, alcohols, terpenes, sterols, steryl esters and glycerols), polar organic extractives (tannins, flavonoids, quinones and lignans) and water-soluble extractives (carbohydrates, alkaloids, proteins and inorganic material).

Trees are characterized botanically as gymnosperms (principally conifers, generally referred to as softwoods) and angiosperms (principally deciduous trees, generally referred to as hardwoods). Roughly two-thirds of the wood used commercially worldwide belongs to the group of softwoods. Hardwoods tend to be somewhat more dense and have a higher content of polar extractives than softwoods.

It is estimated that at least two million people are routinely exposed occupationally to wood dust worldwide. Nonoccupational exposure also occurs. The highest exposures have generally been reported in wood furniture and cabinet manufacture, especially during machine sanding and similar operations (with wood dust levels frequently above 5 mg/m^3). Exposure levels above 1 mg/m^3 have also been measured in the finishing departments of plywood and particle-board mills, where wood is sawn and sanded, and in the workroom air of sawmills and planer mills near chippers, saws and planers. Exposure to wood dust also occurs among workers in joinery shops, window and door manufacture, wooden boat manufacture, installation and refinishing of wood floors, pattern and model making, pulp and paper manufacture, construction carpentry and logging. Measurements are generally available only since the 1970s, and exposures may have been higher in the past because of less efficient (or non-existent) local exhaust ventilation and other measures to control dust.

The wood species used in wood-related industries vary greatly by region and by type of product. Both hardwoods and softwoods (either domestically grown or imported) are used in furniture manufacture. Logging, sawmills and plywood and particle-board manufacture generally involve use of trees grown locally. Most of the wood dust (by mass) found in work environments has a mean aerodynamic diameter of more than 5 μm. Some investigators have reported that the dust generated in operations such as sanding and during the processing of hardwoods results in a higher proportion of smaller particle sizes, but the evidence is not consistent.

Within the furniture manufacturing industry, exposure may occur to solvents and formaldehyde in glues and surface coatings. Such exposures are usually greatest for workers with low or negligible exposure to wood dust and are infrequent or low for workers with high exposure to wood dust. The manufacture of plywood and particle-board may entail exposure to formaldehyde, solvents, phenol, wood preservatives and engine exhausts. Sawmill workers may also be exposed to wood preservatives and fungal spores. Exposures to chemicals in industries where other wood products are manufactured vary but are in many cases similar to those in the furniture manufacturing industry.

5.2 Human carcinogenicity data

The risk for cancer, and particularly cancer of the nasal cavities and paranasal sinuses, among woodworkers has been investigated in many epidemiological studies. Some of the studies provided specific information on cancer risk associated with exposure to wood dust, and those studies were given greatest weight in the evaluation.

Most of the available cohort and case–control studies of cancer of the nasal cavities and paranasal sinuses have shown increased risks associated with exposure to wood dust. These findings are supported by numerous case reports. Very high relative risks for adenocarcinoma at this site, associated with exposure to wood dust, have been observed in many countries, particularly in Europe. The lower risks observed in the studies in the United States may be due to differences in concentration or type of wood dust, but in one of these studies the more heavily exposed groups had significantly increased risks. A pooled analysis of 12 case–control studies revealed a clearly increasing risk with increasing estimated levels of exposure to wood dust, overall and in most individual studies. The excess appears to be attributable to wood dust *per se*, rather than to other exposures in the workplace, since the excess was observed in various countries during different periods and among different occupational groups, and because direct exposures to other chemicals do not produce relative risks of the magnitude associated with exposure to wood dust.

Adenocarcinoma of the nasal cavities and paranasal sinuses is clearly associated with exposure to hardwood dust; in several series of cases of adenocarcinoma from different countries, a high proportion of cases had been exposed to hardwood, and these findings were confirmed in several case–control studies as well. There were too few studies of any type to evaluate cancer risks attributable to exposure to softwood alone. In the few studies in which exposure was primarily to softwood, the risk for cancer of the nasal cavities and paranasal sinuses was elevated but considerably lower than that in studies of exposure to hardwood or to mixed wood types; furthermore, in the studies of exposure to softwood, exposure to hardwood could not clearly be ruled out. It is more difficult to attribute excess risk to any particular species of wood. The concentration of wood dust and the duration of exposure may also contribute to differences in the risks of workers exposed to different types of wood. These studies consistently indicate that occupational exposure to wood dust is causally related to adenocarcinoma of the nasal cavities and paranasal sinuses.

In studies of squamous-cell carcinoma of the nasal cavities and paranasal sinuses, smaller excesses were generally reported than for adenocarcinomas, and a pooled analysis of 12 case–control studies found no association with exposure to wood dust.

A number of case–control studies on nasopharyngeal cancer have reported an association with employment in wood-related occupations; however, confounding was not ruled out from these studies, and the largest study, from Denmark, in which exposure to wood dust was estimated, did not confirm the association. Case–control studies of laryngeal cancer consistently showed an association with exposure to wood dust or woodworking; however, cohort studies of woodworkers gave consistently negative results. Overall, these studies provide suggestive but inconclusive evidence for a causal role of occupational exposure to wood dust in cancers of the nasopharynx.

Studies of the association between exposure to wood dust and cancers of the oropharynx, hypopharynx, lung, lymphatic and haematopoietic systems, stomach, colon or rectum individually gave null or low risk estimates, gave inconsistent results across studies, and did not analyse exposure–response relationships. The evidence for an association between exposure to wood dust and Hodgkin's disease was somewhat more suggestive, in that some case–control studies showed moderately high risks, but these results were not substantiated by the results of cohort studies or some of the well-designed case–control studies. In view of the overall lack of consistent findings, there is no indication that occupational exposure to wood dust has a causal role in cancers of the oropharynx, hypopharynx, lung, lymphatic and haematopoietic systems, stomach, colon or rectum.

5.3 Animal carcinogenicity data

Dust from beech wood was tested for carcinogenicity by inhalation and for enhancement of carcinogenicity when administered with sidestream cigarette smoke or formaldehyde in two studies in rats, or with *N*-nitrosodiethylamine administered by subcutaneous injection in two studies in hamsters. The studies did not show any significant carcinogenic or co-carcinogenic potential of beech wood dust, but each of the studies suffered from various kinds of limitations and had some inadequacies in reporting of data.

The mutagenic fraction of a methanol extract of beech wood dust was tested for carcinogenicity by skin application in one study in mice. Although a significant, dose-dependent increase in the incidence of skin tumours and a marginally significant, dose-dependent increase in the incidence of mammary tumours were observed, these results cannot be used in an evaluation of the carcinogenicity of wood dust *per se*.

In a preliminary study, beech wood dust was tested for local carcinogenicity by intraperitoneal injection in rats; no peritoneal tumours were reported.

5.4 Other relevant data

General knowledge of particle size indicates that wood dust can be deposited in human upper and lower airways, the deposition pattern depending partly on particle size. Heavy exposure to wood dust may result in decreased mucociliary clearance and, sometimes, in muco-stasis. No data were available on clearance of wood dust from the lower airways.

Exposure to wood dust may cause cellular changes in the nasal epithelium. Increased frequencies of cuboidal metaplasia and dysplasia were found in some studies of workers exposed to dust from both hardwood and softwood. These changes can potentially progress to nasal carcinoma.

Impaired respiratory function and increased prevalences of pulmonary symptoms and asthma occur in workers exposed to wood dust, especially that from western red cedar.

There is little reliable information on the effects of wood dusts on the respiratory tract of rodents. One study *in vitro* showed that various wood dusts are cytotoxic and can induce drug metabolizing enzymes.

Constituents of beech that can be extracted with polar organic solvents are genotoxic, as demonstrated by the induction of point mutations in bacteria, DNA single-strand breaks in rat hepatocytes *in vitro* and micronuclei in rodent tissues *in vivo*. Extracts of oak wood showed similar activity, but fewer data were available. Extracts of spruce, the only softwood tested, gave consistently negative results.

5.5 Evaluation[1]

There is *sufficient evidence* in humans for the carcinogenicity of wood dust.

There is *inadequate evidence* in experimental animals for the carcinogenicity of wood dust.

Overall evaluation[2]

Wood dust *is carcinogenic to humans (Group 1)*.

6. References

Abramson, J.H., Pridan, H., Sacks, M.I., Avitzour, M. & Peritz, E. (1978) A case–control study of Hodgkin's disease in Israel. *J. natl Cancer Inst.*, **61**, 307–314

Acheson, E.D. (1967) Hodgkin's disease in woodworkers (Letter to the Editor). *Lancet*, **ii**, 988–989

Acheson, E.D. (1976) Nasal cancer in the furniture and boot and shoe manufacturing industries. *Prev. Med.*, **5**, 295–315

Acheson, E.D., Cowdell, R.H., Hadfield, E. & Macbeth, R.G. (1968) Nasal cancer in woodworkers in the furniture industry. *Br. med. J.*, **ii**, 587–596

Acheson, E.D., Cowdell, R.H. & Rang, E. (1972) Adenocarcinoma of the nasal cavity and sinuses in England and Wales. *Br. J. ind. Med.*, **29**, 21–30

Acheson, E.D., Pippard, E.C. & Winter, P.D. (1984) Mortality of English furniture makers. *Scand. J. Work Environ. Health*, **10**, 211–217

---

[1] For definitions of the italicized terms, see Preamble, pp. 23–27.
[2] This evaluation is based on the observation of a marked increase in the occurrence of cancers of the nasal cavities and paranasal sinuses among workers exposed predominantly to hardwood dusts.

Albracht, G., Bolm-Audorff, U., Grosse-Jäger, A., Manthey, U., Straub, U., Walter, A. & Weisskopf, V. (1989) Results of the governmental wood dust programme in the Federal State of Hessen. *Staub Reinhalt. Luft*, **49**, 381–384 (in German)

Al Zuhair, Y.S., Whitaker, C.J. & Cinkoati, F.F. (1981) Ventilatory function in workers exposed to tea and wood dust. *Br. J. ind. Med.*, **38**, 339–345

American Conference of Governmental Industrial Hygienists (1992) *Industrial Ventilation, A Manual of Recommended Practice*, 21st Ed., Cincinnati, OH

American Conference of Governmental Industrial Hygienists (1993) *1993–1994 Threshold Limit Values for Chemical Substances and Physical Agents and Biological Exposure Indices*, Cincinnati, OH, p. 35

American Society for Testing and Materials (1965) *1965 Book of ASTM Standards with Related Material*, Part 16, *Structural Sandwich Constructions; Wood; Adhesives*, Baltimore

Andersen, H.C. (1975) Exogenous causes of cancer of the nasal cavities. *Ugeskr. Laeg.*, **137**, 2567–2571 (in Danish)

Andersen, H.C., Solgaard, J. & Andersen, I. (1976) Nasal cancer and nasal mucus-transport rates in woodworkers. *Acta otolaryngol.*, **82**, 263–265

Andersen, H.C., Andersen, I. & Solgaard, J. (1977) Nasal cancers, symptoms and upper airway function in woodworkers. *Br. J. ind. Med.*, **34**, 201–207

Anttila, A., Jaakkola, J., Tossavainen, A. & Vainio, H. (1992) *Occupational Exposure to Chemical Agents in Finland* (Exposure at Work No. 34); Helsinki, Finnish Institute of Occupational Health and Finnish Work Environment Fund (in Finnish)

Arbman, G., Axelson, O., Fredriksson, M., Nilsson, E. & Sjödahl, R. (1993) Do occupational factors influence the risk of colon and rectal cancer in different ways? *Cancer*, **72**, 2543–2549

Armstrong, R.W., Armstrong, M.J., Yu, M.C. & Henderson, B.E. (1983) Salted fish and inhalants as risk factors for nasopharyngeal carcinoma in Malaysian Chinese. *Cancer Res.*, **43**, 2967–2970

Ávila, R. (1972) Some aspects of suberosis: respiratory disease in cork workers. *Bronches*, **22**, 121–128

Ayars, G.H., Altman, L.C., Frazier, C.E. & Chi, E.Y. ((1989) The toxicity of constituents of cedar and pine woods to pulmonary epithelium. *J. Allergy clin. Immunol.*, **83**, 610–618

Barefoot, A.C. & Hankins, F.W. (1982) *Identification of Modern and Tertiary Woods*, Oxford, Clarendon Press

Barthel, E. & Dietrich, M. (1989) Retrospective cohort study of cancer morbidity in furniture makers exposed to wood dust. *Z. ges. Hyg.*, **35**, 279–281

Battista, G., Cavallucci, F., Comba, P., Quercia, A., Vindigni, C. & Sartorelli, E. (1983) A case–referent study on nasal cancer and exposure to wood dust in the province of Siena, Italy. *Scand. J. Work Environ. Health*, **9**, 25–29

Bauch, J. (1975) *Dendrologie der Nadelbaume und ubrigen Gymnospermen* [Dendrology of conifers and other gymnosperms], Berlin, Walter de Gruyter

Beaulieu, H.J., Fidino, A.V., Arlington, K.L.B. & Buchan, R.M. (1980) A comparison of aerosol sampling techniques: 'open' versus 'closed-face' filter cassettes. *Am. ind. Hyg. Assoc. J.*, **41**, 758–765

Beck, M.H., Hausen, B.M. & Dave, V.K. (1984) Allergic contact dermatitis from *Machaerium scleroxylum* Tul. (Pao ferro) in a joinery shop. *Clin. exp. Dermatol.*, **9**, 159–166

Becker, N., Kuhn, G., Marschall, B., Angerer, R., Frentzel-Beyme, R. & Wahrendorf, J. (1992) Follow-up study among model and pattern makers in an automobile company in the Federal Republic of Germany. *J. occup. Med.*, **34**, 552–558

Beecher, C.W.W., Farnsworth, N.R. & Gyllenhaal, C. (1989) Pharmacologically active secondary metabolites from wood. In: Rowe J.W., ed., *Natural Products of Woody Plants II, Chemicals Extraneous to the Lignocellular Cell Wall*, Berlin, Springer Verlag, pp. 1059–1164

Belin, L. (1987) Sawmill alveolitis in Sweden. *Int. Arch. Allergy appl. Immunol.*, **82**, 440–443

Bernard, S.M., Cartwright, R.A., Darwin, C.M., Richards, I.D.G., Roberts, B., O'Brien, C. & Bird, C.C. (1987) Hodgkin's disease: case control epidemiological study in Yorkshire. *Br. J. Cancer*, **55**, 85–90

Bhattacharjee, J.W., Dogra, R.K.S., Lal, M.M. & Zaidi, S.H. (1979) Wood dust toxicity: in vivo and in vitro studies. *Environ. Res.*, **20**, 455–464

Bimbi, G., Battista, G., Belli, S., Berrino, F. & Comba, P. (1988) A case–control study of nasal tumors and occupational exposure. *Med. Lav.*, **79**, 280–287 (in Italian)

Bjeldanes, L.F. & Chang, G.W. (1977) Mutagenic activity of quercetin and related compounds. *Science*, **197**, 577–578

Björkman, A. (1956) Studies on finely divided wood. Part I. Extraction of lignin with neutral solvents. *Svensk Papperstid.*, **59**, 477–485

Black, A., Evans, J.C., Hadfield, E.H., MacBeth, R.G., Morgan, A. & Walsh, M. (1974) Impairment of nasal mucociliary clearance in woodworkers in the furniture industry. *Br. J. ind. Med.*, **31**, 10–17

Blair, A., Linos, A., Stewart, P.A., Burmeister, L.F., Gibson, R., Everett, G., Schuman, L. & Cantor, K.P. (1993) Evaluation of risks for non-Hodgkin's lymphoma by occupation and industry exposures from a case–control study. *Am. J. ind. Med.*, **23**, 301–312

Blot, W.J., Davies, J.E., Morris Brown, L., Nordwall, C.W., Buiatti, E., Ng, A. & Fraumeni, J.F., Jr (1982) Occupation and the high risk of lung cancer in northeast Florida. *Cancer*, **50**, 364–371

Boffetta, P., Stellman, S.D. & Garfinkel, L. (1989) A case–control study of multiple myeloma nested in the American Cancer Society prospective study. *Int. J. Cancer*, **43**, 554–559

Bolm-Audorff, U., Vogel, C. & Woitowitz, H.-J. (1989) Occupation and environmental risk factors of nasal and nasopharyngeal cancer. *Staub-Reinhalt. Luft*, **49**, 389–393 (in German)

Bolm-Audorff, U., Vogel, C. & Woitowitz, H.-J. (1990) Occupational and smoking as risk factors in nasal and nasopharyngeal cancer. In: Sakurai, H., Okazaki, I. & Omar, K., eds, *Occupational Epidemiology*, Amsterdam, Elsevier Science Publications, pp. 71–74

Bond, G.C., Flores, G.H., Shellenberger, R.J., Cartmill, J.B., Fishbeck, W.A. & Cook, R.R. (1986) Nested case–control study of lung cancer among chemical workers. *Am. J. Epidemiol.*, **124**, 53–66

Boysen, M. & Solberg, L. (1982) Changes in the nasal mucosa of furniture workers. A pilot study. *Scand. J. Work Environ. Health*, **8**, 273–282

Boysen, M., Voss, R. & Solberg, L.A. (1986) The nasal mucosa in softwood exposed furniture workers. *Acta otolaryngol.*, **101**, 501–508

Brain, J.D. & Valberg, P.A. (1979) Deposition of aerosol in the respiratory tract. *Am. Rev. respir. Dis.*, **120**, 1325–1373

Bravo, M.P., Espinosa, J. & del Rey Calero, J. (1990) Occupational risk factors for cancer of the larynx in Spain. *Neoplasma*, **37**, 477–481

Brinton, L.A., Blot, W.J., Stone, B.J. & Fraumeni, J.F., Jr (1977) A death certificate analysis of nasal cancer among furniture workers in North Carolina. *Cancer Res.*, 37, 3473–3474

Brinton, L.A., Blot, W.J., Becker, J.A., Winn, D.M., Browder, J.P., Farmer, J.C., Jr & Fraumeni, J.F., Jr (1984) A case–control study of cancers of the nasal cavity and paranasal sinuses. *Am. J. Epidemiol.*, 119, 896–906

Brockmeier, U. & Norpoth, K. (1981) Ames test studies to find respirable mutagens in different work places. In: Schaecke, G. & Stollenz, E., eds, *Epidemiologie Ansätze in Bereich der Arbeitsmedizin* [Epidemiology onset in the domain of occupational medicine], Stuttgart, Gentner, pp. 283–287 (in German)

Brooks, S.M., Edwards, J.J. & Henderson Edwards, F. (1981) An epidemiologic study of workers exposed to western red cedar and other wood dusts. *Chest*, 80 (Suppl. 1), 30S–32S

Bross, I.D.J., Viadana, E. & Houten, L. (1978) Occupational cancer in men exposed to dust and other environmental hazards. *Arch. environ. Health*, 33, 300–307

Brownson, R.C. & Reif, J.S. (1988) A cancer registry-based study of occupational risk for lymphoma, multiple myeloma and leukemia. *Int. J. Epidemiol.*, 17, 27–32

Brownson, R.C., Hoar Zahm, S., Chang, J.C. & Blair, A. (1989) Occupational risk of colon cancer, an analysis by anatomic subsite. *Am. J. Epidemiol.*, 130, 675–687

Buiatti, E., Kriebel, D., Geddes, M., Santucci, M. & Pucci, N. (1985) A case control study of lung cancer in Florence, Italy. I. Occupational risk factors. *J. Epidemiol. Community Health*, 39, 244–250

Burkart, J.A. (1982) Leukemia in hospital patients with occupational exposure to the sawmill industry (Letter to the Editor). *West. J. Med.*, 137, 440–441

Burns, P.B. & Swanson, G.M. (1991) The Occupational Cancer Incidence Surveillance Study (OCISS): risk of lung cancer by usual occupation and industry in the Detroit metropolitan area. *Am. J. ind. Med.*, 19, 655–671

Carlin, L.M., Colovos, G., Garland, D., Jamin, M.E., Klenck, M., Long, T.J. & Nealy, C.L. (1981) *Analytical Methods Evaluation and Validation. Arsenic, Nickel, Tungsten, Vanadium, Talc, Wood Dust* (Contract Report No. 210-79-0060/US NTIS PB83-155325), Cincinnati, OH, National Institute for Occupational Safety and Health

Cartier, A., Chan, H., Malo, J.-L., Pineau, L., Tse, K.S. & Chan-Yeung, M. (1986) Occupational asthma caused by eastern white cedar (*Thuja occidentalis*) with demonstration that plicatic acid is present in the wood dust and is the causal agent. *J. Allergy clin. Immunol.*, 77, 639–645

Cartwright, R.A., McKinney, P.A., O'Brien, C., Richards, I.D.G., Roberts, B., Lauder, I., Darwin, C.M., Bernard, S.M. & Bird, C.C. (1988) Non-Hodgkin's lymphoma: case control epidemiological study in Yorkshire. *Leukemia Res.*, 12, 81–88

Cecchi, F., Buiatti, E., Kriebel, D., Nastasi, L. & Santucci, M. (1980) Adenocarcinoma of the nose and paranasal sinuses in shoemakers and woodworkers in the province of Florence, Italy (1963–1977). *Br. J. ind. Med.*, 37, 222–225

Chan-Yeung, M. (1982) Immunologic and nonimmunologic mechanisms in asthma due to western red cedar (*Thuja plicata*). *J. Allergy clin. Immunol.*, 70, 32–37

Chan-Yeung, M. (1994) Mechanism of occupational asthma due to western red cedar (*Thuja plicata*). *Am. J. ind. Med.*, 25, 13–18

Chan-Yeung, M., Vedal, S., Kus, J., MacLean, L., Enarson, D. & Tse, K.S. (1984) Symptoms, pulmonary function, and bronchial hyperreactivity in western red cedar workers compared with those in office workers. *Am. Rev. respir. Dis.*, 130, 1038–1041

Charrier, B., Marques, M. & Haluk, J.P. (1992) HPLC analysis of gallic and ellagic acids in European oakwood (*Quercus robur* L.) and eucalyptus (*Eucalyptus globulus*). *Holzforschung*, 46, 87–89

Chovil, A.C. (1982) Other possible explanations for cancer experience among woodworkers in the auto industry (Letter to the Editor). *J. occup. Med.*, 24, 870

Christensen, L.P., Lam, J. & Sigsgaard, T. (1988) A novel stilbene from the wood of *Chlorophora excelsa*. *Phytochemistry*, 27, 3014–3016

Clayton Environmental Consultants Ltd. (1988) *Final Report on an Industry-wide Study of Wood Dust Exposures for the Inter-industry Wood Dust Coordinating Committee*, Novi, MI

Coggon, D., Pannett, B., Osmond, C. & Acheson, E.D. (1986) A survey of cancer and occupation in young and middle aged men. I. Cancers of the respiratory tract. *Br. J. ind. Med.*, 43, 332–338

Cohn, K.K., Marcero, D.H. & Wojinski, S.F. (1984) The use of GC/MS analysis and fungal culturing in a pulp mill industrial hygiene program. *Am. Ind. Hyg. Assoc. J.*, 45, 594–597

Comba, P., Battista, G., Belli, S., de Capua, B., Merler, E., Orsi, D., Rodella, S., Vindigni, C. & Axelson, O. (1992a) A case–control study of cancer of the nose and paranasal sinuses and occupational exposures. *Am. J. ind. Med.*, 22, 511–520

Comba, P., Barbieri, P.G., Battista, G., Belli, S., Ponterio, F., Zanetti, D. & Axelson, O. (1992b) Cancer of the nose and paranasal sinuses in the metal industry: a case–control study. *Br. J. ind. Med.*, 49, 193–196

Correa, P., Williams Pickle, L., Fontham, E., Dalager, N., Lin, Y., Haenszel, W. & Johnson, W.D. (1984) The causes of lung cancer in Louisiana. In: Mizell, M. & Correa, P., eds, *Lung Cancer: Causes and Prevention. Biomedical Advances in Carcinogenesis*, New York, Verlag Chemie International, pp. 73–82

Cuzik, J. & De Stavola, B. (1988) Multiple myeloma—a case–control study. *Br. J. Cancer*, 57, 516–520

Damber, L.A. & Larsson, L.G. (1987) Occupation and male lung cancer: a case–control study in northern Sweden. *Br. J. ind. Med.*, 44, 446–453

Darcy, F.J. (1982) *Physical Characteristics of Wood Dust*, PhD Thesis, University of Minnesota, MN

Darcy, F.J. (1984) Woodworking operations—furniture manufacturing. In: Cralley, L.J. & Cralley, L.V., eds, *Industrial Hygiene Aspects of Plant Operations*, Vol. 2, *Operations and Product Fabrication*, Toronto, Macmillan Press, pp. 349–362

Davies, J.M. (1983) Cancer morbidity among auto industry woodworkers (Letter to the Editor). *J. occup. Med.*, 25, 355

Debois, J.M. (1969) Tumours of the nasal cavities among woodworkers. *Tijdschr. Geneesk.*, 2, 92–93 (in Flemish)

Delemarre, J.F.M. & Themans, H.H. (1971) Adenocarcinoma of the nasal cavities. *Ned. T. Geneesk.*, 115, 688–690 (in Dutch)

Demers, P.A., Kogevinas, M., Boffetta, P., Leclerc, A., Luce, D., Gerin, M., Battista, G., Belli, S., Bolm-Audorf, U., Brinton, L.A., Colin, D., Comba, P., Hardell, L., Hayes, R.B., Magnani, C., Merler, E., Morcet, J.F., Preston-Martin, S., Matos, E., Rodella, S., Vaughan, T.L., Zheng, W. & Vainio, H. (1995) Wood dust and sino-nasal cancer: a pooled re-analysis of twelve case–control studies. *Am. J. ind. Med.* (in press)

Demirbas, A. (1991) Fatty and resin acids recovered from spruce wood by supercritical acetone extraction. *Holzforschung*, **45**, 337–339

Deutsche Forschungsgemeinschaft (1993) *MAK- und BAT-Werte-Liste 1993* [MAK- and BAT-Values 1993] (Report No. 29), Weinheim, VCH Verlagsgesellschaft, pp. 83, 87, 103

Dockerty, J.D., Marshall, S., Fraser, J. & Pearce, N. (1991) Stomach cancer in New Zealand: time trends, ethnic group differences and a cancer registry-based case–control study. *Int. J. Epidemiol.*, **20**, 45–53

Dykewicz, M.S., Laufer, P., Patterson, R., Roberts, M. & Sommers, H.M. (1988) Woodman's disease: hypersensitivity pneumonitits from cutting live trees. *J. Allergy clin. Immunol.*, **81**, 455–460

Edwards, J.J., Jr, Brooks, S.M., Henderson, F.I. & Apol, A.G. (1978) *Health Hazard Evaluation Report: Weyerhäuser Company, Longview, Washington* (NIOSH Report No. HHE 76-79,80-543; US NTIS PB81-144081), Cincinnati, OH, National Institute for Occupational Safety and Health

Eller, P.M., ed. (1984a) *NIOSH Manual of Analytical Methods*, 3rd Ed., Vol. 2 (DHHS (NIOSH) Publ. No. 84-100), Washington DC, US Government Printing Office, pp. 0500-1–0500-3

Eller, P.M., ed. (1984b) *NIOSH Manual of Analytical Methods*, 3rd Ed., Vol. 2 (DHHS (NIOSH) Publ. No. 84-100), Washington DC, US Government Printing Office, pp. 0600-1–0600-6

Elwood, J.M. (1981) Wood exposure and smoking: association with cancer of the nasal cavity and paranasal sinuses in British Columbia. *Can. med. Assoc. J.*, **124**, 1573–1577

Elwood, J.M., Pearson, J.C.G., Skippen, D.H. & Jackson, S.M. (1984) Alcohol, smoking, social and occupational factors in the aetiology of cancer of the oral cavity, pharynx and larynx. *Int. J. Cancer*, **34**, 603–612

Emanuel, D.A., Lawton, B.R. & Wenzel, F.J. (1962) Maple-bark disease. Pneumonitis due to *Coniosporium corticale*. *New Engl. J. Med.*, **266**, 333–337

Engzell, U., Englund, A. & Westerholm, P. (1978) Nasal cancer associated with occupational exposure to organic dust. *Acta otolaryngol.*, **86**, 437–442

Esping, B. & Axelson, O. (1980) A pilot study on respiratory and digestive tract cancer among woodworkers. *Scand. J. Work Environ. Health*, **6**, 201–205

FAO (1981) *Small and Medium Sawmills in Developing Countries, A Guide to their Planning and Establishment* (FAO Forestry Paper 28), Rome

FAO (1992) *FAO Yearbook: Forest Products 1979–1990* (FAO Forestry Series No. 25), Rome

FAO (1993) *FAO Yearbook: Forest Products 1980–1991* (FAO Forestry Series No. 26), Rome

Fengel, D. (1989) Chemistry and morphology of Quebracho colorado wood. In: *Proceedings of the 5th International Symposium on Wood and Pulping Chemistry, 22–25 May 1989, Raleigh, NC, USA*, pp. P267–P269

Fengel, D. & Wegener, G. (1989) *Wood—Chemistry, Ultrastructure, Reactions*, 2nd Ed., Berlin, Walter de Gruyter

Fincham, S., MacMillan, A., Turner, D. & Berkel, J. (1993) Occupational risks for cancer in Alberta. *Health Rep.*, **5**, 67–72

Finkelstein, M.M. (1989) Nasal cancer among North American woodworkers: another look. *J. occup. Med.*, **31**, 899–901

Flanders, W.D. & Rothman, K.J. (1982) Occupational risk for laryngeal cancer. *Am. J. public Health*, **72**, 369–372

Flodin, U., Fredriksson, M., Axelson, O., Persson, B. & Hardell, L. (1986) Background radiation, electrical work, and some other exposures associated with acute myeloid leukemia in a case-referent study. *Arch. environ. Health*, **41**, 77–84

Flodin, U., Fredriksson, M. & Persson, B. (1987) Multiple myeloma and engine exhausts, fresh wood, and creosote: a case-referent study. *Am. J. ind. Med.* **12**, 519–529

Flodin, U., Fredriksson, M., Persson, B. & Axelson, O. (1988) Chronic lymphatic leukaemia and engine exhausts, fresh wood, and DDT: a case-referent study. *Br. J. ind. Med.*, **45**, 33–38

Fonte, R., Grigis, L., Grigis, P. & Franco, G. (1982) Chemicals and Hodgkin's disease (Letter to the Editor). *Lancet*, **ii**, 50

Franceschi, S., Serraino, D., Bidoli, E., Talamini, R., Tirelli, U., Carbone, A. & La Vecchia, C. (1989) The epidemiology of non-Hodgkin's lymphoma in the north-east of Italy: a hospital-based case-control study. *Leukemia Res.*, **13**, 465–472

Fredriksson, M., Bengtsson, N.-O., Hardell, L. & Axelson, O. (1989) Colon cancer, physical activity, and occupational exposures. A case-control study. *Cancer*, **63**, 1838–1842

Fukuda, K. & Shibata, A. (1988) A case-control study of past history of nasal diseases and maxillary sinus cancer in Hokkaido, Japan. *Cancer Res.*, **48**, 1651–1652

Fukuda, K. & Shibata, A. (1990) Exposure-response relationships between woodworking, smoking or passive smoking, and squamous cell neoplasms of the maxillary sinus. *Cancer Causes Control*, **1**, 165–168

Fukuda, K., Shibata, A. & Harada, K. (1987) Squamous cell cancer of the maxillary sinus in Hokkaido, Japan: a case-control study. *Br. J. ind. Med.*, **44**, 263–266

Gallagher, R.P., Threlfall, W.J., Band, P.R. & Spinelli, J.J. (1985) Cancer mortality experience of woodworkers, loggers, fishermen, farmers and miners in British Columbia. *Natl Cancer Inst. Monogr.*, **69**, 163–167

Gallagher, R.P., Threlfall, W.J., Band, P.R. & Spinelli, J.J. (1989) *Occupational Mortality in British Columbia 1950–1984*, Vancouver, Cancer Control Agency of British Columbia

Gerhardsson, M.R., Norell, S.E., Kiviranta, H.J. & Ahlbom, A. (1985) Respiratory cancers in furniture workers. *Br. J. ind. Med.*, **42**, 403–405

Ghezzi, I., Peasso, R., Cortona, G., Berrino, F., Crosignani, P. & Baldasseroni, A. (1983) Incidence of malignant tumours of the nasal cavity in 91 communes in Brianza. *Med. Lav.*, **74**, 88–96 (in Italian)

Goldsmith, D.F. & Shy, C.M. (1988) An epidemiologic study of respiratory health effects in a group of North Carolina furniture workers. *J. occup. Med.*, **30**, 959–965

González, C.A., Sanz, M., Marcos, G., Pita, S., Brullet, E., Vida, F., Agudo, A. & Hsieh, C.-C. (1991) Occupation and gastric cancer in Spain. *Scand. J. Work Environ. Health*, **17**, 240–247

Greene, M.H., Brinton, L.A., Fraumeni, J.F. & D'Amico, R. (1978) Familial and sporadic Hodgkin's disease associated with occupational wood exposure (Letter to the Editor). *Lancet*, **ii**, 626–627

Grosser, D. (1977) *Die Hölzer Mitteleuropas. Ein Mikrophotographischer Lehratlas* [Woods of Mediterranean Europe. A microphotographic text atlas], Berlin, Springer Verlag

Grufferman, S., Duong, T. & Cole, P. (1976) Occupation and Hodgkin's disease. *J. natl Cancer Inst.*, **57**, 1193–1195

Gülzow, J. (1975) Adenocarcinoma of the paranasal sinsuses in woodworkers, an occupational disease? *Laryngol. Rhinol.*, **54**, 304–310 (in German)

Haden-Guest, S., Wright, J.K. & Teclaff, E.M., eds (1956) *A World Geography of Forest Resources* (American Geographic Society Special Publication No. 33), New York, Ronal Press Co.

Haguenoer, J.M., Cordier, S., Morel, C., Lefebvre, J.L. & Hemon, D. (1990) Occupational risk factors for upper respiratory tract and upper digestive tract cancers. *Br. J. ind. Med.*, **47**, 380–383

Hamill, A., Ingle, J., Searle, S. & Williams, K. (1991) Levels of exposure to wood dust. *Ann. occup. Hyg.*, **35**, 397–403

Hampl, V. & Johnston, O.E. (1985) Control of wood dust from horizontal belt sanding. *Am. ind. Hyg. Assoc. J.*, **46**, 567–577

Hardell, L., Johansson, B. & Axelson, O. (1982) Epidemiological study of nasal and nasopharyngeal cancer and their relation to phenoxy acid or chlorophenol exposure. *Am. J. ind. Med.*, **3**, 247–257

Harrington, J.M., Blot, W.J., Hoover, R.N., Housworth, W.J., Heath, C.W., Jr & Fraumeni, J.F., Jr (1978) Lung cancer in coastal Georgia: a death certificate analysis of occupation: brief communication. *J. natl Cancer Inst.*, **60**, 295–298

Hausen, B.M. (1981) *Woods Injuries to Human Health. A Manual*, Berlin, Walter de Gruyter

Hausen, B.M. (1986) Contact allergy to woods. *Clin. Dermatol.*, **4**, 65–76

Hayes, R.B., Gérin, M., Raatgever, J.W. & de Bruyn, A (1986a) Wood-related occupations, wood dust exposure, and sinonasal cancer. *Am. J. Epidemiol.*, **124**, 569–577

Hayes, R.B., Raatgever, J.W., de Bruyn, A. & Gérin, M. (1986b) Cancer of the nasal cavity and paranasal sinuses, and formaldehyde exposure. *Int. J. Cancer*, **37**, 487–492

Hayes, R.B., Kardaun, J.W.P.F. & de Bruyn, A. (1987) Tobacco use and sino-nasal cancer. *Br. J. Cancer*, **56**, 843–846

Hearn, S., Bond, G.G., Cook, R.R., Schneider, E.J. & Kolesar, R.C. (1983) Pattern makers mortality study. *J. occup. Med.*, **35**, 719–722

Hedenstierna, G., Alexandersson, R., Wimander, K. & Rosén, G. (1983) Exposure to terpenes: effects on pulmonary function. *Int. Arch. occup. environ. Health*, **51**, 191–198

Heineman, E.F., Olsen, J.H., Pottern, L.M., Gomez, M., Raffn, E. & Blair, A. (1992) Occupational risk factors for multiple myeloma among Danish men. *Cancer Causes Control*, **3**, 555–568

Hemingway, R.W. & Karchesy, J.J., eds (1989) *Chemistry and Significance of Condensed Tannins*, New York, Plenum Press

Henschler, D., ed. (1983) Wood dust. In: *Gesundheitsschädliche Arbeitsstoffe. Toxikologischarbeitsmedizinische Begründungen von MAK-Werten* [Noxious substances. Toxicological and occupational MAK values], Vol. 9, Weinheim, Verlag Chemie, pp. 1–8

Hernberg, S., Westerholm, P., Schultz-Larsen, K., Degerth, R., Kuosma, E., Englund, A., Engzell, U., Hansen, H.S. & Mutanen, P. (1983) Nasal and sinonasal cancer. Connection with occupational exposures in Denmark, Finland and Sweden. *Scand. J. Work Environ. Health*, **9**, 315–326

Herold, D.A., Wahl, R., Maasch, H.J., Hausen, B.M. & Kunkel, G. (1991) Occupational wood-dust sensitivity from *Euonymus europaeus* (spindle tree) and investigation of cross reactivity between E.e. wood and *Artemisia vulgaris* pollen (mugwort). *Allergy*, **46**, 186–190

Heston, W.E. (1975) Testing for possible effects of cedar wood shavings and diet on occurrence of mammary gland tumors and hepatomas in $C3H-A^{vy}$ and $C3H-A^{vy}fB$ mice (Brief communication). *J. natl Cancer Inst.*, **54**, 1011–1014

Hildesheim, A., West, S., De Veyra, E., De Guzman, M.F., Jurado, A., Jones, C., Imai, J. & Hinuma, Y. (1992) Herbal medicine use, Epstein–Barr virus and risk of nasopharyngeal carcinoma. *Cancer Res.*, **52**, 3048–3051

Hillis, W.E. (1987) *Heartwood and Tree Exudates*, Berlin, Springer Verlag

Hinds, W.C. (1988) Basis for particle size-selective sampling for wood dust. *Appl. ind. Hyg.*, **3**, 67–72

Hoadley, R.B. (1990) *Identifying Wood. Accurate Results with Simple Tools*, Newtown, CT, Taunton Press

Hoar Zahm, S., Brownson R.C., Chang, J.C. & Davis, J.R. (1989) Study of lung cancer histologic types, occupation, and smoking in Missouri. *Am. J. ind. Med.*, **15**, 565–578

Holliday, M.G., Dranitsaris, P., Strahlendorf, P.W., Contala, A. & Engelhardt, J.J. (1986) *Wood Dust Exposure in Ontario Industry: The Occupational Health Aspects*, Ottawa, Michael Holliday & Assoc.

Holmström, M. & Wilhelmsson, B. (1988) Respiratory symptoms and pathophysiological effects of occupational exposure to formaldehyde and wood dust. *Scand. J. Work Environ. Health*, **14**, 306–311

Holmström, M., Wilhelmsson, B. & Hellquist, H. (1989a) Histological changes in the nasal mucosa in rats after long-term exposure to formaldehyde and wood dust. *Acta otolaryngol.*, **108**, 274–283

Holmström, M., Wilhelmsson, B., Hellquist, H. & Rosén, G. (1989b) Histological changes in the nasal mucosa in persons occupationally exposed to formaldehyde alone and in combination with wood dust. *Acta otolaryngol.*, **107**, 120–129

Holness, D.L., Sass-Kortsak, A.M., Pilger, C.W. & Nethercott, J.R. (1985) Respiratory function and exposure–effect relationships in wood dust-exposed and control workers. *J. occup. Med.*, **27**, 501–506

Hounam, R.F. & Williams, J. (1974) Levels of airborne dust in furniture making factories in the High Wycombe area. *Br. J. ind. Med.*, **31**, 1–9

Huebner, W.W., Schoenberg, J.B., Kelsey, J.L., Wilcox, H.B., McLaughlin, J.K., Greenberg, R.S., Preston-Martin, S., Austin, D.F., Stemhagen, A., Blot, W.J., Winn, D.M. & Fraumeni, J.F., Jr (1992) Oral and pharyngeal cancer and occupation: a case–control study. *Epidemiology*, **3**, 300–309

IARC (1974) *IARC Monographs on the Evaluation of Carcinogenic Risk of Chemicals to Man*, Vol. 25, *Some Organochlorine Pesticides*, Lyon, pp. 25–38

IARC (1981) *IARC Monographs on the Evaluation of the Carcinogenic Risk of Chemicals to Humans*, Vol. 25, *Wood, Leather and Some Associated Industries*, Lyon

IARC (1983a) *IARC Monographs on the Evaluation of the Carcinogenic Risk of Chemicals to Humans*, Vol. 31, *Some Food Additives, Feed Additives and Naturally Occurring Substances*, Lyon, pp. 171–178

IARC (1983b) *IARC Monographs on the Evaluation of the Carcinogenic Risk of Chemicals to Humans*, Vol. 31, *Some Food Additives, Feed Additives and Naturally Occurring Substances*, Lyon, pp. 213–229

IARC (1985a) *IARC Monographs on the Evaluation of the Carcinogenic Risk of Chemicals to Humans*, Vol. 36, *Allyl Compounds, Aldehydes, Epoxides and Peroxides*, Lyon, pp. 75–97

IARC (1985b) *IARC Monographs on the Evaluation of the Carcinogenic Risk of Chemicals to Humans*, Vol. 35, *Polynuclear Aromatic Compounds, Part 4: Bitumens, Coal-tars and Derived Products, Shale-oils and Soots*, Lyon, p. 85

IARC (1986) *IARC Monographs on the Evaluation of the Carcinogenic Risk of Chemicals to Humans*, Vol. 41, *Some Halogenated Hydrocarbons and Pesticide Exposures*, Lyon, pp. 319–356

IARC (1987a) *IARC Monographs on the Evaluation of Carcinogenic Risks to Humans*, Suppl. 7, *Overall Evaluations of Carcinogenicity: An Updating of* IARC Monographs *Volumes 1–42*, Lyon, pp. 100–106

IARC (1987b) *IARC Monographs on the Evaluation of Carcinogenic Risks to Humans*, Suppl. 7, *Overall Evaluations of Carcinogenicity: An Updating of* IARC Monographs *Volumes 1–42*, Lyon, pp. 154–156

IARC (1987c) *IARC Monographs on the Evaluation of Carcinogenic Risks to Humans*, Suppl. 7, *Overall Evaluations of Carcinogenicity: An Updating of* IARC Monographs *Volumes 1–42*, Lyon, pp. 350–354

IARC (1987d) *IARC Monographs on the Evaluation of Carcinogenic Risks to Humans*, Suppl. 7, *Overall Evaluations of Carcinogenicity: An Updating of* IARC Monographs *Volumes 1–42*, Lyon, pp. 220–222

IARC (1987e) *IARC Monographs on the Evaluation of Carcinogenic Risks to Humans*, Suppl. 7, *Overall Evaluations of Carcinogenicity: An Updating of* IARC Monographs *Volumes 1–42*, Lyon, pp. 120–122

IARC (1988) *IARC Monographs on the Evaluation of Carcinogenic Risks to Humans*, Vol. 43, *Man-made Mineral Fibres and Radon*, Lyon, pp. 39–172

IARC (1989a) *IARC Monographs on the Evaluation of Carcinogenic Risks to Humans*, Vol. 45, *Occupational Exposures in Petroleum Refining; Crude Oil and Major Petroleum Fuels*, Lyon, pp. 39–117

IARC (1989b) *IARC Monographs on the Evaluation of Carcinogenic Risks to Humans*, Vol. 47, *Some Organic Solvents, Resin Monomers and Related Compounds, Pigments and Occupational Exposures in Paint Manufacture and Painting*, Lyon, pp. 263–287

IARC (1989c) *IARC Monographs on the Evaluation of Carcinogenic Risks to Humans*, Vol. 47, *Some Organic Solvents, Resin Monomers and Related Compounds, Pigments and Occupational Exposures in Paint Manufacture and Painting*, Lyon, pp. 79–123

IARC (1989d) *IARC Monographs on the Evaluation of Carcinogenic Risks to Humans*, Vol. 47, *Some Organic Solvents, Resin Monomers and Related Compounds, Pigments and Occupational Exposures in Paint Manufacture and Painting*, Lyon, pp. 124–156

IARC (1989e) *IARC Monographs on the Evaluation of Carcinogenic Risks to Humans*, Vol. 46, *Diesel and Gasoline Engine Exhausts and Some Nitroarenes*, Lyon, pp. 41–185

IARC (1989f) *IARC Monographs on the Evaluation of Carcinogenic Risks to Humans*, Vol. 45, *Occupational Exposures in Petroleum Refining; Crude Oil and Major Petroleum Fuels*, Lyon, pp. 159–201

IARC (1990) *IARC Monographs on the Evaluation of Carcinogenic Risks to Humans*, Vol. 49, *Chromium, Nickel and Welding*, Lyon, pp. 49–256

IARC (1991) *IARC Monographs on the Evaluation of Carcinogenic Risks to Humans*, Vol. 53, *Occupational Exposures in Insecticide Application, and Some Pesticides*, Lyon, pp. 115–175

IARC (1993) *IARC Monographs on the Evaluation of Carcinogenic Risk to Humans*, Vol. 56, *Some Naturally Occurring Substances: Food Items and Constituents, Heterocyclic Aromatic Amines and Mycotoxins*, Lyon, pp. 135–162

IARC (1994a) *IARC Monographs on the Evaluation of Carcinogenic Risk to Humans*, Vol. 60, *Some Industrial Chemicals*, Lyon, pp. 233–320

IARC (1994b) *Directory of Agents Being Tested for Carcinogenicity*, No. 16, Lyon

Industrial Accident Prevention Association (1985) *Furniture Manufacturing Safety and Health Guide*, Toronto

Ironside, P. & Matthews, J. (1975) Adenocarcinoma of the nose and paranasal sinuses in woodworkers in the state of Victoria, Australia. *Cancer*, **36**, 1115–1121

Jacobs, B.B. & Dieter, D.K. (1978) Spontaneous hepatomas in mice inbred from Ha:ICR Swiss stock: effects of sex, cedar shavings in bedding, and immunization with fetal liver or hepatoma cells. *J. natl Cancer Inst.*, **61**, 1531–1534

Jagels, R. (1985) Health hazards of natural and introduced chemical components of boatbuilding woods. *Am. J. ind. Med.*, **8**, 241–251

Jane, F.W. (1970) *The Structure of Wood*, 2nd Ed., London, Adam & Charles Black

Jäppinen, P., Pukkala, E. & Tola, S. (1989) Cancer incidence of workers in a Finnish sawmill. *Scand. J. Work Environ. Health*, **15**, 18–23

Jöckel, K.-H., Ahrens, W., Wichmann, H.-E., Becher, H., Bolm-Audorff, U., Jahn, I., Molik, B., Greiser, E. & Timm, J. (1992) Occupational and environmental hazards associated with lung cancer. *Int. J. Epidemiol.*, **21**, 202–213

Jones, P.A. & Smith, L.C. (1986) Personal exposures to wood dust of woodworkers in the furniture industry in the High Wycombe area: a statistical comparison of 1983 and 1976/77 survey results. *Ann. occup. Hyg.*, **30**, 171–184

Kadlec, K. & Hanslian, L. (1983) Wood. In: Parmeggiani, L., ed., *Encyclopaedia of Occupational Health and Safety*, Vol. 2, 3rd rev. Ed., Geneva, International Labour Office, pp. 2308–2316

Kauppinen, T. (1986) Occupational exposure to chemical agents in the plywood industry. *Ann. occup. Hyg.*, **30**, 19–29

Kauppinen, T. & Lindroos, L. (1985) Chlorophenol exposure in sawmills. *Am. ind. Hyg. Assoc. J.*, **46**, 34–38

Kauppinen, T. & Niemelä, R. (1985) Occupational exposure to chemical agents in the particleboard industry. *Scand. J. Work Environ. Health*, **11**, 357–363

Kauppinen, T., Lindroos, L. & Mäkinen, R. (1984) Wood dust in the air of sawmills and plywood factories. *Staub Reinhalt. Luft*, **44**, 322–324 (in German)

Kauppinen, T.P., Partanen, T.J., Nurminen, M.M., Nickels, J.I., Hernberg, S.G., Hakulinen, T.R., Pukkala, E.I. & Savonen, E.T. (1986) Respiratory cancers and chemical exposures in the wood industry: a nested case–control study. *Br. J. ind. Med.*, **43**, 84–90

Kauppinen, T.P., Partanen, T.J., Hernberg, S.G., Nickels, J.I., Luukkonen, R.A., Hakulinen, T.R. & Pukkala, E.I. (1993) Chemical exposures and respiratory cancer among Finnish woodworkers. *Br. J. ind. Med.*, **50**, 143–148

Kawachi, I., Pearce, N. & Fraser, J. (1989) A New Zealand Cancer Registry-based study of cancer in wood workers. *Cancer*, **64**, 2609–2613

Kjuus, H., Skjaerven, R., Langård, S., Lien, J.T. & Aamodt, T. (1986) A case–referent study of lung cancer, occupational exposures and smoking. I. Comparison of title-based and exposure-based occupational information. *Scand. J. Work Environ. Health*, **12**, 193–202

Kleinsasser, O. & Schroeder, H.G. (1989) What's new in tumors of the nasal cavity? Adenocarcinomas arising after exposure to wood dust. *Pathol. Res. Pract.*, **184**, 554–558

Klintenberg, C., Olofsson, J., Hellquist, H. & Sökjer, H. (1984) Adenocarcinoma of the ethmoid sinuses. A review of 28 cases with special reference to wood dust exposure. *Cancer*, **54**, 482–488

Koch, P. (1964) *Wood Machining Processes*, New York, The Ronald Press Co.

Kominsky, J.R. & Anstadt, G.P. (1976) *Health Hazard Evaluation Determination Report: Masonite Corporation, Evendale, Ohio* (NIOSH Report No. HHE 75-19-276), Cincinnati, OH, National Institute for Occupational Safety and Health

Kotimaa, M. (1990) *Bioaerosols* (Exposure at Work No. 8), Helsinki, Finnish Institute of Occupational Health and Finnish Work Environment Fund (in Finnish)

Kubel, H., Weissmann, G. & Lange, W. (1988) Investigations on the carcinogenicity of wood dust. The extractives of beech and spruce. *Holz Roh-Werkstoff*, **46**, 215–220 (in German)

Kubiš, T. (1963) Dust ratios in association with cabinet-makers' band lathes. *Prac. Lék.*, **15**, 435–437 (in Czech)

Kurt, T.L. (1986) Colon cancer in the automobile industry (Letter to the Editor). *J. occup. Med.*, **28**, 264

Kurttio, P., Kalliokoski, P., Lampelo, S. & Jantunen, M.J. (1990) Mutagenic compounds in wood-chip drying fumes. *Mutat. Res.*, **242**, 9–15

Kurttio, P., Norppa, H., Järventaus, H., Sorsa, M. & Kalliokoski, P. (1993) Chromosome aberrations in peripheral lymphocytes of workers employed in the plywood industry. *Scand. J. Work Environ. Health*, **19**, 132–134

Leclerc, A., Martinez Cortes, M., Gérin, M., Luce, D. & Brugère, J. (1994) Sinonasal cancer and wood dust exposure: results from a case–control study. *Am. J. Epidemiol.*, **140**, 340–349

Lehmann, E. & Fröhlich, N. (1987) Dust exposure in wood industries. *Zbl. Arbeitsmed.*, **37**, 315–323 (in German)

Lehmann, E. & Fröhlich, N. (1988) Particle size distribution of wood dust at the workplace. *J. Aerosol Sci.*, **19**, 1433–1436

Lerchen, M.L., Wiggins, C.L. & Samet, J.M. (1987) Lung cancer and occupation in New Mexico. *J. natl Cancer Inst.*, **79**, 639–645

Leroux-Robert, J. (1974) Cancers of the ethmoid sinus in woodworkers. *Cah. ORL*, **9**, 585–594 (in French)

Levin, L.I., Zheng, W., Blot, W.J., Gao, Y.-T. & Fraumeni, J.F., Jr (1988) Occupation and lung cancer in Shanghai: a case–control study. *Br. J. ind. Med.*, **45**, 450–458

Lindroos, L. (1983) Wood dust at a woodworking plant. *Työterveyslaitoksen Tutkimuksia*, **1**, 105–114 (in Finnish)

Linet, M.S., Harlow, S.D. & McLaughlin, J.K. (1987) A case–control study of multiple myeloma in whites: chronic antigenic stimulation, occupation, and drug use. *Cancer Res.*, **47**, 2978–2981

Linet, M.S., Malker, H.S.R., McLaughlin, J.K., Weiner, J.A., Stone, B.J., Blot, W.J., Ericsson, J.L.E. & Fraumeni, J.F., Jr (1988) Leukemias and occupation in Sweden: a registry based analysis. *Am. J. ind. Med.*, **14**, 319–330

Linet, M.S., Malker, H.S.R., McLaughlin, J.K., Weiner, J.A., Blot, W.J., Ericsson, J.L.E. & Fraumeni, J.F., Jr (1993) Non-Hodgkin's lymphoma and occupation in Sweden: a registry based analysis. *Br. J. ind. Med.*, **50**, 79–84

Liu, W.K., Wong, M.H., Tam, N.F.Y. & Choy, A.C.K. (1985) Properties and toxicity of airborne wood dust in woodworking establishments. *Toxicol. Lett.*, **26**, 43–52

Löbe, L.-P. & Ehrhardt, H.-P. (1983) Occupational adenocarcinoma of the nose and the paranasal sinuses in woodworkers. *HNO-Praxis*, **8**, 185–188 (in German)

Loi, A.M., Amram, D.L., Bramanti, L., Roselli, M.G., Giacomini, G., Simi, U., Belli, S. & Comba, P. (1989) Nasal cancer and exposure to wood and leather dust. A case–control study in Pisa area. *J. exp. clin. Cancer Res.*, **8**, 13–19

Loomis, D.P. & Savitz, D.A. (1991) Occupation and leukemia mortality among men in 16 states: 1985–1987. *Am. J. ind. Med.*, **19**, 509–521

López, J.I., Nevado, M., Eizaguirre, B. & Pérez, A. (1990) Intestinal-type adenocarcinoma of the nasal cavity and paranasal sinuses. A clinicopathologic study of 6 cases. *Tumori*, **76**, 250–254

Luboinski, B. & Marandas, P. (1975) Cancer of the ethmoid sinus: occupational etiology. *Arch. Mal. prof.*, **36**, 477–487 (in French)

Lucas, A.D. & Salisbury, S.A. (1992) Industrial hygiene survey in a university art department. *J. environ. Pathol. Toxicol. Oncol.*, **11**, 21–27

Luce, D., Leclerc, A., Marne, M.J., Gérin, M., Casal, A. & Brugère, J. (1991) Sinonasal cancer and occupation: a case–control study. *Rev. Epidémiol. Santé publique*, **39**, 7–16 (in French)

Luce, D., Leclerc, A., Morcet, J.-F., Casal-Lareo, A., Gérin, M., Brugère, J., Haguenoer, J.-M. & Goldberg, M. (1992) Occupational risk factors for sinonasal cancer: a case–control study in France. *Am. J. ind. Med.*, **21**, 163–175

Luce, D., Gérin, M., Leclerc, A., Morcet, J.-F., Brugère, J. & Goldberg, M. (1993) Sinonasal cancer and occupational exposure to formaldehyde and other substances. *Int. J. Cancer*, **53**, 224–231

Macbeth, R. (1965) Malignant disease of the paranasal sinuses. *J. Laryngol.*, **79**, 592–612

Magnani, C., Ciambellotti, E., Salvi, U., Zanetti, R. & Comba, P. (1989) Incidence of cancers of the nasal fossae and paranasal sinuses in Biella: 1970–1986. *Acta otorhinol. ital.*, **9**, 511–519 (in Italian)

Magnani, C., Comba, P., Ferraris, F., Ivaldi, C., Meneghin, M. & Terracini, B. (1993) A case–control study of carcinomas of the nose and paranasal sinuses in the woolen textile manufacturing industry. *Arch. environ. Health*, **48**, 94–97

Maier, H., Dietz, A., Gewelke, U. & Heller, W.-D. (1991) Occupation and risk for oral, pharnygeal and laryngeal cancer. A case–control study. *Laryngo-Rhino-Otolaryngol.*, **70**, 93–98 (in German)

Maier, H., Gewelke, U., Dietz, A. & Heller, W.-D. (1992) Risk factors of cancer of the larynx: results of the Heidelberg case–control study. *Otolaryngol. Head Neck Surg.*, **107**, 577–582

Malker, H.S.R., McLaughlin, J.K., Blot, W.J., Weiner, J.A., Malker, B.K., Ericsson, J.L.E. & Stone, B.J. (1986) Nasal cancer and occupation in Sweden, 1961–1979. *Am. J. ind. Med.*, **9**, 477–485

Maloney, T.M. (1977) *Modern Particleboard and Dry-process Fiberboard Manufacturing*, San Francisco, Miller Freeman

Mark, D. & Vincent, J.H. (1986) A new personal sampler for airborne total dust in workplaces. *Ann. occup. Hyg.*, **30**, 89–102

Mayer, W., Seitz, H. & Jochims, J.C. (1969) Tanning compounds from the wood of chestnut and oak. IV. The structure of castalagine. *Liebigs Ann. Chem.*, **721**, 186–193 (in German)

Mayer, W., Seitz, H., Jochims, J.C., Schauerte, K. & Schilling, G. (1971) Tanning compounds from the wood of chestnut and oak. VI. The structure of vescalagine. *Liebigs Ann. Chem.*, **751**, 60–68 (in German)

McCammon, C.S., Jr, Robinson, C., Waxweiler, R.J. & Roscoe, R. (1985) Industrial hygiene characterization of automotive wood model shops. *Am. ind. Hyg. Assoc. J.*, **46**, 343–349

McGregor, D.B. (1982) Mutagenicity of wood dust. In: *The Carcinogenicity of Wood Dust* (Scientific Report No. 1), Southampton, MRC Environmental Unit, pp. 26–29

Menck, H.R. & Henderson, B.E. (1976) Occupational differences in rates of lung cancer. *J. occup. Med.*, **18**, 797–801

Merler, E., Baldasseroni, A., Laria, R., Faravelli, P., Agostini, R., Pisa, R. & Berrino, F. (1986) On the causal association between exposure to leather dust and nasal cancer: further evidence from a case–control study. *Br. J. ind. Med.*, **43**, 91–95

Merletti, F., Boffetta, P., Ferro, G., Pisani, P. & Terracini, B. (1991) Occupation and cancer of the oral cavity or oropharynx in Turin, Italy. *Scand. J. Work Environ. Health*, **17**, 248–254

Milham, S., Jr (1976) Neoplasia in the wood and pulp industry. *Ann. N.Y. Acad. Sci.*, **271**, 294–300

Milham, S., Jr (1978) *Mortality Experience of the AFL–CIO United Brotherhood of Carpenters and Joiners of America 1969–1970 and 1972–1973* (DHEW (NIOSH) Publication No. 78–152), Washington DC, US Government Printing Office

Milham, S., Jr & Hesser, J.E. (1967) Hodgkin's disease in woodworkers. *Lancet*, **ii**, 136–137

Miller, B.A., Blair, A.E., Raynor, H.L., Stewart, P.A., Zahm, S.H. & Fraumeni, J.F., Jr (1989) Cancer and other mortality patterns among United States furniture workers. *Br. J. ind. Med.*, **46**, 508–515

Miller, B.A., Blair, A. & Reed, E.J. (1994) Extended mortality follow-up among men and women in a US furniture workers union. *Am. J. ind. Med.*, **25**, 537–549

Milne, K.L., Sandler, D.P., Everson, R.B. & Brown, S.M. (1983) Lung cancer and occupation in Alameda County: a death certificate case–control study. *Am. J. ind. Med.*, **4**, 565–575

Minder, C.E. & Vader, J.P. (1987) Sinonasal cancer and furniture workers: update and methodological points. *Méd. soc. prév.*, **32**, 228–229

Mohtashamipur, E. & Norpoth, K. (1984) Non-mutagenicity of some wood-related compounds in the bacterial/microsome plate incorporation and microsuspension assays. *Int. Arch. occup. environ. Health*, **54**, 83–90

Mohtashamipur, E. & Norpoth, K. (1989) Nuclear aberrations in the small intestine of mice and bacterial mutagenicity caused by a fraction isolated from beech wood dusts (Abstract No. 548). *Proc. Am. Assoc. Cancer Res.*, **30**, 139

Mohtashamipur, E. & Norpoth, K. (1990) Release of mutagens after chemical or microbial degradation of beech wood lignin. *Toxicol. Lett.*, **51**, 277–285

Mohtashamipur, E., Brochmeier, U. & Norpoth, K. (1984) Studies on the isolation and identification of a genotoxic principle in wood dusts. In: *Verhandlungen der Deutschen Gesellschaft für Arbeitsmedizin* [Deliberations of the German Society of Occupational Medicine], Vol. 24, Stuttgart, Gentner, pp. 395–399 (in German)

Mohtashamipur, E., Norpoth, K. & Hallerberg, B. (1986) A fraction of beech wood mutagenic in the Salmonella/mammalian microsome assay. *Int. Arch. occup. environ. Health*, **58**, 277–234

Mohtashamipur, E., Norpoth, K. & Lühmann, F. (1989a) Cancer epidemiology of woodworking. *J. Cancer Res. clin. Oncol.*, **115**, 503–515

Mohtashamipur, E., Norpoth, K., Ernst, H. & Mohr, U. (1989b) The mouse-skin carcinogenicity of a mutagenic fraction from beech wood dusts. *Carcinogenesis*, **10**, 483–487

Moore, L.L., Dube, D.J. & Burk, T. (1990) Improved sampling and recovery of wood dust using MSA respirable dust cassettes. *Am. ind. Hyg. Assoc. J.*, **51**, A-475–A-476

Morabia, A., Markowitz, S., Garibaldi, K. & Wynder, E.L. (1992) Lung cancer and occupation: results of a multicentre case–control study. *Br J. ind. Med.*, **49**, 721–727

Morris Brown, L., Mason, T.J., Williams Pickle, L., Stewart, P.A., Buffler, P.A., Burau, K., Ziegler, R.G. & Fraumeni, J.F., Jr (1988) Occupational risk factors for laryngeal cancer on the Texas Gulf Coast. *Cancer Res.*, **48**, 1960–1964

Muscat, J.E. & Wynder, E.L. (1992) Tobacco, alcohol, asbestos, and occupational risk factors for laryngeal cancer. *Cancer*, **69**, 2244–2251

Nabeta, K., Yonekubo, J. & Miyake, M. (1987) Phenolic compounds from the heartwood of European oak (*Quercus robur* L.) and brandy. *Mokuzai Gakkaishi*, **33**, 408–415

Nandakumar, A., Dougan, L.E., English, D.R. & Armstrong, B.K. (1988) Incidence and outcome of multiple myeloma in western Australia, 1960 to 1984. *Aust. N.Z. J. Med.*, **18**, 774–779

Nelson, E., Zhou, Z., Carmichael, P.L., Norpoth, K. & Fu, J. (1993) Genotoxic effects of subacute treatments with wood dust extracts on the nasal epithelium of rats: assessment by the micronucleus and ^{32}P-postlabelling. *Arch. Toxicol.*, **67**, 586–589

Ng, T.P. (1986) A case–referent study of cancer of the nasal cavity and sinuses in Hong Kong. *Int. J. Epidemiol.*, **15**, 171–175

Norrish, A.E., Beasley, R., Hodgkinson, E.J. & Pearce, N. (1992) A study of New Zealand wood workers: exposure to wood dust, respiratory symptoms, and suspected cases of occupational asthma. *N.Z. med. J,*, **105**, 185–187

Notani, P.N., Shah, P., Jayant, K. & Balakrishnan, V. (1993) Occupation and cancers of the lung and bladder: a case–control study in Bombay. *Int. J. Epidemiol.*, **22**, 185–191

Nygren, O., Nilsson, C.-A. & Lindahl, R. (1992) Occupational exposure to chromium, copper and arsenic during work with impregnated wood in joinery shops. *Ann. occup. Hyg.*, **36**, 509–517

Oehling, A. (1963) Occupational allergy in the wood industry. *Allerg. Asthma*, **9**, 312–322 (in German)

Oleske, D., Golomb, H.M., Farber, M.D. & Levy, P.S. (1985) A case–control inquiry into the etiology of hairy cell leukemia. *Am. J. Epidemiol.*, **121**, 675–683

Olsen, J.H. (1988) Occupational risks of sinonasal cancer in Denmark. *Br. J. ind. Med.*, **45**, 329–335

Olsen, J.H. & Asnaes, S. (1986) Formaldehyde and the risk of squamous cell carcinoma of the sinonasal cavities. *Br. J. ind. Med.*, **43**, 769–774

Olsen, J.H. & Møller Jensen, O. (1987) Occupation and risk of cancer in Denmark. An analysis of 93 810 cancer cases 1970–1979. *Scand. J. Work Environ. Health.*, **13** (Suppl. 1), 1–91

Olsen, J. & Sabroe, S. (1979) A follow-up study of non-retired and retired members of the Danish carpenter/cabinet makers' trade union. *Int. J. Epidemiol.*, **8**, 375–382

Olsen, J.H., Plough Jensen, S., Hink, M., Faurbo, K., Breum, N.O. & Møller Jensen, O. (1984) Occupational formaldehyde exposure and increased nasal cancer risk in man. *Int. J. Cancer*, **34**, 639–644

Olsen, J.H., Møller, H. & Møller Jensen, O. (1988) Risks for respiratory and gastric cancer in woodworking occupations in Denmark. *J. Cancer Res. clin. Oncol.*, **114**, 420–424

Panshin, A.J. & de Zeeuw, C. (1970) *Textbook of Wood Technology*, Vol. I, *Structure, Identification, Uses and Properties of the Commercial Woods of the United States and Canada*, 3rd Ed., New York, McGraw-Hill

Partanen, T., Kauppinen, T., Luukkonen, R., Hakulinen, T. & Pukkala, E. (1993) Malignant lymphomas and leukemias, and exposures in the wood industry: an industry-based case–referent study. *Int. Arch. occup. environ. Health*, **64**, 593–596

Pearce, N.E. & Howard, J.K. (1986) Occupation, social class and male cancer mortality in New Zealand, 1974–78. *Int. J. Epidemiol.*, **15**, 456–462

Pearce, N.E., Sheppard, R.A., Howard, J.K., Fraser, J. & Lilley, B.M. (1986) Leukemia among new Zealand agricultural workers. A cancer registry-based study. *Am. J. Epidemiol.*, **124**, 402–409

Persson, B., Dahlander, A.-M., Fredriksson, M., Noorlind Brage, H., Ohlson, C.-G. & Axelson, O. (1989) Malignant lymphomas and occupational exposures. *Br. J. ind. Med.*, **46**, 516–520

Persson, B., Fredriksson, M., Olsen, K., Boeryd, B. & Axelson, O. (1993) Some occupational exposures as risk factors for malignant lymphomas. *Cancer*, **72**, 1773–1778

Peters, R.K., Garabrant, D.H., Yu, M.C. & Mack, T.M. (1989) A case–control study of occupational and dietary factors in colorectal cancer in young men by subsite. *Cancer Res.*, **49**, 5459–5468

Petersen, G.R. & Milham, S., Jr (1974) Hodgkin's disease mortality and occupational exposure to wood (Brief communication). *J. natl Cancer Inst.*, **53**, 957–958

Petersen, G.R. & Milham, S., Jr (1980) *Occupational Mortality in the State of California 1959–1961* (DHEW (NIOSH) publication no. 80-104), Cincinnati, OH, National Institute for Occupational Safety and Health

Petronio, L., Negro, C., Bovenzi, M. & Stanta, G. (1983) Cancer of the nasal cavities and paranasal sinuses in Trieste during 1968 to 1980. *Med. Lav.*, **74**, 97–105 (in Italian)

Phalen, R.F., Hinds, W.C., John, W., Lloy, P.J., Lippmann, M., McCawley, M.A., Raabe, O.G., Soderholm, S.C. & Stuart, B.O. (1986) Rationale and recommendations for particle size-selective sampling in the workplace. *Appl. ind. Hyg.*, **1**, 3–14

Pisaniello, D.L., Connell, K.E. & Muriale, L. (1991) Wood dust exposure during furniture manufacture—Results from an Australian survey and considerations for threshold limit value development. *Am. ind. Hyg. Assoc. J.*, **52**, 485–492

Pisaniello, D.L., Tkaczuk, M.N. & Owen, N. (1992) Occupational wood dust exposures, lifestyle variables, and respiratory symptoms. *J. occup. Med.*, **34**, 788–792

Pott, F., Roller, M., Ziem, U., Reiffer, F.-J., Bellmann, B., Rosenbruch, M. & Huth, F. (1989) Carcinogenicity studies on natural and man-made fibres with the intraperitoneal test in rats. In: Bignon, J., Peto, J. & Saracci, R., eds, *Non-occupational Exposure to Mineral Fibres* (IARC Scientific Publications No. 90), Lyon, IARC, pp. 173–179

Pottern, L.M., Heineman, E.F., Olsen, J.H., Raffn, E. & Blair, A. (1992) Multiple myeloma among Danish women: employment history and workplace exposures. *Cancer Causes Control*, **3**, 427–432

Priha, E., Riipinen, H. & Korhonen, K. (1986) Exposure to formaldehyde and solvents in Finnish furniture factories in 1975–1984. *Ann. occup. Hyg.*, **30**, 289–294

Proctor, D.F. (1982) The mucociliary system. In: Proctor, D.F. & Andersen, I., eds, *The Nose: Upper Airway Physiology and the Atmospheric Environment*, Amsterdam, Elsevier Biomedical Press, pp. 244–278

Pukkala, E., Teppo, L., Hakulinen, T. & Rimpelä, M. (1983) Occupation and smoking as risk determinants of lung cancer. *Int. J. Epidemiol.*, **12**, 290–296

Rang, E.H. & Acheson E.D. (1981) Cancer in furniture workers. *Int. J. Epidemiol.*, **10**, 253–261

Reif, J., Pearce, N., Kawachi, I. & Fraser, J. (1989) Soft-tissue sarcoma, non-Hodgkin's lymphoma and other cancers in New Zealand forestry workers. *Int. J. Cancer,* **43**, 49–54

Robinson, C.F., Fowler, D., Brown, D.P. & Lemen, R.A. (1990) *Plywood Mills Workers Mortality Patterns 1945–1977* (Publ. No. PB90-147056), Springfield, VA, National Technical Information Service

Ronco, G., Ciccone, G., Mirabelli, D., Troia, B. & Vineis, P. (1988) Occupation and lung cancer in two industrialized areas of northern Italy. *Int. J. Cancer,* **41**, 354–358

Roscoe, R.J., Steenland, K., McCammon, C.S., Jr, Schober, S.E., Robinson, C.F., Halperin, W.E. & Fingerhut, M.A. (1992) Colon and stomach cancer mortality among automotive wood model makers. *J. occup. Med.*, **34**, 759–768

Rosenberg, M.J., Landrigan, P.J. & Crowley, S. (1980) Low-level arsenic exposure in wood processing plants. *Am. J. ind. Med.*, **1**, 99–107

Roush, G.C., Meigs, J.W., Kelly, J., Flannery, J.T. & Burdo, H. (1980) Sinonasal cancer and occupation: a case–control study. *Am. J. Epidemiol.*, **111**, 183–193

Ruppe, K. (1973) Diseases and functional disturbance in the respiratory tract of workers in the wood processing industry. *Z. ges. Hyg.*, **19**, 261–264 (in German)

Rüttner, J.R. & Makek, M. (1985) Adenocarcinoma of nose and nasal sinuses, an occupational disease? *Schweiz. med. Wschr.*, **115**, 1838–1842 (in German)

Sabine, J.R. (1975) Exposure to an environment containing the aromatic red cedar, *Juniperus virginiana*: procarcinogenic, enzyme-inducing and insecticidal effects. *Toxicology*, **5**, 221–235

Sabine, J.R., Horton, B.J. & Wicks, M.B. (1973) Spontaneous tumors in C3H-A^{vy} and C3H-A^{vy}fB mice: high incidence in the United States and low incidence in Australia. *J. natl Cancer Inst.*, **50**, 1237–1242

Saka, S. & Goring, D.A.I. (1983) The distribution of inorganic constituents in black spruce wood as determined by TEM-EDXA. *Mokuzai Gakkaishi*, **29**, 648–656

Sass-Kortsak, A.M., Holness, D.L., Pilger, C.W. & Nethercott, J.R. (1986) Wood dust and formaldehyde exposures in the cabinet-making industry. *Am. ind. Hyg. Assoc. J.*, **47**, 747–753

Sass-Kortsak, A.M., Tracy, C. & Purdham, J. (1989) Filter preparation techniques for dust exposure determination by gravimetric analysis. *Appl. ind. Hyg.*, **4**, 222–226

Sass-Kortsak, A.M., O'Brien, C.R., Bozek, P.R. & Purdham, J.T. (1993) Comparison of the 10-mm nylon cyclone, horizontal elutriator, and aluminum cyclone for silica and wood dust measurements. *Appl. occup. Environ. Hyg.*, **8**, 31–37

Scheidt, R., Ehrhardt, H.-P., Bartsch, R. & Schwarz, H. (1989) Wood dust concentration in the furniture industry. *Z. ges. Hyg.*, **35**, 219–222 (in German)

Scherr, P.A., Hutchison, G.B. & Neiman, R.S. (1992) Non-Hodgkin's lymphoma and occupational exposure. *Cancer Res.,* **52**, 5503s–5509s

Schmezer, P., Kuchenmeister, F., Klein, R.G., Pool-Zobel, B.L., Fengel, D., Stehlin, J. & Wolf, J. (1994) Study of the genotoxic potential of different wood extracts and of selected additives in the wood industry. *Arbeitsmed. Sozialmed. Umweltmed.*, **21**, 13–17 (in German)

Schoenberg, J.B., Stemhagen, A., Mason, T.J., Patterson, J., Bill, J. & Altman, R. (1987) Occupation and lung cancer risk among New Jersey white males. *J. natl Cancer Inst.*, **79**, 13–21

Schraub, S., Belon-Leneubre, M., Mercier, M. & Bourgeois, P. (1989) Adenocarcinoma and wood. *Am. J. Epidemiol.*, **130**, 1164–1166

Schroeder, H.-G. (1989) *Adenokarzinome der inneren Nase und Holzstaubexposition. Klinische, morphologische und epidemiologische Aspekte* [Adenocarcinoma of the nasal cavity and exposure to wood. Clinical, morphological and epidemiological aspects], Essen, A. Sutter Druckerei, GmbH

Schulz, H. (1993) The history of wood use in the nineteenth, twentieth and twenty-first centuries. *Holz. Roh. Werkst.*, **51**, 75–82 (in German)

Schumacher, M.C. & Delzell, E. (1988) A death-certificate case–control study of non-Hodgkin's lymphoma and occupation in men in North Carolina. *Am. J. ind. Med.*, **13**, 317–330

Shamssain, M.H. (1992) Pulmonary function and symptoms in workers exposed to wood dust. *Thorax*, **47**, 84–87

Shimizu, H., Hozawa, J., Saito, H., Murai, K., Hirata, H., Takasaka, T., Togawa, K., Konno, A., Kimura, Y., Kikuchi, A., Ohkouchi, Y., Ohtani, I. & Hisamichi, S. (1989) Chronic sinusitis and woodworking as risk factors for cancer of the maxillary sinus in northeast Japan. *Laryngoscope*, **99**, 58–61

Siemiatycki, J. (1991) *Risk Factors for Cancer in the Workplace*, Boca Raton, CRC Press

Siemiatycki, J., Richardson, L., Gérin, M., Goldberg, M., Dewar, R., Désy, M., Campbell, S. & Wacholder, S. (1986) Associations between several sites of cancer and nine organic dusts: results from an hypothesis-generating case–control study in Montreal, 1979–1983. *Am. J. Epidemiol.*, **123**, 235–249

Smetana, R. & Horak, F. (1983) Rhinogenic adenocarcinoma in woodworkers. *Laryngol. Rhinol. Otolaryngol.*, **62**, 74–76 (in German)

Solgaard, J. & Andersen, I. (1975) Airway function and symptoms in woodworkers. *Ugeskr. Laeg.*, **44**, 2593–2599 (in Danish)

Spiegelman, D. & Wegman, D.H. (1985) Occupation-related risks for colorectal cancer. *J. natl Cancer Inst*, **75**, 813–821

Sriamporn, S., Vatanasapt, V., Pisani, P., Yongchaiyudha, S. & Rungpitarangsri, V. (1992) Environmental risk factors for nasopharyngeal carcinoma: a case–control study in northeastern Thailand. *Cancer Epidemiol. Biomarkers Prev.*, **1**, 345–348

Stellman, S.D. & Garfinkel, L. (1984) Cancer mortality among woodworkers. *Am. J. ind. Med.*, **5**, 343–357

Streit, W. & Fengel, D. (1994) On the changes of the extractive composition during heartwood formation in *Quebracho colorado* (*Schinopsis balansae* Engl.). *Holzforschung*, **48** (Suppl.), 15–20

Stumpf, J.M., Blehm, K.D., Buchan, R.M. & Gunter, B.J. (1986) Characterization of particle board aerosol-size distribution and formaldehyde content. *Am. ind. Hyg. Assoc. J.*, **47**, 725–730

Suchsland, O. & Woodson, G.E. (1986) *Fiberboard Manufacturing Practices in the United States* (Agricultural Handbook No. 640), Washington DC, United States Forest Service, United States Department of Agriculture

Suckling, I.D., Gallagher, S.S. & Ede, R.M. (1990) A new method for softwood extractives analysis using high performance liquid chromatography. *Holzforschung*, **44**, 339–345

Swan, E.P. (1989) Health hazards associated with extractives. In: Rowe, J.W., ed., *Natural Products of Woody Plants, II, Chemicals Extraneous to the Lignocellulosic Cell Wall*, Berlin, Springer Verlag, pp. 931–952

Swanson, G.M. & Belle, S.H. (1982) Cancer morbidity among woodworkers in the US automotive industry. *J. occup. Med.*, **24**, 315–319

Swanson, G.M., Belle, S.H. & Burrows, R.W., Jr (1985) Colon cancer incidence among modelmakers and patternmakers in the automobile manufacturing industry. A continuing dilemma. *J. occup. Med.*, **27**, 567–569

Takasaka, T., Kawamoto, K. & Nakamura, K. (1987) A case–control study of nasal cancers. An occupational survey. *Acta otolaryngol.*, **435** (Suppl.), 136–142

Tanaka, I., Haratake, J., Horie, A. & Yoshimura, T. (1991) Cumulative toxicity potential of hardwood dust and sidestream tobacco smoke in rats by repeated inhalation. *Inhal. Toxicol.*, **3**, 101–112

Technical Association of the Pulp and Paper Industry (1985a) *Sampling and Preparing Wood for Analysis* (T257cm-85), Atlanta, GA

Technical Association of the Pulp and Paper Industry (1985b) *Ash in Wood and Pulp* (T211om-85), Atlanta, GA

Technical Committee of the Inhalation Specialty Section, Society of Toxicology (1987) Commentary. Recommendations for the conduct of acute inhalation limit tests. *Fundam. appl. Toxicol.*, **18**, 321–327

Teschke, K., Hertzman, C. & Morrison, B. (1994) Level and distribution of employee exposures to total and respirable wood dust in two Canadian sawmills. *Am. ind. Hyg. Assoc. J.*, **55**, 245–250

Teschke, K., Hertzman, C., Fenske, R.A., Jin, A., Ostry, A., van Netten, C. & Leiss, W. (1995) A history of process and chemical changes for fungicide application in the western Canadian lumber industry: What can we learn? *Appl. occup. Environ. Hyg.* (in press)

Tilley, B.C., Johnson, C.C., Schultz, L.R., Buffler, P.A. & Joseph, C.L.M. (1990) Risk of colorectal cancer among automotive pattern and model makers. *J. occup. Med.*, **32**, 541–546

Tola, S., Hernberg, S., Collan, Y., Linderborg, H. & Korkala, M.-L. (1980) A case control study of the etiology of nasal cancer in Finland. *Int. Arch. occup. environ. Health*, **46**, 79–85

Tollerud, D.J., Brinton, L.A., Stone, B.J., Tobacman, J.K. & Blattner, W.A. (1985) Mortality from multiple myeloma among North Carolina furniture workers. *J. natl Cancer Inst.*, **74**, 799–801

Torén, K., Hörte, L.-G. & Järvholm, B. (1991) Occupation and smoking adjusted mortality due to asthma among Swedish men. *Br. J. ind. Med.*, **48**, 323–326

Törrönen, R., Pelkonen, K. & Kärenlampi, S. (1989) Enzyme-inducing and cytotoxic effects of wood-based materials used as bedding for laboratory animals. Comparison by a cell culture study. *Life Sci.*, **45**, 559–565

Torul, O. & Olcay, A. (1984) Terpene hydrocarbons of soxhlet and supercritical-gas extracts of oriental spruce and oriental beech. *Holzforschung*, **38**, 221–224

Trotel, E. (1976) *Adénocarcinome de l'Ethmoïde et Travail du Bois* [Adenocarcinoma of the ethmoid sinus and woodworking], MD Thesis No. 117, University of Rennes (in French)

Työministeriö [Ministry of Labour] (1993) *Limit Values 1993*, Helsinki, p. 16 (in Finnish)

United Kingdom Health and Safety Executive (1989) *MDHS 14 General Methods for the Gravimetric Determination of Respirable and Total Inhalable Dust*, Bootle, Lancs

United Kingdom Health and Safety Executive (1992) *EH40/92 Occupational Exposure Limits 1992*, London, Her Majesty's Stationery Office, pp. 12, 29

United States National Institute for Occupational Safety and Health (1987) *Health Effects of Exposure to Wood Dust: A Summary of the Literature* (US NTIS PB87-218251), Cincinnati, OH

United States National Institute for Occupational Safety and Health (1990) *National Occupational Exposure Survey 1981–83*, Cincinnati, OH

United States National Institute for Occupational Safety and Health (1992) *NIOSH Recommendations for Occupational Safety and Health. Compendium of Policy Documents and Statements* (DHHS (NIOSH) Publ. No. 92-100), Cincinnati, OH, Division of Standards Development and Technology Transfer, pp. 43, 133

United States Occupational Safety and Health Administration (1993) Air contaminants. *US Code Fed. Regul.*, **Title 29**, Part 1910.1000, pp. 6–19

Vaucher, H., compiler (1986) *Elsevier's Dictionary of Trees and Shrubs*, Amsterdam, Elsevier

Vaughan, T.L. (1989) Occupation and squamous cell cancers of the pharynx and sinonasal cavity. *Am. J. ind. Med.*, **16**, 493–510

Vaughan, T.L. & Davis, S. (1991) Wood dust exposure and squamous cell cancers of the upper respiratory tract. *Am. J. Epidemiol.*, **133**, 560–564

Vaughan, N.P., Chalmers, C.P. & Botham, R.A. (1990) Field comparison of personal samplers for inhalable dust. *Ann. occup. Hyg.*, **34**, 553–573

Vedal, S., Chan-Yeung, M., Enarson, D., Fera, T., Maclean, L., Tse, K.S. & Langille, R. (1986) Symptoms and pulmonary function in western red cedar workers related to duration of employment and dust exposure. *Arch. environ. Health*, **41**, 179–183

Vetrugno, T. & Comba, P. (1987) Collaborative study on nasal cancer and occupation in Italy, 1983–1985. *Acta otorhinol. ital.*, **7**, 485–494 (in Italian)

Vincent, J.H. & Mark, D. (1990) Entry characteristics of practical workplace aerosol samplers in relation to the ISO recommendations. *Ann. occup. Hyg.*, **34**, 249–262

Vinzents, P. & Laursen, B. (1993) A national cross-sectional study of the working environment in the Danish wood and furniture industry—air pollution and noise. *Ann. occup. Hyg.*, **37**, 25–34

Viren, J.R. & Imbus, H.R. (1989) Case–control study of nasal cancer in workers employed in wood-related industries. *J. occup. Med.*, **31**, 35–40

Vlahakis, G. (1977) Possible carcinogenic effects of cedar shavings in bedding of C3H-A^{vy}fB mice (Brief communication). *J. natl Cancer Inst.*, **58**, 149–150

Voss, R., Stenersen, T., Roald Oppedal, B. & Boysen, M. (1985) Sinonasal cancer and exposure to softwood. *Acta otolaryngol.*, **99**, 172–178

Wagenführ, R. & Scheiber, C. (1974) *Holzatlas* [Wood atlas], Leipzig, VEB Fachbuchverlag (in German)

Warheit, D.B. (1989) Interspecies comparisons of lung responses to inhaled particles and gases. *Crit. Rev. Toxicol.*, **20**, 1–29

Wegener, G. & Fengel, D. (1978) Rapid extraction of lignin from ball-milled spruce wood by the use of ultrasonics (Letter to the Editor). *Ultrasonics*, **16**, 186

Weissmann, G., Kubel, H. & Lange, W. (1989) Studies on the carcinogenicity of wood dust. Extractives of oak wood (*Quercus robur* L.). *Holzforschung*, **43**, 75–82 (in German)

Weissmann, G., Lange, W., Kubel, H. & Wenzel-Hartung, R. (1992) Investigations on the carcinogenicity of wood dust. *Holz Roh.-Werkstoff.*, **50**, 421–428 (in German)

Welling, I. & Kallas, T. (1991) *Wood Dust* (Exposure at Work No. 7), Helsinki, Finnish Institute of Occupational Health and Finnish Work Environment Fund (in Finnish)

West, S., Hildesheim, A. & Dosemeci, M. (1993) Non-viral risk factors for nasopharyngeal carcinoma in the Philippines: results from a case–control study. *Int. J. Cancer*, **55**, 722–727

Whitehead, L.W., Freund, T. & Hahn, L.L. (1981a) Suspended dust concentrations and size distributions and quantitative analysis of inorganic particles from woodworking operations. *Am. ind. Hyg. Assoc. J.*, **42**, 461–467

Whitehead, L.W., Ashikaga, T. & Vacek, P. (1981b) Pulmonary function status of workers exposed to hardwood or pine dust. *Am. ind. Hyg. Assoc.*, **42**, 178–186

Whittemore, A.S., Holly, E.A., Lee, I.-M., Abel, E.A., Adams, R.M., Nickoloff, B.J., Bley, L., Peters, J.M. & Gibney, C. (1989) Mycosis fungoides in relation to environmental exposures and immune response: a case–control study. *J. natl Cancer Inst.*, **81**, 1560–1567

Wilhelmsson, B. & Drettner, B. (1984) Nasal problems in wood furniture workers. A study of symptoms and physiological variables. *Acta otolaryngol.*, **98**, 548–555

Wilhelmsson, B. & Lundh, B. (1984) Nasal epithelium in woodworkers in the furniture industry. *Acta otolaryngol.*, **98**, 321–324

Wilhelmsson, B., Lundh, B., Drettner, B. & Stenkvist, B. (1985a) Effects of wood dust exposure and diethylnitrosamine. A pilot study in Syrian golden hamsters. *Acta otolaryngol.*, **99**, 160–171

Wilhelmsson, B., Lundh, B. & Drettner, B. (1985b) Effects of wood dust exposure and diethylnitrosamine in an animal experimental system. *Rhinology*, **23**, 114–117

Wilhelmsson, B., Hellqvist, H., Olofsson, J. & Klingenberg, C. (1985c) Nasal cuboidal metaplasia with dysplasia. *Acta otolaryngol.*, **99**, 641–648

Williston, E.M. (1988) *Lumber Manufacturing. The Design and Operation of Sawmills and Planer Mills*, San Francisco, Miller Freeman Publications

Wolf, J., Hartung M., Schaller, K.H., Kochem, W. & Valentin, H. (1986) The occurrence of adenocarcinomas of the nasal and paranasal sinuses in woodworkers. II. Further investigations. *Arbeitsmed. Sozialmed. Präventivmed.*, **7**, 3–22 (in German)

Wolf, J., Schmelzer, P., Kuchenmeister, F., Klein, R.G., Schroeder, H.G., Pool-Zobel, B., Ziegler, H., Detering, B., Fengel, D., Stehlin, J. & Kleinsasser, O. (1994) Etiology of malignant nasal tumours in people working in the furniture industry. *Arbeitsmed. Sozialmed. Präventivmed.*, **21**, 3–17 (in German)

Woods, B. & Calnan, C.D. (1976) Toxic woods. *Br. J. Dermatol.*, **94** (Suppl. 13), 1–97

Wortley, P., Vaughan, T.L., Davis, S., Morgan, M.S. & Thomas, D.B. (1992) A case–control study of occupational risk factors for laryngeal cancer. *Br. J. ind. Med.*, **49**, 837–844

Wu-Williams, A.H., Xu, Z.Y., Blot, W.J., Dai, X.D., Louie, R., Xiao, H.P., Stone, B.J., Sun, X.W., Yu, S.F., Feng, Y.P., Fraumeni, J.F., Jr & Henderson, B.E. (1993) Occupation and lung cancer risk among women in northern China. *Am. J. ind. Med.*, **24**: 67–79

Wynder, E.L., Covey, L.S., Mabuchi, K. & Mushinski, M. (1976) Environmental factors in cancer of the larynx. A second look. *Cancer*, **38**, 1591–1601

Zagraniski, R.T., Kelsey, J.L. & Walter, S.D. (1986) Occupational risk factors for laryngeal carcinoma: Connecticut, 1975–1980. *Am. J. Epidemiol.*, **124**, 67–76

Zheng, W., Blot, W.J., Shu, X.-O., Diamond, E.L., Gao, Y.-T., Ji, B.-T. & Fraumeni, J.F., Jr (1992a) A population-based case–control study of cancers of the nasal cavity and paranasal sinuses in Shanghai. *Int. J. Cancer*, **52**, 557–561

Zheng, W., Blot, W.J., Shu, X.-O., Gao, Y.-T., Ji, B.-T., Ziegler, R.G. & Fraumeni, J.F., Jr (1992b) Diet and other risk factors for laryngeal cancer in Shanghai, China. *Am. J. Epidemiol.*, **136**, 178–191

Zheng, W., McLaughlin, J.K., Chow, W.-H., Co Chien, H.T. & Blot, W.J. (1993) Risk factors for cancers of the nasal cavity and paranasal sinuses among white men in the United States. *Am. J. Epidemiol.*, **138**, 965–972

Zinkel, D.F. (1983) Quantitative separation of ether-soluble acidic and neutral materials. *J. Wood Chem. Technol.*, **3**, 131–143

FORMALDEHYDE

This substance was considered by previous working groups, in October 1981 (IARC, 1982) and March 1987 (IARC, 1987a). Since that time, new data have become available, and these have been incorporated in the monograph and taken into consideration in the evaluation.

1. Exposure Data

1.1 Chemical and physical data

1.1.1 Nomenclature

Chem. Abstr. Serv. Reg. No.: 50-00-0
Deleted CAS Reg. Nos: 8005-38-7; 8006-07-3; 8013-13-6; 112068-71-0
Chem. Abstr. Name: Formaldehyde
IUPAC Systematic Name: Methanal
Synonyms: Formaldehyde, gas; formic aldehyde; methaldehyde; methyl aldehyde; methyl oxide; methylene oxide; oxomethane; oxymethylene

1.1.2 Structural and molecular formulae and relative molecular mass

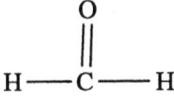

CH_2O Relative molecular mass: 30.03

1.1.3 Chemical and physical properties of the pure substance

From Lide (1993), unless otherwise noted

(a) *Description*: Colourless gas with a pungent odour (Reuss *et al.*, 1988)
(b) *Boiling-point*: −21 °C
(c) *Melting-point*: −92 °C
(d) *Density*: 0.815 at 20 °C/4 °C
(e) *Spectroscopy data*: Infrared [prism, 2538], ultraviolet [3.1] and mass spectral data have been reported (Weast & Astle, 1985; Sadtler Research Laboratories, 1991).
(f) *Solubility*: Very soluble in water, ethanol and diethyl ether
(g) *Stability*: Commercial formaldehyde–alcohol solutions are stable; the gas is stable in absence of water; incompatible with oxidizers, alkalis, acids, phenols and urea

(Gerberich et al., 1980; IARC, 1982; Cosmetic Ingredient Review Expert Panel, 1984; Reuss et al., 1988).

(h) *Reactivity*: Reacts explosively with peroxide, nitrogen oxide and performic acid; can react with hydrogen chloride or other inorganic chlorides to form bis(chloromethyl) ether (see IARC, 1987b) (Gerberich et al., 1980; IARC, 1982; Cosmetic Ingredient Review Expert Panel, 1984; Reuss et al., 1988).

(i) Octanol/water partition coefficient (P): log P = 0.35 (Sangster, 1989)

(j) Conversion factor: $mg/m^3 = 1.23 \times ppm^a$

1.1.4 Technical products and impurities

Trade names: BFV; FA; Fannoform; Floguard 1015; FM 282; Formalin; Formalin 40; Formalith; Formol; Fyde; Hoch; Ivalon; Karsan; Lysoform; Morbicid; Paraform; Superlysoform

Formaldehyde is most commonly available commercially as a 30–50% (by weight) aqueous solution, commonly referred to as 'formalin'. In dilute aqueous solution, the predominant form of formaldehyde is its monomeric hydrate, methylene glycol. In more concentrated aqueous solutions, oligomers and polymers that are mainly polyoxymethylene glycol are formed and may predominate. Methanol and other substances (e.g. various amine derivatives) are usually added to the solutions as stabilizers, in order to reduce intrinsic polymerization. The concentration of methanol can be as high as 15%, while that of other stabilizers is of the order of several hundred milligrams per litre. Concentrated liquid formaldehyde–water systems containing up to 95% formaldehyde are also available, but the temperature necessary to maintain solution and to prevent separation of the polymer increases from room temperature to 120 °C, as the concentration in solution increases. Impurities include formic acid, iron and copper (Cosmetic Ingredient Review Expert Panel, 1984).

Formaldehyde is marketed in solid form as its cyclic trimer, trioxane $(CH_2O)_3$, and its polymer, paraformaldehyde, with 8–100 units of formaldehyde (Cosmetic Ingredient Review Expert Panel, 1984; Reuss et al., 1988; WHO, 1991).

1.1.5 Analysis

The most widely used methods for the determination of formaldehyde are based on spectrophotometry, with which sensitivities of 0.01–0.03 mg/m^3 can be achieved. Other methods include colorimetry, fluorimetry, high-performance liquid chromatography, polarography, gas chromatography, infrared detection and gas detector tubes. High-performance liquid chromatography is the most sensitive method (limit of detection, 0.002 mg/m^3). In all of these methods, other organic and inorganic chemicals, such as sulfur dioxide, other aldehydes and amines, cause interference.

The method of sampling and the treatment of samples before analysis are important in the accuracy of the determination. Gas detector tubes (WHO, 1989) and infrared analysers are often

---

[a] Calculated from: mg/m^3 = (relative molecular mass/24.45) × ppm, assuming normal temperature (25 °C) and pressure (103.5 kPa)

used for monitoring workplace atmospheres, with a sensitivity of about 0.4–0.5 mg/m^3 (Gollob & Wellons, 1980; Heck et al., 1982; Kennedy et al., 1985; Kennedy & Hull, 1986; Stewart et al., 1987a; Bicking et al., 1988; Greenblatt, 1988; WHO, 1989; United States Occupational Safety and Health Administration, 1991; WHO, 1991). Selected methods for the determination of formaldehyde in various matrices are presented in Table 1.

Four methods have been developed to measure formaldehyde emissions from wood products. The large-scale chamber test developed by the Wilhelm Klauditz Institute in Germany is the principal method used in Europe. A value of 0.1 mg/m^3 in this test, used for the German 'E-1' classification, is often applied for approval of the use of formaldehyde-emitting building products. The large-scale chamber formaldehyde test method 2 (FTM-2) is used to test wood panels in Canada and the United States (European Commission, 1989; American Society for Testing and Materials, 1990; Groah et al., 1991; Jann, 1991). A 2-h desiccator test (FTM-1) is a small-scale method for determining formaldehyde emitted from wood products; formaldehyde, absorbed in distilled water, reacts specifically with a chromotropic acid–sulfuric acid solution (National Particleboard Association, 1983; Groah et al., 1991).

In the perforator method for extracting formaldehyde, small samples are boiled in toluene, and the formaldehyde-laden toluene is distilled through distilled/deionized water, which absorbs the formaldehyde; a sample of the water is then analysed photometrically by the acetylacetone or pararosaniline method. In the iodometric method, formaldehyde in water is determined by adding sulfuric acid solution and an excess of iodine; the iodine oxidizes the formaldehyde, and the excess is back-titrated with sodium thiosulfate (British Standards Institution, 1989).

1.2 Production and use

1.2.1 Production

Since 1889 in Germany, formaldehyde has been produced commercially by the catalytic oxidation of methanol. Various methods were used in the past, but only two are widely used currently: the silver catalyst and metal oxide catalyst processes (Reuss et al., 1988; Gerberich & Seaman, 1994).

The silver catalyst process is conducted in one of two ways: (i) partial oxidation and dehydrogenation with air in the presence of silver crystals, steam and excess methanol at 680–720 °C and atmospheric pressure (also called the BASF process; methanol conversion, 97–98%); and (ii) partial oxidation and dehydrogenation with air in the presence of crystalline silver or silver gauze, steam and excess methanol at 600–650 °C (primary conversion of methanol, 77–87%); the conversion is completed by distilling the product and recycling the unreacted methanol. Carbon monoxide, carbon dioxide, methyl formate and formic acid are by-products (Gerberich et al., 1980; Reuss et al., 1988; Gerberich & Seaman, 1994).

In the metal oxide (Formox) process, methanol is oxidized with excess air in the presence of a modified iron–molybdenum–vanadium oxide catalyst at 250–400 °C and atmospheric pressure (methanol conversion, 98–99%). By-products are carbon monoxide and dimethyl ether and small amounts of carbon dioxide and formic acid (Gerberich et al., 1980; Reuss et al., 1988; Gerberich & Seaman, 1994).

Table 1. Methods for the analysis of formaldehyde in air and food

Sample matrix	Sample preparation	Assay procedure	Limit of detection	Reference
Air	Draw air through impinger containing aqueous pararosaniline; treat with acidic pararosaniline and sodium sulfite	S	0.01 mg/m^3	Georghiou et al. (1993)
	Draw air through PTFE filter and impingers, each treated with sodium bisulfite solution; develop colour with chromotropic acid and sulfuric acid; read absorbance at 580 nm	S	0.03 mg/m^3	Eller (1989a) [Method 3500]
	Draw air through solid sorbent tube treated with 10% 2-(hydroxymethyl) piperidine on XAD-2; desorb with toluene	GC/FID	0.3 mg/m^3	Eller (1989b) [Method 2541]
		GC/NSD	0.02 mg/m^3	United States Occupational Safety and Health Administration (1990) [Method 52]
	Draw air through impinger containing hydrochloric acid/2,4-dinitrophenyl-hydrazine reagent and isooctane; extract with hexane/dichloromethane	HPLC/UV	0.002 mg/m^3	United States Environmental Protection Agency (1988a) [Method TO5]
	Draw air through silica gel coated with acidified 2,4-dinitrophenylhydrazine reagent	HPLC/UV	0.002 mg/m^3	United States Environmental Protection Agency (1988b) [Method TO11]
	Expose passive monitor (Du Pont Pro-Tek® Formaldehyde Badge) for at least 2 ppm-h. Analyse according to manufacturer's specifications	Chromotropic acid test	0.1 mg/m^3	Kennedy & Hull (1986); Stewart et al. (1987a)
Food	Distil sample; add 1,8-dihydroxy-naphthalene-3,6-disulfonic acid in H_2SO_4; purple colour indicates presence of formaldehyde	Chromotropic acid test	NR	Helrich (1990) [Method 931.08]
	Distil sample; add to cold H_2SO_4; add aldehyde-free milk; add bromine hydrate solution; purplish-pink colour indicates presence of formaldehyde	Hehner-Fulton test	NR	Helrich (1990) [Method 931.08]

S, spectrometry; PTFE, polytetrafluoroethylene; GC/FID, gas chromatography/flame ionization detection; GC/NSD, gas chromatography/nitrogen selective detection; HPLC/UV, high-performance liquid chromatography/-ultraviolet detection; NR, not reported

Paraformaldehyde, a solid polymer of formaldehyde, consists of a mixture of poly(oxymethylene) glycols [HO–(CH$_2$O)$_n$–H; n = 8–100]. The formaldehyde content is 90–99%, depending on the degree of polymerization, n and product specifications; the remainder is bound or free water. Paraformaldehyde, a convenient source of formaldehyde for certain applications, is prepared commercially by concentrating aqueous formaldehyde solutions under vacuum in the presence of small amounts of formic acid and metal formates. An alternative solid source of formaldehyde, 1,3,5-trioxane, is the cyclic trimer of formaldehyde and is prepared commercially by strong-acid-catalysed condensation of formaldehyde in a continuous process (Reuss *et al.*, 1988; Gerberich & Seaman, 1994).

Formaldehyde is reported to be produced by 11 companies in Japan, eight companies each in China and the United States, seven companies in Italy, six companies each in Brazil, Mexico, Spain and the United Kingdom, five companies in Germany, four companies in India, three companies each in Austria and Canada, two companies each in Argentina, Australia, Finland, France, Israel, the Republic of Korea, Sweden, Turkey and the former Yugoslavia and one company each in Bulgaria, Colombia, the former Czechoslovakia, Ecuador, Greece, Hungary, New Zealand, Norway, Pakistan, Peru, Philippines, Poland, Portugal, South Africa, Switzerland, the Russian Federation and Thailand. Paraformaldehyde is reported to be produced by three companies in Japan and China, two companies in Spain and one company each in Argentina, Brazil, France, Germany, Israel, Mexico, the United Kingdom and the United States. 1,3,5-Trioxane is produced by three companies Germany (Chemical Information Services Ltd, 1991).

Worldwide production of formaldehyde in 1992 was approximately 12 million tonnes (Smith, 1993). Production of formaldehyde in selected years and countries is shown in Table 2. In 1986, most of the worldwide production capacity was based on silver catalyst processes (Reuss *et al.*, 1988).

1.2.2 Use

The widest use of formaldehyde is in the production of resins with urea, phenol and melamine and, to a small extent, their derivatives. Formaldehyde-based resins are used as adhesives and impregnating resins in the manufacture of particle-board, plywood, furniture and other wood products. They are also used for the production of curable moulding materials (appliances, electric controls, telephones, wiring services) and as raw materials for surface coatings and controlled-release nitrogen fertilizers. They are used in the textile, leather, rubber and cement industries. Further uses are as binders for foundry sand, stonewool and glasswool mats in insulating materials, abrasive paper and brake linings. Smaller amounts of urea–formaldehyde resins are used in the manufacture of foamed resins for mining and for building insulation (Reuss *et al.*, 1988; WHO, 1989; Gerberich & Seaman, 1994).

Another major use of formaldehyde is as an intermediate for synthesizing other industrial chemical compounds, such as 1,4-butanediol, trimethylolpropane and neopentyl glycol, which are used in the manufacture of polyurethane and polyester plastics, synthetic resin coatings, synthetic lubricating oils and plasticizers. Other compounds produced from formaldehyde

Table 2. Production of formaldehyde in selected countries (thousand tonnes)

Country or region	1982	1986	1990
Brazil	152	226	NA
Canada	70	117	106
China	286	426	467
Former Czechoslovakia	254	274	NA
Denmark	NA	3	0.3
Finland	NA	5	48
France	79	80	100
Germany	630	714	680
Hungary	13	11	NA
Italy	125	135	114
Japan	NA	1188	1460
Mexico	83	93	NA
Poland	219	154	NA
Portugal	NA	70	NA
Republic of Korea	NA	122	NA
Spain	NA	91	136
Sweden	NA	223	244
Taiwan	NA	204	215
Turkey	NA	21	NA
United Kingdom	107	103	80
United States[a]	2185	2517	3048
Former Yugoslavia	108	99	88

From Anon. (1985, 1989); Japan Chemical Week (1991); Anon. (1993); China National Chemical Information Centre (1993); Anon. (1994); Data-Star/Dialog (1994)
NA, not available
[a] 37% by weight

include pentaerythritol, used primarily in raw materials for surface coatings and explosives, and hexamethylenetetramine, used as a cross-linking agent for phenol–formaldehyde resins and explosives. The complexing agents nitrilotriacetic acid (see IARC, 1990a) and ethylenediaminetetraacetic acid are derived from formaldehyde and are components of some detergents. There is a steadily increasing demand for formaldehyde for the production of 4,4'-diphenylmethane diisocyanate (see IARC, 1979), which is a constituent of polyurethanes used in the production of soft and rigid foams and, more recently, as an adhesive and for bonding particle-board (Reuss et al., 1988; WHO, 1989; Gerberich & Seaman, 1994).

Polyacetal plastics produced by polymerization of formaldehyde are incorporated into automobiles to reduce weight and fuel consumption, and are used to make functional components of audio and video electronics equipment. Formaldehyde is also the basis for products used to manufacture dyes, tanning agents, dispersion and plastics precursors, extraction agents, crop protection agents, animal feeds, perfumes, vitamins, flavourings and drugs (Reuss *et al.*, 1988; WHO, 1989).

Formaldehyde itself is used for preservation and disinfection, for example, in human and veterinary drugs and biological materials (viral vaccines contain 0.05% formalin as an inactivating agent), for disinfecting hospital wards and preserving and embalming biological specimens. It is used as an antimicrobial agent in many cosmetics products, including soaps, shampoos, hair preparations, deodorants, lotions, make-up, mouthwashes and nail products (Cosmetic Ingredient Review Expert Panel, 1984; Reuss *et al.*, 1988). Formaldehyde is also used directly to inhibit corrosion, in mirror finishing and electroplating, in the electrodeposition of printed circuits and in photographic film development (Reuss *et al.*, 1988).

Paraformaldehyde is used in place of aqueous formaldehyde solutions, especially when the presence of water interferes, e.g. in the plastics industry for the preparation of phenol, urea and melamine resins, varnish resins, thermosets and foundry resins. Other uses include the synthesis of chemical and pharmaceutical products (e.g. Prins reaction, chloromethylation, Mannich reaction), the production of textile products (e.g. for crease-resistant finishes), preparation of disinfectants and deodorants (Reuss *et al.*, 1988) and in selected pesticide applications (United States Environmental Protection Agency, 1993).

The pattern of use of formaldehyde in the United States in 1992 (3 million tonnes) was: urea–formaldehyde resins, 24%; phenolic resins, 21%; acetylenic chemicals (precursors to diol monomers), 11%; polyacetal resins, 9%; pentaerythritol, 6%; urea–formaldehyde concentrates, 5%; diphenylmethane diisocyanate, 5%; hexamethylenetetramine, 4%; melamine, 4%; other (including chelating agents, trimethylolpropane, pyridine chemicals), 10% (Anon., 1992). The pattern of use in Japan in 1992 (1.4 million tonnes) was: urea–melamine adhesive, 25.4%; polyacetal resins, 22.7%; pentaerythritol, 7.4%; phenolic resins, 7.3%; paraformaldehyde, 6.0%; diphenylmethane diisocyanate, 4.4%; urea–melamine resin, excluding adhesives, 3.3%; hexamethylenetetramine, 3.1%; other uses, 20.1% (Japan Chemical Week, 1993).

1.3 Occurrence

The natural and man-made sources of formaldehyde in the environment and environmental levels in indoor and outdoor air, water, soil and food have been reviewed (WHO, 1989; see also IARC, 1982). Information on sources and emissions of formaldehyde in the United States has been compiled by the Environmental Protection Agency (Vaught, 1991).

1.3.1 Natural occurrence

Formaldehyde is ubiquitous in the environment; it is an important endogenous chemical that occurs in most life forms, including humans. It is formed naturally in the troposphere during the oxidation of hydrocarbons, which react with hydroxyl radicals and ozone to form formaldehyde and other aldehydes, as intermediates in a series of reactions that ultimately lead to the formation of carbon monoxide and dioxide, hydrogen and water. Of the hydrocarbons found in the troposphere, methane is the single most important source of formaldehyde. Terpenes and isoprene, emitted by foliage, react with hydroxyl radicals, forming formaldehyde as an intermediate product. Because of their short half-life, these potentially important sources of formaldehyde are important only in the vicinity of vegetation. Formaldehyde is one of the volatile compounds formed in the early stages of decomposition of plant residues in the soil (WHO, 1989), and it occurs naturally in fruits and other foods (WHO, 1991).

1.3.2 Occupational exposure

(a) Extent of exposure

As nonoccupational exposure to formaldehyde is ubiquitous, all work, e.g. in offices, contributes to total human exposure. About 1 500 000 workers in the United States in 1981–83 were estimated in the National Occupational Exposure Survey to be exposed to formaldehyde all or part of the time, representing about 0.6% of the population of the country. Industries in which more than 50 000 workers were exposed included health services, business services, printing and publishing, manufacture of chemicals and allied products, apparel and allied products, paper and allied products, personal services, machinery except electrical, transport equipment and furniture and fixtures. The minimal exposure to formaldehyde was not specified in this survey (United States National Institute for Occupational Safety and Health, 1990). In Finland, 9000–10 000 workers, representing about 0.2% of the Finnish population, were estimated to have been exposed to an 8-h time-weighted concentration of at least 0.3 mg/m^3 during at least one day per year. If the minimal level considered is decreased to 0.15 mg/m^3 (0.12 ppm), the number of people exposed increases by many thousands (Heikkilä *et al.*, 1991). It is impossible to estimate accurately the number of people occupationally exposed to formaldehyde worldwide, but it is likely to be several millions in industrialized countries alone.

Formaldehyde occurs in occupational environments mainly as gas. Inhalation of formaldehyde-containing particulates may occur when paraformaldehyde or powdered resins are being used. For example, production of solid resins may include dust-forming operations, such as drying, crushing grinding and screening (Stewart *et al.*, 1987b). Formaldehyde-based resins may also occur in air attached to carrier agents, such as wood dust from sawing of plywood (Kauppinen, 1986). Dermal exposure to formaldehyde is possible when formalin solutions or liquid resins come into contact with the skin.

(b) Manufacture of formaldehyde, formaldehyde-based resins and other chemical products

The concentrations of formaldehyde measured in the 1980s during the manufacture of formaldehyde and formaldehyde-based resins are summarized in Table 3. The mean levels during manufacture were below 1 ppm (1.2 mg/m^3). The values are sometimes reported as geometric means, which give less weight to occasional heavy exposures than arithmetic means. The workers may also be exposed to methanol (raw material), carbon monoxide, carbon dioxide and hydrogen (process gases) (Stewart *et al.*, 1987b).

The reported mean concentrations in the air of factories producing formaldehyde-based resins vary from < 1 to > 10 ppm (< 1.23–> 12.3 mg/m^3) . There are obvious differences between the factories but no consistent seasonal variation. The earliest measurements are from 1979. The chemicals other than formaldehyde to which exposure may occur depend on the types of resins manufactured: Urea, phenol, melamine and furfural alcohol are the chemicals most commonly reacted with liquid formaldehyde (formalin) or hexamethylenetetramine. Some processes require addition of ammonia. Alcohols are used as solvents in the production of liquid resins (Stewart et al., 1987b).

Some potential occupational exposures to formaldehyde

Agricultural workers	Fur processors (see IARC, 1981)
Anatomists	Furniture makers (see IARC, 1981)
Beauticians	Glue and adhesive makers
Biologists	Hide preservers (see IARC, 1981)
Bookbinders	Histology technicians (including necropsy and autopsy technicians)
Botanists	
Chemical production workers	Ink makers
Cosmetic formulators	Lacquerers and lacquer makers
Crease-resistant textile finishers	Medical personnel (including pathologists)
Disinfectant makers	Mirror manufacturers
Disinfectors	Paper makers (see IARC, 1981)
Dress-goods shop personnel	Particle-board makers (see IARC, 1981)
Electrical insulation makers	Photographic film makers
Embalmers	Plastics workers
Embalming-fluid makers	Plywood makers
Fireproofers	Rubber makers (see IARC, 1982)
Formaldehyde production workers	Taxidermists
Formaldehyde resin makers	Textile mordanters and printers
Foundry employees (see IARC, 1984)	Textile waterproofers
Fumigators (see IARC, 1991)	Varnish workers (see IARC, 1981)
	Wood preservers (see IARC, 1981)

From United States National Institute for Occupational Safety and Health (1976)

No measurements were available to the Working Group of exposure to formaldehyde in other chemical plants where it is used, e.g. in the production of pentaerythritol, hexamethylenetetramine or ethylene glycol.

(c) Manufacture of wood products and paper

Table 4 is a summary of the formaldehyde concentrations in wood and pulp and paper industries. Formaldehyde-based glues have been used in the assembly of plywood for over 30 years. The highest mean concentrations are usually measured in glueing departments, where the glue mixture is prepared, veneers are glued to form plywood and the plywood is cured in hot presses; the mean levels were usually > 1 ppm (1.2 mg/m^3) before the mid-1970s but have been below that level more recently. This decrease in levels is consistent in each operation and is due

Table 3. Concentrations of formaldehyde (in ppm [mg/m^3]) in the workroom air in formaldehyde and resin manufacturing plants

Industry and operation	No. of measurements	Mean[a]	Range	Year	Reference
Chemical factory producing formaldehyde and formaldehyde resins (Sweden)	62	0.2 [0.3]	0.04–0.4 [0.05–0.5]	1979–85	Holmström et al. (1989a)
Production of formaldehyde (Sweden)	9	0.3 [0.34]		1980s	Rosén et al. (1984)
Formaldehyde manufacture (USA)				1983	Stewart et al. (1987b)
Plant no. 2, summer	15	0.6[b] [0.7]	0.03–1.9 [0.04–2.3]		
Plant no. 10, summer	9	0.7[b] [0.9]	0.6–0.8 [0.7–1.0]		
Resin plant (Finland)					Heikkilä et al. (1991)
Furan resin production	3	2.3 [2.9]	1.0–3.4 [1.3–4.2]	1982	
Maintenance	4	2.9 [3.6]	1.4–5.5 [1.8–6.9]	1981	
Urea–formaldehyde resin production	7	0.7 [0.87]	0.6–0.8 [0.7–1.1]	1981	
Resin manufacture (Sweden)	22	0.5 [0.6]		1980s	Rosén et al. (1984)
Resin manufacture (USA)				1983–84	Stewart et al. (1987b)
Plant no. 1, summer	24	3.4[b] [4.2]	0.2–13.2 [0.3–16.2]		
Plant no. 6, summer	6	0.2[b,c] [0.3]	0.1–0.2 [0.1–0.3]		
Plant no. 7, summer	9	0.2[b] [0.3]	0.1–0.3 [0.1–0.4]		
Plant no. 7, winter	9	0.6[b] [0.7]	0.4–0.9 [0.5–1.1]		
Plant no. 8, summer	13	0.4[b-d] [0.7]	0.2–0.8 [0.3–1.0]		
Plant no. 8, winter	9	0.1[b-d] [0.1]	0.1–0.2 [0.1–0.3]		
Plant no. 9, summer	8	14.2[b-d] [17.5]	4.1–30.5 [5.0–37.5]		
Plant no. 9, winter	9	1.7[b] [2.1]	1.1–2.5 [1.4–3.1]		
Plant no. 10, summer	23	0.7[b,d] [0.9]	0.3–1.2 [0.4–1.5]		
Special chemical manufacturing plant (USA)	8		<0.03–1.6 [0.04–2.0]		Blade (1983)

[a] Arithmetic mean unless otherwise specified
[b] Mean and range of geometric means
Some of the results were affected by the simultaneous occurrence in the samples (Stewart et al., 1987b) of:
[c] phenol (leading to low values)
[d] particulates containing nascent formaldehyde (leading to high values).

Table 4. Concentrations of formaldehyde (in ppm [mg/m^3]) in the workroom air of plywood mills, particle-board mills, furniture factories, other wood product plants and paper mills

Industry and operation	No. of measurements	Mean[a]	Range	Year	Reference
Plywood mills					
Plywood factory (Italy)					Ballarin et al. (1992)
Warehouse	3	0.3 [0.4]	0.1–0.5 [0.2–0.6]	NR	
Shearing press	8	0.08 [0.1]	0.06–0.11 [0.08–0.14]		
Plywood mills (Finland)					Kauppinen (1986)
Glue preparation, short-term	15	2.2 [2.7]	0.6–5.0 [0.7–6.2]	1965–74	
Glue preparation, short-term	19	0.7 [0.9]	0.1–2.3 [0.1–2.8]	1975–84	
Assembling	32	1.5 [1.9]	<0.1–4.4 [<0.1–5.4]	1965–74	
Assembling	55	0.6 [0.7]	0.02–6.8 [0.03–8.3]	1975–84	
Hot pressing	41	2.0 [2.5]	<0.1–7.7 [<0.1–9.5]	1965–74	
Hot pressing	43	0.5 [0.6]	0.06–2.1 [0.07–2.6]	1975–84	
Sawing of plywood	5	0.5 [0.6]	0.3–0.8 [0.4–1.0]	1965–74	
Sawing of plywood	12	0.1 [0.1]	0.02–0.2 [0.03–0.3]	1975–84	
Coating of plywood	7	1.0 [1.2]	0.5–1.8 [0.6–2.2]	1965–74	
Coating of plywood	28	0.3 [0.4]	0.02–0.6 [0.03–0.7]	1975–84	
Plywood mill (Indonesia)	40	0.6 [0.8]	0.2–2.3 [0.3–2.8]	NR	Malaka & Kodama (1990)
Plywood production (Sweden)	47	0.3 [0.36]		1980s	Rosén et al. (1984)
Plywood panelling manufacture (USA)				1983–84	Stewart et al. (1987b)
Plant no 3, winter	27	0.2[b] [0.3]	0.08–0.4 [1.0–0.5]		
Plant no 3, summer	26	0.1[b] [0.1]	0.01–0.5 [0.01–0.6]		
Particle-board mills					
Particle-board mills (Finland)					Kauppinen & Niemelä (1985)
Glue preparation	10	2.2 [2.7]	0.3–4.9 [0.4–6.0]	1975–84	
Blending	10	1.0 [1.2]	0.1–2.0 [0.1–2.5]	1965–74	
Blending	8	0.7 [0.9]	<0.1–1.4 [<0.1–1.7]	1975–84	
Forming	26	1.7 [2.1]	<0.5–4.6 [<0.6–5.7]	1965–74	

Table 4 (contd)

Industry and operation	No. of measurements	Mean[a]	Range	Year	Reference
Particle-board mills (contd)					
Particle-board mills (Finland) (contd)					
Forming	32	1.4 [1.7]	0.1–4.8 [0.1–5.9]	1975–84	
Hot pressing	35	3.4 [4.2]	1.1–9.5 [1.4–11.7]	1965–74	
Hot pressing	61	1.7 [2.1]	0.2–4.6 [0.25–5.7]	1975–84	
Sawing	17	4.8 [5.9]	0.7–9.2 [0.9–11.3]	1965–74	
Sawing	36	1.0 [1.2]	< 0.1–3.3 [< 0.1–4.1]	1975–84	
Coating	7	1.0 [1.2]	0.5–1.8 [0.6–2.2]	1965–74	
Coating	12	0.4 [0.5]	0.1–1.2 [0.1–1.5]	1975–84	
Particle-board mill (Indonesia)	9	2.4 [3.0]	1.2–3.5 [1.5–4.3]	NR	Malaka & Kodama (1990)
Particle-board production (Sweden)	21	0.3 [0.4]		1980s	Rosén et al. (1984)
Chip-board production (Germany)	24	1.5 [1.9]	< 0.01–8.4 [< 0.01–10]	1980–88	Triebig et al. (1989)
Block-board mill (Indonesia)	6	0.5 [0.6]	0.4–0.6 [0.5–0.7]	NR	Malaka & Kodama (1990)
Medium-density fibre-board production (Sweden)	19	0.2 [0.3]		1980s	Rosén et al. (1984)
Furniture factories					
Furniture factories, surface finishing with acid curing paints (Sweden)				NR	Alexandersson & Hedenstierna (1988)
Paint mixer/supervisor	6	0.2 [0.3]	0.1–0.4 [0.2–0.5]		
Mixed duties on the line	5	0.4 [0.5]	0.3–0.5 [0.3–0.6]		
Assistant painters	3	0.5 [0.6]	0.2–0.7 [0.2–0.9]		
Spray painters	10	0.4 [0.5]	0.1–1.1 [0.2–1.3]		
Feeder/receiver	13	0.2 [0.3]	0.1–0.8 [0.1–0.9]		
Furniture factories (Finland)				1981–86	Heikkilä et al. (1991)
Glueing	73	0.3 [0.4]	0.07–1.0 [0.09–1.2]		
Machining in finishing department	9	0.3 [0.5]	0.1–0.9 [0.1–1.1]		
Varnishing	150	1.1 [1.4]	0.1–6.3 [0.1–7.9]		

Table 4 (contd)

Industry and operation	No. of measurements	Mean[a]	Range	Year	Reference
Furniture factories (contd)					
Furniture factories (Finland) (contd)				1975–84	Priha et al. (1986)
Feeding painting machine	14	1.1 [1.4]	0.3–2.7 [0.4–3.3]		
Spray painting	60	1.0 [1.2]	0.2–4.0 [0.3–5.0]		
Spray painting assistance	10	1.0 [1.2]	0.2–1.6 [0.3–2.0]		
Curtain painting	18	1.1 [1.4]	0.2–6.1 [0.3–7.5]		
Before drying of varnished furniture	34	1.5 [1.8]	0.1–4.2 [0.1–5.2]		
After drying of varnished furniture	14	1.4 [1.7]	0.2–5.4 [0.3–6.6]		
Furniture factories (Sweden)				1980s	Rosén et al. (1984)
Varnishing with acid-cured varnishes	32	0.7 [0.9]			
Manufacture of furniture (Denmark)				NR	Vinzents & Laursen (1993)
Painting	43	0.2 [0.3]	0.15–0.2 [0.2–0.24]		
Glueing	68	0.1 [0.2]	0.1–0.14 [0.1–0.2]		
Cabinet-making (Canada)	48		<0.1 [<0.1]	NR	Sass-Kortsak et al. (1986)
Other wood product plants					
Match mill, impregnation of matchbox parts, short-term (Finland)	2	2.0 [2.5]	1.9–2.1 [2.3–2.6]	1963	FIOH (1994)
Wooden container mill, glueing and sawing (Finland)	6	0.3 [0.4]	0.2–0.4 [0.3–0.5]	1961	FIOH (1994)
Manufacture of wooden bars (Finland)				1983	Heikkilä et al. (1991)
Glueing	33	0.6 [0.7]	0.16–1.9 [0.2–2.4]		
Machining	7	1.2 [1.5]	0.2–2.2 [0.3–2.7]		
Parquet plant (Finland)				1981	Heikkilä et al. (1991)
Machining	3	0.3 [0.4]	0.16–0.5 [0.2–0.6]		
Varnishing	5	0.8 [1.0]	0.2–1.4 [0.3–1.7]		
Production of wooden structures (Finland)				1981–86	Heikkilä et al. (1991)
Glueing	36	0.7 [0.8]	0.07–1.8 [0.1–2.2]		
Machining	19	0.4 [0.44]	0.1–0.8 [0.1–0.9]		
Glueing in wood industry (Sweden)	65	0.2 [0.26]		1980s	Rosén et al. (1984)

Table 4 (contd)

Industry and operation	No. of measurements	Mean[a]	Range	Year	Reference
Paper mills					
Paper mill (Finland)					Finnish Institute of Occupational Health (1994)
Glueing, hardening, lamination and rolling of special paper	12	0.9 [1.1]	0.3–2.5 [0.4–3.1]	1971–73	
Impregnation of paper with phenol resin, partly short-term	38	7.4 [9.1]	<1.0–33.0 [<1.1–40.6]	1968–69	
Paper storage, diesel truck traffic	5	0.3 [0.4]	0.2–0.4 [0.25–0.5]	1969	
Paper mill (Finland)				1975–84	Heikkilä et al. (1991)
Coating of paper	30	0.7 [0.9]	0.4–31 [0.5–39]		
Gum paper production	4	0.4 [0.5]	0.3–0.6 [0.3–0.8]		
Impregnation of paper with amino resin	6	3.1 [3.9]	0.5–13 [0.6–16]		
Impregnation of paper with phenol resin	20	0.1 [0.1]	0.05–0.3 [0.06–0.4]		
Laminated paper production (Sweden)	23	0.3 [0.39]		1980s	Rosén et al. (1984)
Manufacture of offset paper (Sweden)	8	0.2 [0.21]		1980s	Rosén et al. (1984)
Lamination and impregnation of paper with melamine and phenol resins (USA)				1983	Stewart et al. (1987b)
Plant no 6, summer	53	0.7[b-d] [0.9]	<0.01–7.4 [<0.01–9.1]		
Plant no 6, winter	39	0.3[b,d] [0.4]	0.05–0.7 [0.06–0.9]		

NR, not reported
[a] Arithmetic mean unless otherwise specified
[b] Mean and range of geometric means
Some of the results were affected by the simultaneous occurrence in the samples (Stewart et al., 1987) of:
[c] phenol (leading to low values)
[d] particulates containing nascent formaldehyde (leading to high values).

mainly to the introduction of glues that release less formaldehyde (Kauppinen, 1986). Other exposures in plywood mills include wood dust (see monograph, p. 74), phenol (see IARC, 1989a), pesticides (see IARC, 1991), heating products from coniferous veneers, solvents from coating materials and engine exhaust from forklift trucks (see IARC, 1989b). These exposures are described in more detail in the monograph on wood dust.

Particle-board mills, which started operating in many countries in the 1950s, use urea–formaldehyde resins, which are mixed with wood particles to form particle-board and then cured in hot presses. Phenol–, melamine– and resorcinol–formaldehyde resins are less commonly used in particle-board mills than in plywood mills, mainly for economic reasons. The levels of formaldehyde measured in particle-board mills before the mid-1970s were high—often well over 2 ppm (2.5 mg/m^3)—but the development of glues with a lower formaldehyde content and ventilation systems have decreased the levels to about 1 ppm (1.2 mg/m^3) or below (Kauppinen & Niemelä, 1985). Other exposures in particle-board and other reconstituted-board mills are usually similar to those in plywood mills. Exposure to wood dust is described in detail in the monograph on wood dust.

Furniture varnishes may contain urea–formaldehyde (carbamide) resins dissolved in organic solvents. Finnish workers varnishing, lacquering or painting wooden furniture in 1975–84 were continuously exposed to an average level of about 1 ppm (1.2 mg/m^3) formaldehyde (Priha *et al.*, 1986), but the levels decreased slightly during this period. The levels in Sweden in the 1980s were somewhat lower (Rosén *et al.*, 1984; Alexandersson & Hedenstierna, 1988). Formaldehyde-based glues are also used occasionally in veneering wood-based boards, but the levels associated with glueing are generally lower, 0.1–0.3 ppm [0.12–0.37 mg/m^3] (Heikkilä *et al.*, 1991; Vinzents & Laursen, 1993). Other exposures in furniture factories are to wood dust, organic solvents and pigments, described in detail in the monograph on wood dust.

Some paper mills produce special products coated with formaldehyde-based phenol or amino (urea or melamine) resins. Coating agents and other chemicals used in paper mills may also contain formaldehyde as a bactericide. The average levels related to lamination and impregnation of paper in a mill in the United States in the 1980s were below 1 ppm (Stewart *et al.*, 1987b). In Sweden and Finland, the levels of formaldehyde are also usually below 1 ppm, but there is considerable variation, depending on the resin used and the product manufactured. The earliest measurements, from the late 1960s, suggest that much higher exposure may occur in some circumstances (Table 4). Other exposures in paper-coating mills include phenol, urea, melamine, paper dust, solvents and engine exhaust from factory trucks.

(d) Manufacture of textiles and garments

The use of formaldehyde-based resins to produce crease-resistant fabrics started in the 1950s. The early resins contained substantial amounts of extractable formaldehyde: over 0.4% by weight of fabric. Introduction of dimethyloldihydroxyethyleneurea resins in 1970 reduced the levels of free formaldehyde in fabrics to 0.15–0.2%. Since then, methylation of dimethyloldihydroxyethyleneurea and other modifications of the resin have decreased the level of formaldehyde gradually to 0.01–0.02% (Elliott *et al.*, 1987). Some flame-retardants, such as Pyrovatex CP, however, contain agents that release formaldehyde (Heikkilä *et al.*, 1991).

Measurements of formaldehyde in the air of textile mills in the late 1970s and 1980s show average levels of 0.2–2 ppm (0.25–2.5 mg/m^3) (Table 5). Levels were probably higher in the 1950s and 1960s because the content of free formaldehyde in resins was higher (Elliott et al., 1987). Finishing workers in textile mills may also be exposed to textile dyes, flame retardants, carrier agents, textile finishing agents and solvents. The exposures of textile workers are described in an earlier monograph (IARC, 1990b).

Measurements from the 1980s indicated that the formaldehyde levels in the garment industry were relatively low, usually averaging 0.1–0.2 ppm (0.12–0.25 mg/m^3) (Table 5). Exposures in the past were probably higher owing to the higher content of formaldehyde in fabrics. For example, the mean formaldehyde concentration in air increased from 0.1 to 1.0 ppm (0.12–1.23 mg/m^3) in a study in the United States when the formaldehyde content of the fabric increased from 0.015 to 0.04% (Luker & Van Houten, 1990). The concentration of formaldehyde was reported to have been 0.9–2.7 ppm (1.1–3.3 mg/m^3) in a post-cure garment manufacturing plant and 0.3–2.7 ppm (0.4–3.3 mg/m^3) in eight other garment plants in the United States in 1966. Few chemicals other than formaldehyde are used in garment factories. Cutting and sewing of fabrics release low levels of textile dust, and small amounts of chlorinated organic solvents are used in cleaning spots. Pattern copying machines may emit ammonia and dimethylthiourea in some plants (Elliott et al., 1987).

(e) Manufacture of metal products, mineral wool and other products

Formaldehyde-based resins are commonly used as core binders in foundries. Urea–formaldehyde resin is usually blended with oleoresin or phenol–formaldehyde resin and mixed with sand to form a core, which is then cured by baking in an oven or by heating from inside the core box (hot box method). The original hot box binder was a mixture of urea–formaldehyde resin and furfuryl alcohol, commonly referred to as furan resin. The furan resins were then modified with phenol to produce urea–formaldehyde/furfuryl alcohol, phenol–formaldehyde/-furfuryl alcohol and phenol–formaldehyde/urea–formaldehyde resins. The mean levels of formaldehyde measured in core-making and operations following core-making in the 1980s in Sweden and Finland were usually below 1 ppm (Table 6); however, measurements made before 1975 suggest that past exposures may have been considerably higher (Heikkilä et al., 1991). Many other chemicals occur in foundries, e.g. silica (IARC, 1987c) and other mineral dusts, polynuclear aromatic hydrocarbons (IARC, 1983), asbestos (IARC, 1987d), metal fumes and dusts, carbon monoxide, isocyanates (IARC, 1986a), phenols (IARC, 1989a), organic solvents and amines. These have been described in a previous monograph (IARC, 1984).

Phenol–formaldehyde resins are commonly used to bind man-made mineral fibre products. Measurements in glass-wool and stone-wool plants in the 1980s showed mean concentrations of 0.1–0.2 ppm (0.12–0.25 mg/m^3) formaldehyde (Table 6). Highest levels were measured occasionally in Finnish factories close to cupola ovens and hardening chambers (Heikkilä et al., 1991). Other exposures in man-made mineral fibre production were described in a previous monograph (IARC, 1988).

Formaldehyde-based plastics are used in the production of electrical parts, dishware and various other plastic products. The concentrations of formaldehyde measured in such industries

Table 5. Concentrations of formaldehyde (in ppm [mg/m^3]) in the workroom air of textile mills and garment factories

Industry and operation	No. of measurements	Mean[a]	Range	Year	Reference
Textile mills					
Textile plants (Finland)				1975–78	Nousiainen & Lindqvist (1979)
Finishing department, mixing	8	0.8 [1.1]	<0.2–>5 [<0.2–>6]		
Crease-resistant treatment	52	0.4 [0.5]	<0.2–>3 [<0.2–>4]		
Finishing department (excluding crease-resistant and flame-retardant treatment)	17	0.3 [0.4]	<1.3 [<1.5]		
Flame-retardant treatment	67	1.9 [2.5]	<0.2–>10 [<0.2–>11]		
Fabric store	6	0.8 [1.1]	0.1–1.3 [0.1–1.6]		
Textile mills (Sweden)				1980s	Rosén et al. (1984)
Crease-resistant treatment	29	0.2 [0.23]			
Flame-retardant treatment	2	1.2 [1.5]			
Garment factories					
Manufacture from crease-resistant cloth (USA)	181	~0.2 [~0.25]	<0.1–0.9 [<0.1–1.1]	1980s	Blade (1983)
Manufacture of shirts from fabric treated with formaldehyde-based resins (USA)	326		<0.1–0.4 [<0.1–0.5]		Elliott et al. (1987)
Garment industry (Finland)				1981–86	Heikkilä et al. (1991)
Handling of leather	3	0.1 [0.1]	0.02–0.1 [0.03–0.13]		
Pressing	32	0.2 [0.3]	0.02–0.7 [0.03–0.86]		
Sewing	15	0.1 [0.1]	0.02–0.3 [0.05–0.34]		
Sewing plant (USA)				NR	Luker & Van Houten (1990)
Processing of 0.04% formaldehyde fabric	9	1.0 [1.2]	0.5–1.1 [0.6–1.4]		
Processing of 0.016% formaldehyde fabric	9	0.1 [0.1]	<0.1–0.2 [<0.1–0.25]		

NR, not reported

Table 6. Concentrations of formaldehyde (in ppm [mg/m^3]) in the workroom air of foundries and other industrial facilities

Industry and operation	No. of measurements	Mean[a]	Range	Year	Reference
Foundries					
Foundries (Finland)					Heikkilä et al. (1991)
Coremaking	43	2.8 [3.4]	<0.1–>10 [<0.1–>11]	Before 1975	
Coremaking	17	0.3 [0.4]	0.02–1.4 [0.03–1.8]	1981–86	
Casting	10	0.15 [0.19]	0.02–0.2 [0.03–0.8]	1981–86	
Moulding	25	0.3 [0.39]	0.04–2.0 [0.05–2.5]	1981–86	
Foundries (Sweden)					Åhman et al. (1991)
Moulders and coremaker handling furan resin sand (8-h TWA)	36	0.1 [0.1]	0.02–0.2 [0.02–0.27]	NR	
Foundry (Sweden)				1980s	Rosén et al. (1984)
Hotbox method	5	1.5 [1.9]			
Moulding	17	0.1 [0.12]			
Man-made mineral fibre plants					
(Finland)				1981–86	Heikkilä et al. (1991)
Production	36	0.2 [0.25]	0.02–1.5 [0.03–1.7]		
Form pressing	24	0.1 [0.11]	0.01–0.3 [0.01–0.44]		
(Sweden)				1980s	Rosén et al. (1984)
Production	16	0.15 [0.19]			
Form pressing	4	0.16 [0.20]			
Plastics production					
Plastics production (Finland)				1981–86	Heikkilä et al. (1991)
Casting of polyacetal resin	10	0.3 [0.36]	0.06–0.7 [0.08–0.82]		
Casting of urea–formaldehyde resin	4	0.4 [0.46]	0.2–0.5 [0.27–0.59]		
Casting of other plastics	29	<0.1	<0.1–0.2 [<0.1–0.3]		

Table 6 (contd)

Industry and operation	No. of measurements	Mean[a]	Range	Year	Reference
Plastics production (contd)					
Production of moulded plastic products (USA)				1983–84	Stewart et al. (1987b)
Plant no. 8, phenol resin, summer	10	0.5[b] [0.6]	0.1–0.9 [0.1–1.1]		
Plant no. 9, melamine resin, summer	13	9.2[b] [11.3]	<0.01–26.5 [<0.01–32.6]		
Moulding compound manufacture (USA)				1983–84	Stewart et al. (1987b)
Plant no. 9, winter	9	2.8[b] [3.4]	0.04–6.7 [0.05–8.2]		
Plant no. 9, summer	18	38.2[b,c] [45]	9.5–60.8 [11.7–74.8]		
Plant no. 1, winter	12	1.5[b] [1.8]	0.9–2.0 [1.1–2.1]		
Plant no. 1, summer	24	9.7[b] [11.9]	3.8–14.4 [4.7–17.7]		
Plant no. 8, winter	13	0.3[b] [0.4]	0.07–0.7 [0.09–0.9]		
Plant no. 7, summer	43	0.3[b] [0.4]	0.05–0.6 [0.06–0.8]		
Plant no. 2, summer	15	6.5[b] [8.0]	0.3–20.6 [0.4–25.3]		
Metalware plant, bake painting (Finland)	18	0.3 [0.4]	0.03–0.7 [0.04–0.9]	1981–86	Heikkilä et al. (1991)
Electrical machinery manufacture (Finland)				1977–79	Niemelä & Vainio (1981)
Soldering	47	<0.1			
Lacquering and treatment of melamine plastics	8	0.35 [0.4]			
Painting with bake-drying paints (Sweden)	13	<0.1		1980s	Rosén et al. (1984)
Miscellaneous					
Photographic film manufacture (USA)				1983–84	Stewart et al. (1987b)
Plants no. 4 and 5, summer	49	0.1[b]	<0.01–0.4 [<0.01–0.5]		
Plants no. 4 and 5, winter	29	0.3[b] [0.4]	0.02–0.9 [0.03–1.1]		
Print (Finland)				1981–86	Heikkilä et al. (1991)
Development of photographs	11	0.04 [0.05]	0.02–0.1 [0.03–0.13]		

Table 6 (contd)

Industry and operation	No. of measurements	Mean[a]	Range	Year	
Miscellaneous (contd)					
Photographic laboratories (Finland)	10	0.07 [0.09]	0.02–0.3 [0.03–0.40]	1981–86	Heikkilä et al. (1991)
Abrasive production (Sweden)	20	0.2 [0.3]		1980s	Rosén et al. (1984)
Coal coking plant (former Czechoslovakia)	NR	0.05[d] [0.06]	< 0.01–0.25 [< 0.01–0.3]	NR	Mašek (1972)
Pitch coking plant (former Czechoslovakia)	NR	0.4[d] [0.5]	0.05–1.6 [0.07–2.0]	NR	Mašek (1972)
Rubber processing (USA)	NR	NR	0.4–0.8 [0.5–0.98]	1975	IARC (1982)
Sugar mill (Sweden) Preservation of sugar beets	26	0.4 [0.5]	NR	1980s	Rosén et al. (1984)
Malt barley production (Finland) Preservation of malt barley	6	0.7 [0.9]	0.4–1.5 [0.5–1.8]	1981	Heikkilä et al. (1991)

[a] Arithmetic mean unless otherwise specified
[b] Mean of geometric mean
[c] Some of the results were affected by the simultaneous occurrence in the samples (Stewart et al., 1987b) of particulates containing formaldehyde (leading to high values)
[d] Mean of arithmetic means

have usually been < 1 ppm, but much higher exposures may occur (Table 6). Plastic dust and fumes may be present in the atmospheres of moulded plastics product plants, and exposures in these facilities are usually considerably higher than those in facilities where the products are used. The mean concentration of formaldehyde was > 1 ppm in many plants in the United States where moulding compounds were used. Some workers may be exposed to pigments, lubricants and fillers (e.g. asbestos and wood flour) used as constituents of moulding compounds (Stewart *et al.*, 1987b).

Heating of bake-drying paints and soldering may release some formaldehyde in plants where metalware and electrical equipment are produced, but the measured levels are usually well below 1 ppm (Table 6).

The mean concentrations of formaldehyde measured during coating of photographic films and during development of photographs are usually well below 1 ppm (Table 6). Methanol, ethanol, acetone and ammonia are other volatile agents that may occur in film manufacturing facilities (Stewart *et al.*, 1987b). Skin contact with numerous photographic chemicals occurs occasionally in photographic laboratories.

Formaldehyde is also used or formed during many other industrial operations, such as preservation of fur, leather, barley and sugar beets, coal and pitch coking, rubber processing and abrasive production. Some of these activities may entail heavy exposure. For example, treatment of furs with formaldehyde resulted in the highest exposure to formaldehyde of all jobs and industries studied in a large Swedish survey in the early 1980s. The 8-h time-weighted average concentration of formaldehyde was assessed to be 0.8–1.6 ppm (1.0–2.0 mg/m^3), and high peak exposures occurred many times per day (Rosén *et al.*, 1984).

(f) Mortuaries, hospitals and laboratories

Formaldehyde is used as a tissue preservative and disinfectant in embalming fluids. Some parts of bodies to be embalmed are also cauterized and sealed with a hardening compound that contains paraformaldehyde powder. The concentration of formaldehyde in the air during embalming depends on the content of embalming fluid, type of the body, ventilation and work practices; the mean level is about 1 ppm (Table 7). Embalming of a normal intact body usually takes about 1 h. A survey in West Virginia (United States of America) in 1979 showed that undertakers prepare an average of 75 bodies per year, 15 of which have been autopsied; the 8-h time-weighted average exposure is therefore likely to be 0.1–0.4 ppm (0.1–0.5 mg/m^3). Disinfectant sprays are occasionally used which may release small amounts of solvents, such as isopropanol (Williams *et al.*, 1984). Methanol is used as a stabilizer in embalming fluids, and levels of 0.5–22 ppm (0.7–28.4 mg/m^3) have been measured during embalming. Low levels of phenol have also been detected in embalming rooms (Stewart *et al.*, 1992).

Formaldehyde is widely used in hospitals for disinfection. The mean concentrations, summarized in Table 7, range from 0.1 to 0.8 ppm (0.1–1.0 mg/m^3), but many of the measurements were made during disinfection, which usually takes a relatively short time. Low levels are found when detergents are used for cleaning; higher levels, which may occasionally exceed 1 ppm, occur when more concentrated formalin solutions are used, e.g. during the disinfection of operating theatres and dialysers (Salisbury, 1983).

Table 7. Concentrations of formaldehyde (in ppm [mg/m³]) in the workroom air of mortuaries, hospitals and laboratories

Industry and operation (type of sample)	No. of measurements	Mean[a]	Range	Year	Reference
Embalming, six funeral homes (USA)	NR	0.7 [0.9]	0.09–5.3 [0.1–6.5]	NR	Kerfoot & Mooney (1975)
Embalming, 23 mortuaries (USA) 8-h TWA	NR	1.1 [1.4]	0.03–3.15 [0.04–3.9]	NR	Lamont Moore & Ogrodnik (1986)
		0.16 [0.2]	0.01–0.49 [0.01–0.6]		
Embalming, seven funeral homes				1980	Williams et al. (1984)
Intact bodies (personal samples)	8	0.3 [0.4]	0.18–0.3 [0.2–0.4]		
Autopsied bodies (personal samples)	15	0.9 [1.1]	2.1 [0–2.6]		
Embalming (USA)				NR	Stewart et al. (1992)
Personal samples	25	2.58 [3.2]	0.31–8.72 [0.4–10.7]		
Area 1	25	2.03 [3.0]	0.23–7.52 [0.3–9.2]		
Area 2	25	2.16 [2.7]	0.28–8.15 [0.3–10.0]		
Embalming (Canada)				NR	Korczynski (1994)
Intact bodies (personal samples)	24	0.6 [0.8]	0.10–4.57 [0.12–5.64]		
Autopsied bodies (personal samples)	24	0.6 [0.8]	0.09–3.35 [0.11–4.13]		
Area samples	72	0.5 [0.6]	0.04–6.79 [0.05–8.37]		
Autopsy service (USA)[b]				NR	Coldiron et al. (1983)
Personal samples	27	1.3[c] [1.66]	0.4–3.28 [0.5–4.0]		
Area samples	23	4.2 [5]	0.1–13.6 [0.1–16.7]		
Autopsy (Finland)	5	0.7 [0.8]	<0.1–1.4 [<0.1–1.7]	1981–86	Heikkilä et al. (1991)
Anatomical theatre (Germany)	29	1.1[d] [1.4]	0.7–1.7 [0.9–2.2]	1980–88	Triebig et al. (1989)
Cleaning hospital floors with detergent containing formaldehyde (38–74 min) (Italy) (personal)	4	0.18 [0.22]	0.15–0.21 [0.18–0.26]		Bernardini et al. (1983)
Disinfecting operating theatres (Germany)[b]				NR	Binding & Witting (1990)
3% cleaning solution	43	0.8 [1.1]	0.01–5.1 [0.01–6.3]		
0.5% cleaning solution	26	0.2 [0.22]	0.01–0.43 [0.01–0.53]		
Disinfecting operating theatres (Germany)	43	0.4[c] [0.5]	0.04–1.4 [0.05–1.7]	NR	Elias (1987)

Table 7 contd

Industry and operation (type of sample)	No. of measurements	Mean[a]	Range	Year	Reference
Disinfection of dialysis clinic (USA), personal samples (37–63 min)	7	0.6 [0.8]	0.09–1.8 [0.12–2.2]	1983	Salisbury (1983)
Disinfection in hospitals (Finland)	18	0.1 [0.14]	0.03–0.2 [0.04–0.3]	1981–86	Heikkilä et al. (1991)
Bedrooms in hospital (Germany)	14	0.05 [0.06]	<0.01–0.7 [<0.01–0.9]	1980–88	Triebig et al. (1989)
Pathology laboratory (Sweden)	13	0.5 [0.66]	NR	1980s	Rosén et al. (1984)
Pathology laboratories (Germany)	21	0.5[d] [0.6]	<0.01–1.2 [<0.01–1.6]	1980–88	Triebig et al. (1989)
Hospital laboratories (Finland)	80	0.5 [0.6]	0.01–7.3 [0.01–9.1]	1981–86	Heikkilä et al. (1991)

[a] Arithmetic means unless otherwise specified
[b] Mean is 8-h time-weighted average
[c] Mean of arithmetic means
[d] Median

Formalin solution is commonly used to preserve tissue samples in histopathology laboratories. The concentrations of formaldehyde are sometimes high, e.g. during tissue disposal, formalin preparation and changing of tissue processor solutions (Belanger & Kilburn, 1981). The mean level during exposure is usually about 0.5 ppm (0.6 mg/m^3) (Table 7). Other agents to which pathologists and histology technicians may be exposed include xylene (see IARC, 1989c), toluene (see IARC, 1989d), chloroform (see IARC, 1987e) and methyl methacrylate (see IARC, 1994). The 8-h time-weighted average concentrations were from nondetectable to 22 ppm (95.5 mg/m^3) for xylene, 9–13 ppm (34–49 mg/m^3) for toluene and 0.4–7 ppm (2.0–34.3 mg/m^3) for chloroform in a study of the exposure and symptoms of histology laboratory technicians in the United States (Belanger & Kilburn, 1981).

(g) Building, agriculture, forestry and other activities

Exposure to formaldehyde may also occur in the construction industry, agriculture, forestry and the service sector. Specialized construction workers who varnish wooden parquet floors may have relatively high exposure. The mean levels of formaldehyde in the air during varnishing with urea–formaldehyde varnishes were 2–5 ppm (2.5–6.2 mg/m^3) (Table 8). One coat of varnish takes only about 30 min to apply (Riala & Riihimäki, 1991), but the same worker may apply five or even 10 coats per day. Use of water-based polyurethane varnishes that do not release formaldehyde reduces exposure. Other chemical agents to which parquetry workers are usually exposed include wood dust from sanding (see the monograph on wood dust) and solvent vapours from varnishes, putties and adhesives. The concentrations of solvents measured during varnishing were 6.5 times the exposure limit of the solvent mixture during nitrocellulose varnishing and 3.4 times the exposure limit during urea–formaldehyde varnishing in a Finnish study. The main solvents released from nitrocellulose varnishes were acetone, ethyl alcohol, ethyl acetate, hexane and other aliphatic hydrocarbons. The solvents in the urea–formaldehyde varnishes were mainly ethanol, acetone, isobutyl alcohol, ethyl acetate and propylene glycol monomethyl ether (Riala & Riihimäki, 1991). Other operations that may result in exposure to formaldehyde in the building trades are insulation with urea–formaldehyde foam and machining of particle-board. Various levels of formaldehyde have been measured during insulation with urea–formaldehyde foam, but exposure during handling and sawing of particle-board seems to be consistently low (Table 8).

Formaldehyde is used in agriculture as a preservative for fodder and as a disinfectant. For example, fodder was preserved with a 2% formalin solution for several days per year from the late 1960s until the early 1980s on farms in Finland. As the concentration during preservation was < 0.5 ppm (0.6 mg/m^3), the annual mean exposure is probably very low. Formaldehyde gas is also used 5–10 times a year to disinfect eggs in brooding houses. The concentration of formaldehyde in front of the disinfection chamber immediately after disinfection was as high as 7–8 ppm (8.6–9.8 mg/m^3), but annual exposure from this source probably remains very low (Heikkilä *et al.*, 1991).

Engine exhausts contain a small amount of formaldehyde (see section 1.3.3). The average exposure of lumberjacks using chainsaws in Sweden and Finland was, however, < 0.1 ppm (Table 8).

Table 8. Concentrations of formaldehyde (in ppm [mg/m^3]) in the workroom air in building sites, agriculture, forestry and miscellaneous other activities

Industry and operation	No. of measurements	Mean[a]	Range	Year	Reference
Varnishing parquet with urea–formaldehyde varnish (Finland)	10	2.9 [3.6]	0.3–6.6 [0.4–8.1]	1976	Heikkilä et al. (1991)
Varnishing parquet with urea–formaldehyde varnish (Finland)	6	4.3 [5.3]	2.6–6.1 [3.2–7.5]	1987	Riala & Riihimäki (1991)
Insulating buildings with urea–formaldehyde foam (USA)	66	1.3[b] [1.6]	0.3–3.1 [0.4–3.8]	NR	WHO (1989)
Insulating buildings with urea–formaldehyde foam (Sweden)	6	0.1 [0.18]	NR	1980s	Rosén et al. (1984)
Sawing particle-board at construction site (Finland)	5	<0.5 [<0.6]	NR	1967	Finnish Institute of Occupational Health (1994)
Agriculture (Finland)				1982	Heikkilä et al. (1991)
Handling of fodder	NR	NR	0.02–0.4 [0.03–0.46]		
Disinfection of eggs	11	2.6 [3.2]	0.2–7.8 [0.3–9.6]		
Chain-sawing (Sweden)	NR	0.05 [0.06]	0.02–0.1 [0.03–0.13]	1981–86	Hagberg et al. (1985)
Chain-sawing (Finland) (8-h TWA)	NR	<0.1	<0.1–0.5 [0.1–0.6]	NR	Heikkilä et al. (1991)
Retail dress shops (USA)	NR	NR	0.1–0.5 [0.15–0.6]	1959	Elliott et al. (1987)
Fabric shops (Finland)	3	0.16 [0.2]	0.1–0.2 [0.15–0.3]	1985–87	Priha et al. (1988)
Fire-fighting (USA)	30	0.1 [0.16]	0.04–0.3 [0.05–0.4]	1989	Materna et al. (1992)
Museum, taxidermy (Sweden)	8	0.2 [0.3]	NR	1980s	Rosén et al. (1984)

NR, not reported; TWA, time-weighted average
[a] Arithmetic mean unless otherwise specified
[b] Mean of arithmetic means

The use of formaldehyde-based resin in finishing textiles and some garments may also result in exposure in retail shops. Measurements in dress shops in the United States of America in the 1950s showed levels up to 0.5 ppm [0.62 mg/m^3]. The air in three Finnish fabric shops in the 1980s contained 0.1–0.2 ppm [0.12–0.25 mg/m^3] formaldehyde (Table 8).

Low concentrations of formaldehyde may occur also during firefighting and taxidermy (Table 8).

1.3.3 Ambient air

Although formaldehyde is a natural component of ambient air, anthropogenic sources usually contribute the most formaldehyde in populated regions, since the ambient levels are generally < 1 µg/m^3 in remote areas. For example, in the unpopulated Eniwetok Atoll in the Pacific Ocean, a mean of 0.5 µg/m^3 and a maximum of 1.0 µg/m^3 formaldehyde were measured in outdoor air (Preuss *et al.*, 1985). Other authors have reported similar levels in remote, unpopulated areas (Gammage & Travis, 1989; WHO, 1989).

Outdoor air concentrations in urban environments are more variable and depend on local conditions. They are usually 1–20 µg/m^3 (United States National Research Council, 1980, 1981; Preuss *et al.*, 1985; Gammage & Travis, 1989; WHO, 1989). A major source of formaldehyde in urban air is incomplete combustion of hydrocarbon fuels, especially from vehicle emissions (Sittig, 1985; WHO, 1989; Vaught, 1991). Urban air concentrations in heavy traffic or during severe inversions can range up to 100 µg/m^3 (United States National Research Council, 1980, 1981; Preuss *et al.*, 1985; WHO, 1989).

1.3.4 Residential indoor air

The levels of formaldehyde in indoor air are often higher than those outside (Gammage & Gupta, 1984). The concentrations in dwellings depend on the sources of formaldehyde that are present, the age of the source materials, ventilation, temperature and humidity. Major sources of formaldehyde in some dwellings have been reported to be off-gassing of urea–formaldehyde foam insulation and particle-board. In studies summarized by Preuss *et al.* (1985), the mean levels in conventional homes with no urea–formaldehyde foam insulation were 25–60 µg/m^3.

Many studies have been reported since the late 1970s of formaldehyde levels in 'mobile homes' (caravans) (see, for example, the review of Gammage & Travis, 1989). The levels appear to decrease as the mobile home (and its formaldehyde-based resins) age, with a half-life of four to five years (Preuss *et al.*, 1985). In the early 1980s, a mean levels of 0.4 ppm [0.49 mg/m^3] and individual measurements as high as several parts per million were measured in new mobile homes. As a result of new standards set in the mid-1980s for building materials used in mobile homes and voluntary reductions by the manufacturers, formaldehyde levels in mobile homes are now typically around 0.1 ppm [0.12 mg/m^3] or less (Gammage & Travis, 1989; Sexton *et al.*, 1989; Gylseth & Digernes, 1992; Lehmann & Roffael, 1992).

1.3.5 Other exposures

Several studies have been conducted to determine exposures of students in laboratories. Skisak (1983) measured formaldehyde in the breathing zone at dissecting tables and in the ambient air in a medical school in the United States for 12 weeks. Concentrations > 1.2 mg/m^3 were found in 44% of the breathing zone samples and 11 ambient air samples; 50% of the breathing zone samples contained 0.7–1.2 mg/m^3, with a range of 0.4–3.2 mg/m^3. Korky *et al.* (1987) studied the dissecting facilities at a university in the United States during the 1982–83 academic year. The airborne concentration of formaldehyde was 7–16.5 ppm (8.6–20.3 mg/m^3) in the laboratory, 1.97–2.62 ppm (2.4–3.2 mg/m^3) in the stockroom and < 1 ppm (< 1.2 mg/m^3) in the public hallway. In another study, of 253 samples of air taken during laboratory dissection classes at a university in the United States, 97 contained concentrations above the detection limit of 0.01 mg/m^3; all but four samples had levels < 1.2 mg/m^3. The average concentration detected was 0.5 mg/m^3 (Poslusny *et al.*, 1992).

Cigarette smoke has been reported to contain levels of a few micrograms to several milligrams of formaldehyde per cigarette. A 'pack-a-day' smoker may inhale as much as 0.4–2.0 mg formaldehyde (IARC, 1986b; WHO, 1989; American Conference of Governmental Industrial Hygienists, 1991).

Cosmetic products containing formaldehyde, formalin and/or paraformaldehyde may come into contact with hair (e.g. shampoos and hair preparations), skin (deodorants, bath products, skin preparations and lotions), eyes (mascara and eye make-up), oral mucosa (mouthwashes and breath fresheners), vaginal mucosa (vaginal deodorants) and nails (cuticle softeners and nail creams and lotions). Aerosol products (e.g. shaving creams) result in potential inhalation of formaldehyde (Cosmetic Ingredient Review Expert Panel, 1984).

Formaldehyde occurs naturally in foods, and foods may be contaminated as a result of fumigation (of e.g. grain), cooking (as a combustion product) and release from formaldehyde resin-based tableware (WHO, 1989). It has been used as a bacteriostatic agent in some foods, such as cheese (Restani *et al.*, 1992). Fruits and vegetables typically contain 3–60 mg/kg, milk and milk products about 1 mg/kg, meat and fish 6–20 mg/kg and shellfish 1–100 mg/kg. Drinking-water normally contains < 0.1 mg/L (WHO, 1989).

Other exposures to formaldehyde are reviewed by IARC (1982) and WHO (1989).

1.4 Regulations and guidelines

Occupational exposure limits and guidelines for formaldehyde are presented in Table 9. International regulations and guidelines related to emissions of and exposures to formaldehyde in occupational settings, indoor air and building materials have been reviewed (Scheuplein, 1985; Sundin, 1985; Coutrot, 1986; Meyer, 1986; McCredie, 1988; Gylseth & Digernes, 1992; Halligan, 1992; Lehmann & Roffael, 1992; McCredie, 1992).

Table 9. Occupational exposure limits and guidelines for formaldehyde

Country or region	Year	Concentration (mg/m^3)	Interpretation
Australia	1991	1.5	TWA; probable human carcinogen, sensitizer
		3	STEL
Austria	1982	1.2	TWA
Brazil	1978	2.3	TWA
Belgium	1991	1.2	TWA; probable human carcinogen
		2.5	STEL
Bulgaria	1984	1	TWA
Chile	1983	2.4	Ceiling
China	1982	3	TWA
Czech Republic	1991	0.5	TWA
		1	STEL
Denmark	1991	0.4	STEL; suspected carcinoegn
Egypt	1959	6.2	TWA
Finland	1993	1.3	STEL, 15 min; significant absorption through skin
France	1991	3	STEL
Germany	1993	0.6	TWA; suspected carcinogenic potential; local irritant; sensitizer
Hungary	1991	0.6	Ceiling; probable human carcinogen; irritant; sensitizer
India	1983	3	Ceiling
Indonesia	1978	6	Ceiling
Italy	1978	1.2	TWA
Japan	1991	0.61	TWA; suspected carcinogenic potential
Mexico	1983	3	TWA
Netherlands	1986	1.5	TWA
		3	Ceiling, 15 min
Norway	1990	0.6	TWA; allergen; suspected carcinogen
		1.2	Ceiling
Poland	1991	2	TWA
Romania	1975	4	MAX
Russian Federation	1991	0.5	STEL; significant absorption through skin; allergen
Sweden	1991	0.6	TWA; sensitizer
		1.2	Ceiling
Switzerland	1991	0.6	TWA; sensitizer
		1.2	STEL
Taiwan	1981	6	TWA; significant absorption through skin
United Kingdom	1992	2.5	TWA; maximum exposure limit
		2.5	STEL, 10 min
USA			
ACGIH	1993	0.37	Ceiling; suspected human carcinogen
NIOSH	1992	0.02	TWA; potential human carcinogen
		0.12	Ceiling, 15 min
OSHA	1993	0.9	TWA
		2.5	STEL

Table 9 (contd)

Country or region	Year	Concentration (mg/m^3)	Interpretation
Venezuela	1978	3	TWA, ceiling
Former Yugoslavia	1971	1	TWA

From Arbeidsinspectie (1986); Cook (1987); Direktoratet for Arbeidstilsynet (1990); International Labour Office (1991); United Kingdom Health and Safety Executive (1992); United States National Institute for Occupational Safety and Health (NIOSH) (1992); American Conference of Governmental Industrial Hygienists (ACGIH) (1993); Deutsche Forschungsgemeinschaft (1993); Työministeriö (1993); United States Occupational Safety and Health Administration (OSHA) (1993)

TWA, time-weighted average; STEL, short-term exposure limit; MAX, maximum

The European Union has adopted a Directive that imposes concentration limits for formaldehyde and paraformaldehyde in cosmetics. These substances are permitted at a maximal concentration of 0.2% (expressed as free formaldehyde) in all cosmetic formulations except nail hardeners, oral hygiene products and aerosol dispensers. Nail hardeners and oral hygiene products may contain maximal concentrations of 5 and 0.1%, respectively, whereas formaldehyde and paraformaldehyde are prohibited for use in aerosol dispensers (except for foams). Cosmetic product labels are required to list formaldehyde and paraformaldehyde as ingredients when the concentration of either exceeds 0.05% (Cosmetic Ingredient Review Expert Panel, 1984; European Commission, 1990).

Guidelines for ambient air levels of formaldehyde in living spaces have been set in several countries and range from 0.05–0.4 ppm (0.06–0.5 mg/m^3), with a preference for 0.1 ppm (0.12 mg/m^3) (Lehmann & Roffael, 1992).

In the United States, all plywood and particle-board materials bonded with a resin system or coated with a surface finish containing formaldehyde cannot exceed the following formaldehyde emission levels when installed in manufactured homes: plywood materials and particle-board flooring products (including urea–formaldehyde bonded particle-board) cannot emit more than 0.25 mg/m^3 formaldehyde, and particle-board materials and medium-density fibre-board cannot emit more than 0.37 mg/m^3 (National Particleboard Association, 1992, 1993; National Particleboard Association, 1994; United States Department of Housing and Urban Development, 1994). Several other countries have similar regulations (Lehmann & Roffael, 1992).

2. Studies of Cancer in Humans

2.1 Case reports

Halperin *et al.* (1983) and Brandwein *et al.* (1987) reported cases of squamous-cell carcinoma of the sinonasal cavities associated with exposure to formaldehyde at work or at home. Holmstrom and Lund (1991) drew attention to the possibility of a causal relationship with malignant melanoma of the nasal cavity on the basis of three cases seen after occupational exposure to formaldehyde.

2.2 Descriptive studies

Gallagher *et al.* (1989) calculated proportionate mortality ratios (PMRs) by occupation for 320 423 men who died in British Columbia, Canada, during 1950–84. One of the 79 funeral directors included in the study had died from sinonasal cancer (PMR, 16; 95% confidence interval [CI], 0.4–87).

A similar analysis was conducted by Petersen and Milham (1980) on about 200 000 white male residents of California, USA, for the period 1959–61. Funeral directors and embalmers accounted for 130 deaths, none of which were from cancer of the buccal cavity or pharynx (one expected) or from sinonasal cancer (< 0.1 expected).

Malker *et al.* (1990) used data from the Swedish Cancer–Environment Registry, which combines data from the 1960 census with those from the National Cancer Registry through 1979, to calculate standardized incidence ratios (SIRs) for nasopharyngeal cancer (471 cases) in various occupational and industrial groups in 1961–79. Significant excesses were seen for glassmakers (SIR, 6.2; 3 cases), bookbinders (6.1; 3 cases), shoemakers (3.8; 5 cases) and workers in shoe repair (4.0; 5 cases) and fibre-board manufacture (3.9; 4 cases). No significantly high risks for nasopharyngeal cancer were seen in other occupations in which exposure to formaldehyde probably occurs [expected numbers not given].

2.3 Cohort studies

The relationship between exposure to formaldehyde and cancer has been investigated in over 25 cohort studies of professional (pathologists, anatomists and embalmers) and industrial groups (formaldehyde producers, formaldehyde resin makers, plywood and particle-board manufacturers, garment workers and workers in the abrasives industry). Relative risks have been estimated as standardized mortality ratios (SMRs), PMRs, proportionate cancer mortality ratios (PCMRs) and SIRs. In some studies, exposure was not assessed but was assumed on the basis of the subject's occupation or industry; in others, it was based on duration of exposure and quantitative estimates of historical exposure levels. Mortality in several of the cohorts was followed beyond the period covered by the original report; only the latest results are reviewed below, unless there were important differences in the analyses performed or changes in the cohort definition. Reviews are available which summarize the epidemiological data (Blair et al., 1990a; Partanen, 1993; McLaughlin, 1995); in the first two, the technique of meta-analysis was used.

For each point estimate expressed as an SMR, PMR, PCMR or SIR, a 95% CI is given, even if the original authors did not report one. The 95% CI bears a relationship to the usual judgement of statistical significance, in that a CI that does not include the value of 1.0 occurs when the point estimate is significant at the traditional 5% level. The 95% CI provides more information, however, in that it shows the magnitude of the estimated random variation around the point estimate.

2.3.1 Professional groups

Pathologists, anatomists, embalmers and funeral directors were studied because they use formaldehyde as a tissue preservative. Investigations of these occupations have several methodological problems. Use of national statistics to generate expected numbers may bias estimates of relative risks toward the null for some cancers and away from the null for others because these groups have a higher socioeconomic level than the general population; only a few investigations included a special referent population designed to diminish potential socio-economic confounding. None of these studies had the data necessary to adjust for tobacco use. Since anatomists and pathologists in the United States generally smoke less than the general population (Sterling & Weinkam, 1976), estimates of relative risks for smoking-related cancers will be artificially low. Without adjustments, the biases introduced by socioeconomic factors and smoking may be strong enough to preclude any possibility of detecting excess occurrence of tobacco-related cancers. This may be less of a problem for embalmers, however, because their smoking habits may not differ from those of the general population (Sterling & Weinkam, 1976). In no study were risk estimates developed by level of exposure, and in only a few studies were risks evaluated by duration of exposure. When exposure estimates are not summarized in the following text, they were not provided in the original study. Non-differential error in exposure assessment, which occurs when the measures of exposure are about equally inaccurate for study subjects with and without the cancer of interest, diminishes the chances of uncovering an underlying association, as it biases estimates of the relative risk toward the null.

Harrington and Shannon (1975) evaluated the mortality of pathologists and medical laboratory technicians in the United Kingdom. Members of the Royal College of Pathologists and the Pathological Society who were alive in 1955 were enrolled and followed through 1973. Of the 2079 pathologists included, only 13 could not be traced successfully; 156 deaths were identified. The Council for Professions Supplementary to Medicine was used to identify 12 944 technicians; 154 subsequently died and 199 could not be traced. Ten of the pathologists who died and 20 of the medical technicians who died were women, but the number of women included in the cohort was not provided. Expected numbers of deaths were calculated from sex-, five-year calendar period- and five-year age group-specific rates for England and Wales or Scotland, as appropriate. The SMRs for all causes of deaths were 0.6 for pathologists and 0.7 for medical technicians; the numbers of deaths from all cancers and from ischaemic heart disease were also fewer than expected. The SMR for lymphatic and haematopoietic cancer was elevated among pathologists ([2.0; 95% CI, 0.9–3.9]; 8 observed), but not among technicians ([0.6; 0.1–1.6]; 3 observed). The SMRs for cancers at other sites were below 1.0.

The study of British pathologists was extended and expanded by Harrington and Oakes (1984), who added new entrants and traced new and previously studied subjects from 1974 through 1980. The population now included 2307 men and 413 women. Vital status was confirmed for 99.9% of the men and 99.5% of the women. SMRs were calculated using expected rates based on age-, sex- and calendar time-specific data from England and Wales. The SMRs for all causes and for all cancers among men were both 0.6; among women, the SMR for all causes was 1.0 and that for all cancers, 1.4. Mortality from brain cancer was elevated among men (SMR, 3.3 [95% CI, 0.9–8.5]; 4 deaths). No nasal cancer and no cancer of the nasal sinuses was seen.

This cohort was further evaluated by Hall *et al.* (1991), who extended follow-up of mortality from 1980 through 1986 and added new members of the Pathological Society, resulting in 4512 individuals available for study (3478 men; 803 women; and 231 unaccounted subjects [who may have been women from Scotland, but it was not clear in the article]). Sex-specific SMRs were based on expected rates for England and Wales or Scotland, as appropriate, and were adjusted for age and calendar time. The SMRs for all causes of death were all considerably below 1.0: men from England and Wales, 0.4 (95% CI, 0.4–0.5); women from England and Wales, 0.7 (0.4–1.0); men from Scotland, 0.5 (0.3–0.7). The SMRs for cancers at all sites were 0.4 (0.3–0.6) and 0.6 (0.3–1.1) among men from England and Wales and Scotland, respectively, and 1.0 (0.5–1.9) among women from England and Wales. No significant excess was seen for cancer at any site. Nonsignificant excesses occurred for brain cancer (SMR, 2.4; 0.9–5.2) and lymphatic and haematopoietic cancer (1.4; 0.7–2.7) among men from England and Wales, breast cancer (1.6; 0.4–4.1) among women from England and Wales and prostatic cancer (3.3; 0.4–12) among men from Scotland.

Walrath and Fraumeni (1983) used licensure records from the New York State (United States) Department of Health, Bureau of Funeral Directing and Embalming to identify 1678 embalmers who had died between 1925 and 1980. Death certificates were obtained for 1263 (75%) decedents (1132 white men, 79 nonwhite men, 42 men of unknown race and 10 women), and PMRs and PCMRs were calculated for white men and nonwhite men on the basis of age-, race-, sex- and calendar time-specific proportions in the general population. Observed and expected numbers were generally not provided for nonwhite men, but the paper indicated that there was significant excess mortality from arteriosclerotic heart disease (PMR, 1.5 [95% CI, 1.1–2.2]; 33 observed) and from cancers of the larynx (2 observed) and lymphatic and haematopoietic system (3 observed). Among white men, the PMRs were 1.1 [1.0–1.3] for all cancer combined, 1.1 [1.0–1.3] for arteriosclerotic heart disease, 0.7 [0.3–1.2] for emphysema, and 1.3 [0.9–1.9] for cirrhosis of the liver. The PCMRs for other cancers were 1.0 [0.4–2.0] for buccal cavity and pharynx, 1.3 [0.9–1.9] for colon, 1.1 [0.9–1.4] for lung, 0.8 [0.4–1.4] for prostate, 1.4 [0.6–2.7] for brain, 1.2 [0.8–1.8] for lymphatic and haematopoietic system (PMR), 0.8 [0.3–1.9] for lymphoma and 1.2 [0.6–2.1] for leukaemia. No deaths occurred from cancer of the nasal sinuses or nasopharynx. There was little difference in PMR by time since first licensure. Although the subjects had been licensed as either embalmers or embalmers/funeral directors, these two groups were analysed separately because the authors assumed that embalmers would have had more exposure to formaldehyde than embalmers/funeral directors. The PMR for brain cancer was increased among people licensed only as embalmers (2.3 [0.8–5.0]; 6 observed) but not among those also licensed as a funeral director (0.9 [0.2–2.7]; 3 observed). A difference was also observed for mortality from cancer of the buccal cavity and pharynx; the PMR for embalmers was 2.0 ([0.8–4.1] 7 observed), and that for embalmers/funeral directors was 0.3 ([0.0–1.5] 1 observed).

Walrath and Fraumeni (1984) used the records of the California (United States) Bureau of Funeral Directing and Embalming to examine mortality among embalmers first licensed in California between 1916 and 1978. They identified 1109 embalmers who died between 1925 and 1980, comprising 1007 white men, 39 nonwhite men, 58 white women and five nonwhite women. Only mortality of white men was analysed. The expected numbers of deaths were calculated on the basis of age-, race-, sex- and calendar year-specific proportions from the general population. The PMRs for major categories of death were 1.2 [95% CI, 1.0–1.4] for all cancers combined (205 observed), 1.2 [1.1–1.3] for ischaemic heart disease (355 observed), 0.4 [0.1–1.0] for emphysema (4 observed) and 1.8 [1.3–2.4] for suicide (44 observed). The PCMRs for specific cancers were 1.0 [0.4–2.0] for buccal cavity and pharynx (8 observed), 1.4 [0.9–2.0] for colon (30 observed), 0.9 [0.6–1.2] for lung (41 observed), 1.3 [0.8–2.0] for prostate (23 observed), 1.7 [0.8–3.2] for brain (9 observed), 1.2 (PMR) [0.7–1.9] for lymphatic and haematopoietic system (19 observed), [1.0 (PMR); 0.2–2.8] for lymphoma (3 observed) and 1.4 [0.7–2.4] for leukaemia (12 observed). There was no death from cancer of the nasal passages (0.6 expected). The PMRs by length of licensure (< 20 years and ≥ 20 years) were 2.0 and 1.9 for brain cancer and 1.2 and 2.2 for leukaemia.

Mortality among 1477 male embalmers licensed by the Ontario (Canada) Board of Funeral Services between 1928 and 1957 was evaluated by Levine *et al.* (1984a) by following-up the cohort through 1977: 359 deaths were identified; 54 (4%) could not be traced. The expected numbers were derived from the mortality rates for men in Ontario in 1950–77, adjusted for age and calendar year. Since mortality rates for Ontario were not available before 1950, person-years and deaths in the cohort before that time were excluded from the analysis, leaving 1413 men known to be alive in 1950. The SMRs for all causes and for all cancers were 1.0 [95% CI, 0.9–1.1] and 0.9 [0.7–1.1], respectively. Excesses were seen for chronic rheumatic heart disease (2.0 [0.9–3.9]; 8 observed) and cirrhosis of the liver (2.4 [1.4–3.7]; 18 observed). For specific cancers, the numbers of deaths and those expected were as follows: buccal cavity and pharynx (1/2.1), nose, middle ear, sinuses (0/0.2), lung (19/20.2), prostate (3/3.4), brain (3/2.6), lymphatic and haematopoietic system (SMR, 1.2; 8/6.5) and leukaemia (4/2.5).

Stroup *et al.* (1986) evaluated mortality among members of the American Association of Anatomists. A total of 2317 men had joined the Association between 1888 and 1969; because only 299 women had joined during this time period, they were not included. Follow-up of the cohort for vital status through 1979 resulted in 738 deaths; 39 individuals could not be traced. The expected numbers of deaths were calculated from age-, race-, sex- and calendar time-specific rates for the general population of the United States for the period 1925–79 or for male members of the American Psychiatric Association, a population that should be similar to anatomists with regard to socioeconomic status, in 1900–69. Between 1925 and 1979, 738 anatomists died and only 2% were of unknown vital status at the close of the follow-up. In comparison with the general population, the cohort showed a very large 'healthy worker effect', with SMRs of 0.7 for all causes (738 observed), 0.6 (95% CI, 0.5–0.8) for cancer at all sites (118 observed) and 0.8 (0.7–0.9) for ischaemic heart disease (271 observed). The SMRs were less than 1.0 for individual cancers (e.g. lung cancer, SMR, 0.3; 0.1–0.5; 12 observed; oral and pharyngeal cancer, 0.2; 0.0–0.8; 1 observed), except for cancers of the brain (2.7; 1.3–5.0; 10 observed), lymphatic and haematopoietic system (2.0; 0.7–4.4; 6 observed) and leukaemia (1.5; 0.7–2.7; 10 observed). No death from nasal cancer occurred (0.5 expected). The risk for brain cancer increased with duration of membership, from 2.0 for < 20 years, to 2.8 for 20–39 years and to 7.0 for ≥ 40 years [trend cannot be calculated]; no such pattern was seen for lung cancer. Anatomists had deficits of lung cancer (0.5; 0.2–1.1) and leukaemia (0.8; 0.2–2.9) when compared with members of the American Psychiatric Association, but they still had an excess of brain cancer (6.0; 2.3–16).

Logue *et al.* (1986) evaluated mortality among 5585 members of the College of American Pathologists listed in the Radiation Registry of Physicians; 496 deaths were identified. The cohort was established by enrolling members between 1962 and 1972 and following them up through 1977. SMRs were calculated on the basis of the mortality rates for white men in the United States in 1970, but information on age and calendar-year categories was not provided. The SMRs were 0.7 [observed number of cases calculated from rates, 4] for cancer of the buccal cavity and pharynx, 0.2 [14] for cancer of the respiratory system ($p < 0.01$), 0.8 [36] for cancer of the digestive organs, 0.5 [5] for cancers of the lymphatic and haematopoietic system and 1.1 [7] for leukaemia. The age-adjusted mortality rates for these cancers were similar to those calculated for members of the American College of Radiology who were also listed in the Registry.

Hayes *et al.* (1990), in the United States, identified 6651 deceased embalmers/funeral directors from the records of licensing boards and state funeral directors' associations in 32 states and the District of Columbia and from the vital statistics offices of nine states and New York City between 1975 and 1985. Decedents included in studies in New York (Walrath & Fraumeni, 1983) and California (Walrath & Fraumeni, 1984) were excluded. Death certificates were received for 5265. Exclusion of 449 decedents included in previous studies of embalmers, 376 subjects who probably did not work in the funeral industry, eight subjects of unknown race or age and 386 women left 4046 decedents available for analysis (3649 whites and 397 nonwhites). PMRs and PCMRs were calculated on the basis of expected numbers from race- and sex-specific groups of the general population, adjusted for five-year age and calendar-time categories. The PMR for all cancers was 1.1 (95% CI, 1.0–1.2; 900 observed) for whites and 1.1 (0.9–1.3; 102 observed) for nonwhites. The PMR for ischaemic heart disease was elevated in both whites (1.1; 1.1–1.2; 1418 observed) and nonwhites (1.5; 1.2–1.7; 135 observed). That for emphysema was about as expected among whites (1.0; 0.8–1.4; 48 observed), but only one death from this cause occurred among nonwhites (0.5; 0.1–2.6; 1 observed). The PMRs for specific cancers were: buccal cavity and pharynx (whites: 1.2; 0.8–1.7; 26 observed; nonwhites: 1.3; 0.3–3.2; 4 observed), nasopharynx (whites: 1.9; 0.4–5.5; 3 observed; nonwhites: 4.0; 0.1–22;

1 observed), colon (whites: 1.2; 1.0–1.4; 95 observed; nonwhites: 2.3; 1.3–3.8; 16 observed), nasal sinuses (whites and nonwhites: 0 observed, 1.7 expected), lung (whites: 1.0; 0.9–1.1; 285 observed; nonwhites: 0.8; 0.5–1.1; 23 observed), prostate (whites: 1.1; 0.8–1.3; 79 observed; nonwhites: 1.4; 0.8–2.1; 19 observed), brain (whites: 1.2; 0.8–1.8; 24 observed; nonwhites: 0 observed, 0.8 expected) and lymphatic and haematopoietic system (whites: 1.3; 1.1–1.6; 100 observed; nonwhites: 2.4; 1.4–4.0; 15 observed). The risks for cancers of the lymphatic and haematopoietic system and brain did not vary substantially by licensing category (embalmer versus funeral director), by geographic region, by age at death or by source of data on mortality. Among the lymphatic and haematopoietic cancers, the PMRs were significantly elevated for myeloid leukaemia (1.6; 1.0–2.3; 24 observed) and other and unspecified leukaemia (2.3; 1.4–3.5; 20 observed); nonsignificant excesses were observed for several other histological types.

In a study of users of various medicinal drugs based on computer-stored hospitalization records of the out-patient pharmacy at the Kaiser–Permanente Medical Center in San Francisco (CA, United States), Friedman and Ury (1983) evaluated cancer incidence in a cohort of 143 574 pharmacy users from July 1969 through August 1973 and followed them up to the end of 1978. The number of cases among users of specific drugs was compared with the number expected on the basis of rates for all pharmacy users, adjusted for age and sex. Since many analyses were performed (56 cancers and 120 drugs, for 6720 combinations), chance findings would be expected. Five cancers were associated with use of formaldehyde solution (topically for warts) (morbidity ratio, 0.8 [95% CI, 0.3–2.0]). The morbidity ratio for lung cancer was significantly elevated (5.7 [1.6–15]) for people using formaldehyde, with four cases observed. Information on smoking was not provided. [The Working Group noted the short period of follow-up.]

Cohort and proportionate mortality studies of professional groups are summarized in Table 10.

2.3.2 Industrial groups

Several studies of industrial groups included evaluations by duration of exposure or employment, but only four contained assessments by level of exposure: a study in the United Kingdom (Acheson *et al.*, 1984a; Gardner *et al.*, 1993) and three in the United States (Blair *et al.* 1986; Stewart *et al.*, 1986; Blair *et al.*, 1990b) (Andjelkovich *et al.*, 1995) (Marsh *et al.*, 1994a). Information on tobacco use was generally absent, although Blair *et al.* (1990b) and Andjelkovich *et al.* (1995) obtained some information. The reports on industrial cohorts are not entirely independent; some publications are based on extended follow-ups (reports on the British cohort by Acheson *et al.* (1984a) and Gardner *et al.* (1993) and the Italian cohort by Bertazzi *et al.* (1986, 1989)). There is also partial overlap because of inclusion of the same workers in several studies: the two reports on the garment industry by Stayner *et al.* (1985, 1988) included two common facilities; the 10-plant study by Blair *et al.* (1986, 1990b) included workers also reported in cohort studies by Marsh (1982), Wong (1983), Liebling *et al.* (1984) and Marsh *et al.* (1994a) and in a case–control study by Fayerweather *et al.* (1983). Analyses of the data from the cohort study by Blair *et al.* (1986, 1990b) have also been published by others (Robins

Table 10. Cohort and proportionate mortality studies of cancer in professionals exposed to formaldehyde

Country (reference)	Population, design (number), date	Exposure estimates	Cancer	Relative risk (95% CI)	Comments
United Kingdom (Hall et al., 1991) (update of Harrington et al. (1984) plus new members since 1973)	Pathologists, SMR (4512), 1974–87	None	All causes	0.4 [0.4–0.5]	194 deaths
			All cancers	0.5 [0.4–0.6]	55 deaths
			Colon	1.0 [0.4–2.0]	Seven deaths
			Lung	0.2 [0.1–0.4]	Nine deaths
			Brain	2.2 [0.8–4.8]	Six deaths
			Lymphatic/haematopoietic	1.4 [0.7–2.7]	10 deaths
			Leukaemia	1.5 [0.4–3.9]	Four deaths
			Breast	1.6 [0.4–4.1]	Four deaths among women
			Prostate	3.3 [0.4–12]	Two deaths among men in Scotland
New York, USA (Walrath & Fraumeni, 1983)	Embalmers, PMR, PCMR (1132 men), 1925–80	None	All cancers	1.1 [1.0–1.3]	243 deaths, PMR
			Buccal/pharynx	1.0 [0.4–2.0]	Eight deaths, PCMR
				2.0 [0.8–4.1]	Embalmers only, seven deaths, PMR
				0.3 [0.0–1.5]	Funeral directors, one death, PMR
			Colon	1.3 [0.9–1.9]	29 deaths, PCMR
			Lung	1.1 [0.9–1.4]	70 deaths, PCMR
			Brain	1.4 [0.6–2.7]	Nine deaths, PCMR
				2.3 [0.8–5.0]	Embalmers only, six deaths, PMR
				0.9 [0.2–2.7]	Funeral directors, three deaths, PMR
			Lymphatic/haematopoietic	1.2 [0.8–1.8]	25 deaths, PMR
			Lymphoma	0.8 [0.3–1.9]	Five deaths, PCMR
			Leukaemia	1.2 [0.6–2.1]	12 deaths, PCMR
California, USA (Walrath & Fraumeni, 1984)	Embalmers, PMR, PCMR (1007 white men), 1925–80	Duration	All cancer	1.2 [1.0–1.4]	205 deaths, PMR
			Buccal/pharynx	1.3 [0.6–2.6]	Eight deaths, PMR, inverse trend with duration
			Colon	1.4 [0.9–2.0]	30 deaths, PCMR, no trend
			Lung	0.9 [0.6–1.2]	41 deaths, PCMR, no trend
			Nasal	–	0 deaths
			Prostate	1.3 [0.8–2.0]	23 deaths, PCMR, no trend
			Brain	1.9 ($p < 0.05$)	Nine deaths, PMR, no trend
			Lymphatic/haematopoietic	1.2 [0.7–1.9]	19 deaths, PMR
			Lymphoma	[1.0] [0.2–2.8]	Three deaths, PMR
			Leukaemia	1.4 [0.7–2.4]	12 deaths, PCMR, trend with duration

Table 10 (contd)

Country (reference)	Population, design (number), date	Exposure estimates	Cancer	Relative risk (95% CI)	Comments
Canada (Levine et al., 1984a)	Embalmers, SMR (1477 men), 1950–77	None	All causes	1.0 [0.9–1.1]	319 deaths
			All cancers	0.9 [0.7–1.1]	58 deaths
			Nasal	—	0 deaths (0.2 expected)
			Buccal/pharynx	[0.5] (NA)	One death
			Lung	0.9 [0.6–1.5]	19 deaths
			Brain	[1.2] [0.2–3.4]	Three deaths
			Prostate	[0.9] [0.2–2.6]	Three deaths
			Lymphatic/haematopoietic	1.2 [0.5–2.4]	Eight deaths
			Leukaemia	[1.6] [0.4–4.1]	Four deaths
USA (Stroup et al., 1986)	Anatomists, SMR (2239 men), 1925–79	Duration	All causes	0.7 (0.6–0.7)	738 deaths
			All cancers	0.6 (0.5–0.8)	118 deaths
			Buccal/pharynx	0.2 (0.0–0.8)	One death
			Colon	1.1 (0.7–1.7)	20 deaths
			Nasal	–(0.0–7.2)	0 deaths, 0.5 expected
			Lung	0.3 (0.1–0.5)	12 deaths, no trend with duration
			Prostate	1.0 (0.6–1.6)	19 deaths
			Brain	2.7 (1.3–5.0)	10 deaths, trend with duration
			Lymphatic/haematopoietic	1.2 (0.7–2.0)	18 deaths
			Lymphoma	0.7 (0.1–2.5)	Two deaths
			Leukaemia	1.5 (0.7–2.7)	10 deaths
			Other lymphatic tissue	2.0 (0.7–4.4)	Six deaths
USA (Logue et al., 1986)	Pathologists, SMR (5585 men), 1962–77	None	Buccal/pharynx	0.7 (NA)	
			Digestive organs	0.8 (NA)	
			Respiratory system	0.2 ($p < 0.01$)	
			Lymphatic/haematopoietic	0.5 (NA)	
			Leukaemia	1.1 (NA)	

Table 10 (contd)

Country (reference)	Population, design (number), date	Exposure estimates	Cancer	Relative risk (95% CI)	Comments
USA (Hayes et al., 1990)	Embalmers, PMR (3649 white men, 397 nonwhite men), 1975–85	None	All cancers	1.1 (1.0–1.2)	900 deaths among white men
				1.1 (0.9–1.3)	102 deaths among nonwhite men
			Buccal/pharynx	1.2 (0.8–1.7)	26 deaths among white men
				1.3 (0.3–3.2)	Four deaths among nonwhite men
			Nasopharynx	1.9 (0.4–5.5)	Three deaths among white men
				4.0 (0.1–22)	One death among nonwhite men
			Colon	1.2 (1.0–1.4)	95 deaths among white men
				2.3 (1.3–3.8)	16 deaths among nonwhite men
			Nasal	–	0 deaths among white and nonwhite men
			Lung	1.0 (0.9–1.1)	285 deaths among white men
				0.8 (0.5–1.1)	23 deaths among nonwhite men
			Prostate	1.1 (0.8–1.3)	79 deaths among white men
				1.4 (0.8–2.1)	19 deaths among nonwhite men
			Brain	1.2 (0.8–1.8)	24 deaths among white men
				–	0 deaths among nonwhite men
			Lymphatic/haematopoietic	1.3 (1.1–1.6)	100 deaths among white men
				2.4 (1.4–4.0)	15 deaths among nonwhite men
			Lymphoma	1.1 (0.5–1.9)	11 deaths among white men
				1.9 (0.1–11)	One death among nonwhite men
			Lymphatic leukaemia	0.6 (0.2–1.3)	Five deaths among white men
				3.0 (0.4–11)	Two deaths among nonwhite men
			Myeloid leukaemia	1.6 (1.0–2.4)	23 deaths among white men
				1.1 (0.1–5.9)	One death among nonwhite men
California, USA (Friedman & Ury, 1983)	Patients, SIR (143 574 pharmacy users), 1969–78	Use as a drug	All cancers	0.8 [0.3–2.0]	Five cases
			Lung	5.6 [1.6–15]	Four cases

CI, confidence interval; SMR, standardized mortality ratio; PMR, proportionate mortality ratio; PCMR, proportionate cancer mortality ratio; NA, not available; SIR, standardized incidence ratio *et al.*, 1988; Sterling & Weinkam, 1988, 1989a,b; Marsh *et al.*, 1992a,b, 1994a; Sterling & Weinkam, 1994).

Studies that provided detailed information indicate that workers had a range of levels of exposure to formaldehyde. Blair *et al.* (1986) found that 4% of their cohort was exposed to ≥ 2 ppm (≥ 2.5 mg/m^3); Acheson *et al.* (1984a) found that 35% were exposed to > 2 ppm; Andjelkovich *et al.* (1995) found 25% exposed to > 1.5 ppm (> 1.8 mg/m^3) and Marsh *et al.* (1994a) found 25% exposed to > 0.7 ppm (> 0.9 mg/m^3).

Cohort mortality studies in which cancers of the upper and lower respiratory tract are addressed may be biased by differences in the prevalence of tobacco smoking between the cohort and the referent population. Axelson and Steenland (1988), Blair *et al.* (1988), Siemiatycki *et al.* (1988) and others have shown that this potential bias is not a major problem, because the distribution of smoking habits between most occupational cohorts and referent populations differs little, if at all. Furthermore, when respiratory tract cancer rates are evaluated across a gradient of occupational exposures within the same cohort, the prevalence of smoking is generally so similar among the groups that tobacco smoking does not confound the relationship between occupation and cancer. In the present context, this theoretical lack of confounding was confirmed by Blair *et al.* (1990b) and Andjelkovich *et al.* (1995), who obtained data on the smoking habits of individual workers and found no effect on the risk estimates.

Three groups of workers were studied, which were totally or partially subsumed in the study of Blair *et al.* (1986, 1990b), described below. Marsh (1982) evaluated proportionate mortality patterns among workers engaged in the production of phenolic resins, urea–formaldehyde resins, melamine–formaldehyde resins, hexamethylenetetramine and resorcinol in the United States. He identified 603 deaths that occurred among men in 1950–76 and included 580 (132 exposed to formaldehyde for one month or more and 448 others) in the analysis. Wong (1983) studied workers employed at a formaldehyde production plant between the early 1940s and 1977. After exclusion of about 200 women, 12 blacks and two orientals, 2026 white men were included in the analysis. Tracing through 1977 was successful for all but 51 workers (2.5%), and 146 deaths were identified (death certificates were obtained for 136). Approximately 800 workers who were exposed to formaldehyde were included in the investigation of Blair *et al.* (1986, 1990b). Liebling *et al.* (1984) evaluated the proportionate mortality of 24 workers in the formaldehyde resin plant studied by Marsh (1982), who were also included by Blair *et al.* (1986, 1990b).

Blair *et al.* (1986) conducted a cohort mortality study of workers employed at 10 plants in the United States where formaldehyde was produced and used before 1966 and followed the workers up through 1979; some were included in the three studies mentioned above. The 10 plants were selected from a survey of about 200 companies because they had the most workers, the longest use of formaldehyde and the records necessary for a study. The cohort was assembled from company personnel records and verified for completeness from Social Security Quarterly Earnings reports. Relative risks were estimated from SMRs and directly standardized rate ratios. The expected numbers were calculated from rates for the general population and for the populations of the 10 counties in which the plants were located and were adjusted for race, sex, age and calendar time. Directly standardized rate ratios were adjusted to the distribution of age and calendar-time person-years of the entire cohort for internal comparisons. Quantitative estimates of exposure were made on the basis of monitoring data available from the companies, from monitoring conducted by the study investigators and from information on tasks, plant operations, effects of controls and production levels (Stewart *et al.* (1986). On the basis of the job held with the highest exposure, 11% of the workers were in the background/unexposed category, 12% were exposed to < 0.1 ppm (< 0.12 mg/m^3), 34% to 0.1–0.5 ppm (0.12–0.6 mg/m^3), 40% to 0.5–2.0 ppm (0.6–2.5 mg/m^3) and 4% to ≥ 2.0 ppm (≥ 2.5 mg/m^3). The vital status of the 26 561 people in the cohort (20 714 white men, 1839 black men, 3104 white women, 26 black women and 878 workers of unknown race or sex) was determined successfully as of 1980 for 96% of the men and 83% of the women, yielding 4396 deaths. Death certificates were obtained for 4059 of the decedents (92%). The SMRs for all causes for workers exposed to formaldehyde were 1.0 (95% CI, 0.9–1.0) for white men, 0.9 (0.8–1.0) for white women and 0.8 (0.7–0.9) for black men. The SMRs for all cancers combined were 1.0 (0.9–1.1) for white men, 0.8 (0.6–1.0) for white women and 0.7 (0.5–1.0) for black men. Neither emphysema (SMR, 0.9; 0.7–1.3) nor cirrhosis of the liver (0.9; 0.7–1.1) occurred in excess among white men. No significant excess of mortality from any cancer was seen among exposed white men, white women or black men. Among exposed white men (the only group for which there was sufficient information), there were fewer deaths from leukaemia (0.8 [0.5–1.2]; 19 deaths) and brain cancer (0.8 [0.5–1.3]; 17 deaths) than expected, while the number of deaths from prostate cancer was about that expected (1.2 [0.8–1.6]; 33 deaths). Two nasal cancers occurred among white men, with 2.2 expected; none was observed among white women or black men. Although based on small numbers, the risk for cancer of the nasopharynx was increased ([3.2; 1.3–6.6] 7 deaths). One cancer of the nasopharynx occurred in a person who was not exposed to formaldehyde and one in a person not exposed to particulates, i.e. work environments in the formaldehyde-resin industry where the particles include urea–, phenol– and melamine–formaldehyde resins. The risk for death from nasopharyngeal cancer (5 deaths) among white men exposed to particulates rose with cumulative exposure to formaldehyde (0 deaths among unexposed; 1.9 (1 death) among those exposed for < 0.5 ppm-years; 4.0 (2 deaths) exposed for 0.5–5.5 ppm-years; and 7.5 (2 deaths) exposed for ≥ 5.5 ppm-years) (Blair *et al.*, 1987). In the same group, the SMRs for cancer of the oropharynx by cumulative exposure were 0 (no deaths) for the unexposed, 4.6 (3 deaths) for those exposed for < 0.5 ppm-years, 0 for exposure for 0.5–< 5.5 ppm-years and ≥ 5.5 ppm-years. Collins *et al.* (1988) pointed out that the excess occurred primarily at one plant and that subjects included in the analysis of Blair *et al.* (1987) were not required to have had simultaneous exposure to formaldehyde and particulates. They extended the follow-up of workers at the plant where four of the seven nasopharyngeal cancer deaths occurred and found no additional death for 13 656 person-years of follow-up. Tamburro and Waddell (1987) objected to the interpretation of the pattern of nasopharyngeal cancers as a trend in the absence of a significant exposure–response gradient. Lucas (1994) compared death certificate diagnoses of four nasopharyngeal cancers with information from a cancer registry. Because one of the four cancers was incorrectly labelled as nasopharyngeal cancer on the death certificate, Lucas (1994) suggested that corrected diagnoses should be used. Marsh *et al.* (1994b) and Blair and Stewart (1994) discussed the appropriateness of this recommendation. [The Working Group noted that correction of the diagnoses in the cohort and not in the comparison population would bias estimates of relative risks towards the null. This bias would occur because there are deaths in the comparison population that are also incorrectly diagnosed as nasopharyngeal cancer. In fact, large surveys indicate that about 25% of death certificates coded as nasopharyngeal cancer are incorrect.] The SMRs for lung cancer among exposed and unexposed workers were 1.1 (95% CI, 1.0–1.3) and 0.9 (0.7–1.2) among white men, 1.3 (0.6–2.5) and 1.7 (0.7–3.5) among white women and 0.7 (0.4–1.3) and 0.6 (0.1–1.7) among black men. Internal comparisons resulted in directly adjusted rate ratios (compared with the unexposed category) for white men of 1.7 (35 deaths) for exposure to < 0.1 ppm (< 0.12 mg/m^3), 1.6 (70 deaths) for 0.1–0.4 ppm (0.12–0.5 mg/m^3), 1.8 (125 deaths) for 0.5–1.9 ppm (0.6–2.3 mg/m^3) and 0.8 (6 deaths) for ≥ 2.0 ppm (≥ 2.5 mg/m^3). There was a significant excess of deaths from lung cancer (1.2 [1.0–1.4] 219 observed) among white male wage workers (mainly non-managerial). Restricting analyses to wage (non-managerial) workers is valuable because it focuses on those employees likely to have had more intense exposures. Combining wage and salaried (managerial) workers in the same exposure–response analysis may introduce socioeconomic confounding, because salaried workers who have lower exposures to formaldehyde also have lower lung cancer rates. The risk was slightly higher when the analysis was restricted to events occurring 20 or more years after first exposure (1.3 [1.1–1.6] 151 observed). The SMRs for workers with a 20-year latency did not rise with cumulative exposure categories: 1.0 [0.3–2.3] (5 deaths) among unexposed; 1.4 [1.0–1.8] (49 deaths) for < 0.5 ppm-years, 1.4 [1.0–1.8] (53 deaths) for 0.5–5.5 ppm-years and 1.3 [0.9–1.7] (44 deaths) for ≥ 5.5 ppm-years.

In further analyses of the deaths from lung cancer in this cohort, Blair *et al.* (1990b) found no exposure–response gradient between the SMRs or directly-adjusted rate ratios and a variety of exposure indicators, including duration, intensity, cumulative exposure, peak, average and cumulative exposure restricted to lagged exposures (5, 10, 20 and 30 years). No exposure–response pattern was observed by duration of employment in various cumulative exposure categories or by cumulative exposure with duration of exposure categories. No increased risk for lung cancer was seen for workers exposed to formaldehyde alone (SMR, 1.0 [0.8–1.2] 88 observed). When exposures other than formaldehyde were considered, the risk for lung cancer was elevated (1.4 [1.2–1.7] 124 deaths) for workers in contact with asbestos, antioxidants, carbon black, dyes, melamine, phenol, urea and wood dust. Significant exposure–response trends were observed between mortality from lung cancer and duration of exposure to melamine and urea. Information on smoking was sought from medical records for 190 subjects with cancer and 950 controls. Although information was found for only about one-third of the subjects, the prevalence of smoking in this small sample did not appear to be associated with exposure to formaldehyde (80% who had ever smoked among the unexposed; 67% among those with a cumulative exposure of < 0.5 ppm-years, 84% with cumulative exposure of 0.5– < 5.5 ppm-years and 70% with cumulative exposure of ≥ 5.5 ppm-years).

Short-term workers sometimes have different mortality patterns from longer-term workers. Stewart *et al.* (1990) compared mortality among short-term (employed in the plants studied for one year or less) and long-term workers (employed for more than one year) in the cohort developed by Blair *et al.* (1986). Short-term workers had higher total mortality (SMR, 1.3;

95% CI, 1.2–1.3) than long-term workers (1.0; 0.9–1.0), and this overall excess was due to elevated rates of deaths from arteriosclerotic heart disease (1.1; 1.0–1.3), emphysema (1.7; 1.0–2.8) and cancers at all sites (1.3; 1.1–1.4). Excess rates were seen for cancers at several sites, including the stomach (1.4; 0.7–2.4), lung (1.4; 1.1–1.7) and brain (1.4; 0.7–2.5). The long-term workers had no cancer excesses. Data on nasal and nasopharyngeal cancers were not presented.

Others have re-analysed the study of Blair et al. (1986, 1990b). Robins et al. (1988), using a G-null test to adjust for the 'healthy worker survivor effect', found no indication of an association between exposure to formaldehyde and lung cancer but found a positive association with non-malignant respiratory disease. They adjusted the analysis for bias that may be created when ill workers leave the workforce: healthy workers continue to have the opportunity for exposure, while ill workers do not. This problem is most likely to occur in connection with debilitating diseases that do not lead to immediate death, such as emphysema.

Marsh et al. (1992a) used Poisson regression to analyse the data from the study of Blair et al. (1986, 1990b). They found excess mortality from lung cancer, which did not increase with level of cumulative exposure: relative risk, 1.0 for < 0.1 ppm-years, 1.4 (95% CI, 0.9–2.0) for 0.1–0.5 ppm-years, 1.2 (0.7–1.9) for 0.5–2.0 ppm-years and 1.3 (0.8–2.3) for ≥ 2.0 ppm-years, and no trend with duration of exposure. In a second report of their analyses, Marsh et al. (1992b) found no significant associations between lung cancer and cumulative, average or duration of exposure to formaldehyde. Significant positive associations with lung cancer were found with exposure to formaldehyde in the presence of other agents (antioxidants, hexamethylene-tetramine, melamine and urea), but not in the absence of these cofactors. Finally, Marsh et al. (1994a) re-assessed the exposures in a five-year update of one of the plants in the study of Blair et al. (1986) and found a significant excess of lung cancer among short-term workers (SMR, 1.3 [95% CI, 1.1–1.8]; 63 deaths) but not among long-term workers (1.2 [0.9–1.6]; 50 deaths). Poisson regression analysis also showed larger relative risks in association with exposure to formaldehyde among short-term workers than long-term workers. No additional case of nasopharyngeal cancer was observed. Two nasopharyngeal cancers occurred among short-term workers.

Sterling and Weinkam (1988, 1989a,b, 1994) performed three re-analyses of the data and reported an exposure–response relationship between mortality from lung cancer and exposure to formaldehyde. The first two re-analyses contained errors (Blair & Stewart, 1989; Sterling & Weinkam, 1994).

Acheson et al. (1984a) studied 7680 British workers in six factories producing formaldehyde or formaldehyde-based resins, after excluding 7898 men who had begun work in 1965 or later, 1326 women and 689 workers for whom essential information was lacking. The date of first use of formaldehyde in the six factories ranged from the 1920s to the 1950s. Each worker was traced through 31 December 1981 from national mortality resources, and more than 98% of the cohort was successfully traced, yielding 1619 deaths. SMRs were used to estimate relative risks, and the expected numbers were based on death rates in England and Wales, adjusted for age and calendar time. Mortality rates in the areas where the factories were located were also used to generate expected numbers. Exposures were estimated on the basis of available data from monitoring (none before 1970) and information from management and labour. Exposure was quantified in consultation with the staff of the six plants and placed in one of four categories: high (> 2.0 ppm; > 2.5 mg/m^3), moderate (0.6–2.0 ppm; 0.7–2.5 mg/m^3), low (0.1–0.5 ppm; 0.12–0.6 mg/m^3) and background/nil (< 0.1 ppm; < 0.12 mg/m^3). According to the job held with the highest exposure, 25% of the cohort were classified as having had background/nil exposure, 24% as low exposure, 9% as moderate exposure, 35% as high exposure and 6% unknown. Although the authors attempted to obtain information on tobacco use from company medical records, they were unsuccessful (Acheson et al., 1984b). They point out that the workers may have had contact with other chemicals, including asbestos. Additional analyses using different approaches for dealing with exposure and death (cumulative and mortality over entire follow-up period, cumulative and mortality after leaving the factory, and cumulative at various calendar periods and subsequent mortality) (Acheson et al., 1984c) yielded no further patterns.

The cohort of Acheson et al. (1984a) was further followed-up from 1981 through 1989 by Gardner et al. (1993). This follow-up includes 7660 people employed before 1965 rather than the 7680 in the original publication, because additional eligibility checks resulted in exclusion of 20 workers. The cohort also included 6357 workers who were first employed from 1965 onwards and who were thus excluded from the original report. The distribution of workers by highest formaldehyde exposure category was nil (25%), low (24%), moderate (9%), high (35%) and unknown (6%) among workers first employed before 1965 and 30%, 31%, 10%, 21% and 8%, respectively, for those first employed in 1965 or later. A further 1582 deaths were identified in the extended follow-up, for a total of 3201. SMRs are presented for workers first employed before 1965 and those first employed later. The SMRs for all causes were 1.0 (95% CI, 1.0–1.1) for workers first employed before 1965 and 1.0 (0.9–1.0) for those employed later, and the respective SMRs for all cancers were 1.1 (1.1–1.2; 802 deaths) and 1.0 (0.8–1.2; 128 deaths). One death from nasal cancer occurred (1.4 expected) in the group first employed before 1965 and none (0.3 expected) in those employed later. No deaths occurred from nasopharyngeal cancer, whereas 1.3 were expected, and no non-fatal cases were reported to the National Cancer Registry [expected number not reported]. Significant excess mortality was noted for cancers of the stomach (1.4; 1.2–1.7) and rectum (1.4; 1.0–1.9) among workers employed before 1965, although both decreased with adjustment for local rates. Non-significant SMRs greater than 2.0 were noted for cancer of the gall-bladder (2.9; 2 observed), breast (6.0; 1 observed) and other genital tumours (4.5; 1 observed) in workers employed in 1965 onwards and for bone cancer (2.4; 5 observed) in workers employed before 1965. For sites of special interest by period of first employment, the SMRs were 1.2 (1.1–1.4) and 1.1 (0.9–1.5) for lung cancer, 1.5 (0.6–3.0) and 0 deaths (1.1 expected) for pharyngeal cancer, 0.9 (0.5–1.5) and 0.9 (0.3–2.1) for brain cancer, 0.9 (0.5–1.6) and 1.9 (0.8–3.9) for non-Hodgkin's lymphoma and 0.9 (0.5–1.5) and 0.9 (0.3–2.3) for leukaemia. For background, low, moderate and high exposure categories, the relative risks for lung cancer were 1.0 (95% CI, 0.8–1.3), 1.1 (0.9–1.4), 0.9 (0.6–1.3) and 1.2 (1.1–1.4) for those first employed before 1965 and 1.4 (0.8–2.1), 1.0 (0.5–1.6), 1.0 (0.4–2.2) and 1.4 (0.8–2.3) for those first employed after 1964. There were no trends in lung cancer mortality that could be associated with level or duration of exposure.

Mortality among workers at a formaldehyde resin plant in Italy was studied by Bertazzi et al. (1986, 1989), who included 1330 male workers who were ever employed for at least 30 days between the start-up of the plant in 1959 and 1980. Vital status was determined as of December 1986, and 179 deaths were identified. Work histories of past employees were reconstructed from interviews with retired workers, current workers and foremen; and actual or reconstructed work histories were available for all but 16% of the cohort. Job mobility was low, and 79% of the workers had held a single job throughout their career. On the basis of their work histories, workers were placed into one of three categories: exposed to formaldehyde, exposed to compounds other than formaldehyde and exposure unknown. Individual exposures could not be estimated, but the mean levels in measurements taken between 1974 and 1979 were 0.2–3.8 mg/m^3. SMRs were used to estimate relative risks. The expected numbers were based on national and local mortality rates, adjusted for age and calendar time. The SMRs based on local rates were 0.9 ([95% CI, 0.7–1.0] 179 deaths) for all causes, 0.9 ([0.7–1.1] 62 deaths) for all cancers, 2.4 ([0.8–5.8] 5 deaths) for liver tumours, 1.0 ([0.6–1.5] 24 deaths) for lung cancer and 1.4 ([0.6–2.9] 7 deaths) for cancers of the lymphatic and haematopoietic system. The rates for lung cancer were not associated with duration of exposure, latency or age at first exposure. The SMRs for lung cancer were 0.7 ([0.3–1.5] 6 deaths) for workers exposed to formaldehyde, 0.8 ([0.4–1.5] 9 deaths) for those exposed to other chemicals and 2.1 ([1.0–4.0] 9 deaths) for those whose exposure to formaldehyde could not be determined. The risk for liver cancer was greater among workers with a latency of 20 or more years (SMR, 4.0) and among those who were first exposed at 45 years of age or older (3.8). The excess of liver cancer was seen in all three exposure categories: formaldehyde (2.4; 2 deaths), other chemicals (2.3; 2 deaths) and exposure unknown (2.9; 2 deaths). In the first report (Bertazzi et al., 1986), no nasal cancer was seen (0.03 expected).

Mortality among workers in the abrasives industry in Sweden was evaluated by Edling et al. (1987a) in plants where grinding wheels were manufactured from abrasives held together by formaldehyde resins. The levels of formaldehyde were reported to be 0.1–1.0 mg/m^3. A cohort of 911 workers (211 women and 700 men; 521 were blue-collar workers) employed between 1955 and 1983 was traced for mortality through 1983 and cancer incidence through 1981, yielding 79 deaths and 24 incident cancers. Deaths and events occurring at the age of 75 or more were excluded because of concerns about diagnostic validity. Loss to follow-up was 2%. The expected numbers were based on rates for the general population, stratified for age, calendar year and sex. No significant excesses were seen. The relative risks for mortality were 1.0 (95% CI, 0.8–1.2; 79 deaths) for all causes, 0.9 (0.5–1.5; 17 deaths) for all cancers and 1.3 (0.3–3.2; 4 deaths) for respiratory diseases. The relative risks for cancer incidence were 0.6 (0.1–2.1; 2 cases) for lung cancer, 0.9 (0.2–2.2; 4 cases) for prostatic cancer and 2.0 (0.2–7.2; 2 cases) for non-Hodgkin's lymphoma.

Stayner et al. (1988) conducted a cohort study of mortality among workers who had been employed for at least three months in three garment manufacturing plants in the United States between 1955 and 1977. The cohort consisted of 11 030 workers and comprised 1602 white men, 406 nonwhite men, 6741 white women and 2281 nonwhite women. Vital status was determined through 1982, and 609 deaths were uncovered. Vital status was determined for 96%

of the cohort. The expected numbers for determining SMRs were derived from the age-, calendar time-, race- and sex-specific rates of the general population. Formaldehyde levels were monitored in each of the plants, but the data were not used in the epidemiological analyses. The geometric mean levels in various departments ranged from 0.14 to 0.17 ppm [0.17–0.21 mg/m^3]. The SMRs for most major causes of death were low, including those for all causes (0.7; 95% CI [0.7–0.8]), all cancers (0.8 [0.7–1.0]), diseases of the circulatory system (0.7 [0.6–0.8]) and diseases of the respiratory system (0.7 [0.5–1.1]). Significant excesses occurred for cancers of the buccal cavity (3.4 [0.9-8.5]) and connective tissue (3.6 [1.0-9.3]). The SMRs for other cancers were 1.1 [0.8–1.6] for lung, 0.6 [0.2–1.6] for lymphosarcoma and 1.1 [0.5–2.2] for leukaemia. The four deaths from cancer of the buccal cavity occurred among white women. Some cancers were evaluated by duration of exposure (< 4 years, 4–9 years and ≥ 10 years), resulting in SMRs of 0, 2.8 (1 death) and 7.6 [1.6–22] (3 deaths) for cancer of the buccal cavity [linear trend, $p < 0.01$]; and 1.5 [0.9–2.4] (18 deaths), 1.1 [0.5–2.0] (11 deaths) and 0.8 [0.4–1.5] (10 deaths) for cancer of the lung [linear trend, $p < 0.05$]. In a previous report, Stayner *et al.* (1985) gave the findings of a proportionate mortality study of workers in three garment plants (two of which were included in the later cohort study). Significantly increased PMRs were observed for malignant neoplasms of the buccal cavity, biliary tract and liver and 'other lymphatic and haematopoietic sites'.

Andjelkovich *et al.* (1995) reported on the mortality of a subset of a previously studied cohort of workers with potential exposure to formaldehyde in an automotive iron foundry in the United States. A cohort of 3929 men (2635 white, 1294 nonwhite) was followed from 1 January 1960 to 31 December 1989 (83 064 person-years of follow-up). For comparison, an unexposed group was also analysed, which consisted of 2032 men (1629 white, 403 nonwhite) who had worked during the same period but had not been exposed to formaldehyde (40 719 person-years of follow-up). The expected numbers of deaths were derived from rates for the general population. Men who worked predominantly in core-making and related operations were exposed to formaldehyde from 1960 through 1987, while all workers in the foundry were exposed to silica. Detailed work histories and an evaluation of occupational exposures by an industrial hygienist permitted categorization of the levels of exposure to formaldehyde (none, low, medium, high) and silica (low, medium, high) for every occupational title. Values for these exposure levels were assigned on the basis of sparse sampling data (low, 0.05; medium, 0.55; high, 1.5 ppm (formaldehyde) or mg/m^3 (silica)). Quartiles were based on each person's cumulative exposure. Basic information was obtained on the smoking habits of the exposed (ever, 1934; never, 637; unknown, 1358) and unexposed (ever, 811; never, 309; unknown, 912) men. In the exposed cohort, 608 men died during the observation period, 3263 lived, and vital status was unknown for 58. In the unexposed cohort, 422 died, 1583 lived and 27 were of unknown vital status. For the men exposed to formaldehyde, the SMRs were 0.93 (95% CI, 0.86–1.0) for all causes, 0.99 (0.82–1.2) for all cancers, 1.3 (0.48–2.9) for cancer of the buccal cavity and pharynx (no deaths from cancer of the nasopharynx [0.5 expected]), 1.2 (0.89–1.6) for lung cancer, 0.62 (0.07–2.2) for brain cancer, 0.43 (0.05–1.6) for leukaemia and 1.3 (0.47–2.8) for emphysema. No death occurred from cancer of the nasal cavity. A similar mortality pattern was seen for unexposed men, with SMRs of 0.91 (0.82–1.0) for all causes, 0.97 (0.79–

1.2) for all cancers, 1.7 (0.54–4.0) for cancer of the buccal cavity and pharynx (one death from cancer of the nasopharynx), 1.2 (0.84–1.6) for lung cancer, 0.41 (0.01–2.3) for brain cancer, 0.86 (0.17–2.5) for leukaemia and 2.3 (1.2–4.1) for emphysema. Race-specific analyses revealed no major differences from the analyses of all races, except that an excess of cancers of the buccal cavity and pharynx was seen only among nonwhites. Several comparisons, including rate ratios for exposed and unexposed men, SMRs with expected numbers based on rates in a working population and Poisson regression with adjustment for smoking, revealed no association between lung cancer and exposure to formaldehyde.

The cohort and proportionate mortality studies of cancer among industrial workers exposed to formaldehyde are summarized in Table 11. Table 12 gives an overview of the occurrence of nasal and nasopharyngeal cancer in the cohort studies of both professionals and industrial workers.

2.4 Case–control studies

The Working Group systematically reviewed case–control studies of cancers of the oral cavity, pharynx and respiratory tract. Although case–control studies of cancer at other sites sporadically contained information on exposure to formaldehyde, these studies were not reviewed systematically.

2.4.1 *Cancers of the nasal cavity, paranasal sinuses, nasopharynx, oropharynx and pharynx (unclassified)*

With the purpose of investigating the carcinogenic effects of exposures to wood dust, Hernberg *et al.* (1983) conducted a joint Nordic case–control study of 167 patients in Finland, Sweden and most of Denmark in whom primary malignant tumours of the nasal cavity and paranasal sinuses had been diagnosed between July 1977 and December 1980. There were 167 country-, age- and sex-matched controls in whom cancers of the colon and rectum had been diagnosed. The study subjects represented 58% of all cancers identified at these anatomical sites; the exclusions were due to early deaths or to non-responding or missing controls. Information on the occupations and smoking habits of the study subjects was obtained by telephone interview on a standardized form. A review of occupations in which exposure to formaldehyde can occur [exposure frequencies not stated] gave no indication of any association with sinonasal cancer. None of the cases or controls had worked in the particle-board or plywood industry or in the production of formaldehyde or formaldehyde-based glues. The authors considered the category 'painting, lacquering and glueing' as a possible exception, as minimal exposure to formaldehyde may occur; 18 cases and six controls had had this exposure [odds ratio, 3.2; 95% CI, 1.3–8.1]. When people with exposure to wood dust were excluded, however, the difference was in the opposite direction (three cases, six controls [odds ratio, 0.5; 0.1–1.9]. [The Working Group noted that the study was not designed to address exposure to formaldehyde and that all the cases in Denmark were also included in the study of Olsen *et al.* (1984).]

Table 11. Cohort and proportionate mortality studies of cancer and exposure to formaldehyde among industrial workers

Country (reference)	Population, design (number), date	Exposure estimates	Cancer	Relative risk (95% CI)	Comment
United States (Blair et al., 1986, 1987)	Formaldehyde producers, resin makers, other users, SMR (26 561; 20 714 white men, 1839 black men, 3104 white women, 904 others), 1934–80	Quantitative estimate, duration	All causes	1.0 (0.9–1.0)	2836 deaths, exposed white men
				0.9 (0.8–1.0)	200 deaths, exposed white women
				0.8 (0.7–0.9)	232 deaths, exposed black men
			All cancers	1.0 (0.9–1.1)	570 deaths, exposed white men
				0.8 (0.6–1.0)	50 deaths, exposed white women
				0.7 (0.5–1.0)	31 deaths, exposed black men
			Buccal/pharynx	1.0 (0.6–1.5)	18 deaths, exposed white men
				[0] (NA)	No deaths, exposed white women
				[1.5]	Three deaths, exposed black men
					With particulate exposure, white men:
			Nasopharynx	0 (0 death)	No formaldehyde
				1.9 (1 death)	< 0.5 ppm-years formaldehyde
				4.0 (2 deaths)	0.5–< 5.5 ppm-years formaldehyde
				7.5 (2 deaths)	≥ 5.5 ppm-years formaldehyde
			Colon	0.9 (0.6–1.2)	42 deaths, exposed white men
				1.1 (0.4–2.2)	Seven deaths, exposed white women
				1.9 (0.6–4.3)	Five deaths, exposed black men
			Nasal	[0.9]	Two deaths, exposed white men
				[0]	No deaths, exposed white women/black men
			Lung	1.1 (1.0–1.3)	201 deaths, exposed white men; no trend with level or duration
				1.3 (0.6–2.5)	Eight deaths, exposed white women
				0.7 (0.4–1.3)	11 deaths, exposed black men
			Breast	0.8 (0.4–1.4)	12 deaths, exposed white women
			Prostate	1.2 (0.8–1.6)	33 deaths, exposed white men
				[0.7]	Two deaths, exposed black men

Table 11 (contd)

Country (reference)	Population, design (number), date	Exposure estimates	Cancer	Relative risk (95% CI)	Comment
Blair et al. (contd)			Brain	0.8 (0.5–1.3)	17 deaths, exposed white men
				—	No deaths, 2 expected, exposed white women
				[1.0] (NA)	One death, exposed black man
			Lymphoma	0.5 (0.2–1.1)	Seven deaths, exposed white men
				—	No deaths, exposed white women/black men
			Leukaemia	0.8 (0.5–1.2)	19 deaths, exposed white men
				—	No deaths, exposed white men
				[1.0] (NA)	One death, exposed black man
United Kingdom (Gardner et al., 1993; update of Acheson et al., 1984a)	Chemical industry, SMR (7660 men employed < 1965, 6357 men employed ≥ 1965), 1941–89	Quantitative estimation, duration	Employed < 1965		
			All causes	1.0 (1.0–1.1)	2744 deaths
			All cancers	1.1 (1.1–1.2)	802 deaths
			Mouth	1.4 (0.3–4.0)	Three deaths
			Pharynx	1.5 (0.6–3.0)	Seven deaths
			Nasal	0.7	One death, 1.4 expected
			Lung	1.2 (1.1–1.4)	348 deaths, no trend with level or duration
			Prostate	0.7 (0.4–1.0)	26 deaths
			Brain	0.9 (0.5–1.5)	16 deaths
			Lymphoma	0.9 (0.5–1.6)	12 deaths
			Leukaemia	0.9 (0.5–1.5)	15 deaths
			Employed ≥ 1965		
			All causes	1.0 (0.9–1.0)	457 deaths
			All cancers	1.0 (0.8–1.2)	128 deaths
			Mouth	1.9 (0.1–11)	One death
			Pharynx	—	0 deaths, 1.1 expected
			Nasal	0	No death, 0.3 expected
			Lung	1.1 (0.9–1.5)	54 deaths, no trend with level or duration, slight trend with years of follow-up

Table 11 (contd)

Country (reference)	Population, design (number), date	Exposure estimates	Cancer	Relative risk (95% CI)	Comment
United Kingdom (Gardner et al., 1993; update of Acheson et al., 1984a) (contd)			Prostate	1.2 [0.5–2.7]	Six deaths
			Brain	0.9 (0.3–2.1)	Five deaths
			Lymphoma	1.9 (0.8–3.9)	Seven deaths
			Leukaemia	0.9 (0.3–2.3)	Four deaths
Italy (Bertazzi et al., 1986, 1989)	Formaldehyde resin makers, SMR (1330 men), 1959–86	Duration of employment	All causes	0.9 [0.7–1.0]	179 deaths, local rates
			All cancers	0.9 [0.7–1.1]	62 deaths, local rates
			Alimentary tract	1.0 [0.6–1.5]	21 deaths, local rates
			Lung	1.0 [0.6–1.5]	24 deaths, local rates; no positive trend with duration
			Liver	2.4 [0.8–5.8]	Five deaths, local rates
			Lymphatic/haematopoietic	1.4 [0.6–2.9]	Seven deaths, local rates
Sweden (Edling et al., 1987a)	Abrasives industry, SMR, SIR (521 men), 1955–83	None	All causes	1.0 (0.8–1.2)	79 deaths
			All cancers	0.9 (0.5–1.5)	17 deaths
			All cancers	0.8 (0.5–1.3	24 incident cases
			Nasopharynx	– (NA)	One incident case
			Colon	1.0 (0.1–2.9)	Two incident cases
			Lung	0.6 (0.1–2.1)	Two incident cases
			Prostate	0.9 (0.2–2.2)	Four incident cases
			Lymphoma	2.0 (0.2–7.2)	Two incident cases
			Myeloma	4.0 (0.5–14)	Two incident cases

Table 11 (contd)

Country (reference)	Population, design (number), date	Exposure estimates	Cancer	Relative risk (95% CI)	Comment
United States (Stayner et al., 1988)	Garment industry, SMR (11 030: 1602 white men, 406 nonwhite men, 6741 white women, 2281 nonwhite women), > 3 months, 1955–82	Duration	All causes	0.7 [0.7–0.8]	609 deaths
			All cancers	0.8 [0.7–1.0]	186 deaths
			Buccal cavity	3.4 [0.9–8.5]	Four deaths, all in white women, trend with duration
			Pharynx	1.1 [0.1–4.0]	Two deaths
			Intestine	0.7 [0.4–1.1]	15 deaths
			Lung	1.1 [0.8–1.6]	39 deaths, inverse trend with duration
			Breast	0.7 [0.5–1.0]	33 deaths
			Brain	0.7 [0.2–1.7]	Five deaths
			Connective tissue	3.6 [1.0–9.3]	Four deaths
			Lymphatic/haematopoietic	0.9 [0.5–1.4]	18 deaths
			Lymphoma	0.6 [0.2–1.6]	Four deaths
			Leukaemia	1.1 [0.5–2.2]	Nine deaths
United States (Andjelkovich et al., 1995)	Cohort in foundry, RR (2635 exposed white men, 1294 exposed nonwhite men; 1629 unexposed white men, 403 unexposed nonwhite men), 1960–89	Duration, quantitative estimate	Buccal cavity, pharynx	0.59 (0.14–2.9)	Any exposure
				1.2 (0.2–6.5)	3rd–4th quartile of exposure
			Lung	0.71 (0.43–1.2)	Any exposure
				0.59 (0.28–1.2)	3rd–4th quartile of exposure

CI, confidence interval; SMR, standardized mortality ratio; NA, not available; SIR, standardized incidence ratio; RR, relative risk

Table 12. Results for nasal and nasopharyngeal cancers in cohort studies of professionals and industrial workers exposed to formaldehyde

Reference	Study size	Nasal			Nasopharyngeal		
		RR	Obs	Exp	RR	Obs	Exp
Professional workers							
Harrington & Oakes (1984)	2 720	-	0	NR	-	0	NR
Hall et al. (1991)	4 512	NR			NR		
Walrath & Fraumeni (1983)	1 263	-	0	NR	-	0	NR
Walrath & Fraumeni (1984)	1 007	-	0	0.6	NR	NR	NR
Levine et al. (1984a)	1 477	-	0	0.2	NR	NR	NR
Stroup et al. (1986)	2 317	-	0	0.5	NR	NR	NR
Hayes et al. (1990)	4 046	-	0	1.7	2.2	4	1.9
Friedman & Ury (1983)	143 574	NR			NR		
Industrial workers							
Blair et al. (1986)	20 714	0.9	2	2.2	3^a	6	2.0
Gardner et al. (1993)	14 017	[0.6]	1	1.7	-	0	1.3
Bertazzi et al. (1986)	1 330	-	0	0.03	NR		
Stayner et al. (1988)	11 030	NR			NR		
Edling et al. (1987a)	911	NR			NR		
Andjelkovich et al. (1995)	3 929	-	0	NR	-	0	[0.5]

NR, not reported
^{a} Trend with level among those also exposed to particulates

In a case–control study conducted in four hospitals in North Carolina and Virginia, United States, in 1970–80, 193 men and women with primary malignancies of the nasal cavity and sinuses were identified (Brinton et al., 1984). Two hospital controls who were alive at the date of the interview were selected for each living patient and matched on hospital, year of admission, age, sex, race and administrative area; for deceased patients, two similarly matched controls were chosen: one patient who had attended the same hospital but who was not necessarily alive at the date of the interview, and one deceased person from records of the state vital statistics offices. Patients with oesophageal and sinonasal cancers and various nasal disorders were excluded from the control group. Telephone interviews were completed with 160 of the nasal cancer patients (83%) and 290 of the controls (78%), either directly with the patients themselves (33% of cases and 39% of controls) or with their next-of-kin. Occupational exposures were assessed by the subjects' responses to a checklist of potentially high-risk industries and exposures, including formaldehyde. Exposure to formaldehyde was reported for two nasal cancer patients (one man and one woman), yielding an odds ratio of 0.35 (95% CI, 0.1–1.8). [The Working Group noted that the power of the study may have been decreased by the high proportion of interviews that were with next-of-kin and that exposure to formaldehyde was reported by subjects or their next-of-kin.]

In a population-based study in Denmark (Olsen *et al.*, 1984), 839 men and women in whom cancer of the sinonasal cavities (525) and nasopharynx (314) was diagnosed during the period 1970–82 and reported to the national Cancer Registry were matched to 2465 controls for sex, and age and year of diagnosis, who were selected among all patients in whom cancer of the colon, rectum, prostate and breast was diagnosed during the same period. Histories of exposure to formaldehyde, wood dust and 10 other specified compounds or industrial procedures were assessed by industrial hygienists, who were unaware of the case or control status of the study subjects, on the basis of individual employment histories obtained from a national pension scheme in operation since 1964. With regard to individual compounds, the industrial hygienists determined whether a subject had not been exposed, had definitely been exposed or had probably been exposed, or whether no information could be obtained. Of the controls, 4.2% of males and 0.1% of females had had occupations with presumed exposure to formaldehyde. The results were presented only for the carcinoma subgroup of sinonasal cancer (93% of cases). The odds ratios for definite exposures to formaldehyde (unadjusted for any other occupational exposure and using the no exposure category as the reference level) were 2.8 for both men and women (95% CI, 1.8–4.3 and 0.5–14, respectively) and 3.1 (1.8–5.3) for men in whom the diagnosis was made more than 10 years after first exposure. Adjustment for exposure to wood dust reduced both risk estimates for men to 1.6, which were no longer significant. Only five men in the group of 33 workers with definite exposure to formaldehyde had not been exposed to wood dust. Probable exposure to formaldehyde was associated with a slightly increased risk for sinonasal cancer in men (odds ratio, 1.2; 0.8–1.7). There was no increase in the risk for carcinoma of the nasopharynx among men with definite exposure to formaldehyde (0.7; 0.3–1.7). [The Working Group noted that the employment histories of study subjects were restricted to 1964 or later and that the study is limited by the fact that the formaldehyde-using industries in Denmark seem to be dominated by exposure to wood dust, which makes it difficult to assess the separate effect of exposure to formaldehyde on the risk for sinonasal cancer.]

A re-analysis was performed (Olsen & Asnaes, 1986) in which data on men with squamous-cell carcinoma (215) and adenocarcinoma (39) of the sinonasal cavities were examined separately. An odds ratio (adjusted for exposure to wood dust) of 2.3 (95% CI, 0.9–5.8) for squamous-cell carcinoma was found for 13 subjects who had ever been exposed to formaldehyde; of these, four had not been exposed to wood dust (2.0; 0.7–5.9). Introduction of a 10-year latent period into the analysis yielded odds ratios of 2.4 (0.8–7.4) and 1.4 (0.3–6.4), respectively. The analysis confirmed that the risk associated with exposure to wood dust is high for adenocarcinoma (odds ratios, 16 for any exposure and 30 for exposure 10 or more years before diagnosis) and small or non-existent for squamous-cell carcinoma (odds ratio, 1.3 irrespective of period). For the 17 cases of adenocarcinoma in men who had ever been exposed to formaldehyde, the odds ratio, after adjustment for exposure to wood dust, was 2.2 (0.7–7.2), and that among men who had been exposed 10 or more years before diagnosis was 1.8 (0.5–6.0); however, only one man with an adenocarcinoma had been exposed to formaldehyde alone. Analysis of the risk for histologically specified carcinomas of the nasopharynx showed no association with exposure to either formaldehyde or wood dust. [The Working Group noted possibly incomplete adjustment for confounding from exposure to wood dust in the assessment of the risk for adenocarcinoma associated with exposure to formaldehyde, but also noted that the assessment of risk for squamous-cell carcinoma was less likely to have been affected because squamous-cell carcinoma is not clearly associated with exposure to wood dust.]

From an examination of medical records in the six major institutions in the Netherlands for surgical and radiographic treatment of tumours of the head and neck, Hayes *et al.* (1986) identified 116 men, aged 35–79, in whom a histologically confirmed epithelial cancer of the nasal cavity and paranasal sinuses had been diagnosed in 1978–81. The cases were frequency matched on age with 259 population controls chosen randomly from among living male residents in 1982 (in a ratio of 2:1 for all patients) and from among deceased men in 1980 (in an approximate ratio of 1:1 for dead cases). Detailed histories, including information on exposure to a selected list of substances in the workplace and subjects' smoking habits, were obtained by personal interview of study subjects or their next-of-kin, with response rates of 78% for case patients and 75% for controls. Independently of one another, two industrial hygienists (A and B) evaluated possible exposure to formaldehyde on a 10-point scale, and subsequently used three categories. Exposure to wood dust was assessed similarly by one hygienist. At least some potential occupational exposure to formaldehyde was considered to have occurred for 23% of all study subjects by assessment A and 44% by assessment B; little or no exposure to wood dust was considered to have occurred in 15 and 30%, respectively. A large excess risk for adenocarcinoma (odds ratio, 26 [95% CI, 7.0-99]) was associated with high levels of exposure to wood dust; thus, there were too few cases (six) with no or limited exposure to wood dust to allow a meaningful assessment of risks associated with exposure to formaldehyde alone. Separate analyses among the 45 men with squamous-type sinonasal cancer who had had little or no exposure to wood dust showed an increase in risk with increasing level of exposure to formaldehyde, with odds ratios for moderate and high exposures of 2.7 [95% CI, 0.8–8.8] and 3.1 [0.7–13] in assessment A and 1.4 [0.4–4.4] and 2.4 [1.0–6.0] in assessment B, respectively. The overall relative risks for squamous-cell carcinoma associated with any exposure to formaldehyde were 3.0 [1.2–7.8] in assessment A and 1.9 [0.9–4.1] in assessment B. [The Working Group noted that a greater proportion of case patients than controls were dead (36% versus 14%) and variable numbers of next-of-kin were interviewed; besides, 10% of controls, but none of the case patients, were interviewed by telephone. The Group noted, however, that although assessments A and B were different both gave positive results.]

In the population of a 13-county area in western Washington State, United States, Vaughan *et al.* (1986a) studied 415 patients, aged 20–74 years, in whom cancer of the sinonasal cavities and pharynx (53 patients with sinonasal cancer, 27 with nasopharyngeal cancer and 205 with oro- or hypopharyngeal cancer) had been newly diagnosed, and 690 control subjects who were identified by random-digit dialing and were of similar age and the same sex as the cases. Medical, smoking, alcohol use, residential and occupational histories were collected in a telephone interview with study subjects or their next-of-kin. The response rates were 69% for cases and 80% for controls; interviews were conducted with next-of-kin for about half of the cases and none of the controls. Occupational exposure to formaldehyde was assessed by means of a job–exposure linkage system, in which each job within each industry was related to the likelihood and intensity of exposure and was categorized as background, low, medium or high exposure. In addition, exposure scores were calculated for maximal and usual exposures weighted by the time spent in the relevant job. The odds ratios for sinonasal cancer, adjusted for sex, age, cigarette smoking and alcohol consumption, were 0.8 (95% CI, 0.4–1.7; nine cases) for low exposure, 0.3 (0.0–1.3; three cases) for medium or high exposure and 0.3 (0.0–2.3) for a high exposure score. The corresponding odds ratios for nasopharyngeal cancer were 1.2 (0.5–3.3), 1.4 (0.4–4.7) and 2.1 (0.6–7.8), and those for oropharyngeal cancer were 0.8 (0.5–1.4), 0.6 (0.1–2.7) [high exposure only] and 1.5 (0.7–3.0). When a latent period of 15 years or more was introduced into the analysis, the odds ratio associated with the highest exposure to formaldehyde was unchanged (2.1; 0.4–10) for nasopharyngeal cancer and was slightly reduced (1.3; 0.6–3.1) for oropharyngeal cancer; no exposed cases of sinonasal cancer remained in this latency category. [The Working Group noted that the different proportions of interviews conducted with next-of-kin of cases and controls may have affected the odds ratios.]

Vaughan *et al.* (1986b) also explored the relationships between these types of tumours and residential exposure to formaldehyde. Living in a mobile home and the presence of urea–formaldehyde foam insulation and particle-board or plywood in residences were taken as indirect measures of residential exposure. Five of the patients with sinonasal cancer had lived in a mobile home (odds ratio, 0.6; 95% CI, 0.2–1.7), all for fewer than 10 years, 25 had lived in residences constructed with particle-board or plywood, yielding odds ratios of 1.8 (0.9–3.8) for periods of < 10 years and 1.5 (0.7–3.2) for ≥ 10 years. An association was found between living in a mobile home and risk for nasopharyngeal cancer, with odds ratios of 2.1 (0.7–6.6; four exposed cases) for < 10 years and 5.5 (1.6–19; four exposed cases) for ≥ 10 years. No association was found with the risk for oropharyngeal cancer (0.9; 0.5–1.8 and 0.8; 0.2–2.7, respectively). No association was seen between the risk for oro- or hypopharyngeal cancer and reported exposure to particle-board or plywood. The risks associated with exposure to formaldehyde foam insulation could not be estimated, owing to low exposure frequencies. [The Working Group considered that living in a mobile home was a poor proxy for exposure to formaldehyde because of large variations in the use of formaldehyde-containing foams and the sharply declining release of formaldehyde to indoor air with time.]

Roush *et al.* (1987) reported on a population-based case–control study of 371 men registered at the Connecticut (United States) Tumor Registry with a diagnosis of sinonasal cancer (198) or nasopharyngeal cancer (173) and who had died (of any cause) in Connecticut in 1935–75, and 605 male controls who died in the same period and were selected randomly from the files of Connecticut death certificates, without stratification or matching. Information on the occupations of the study subjects was derived from death certificates and from annual city directories; the latter were consulted 1, 10, 20, 25, 30, 40 and 50 years before death, when available. Each occupation held by case patients and controls was assessed by an industrial hygienist with regard to the likelihood and level of work-place exposure to formaldehyde, and study subjects were subsequently categorized into one unexposed and four exposed groups according to four degrees of probable exposure to formaldehyde. The odds ratio, adjusted for age at death, year of death and availability of information on occupation for case patients with sinonasal cancer who had probably been exposed to the same level of formaldehyde for most of their working life was 0.8 (95% CI, 0.5–1.3); for those who fulfilled the more restricted exposure criteria of being probably exposed to the same level for most of their working life and probably exposed to high levels for some years, the odds ratio was 1.0 (0.5–2.2); that for men who had probably been exposed to the same level for most of their working life and probably exposed to high levels at some point 20 or more years before death, it was 1.5 (0.6–3.9). The corresponding odds ratios for men with nasopharyngeal cancer were 1.0 (0.6–1.7), 1.4 (0.6–3.1) and 2.3 (0.9–6.0). There was no excess risk for sinonasal cancer among formaldehyde-exposed men who had also been exposed wood dust (0.9; 0.4–1.9).

Luce *et al.* (1993) conducted a case–control study of 303 men and women with primary malignancies of the nasal cavities and paranasal sinuses diagnosed in one of 27 hospitals in France between January 1986 and February 1988, and 443 control subjects selected by frequency matching for age and sex among patients in whom another cancer had been diagnosed during the same period at the same or a nearby hospital (340) or from a list of names of healthy individuals provided by the cases (103). Occupational exposures to formaldehyde and 14 other substances or groups of substances were assessed by an industrial hygienist on the basis of information obtained during a personal interview at the hospital (for the cancer patients) or at home (for the healthy controls) on job histories, a number of pre-defined occupational exposures, socioeconomic variables and smoking habits. The response rates were 68% of cases and 92% of controls. Histological confirmation was available in the medical records of all but one of the remaining 207 case patients. Study subjects were classified according to the likelihood of exposure to each of the suspected determinants of sinonasal cancer and grouped into one of four categories: none, possible, probable or definite exposure; the latter two were further split into a number of subgroups according to levels and calendar periods of exposure and combinations thereof. The risks associated with exposure to formaldehyde were reported for men only. Possible exposure to formaldehyde was considered to have occurred in 12% of the 59 men with squamous-cell carcinoma of the sinonasal cavities and 11% of the 320 control subjects (odds ratio, 1.0; 95% CI, 0.4–2.4). The corresponding proportions with probable or definite exposure to formaldehyde were 27% and 25%, respectively [1.1; 0.6–2.1]. No relationship was observed between any of the measures of exposure to formaldehyde and risk for squamous-cell carcinoma. Nearly all of the adenocarcinomas occurred in men with medium to high exposure to wood dust (77/82). For those exposed to both wood dust and formaldehyde, the odds ratio was 692 (92–5210), and for those exposed to wood dust but not formaldehyde the odds ratio was 130 (14–1191); however, the latter estimate was based on only six exposed cases. For four men who were exposed to formaldehyde but who had had no or little exposure to wood dust, the odds ratio for adenocarcinoma was 8.1 (0.9–73). [The Working Group noted that residual confounding by exposure to wood dust may have occurred.]

The etiology of nasopharyngeal carcinoma was studied in the Philippines; both viral (Hildesheim *et al.*, 1992) and non-viral (West *et al.*, 1993) risk factors were addressed. West *et al.* (1993) conducted a case–control study of 104 histologically confirmed cases of nasopharyngeal carcinoma in Rizal Province, where the incidence rates of this tumour (4.7/100 000 men and 2.6/100 000 women) are intermediate between those in China and those in western countries. The case patients (100% response rate) were identified at the Philippines General Hospital, as were 104 hospital controls (100% response rate), who were matched to cases on sex, age and type of hospital ward, and 101 community controls (77% response rate), who were matched on sex, age and neighbourhood. A personal interview included questions on smoking habits, adult diet, demographic variables and occupational history. An industrial hygienist classified each job held by the study subjects as likely or unlikely to involve exposure to formaldehyde, solvents, exhaust fumes, wood dust, dust in general and pesticides and combined the classification with information on duration of employment in such occupations. The risk for nasopharyngeal carcinoma was associated with likely exposure to formaldehyde; the odds ratios, adjusted for the effects of dusts and exhaust fumes and other suspected risk factors, were 1.2 (0.4–3.6; 12 exposed cases) for subjects first exposed < 25 years before diagnosis and 4.0 (1.3–12; 14 exposed cases) for those first exposed ≥ 25 years before diagnosis. In the subgroup of subjects who were first exposed to formaldehyde ≥ 25 years before diagnosis and first exposed to dust and/or exhaust fumes ≥ 35 years before diagnosis, an odds ratio of 16 (2.7–91) was found relative to people exposed to neither factor [numbers exposed not given]. A reverse trend was seen, however, with increasing duration of exposure to formaldehyde, with odds ratios of 2.7 (1.1–6.6) for < 15 years and 1.2 (0.5–3.2) for ≥ 15 years of exposure. [The Working Group noted that the authors did not control for the presence of Epstein–Barr viral antibodies, which showed a strong association with nasopharyngeal cancer (odds ratio, 21) in the study of Hildesheim *et al.* (1992).]

2.4.2 *Cancers of the lung and larynx*

Andersen *et al.* (1982) conducted a case–control study of 79 male and five female Danish doctors with a notification of lung cancer in the files of the nationwide Danish Cancer Registry during the period 1943–77. Three times as many control subjects, matched individually on sex and age, were selected at random from among individuals on official lists of Danish doctors. Information on postgraduate specialization and places of work during the professional career of cases and controls was obtained from medical directories and supplementary files at the Danish Medical Society. Potential exposure to formaldehyde was assumed to be associated with working in pathology, forensic medicine and anatomy. None of the doctors with lung cancer had specialized in any of these fields, but one control doctor was a pathologist. Eight male case patients and 23 controls had been employed at some time in pathology, forensic medicine or anatomy, giving an odds ratio of 1.0 (95% CI, 0.4–2.4).

Fayerweather *et al.* (1983) reported on a case–control study of mortality from cancer among chemical workers in eight plants in the United States where formaldehyde was manufactured or used. All 481 active or pensioned men who were known to have died of cancer in 1957–79 were individually matched on age, pay class, sex and date of first employment to 481 men selected at random from annual payroll rosters at the plants. The cases included 181 lung cancers, 12 brain cancers and 7 cancers of the buccal cavity and pharynx. The work histories of both case and control subjects were supplied by the plants, and the categories of exposure to formaldehyde were defined, on the basis of frequency and intensity of exposure, as 'continuous direct', 'intermittent' and 'background'. Smoking histories were obtained for about 90% of subjects, primarily by interviewing living co-workers. Of the 481 cases, 142 (30%) had had potential exposure to formaldehyde. For no cancer site examined was the odds ratio significantly greater than 1.0 in relation to any of the defined categories of exposure to formaldehyde. After allowance for a cancer induction period of 20 years, 39 patients with lung cancer and 39 controls had potentially been exposed to formaldehyde [odds ratio, 1.0; 95% CI, 0.6–1.7; unadjusted for smoking habits]; the odds ratios were 1.2 [0.6–2.8] and 0.8 [0.4–1.6] for subgroups with < 5 years and ≥ 5 years of exposure, respectively. Only one of the seven patients who died of cancer of the buccal cavity or pharynx was thought to have been exposed to formaldehyde; this was equivalent to the frequency observed among the matched controls. No death from nasal cancer was identified.

In a population-based case–control study, Coggon *et al.* (1984) used death certificates to obtain information about the occupations of all men under the age of 40 years in England and Wales who had died of bronchial carcinoma during 1975–79; 598 cases were identified, 582 of which were matched for sex, year of death, local authority district of residence and year of birth with two controls and the rest with one control who had died from any other cause. Occupations were coded according to the Office of Population Census and Surveys 1970 classification, and a job–exposure matrix was constructed by an occupational hygienist, in which the occupations were grouped according to three levels (high, low and none) of exposure to nine known or putative carcinogens, including formaldehyde. All occupations that entail exposure to formaldehyde were associated with an elevated, crude odds ratio for bronchial carcinoma of 1.5 (95% CI, 1.2–1.8); however, for occupations in which exposure was presumed to be high, the odds ratio was 0.9 (0.6–1.4). [The Working Group noted that information on occupation from death certificates is limited; they also noted the young age of the subjects and the consequent short exposure and latency.]

Partanen *et al.* (1985) conducted a case–control study in a cohort of 3805 male production workers who had been employed for at least a year in one of three particle-board factories, seven plywood factories, eight sawmills and one formaldehyde glue factory between 1944 and 1966. Of these, 57 subjects were notified to the Finnish Cancer Registry with cancer of the respiratory tract (including at least 51 cases of lung cancer), oral cavity or pharynx in 1957–80. Three controls were selected at random from the same cohort and were individually matched to the case by year of birth. Plant- and time-specific job–exposure matrices were constructed for 12 chemicals, including formaldehyde (Kauppinen & Partanen, 1988), and combined with the work histories of the subjects to yield several indicators of exposure; supplementary information on smoking was collected for 68% of cases and 76% of controls by means of a questionnaire posted to study subjects or their relatives. A slight, nonsignificant increase in risk for all cancers combined was seen among workers with any exposure to at least 0.1 ppm (0.12 mg/m^3) formaldehyde, as contrasted to workers with no exposure to formaldehyde, to give an odds ratio of 1.4 [95% CI, 0.6–3.5]; an odds ratio of 1.3 [0.5–3.5] was seen when a minimal latency of 10 years before diagnosis was assumed. No significant association was found with other indicators of exposure to formaldehyde (mean level and cumulative exposure, repeated peak exposures and 'formaldehyde in wood dust'). Adjustment for cigarette smoking did not change the overall picture.

In an expansion of the study to include a total of 35 Finnish factories and 7307 woodworkers employed in 1944–65, Partanen et al. (1990) identified 136 cases of newly diagnosed cancer of the respiratory tract (118 lung cancers, 12 laryngeal cancers and one sinonasal cancer), oral cavity (four cases) and pharynx (one case) from the files of the cancer registry, 1957–82. The additional factories were mainly involved in construction carpentry and furniture manufacture. Three controls were provided for each of the new cancer cases, and exposure to formaldehyde and 11 other occupational agents was assessed by the same methods as those described in the initial study (Partanen et al., 1985; Kauppinen & Partanen, 1988). Of 20 cases with any exposure to formaldehyde (odds ratio, 1.4 [95% CI, 0.6–3.1]), 18 were cancers of the lung (1.3 [0.5–3.0]) and two were cancers at other sites (2.4 [0.3–18]). Adjustment for smoking reduced the odds ratios to 1.1 for all cancers combined and to 0.7 for lung cancer separately and made the odds ratio for cancers at other sites unassessable. The unadjusted odds ratios for all cancers were 1.5 [0.7–3.6] for an estimated mean level of formaldehyde of 0.1–1 ppm [0.12–1.23 mg/m^3] and 1.0 [0.1–8.2] for > 1 ppm, in comparison with no exposure. Other indicators of exposure to formaldehyde showed similar inverse dose–response relationships, i.e. the lowest risks in the highest exposure categories. Allowance for a minimal latency of 10 years further reduced the risk estimates of the subgroups with presumably the highest exposures, with odds ratios generally below 1.0. [The Working Group noted that there were too few cancers at sites other than the lung to allow meaningful analysis; consequently, this is essentially a study of lung cancer.]

Bond et al. (1986) conducted a case–control study in a cohort of 19 608 men employed for one year or more at a large chemical production facility in Texas, United States, between 1940 and 1980, including all 308 workers who had died from lung cancer and 588 controls chosen at random from among men in the same cohort. Two series of controls, individually matched to cases on race, year of birth and year of hire, were selected: one among men still alive when the matched subjects died of lung cancer, and one among men who had died ≤ 5 years after the matched subjects. Exposures (ever or never) to 171 chemical and physical agents, including formaldehyde, were assessed by an industrial hygienist on the basis of a review of documentation on the subject's employment history at the facility and industrial hygiene records; six exposures, excluding formaldehyde, were assessed in greater detail. Only nine men with lung cancer (3%) were judged ever to have been exposed to formaldehyde, and a negative association was seen between this exposure and mortality from lung cancer (not adjusted for other exposure variables), with an odds ratio of 0.6 (95% CI, 0.3–1.3); incorporation into the analysis of a 15-year minimal latency gave an odds ratio of 0.3 (0.1–0.9).

In a population-based case–control study in the area of Montréal, Canada, 857 men with histologically confirmed primary lung cancer diagnosed during 1979–85, were identified (Gérin et al., 1989). Two groups of control subjects were established, one composed of 1523 men diagnosed during the same years with cancers of other organs (oesophagus, stomach, colorectum, liver, pancreas, prostate, bladder, kidney, melanoma and lymphoid tissue), the other composed of 533 men selected from electoral lists of the Montréal area. Interviews or completed questionnaires, yielding lifelong job history and information on potential nonoccupational confounders, were obtained from the cancer patients or their next-of-kin and from the population controls, with response rates at 82 and 72%, respectively. Each job was translated by a group of chemists and hygienists into a list of 300 potential exposures, including formaldehyde, which were categorized according to the likelihood, intensity and frequency of exposure. Nearly one-quarter of all men had been exposed to formaldehyde in at least one of the jobs they had held during their working life; however, only 3.7% were considered to be definitely exposed and only 0.2% were considered to have had high exposure, defined as more than 1.0 ppm [1.23 mg/m^3] of formaldehyde in the ambient air. Odds ratios, adjusted for age, ethnic group, socioeconomic status, cigarette smoking, the 'dirtiness' of the jobs held and other potential workplace exposure, were 0.8 (95% CI, 0.6–1.2) for < 10 years of exposure to formaldehyde, 0.5 (0.3–0.8) for ≥ 10 years of presumed exposure to < 0.1 ppm [0.12 mg/m^3], 1.0 (0.7–1.4) for ≥ 10 years of presumed exposure to 0.1–1.0 ppm [0.12–1.23 mg/m^3] and 1.5 (0.8–2.8), for ≥ 10 years of presumed exposure to ≥ 1.0 ppm formaldehyde, in relation to controls with other cancers. In comparison with the population controls, the equivalent odds ratios were 1.0 (0.6–1.8), 0.5 (0.3–0.8), 0.9 (0.5–1.6), and 1.0 (0.4–2.4), respectively. Marginally increased risks were seen for subjects with the adenocarcinoma subtype of lung cancer who had had long exposure to a high concentration of formaldehyde, with odds ratios of 2.3 (0.9–6–0) and 2.2 (0.7–7.6) in comparison with the cancer and population control groups, respectively; however, the estimates were based on only seven exposed cases.

Wortley *et al.* (1992) studied 291 male and female residents of a 13-county area of western Washington State, United States, in whom laryngeal cancer was diagnosed in 1983–87 and notified to a population-based cancer registry in the area; 81% were interviewed. Control subjects were identified by random-digit dialing and selected to be similar in age and sex to the cases; 80% of eligible subjects were interviewed, leaving 547 for analysis. Lifetime occupational, smoking and drinking histories were obtained by personal interview, and each job held for least six months was coded according to the United States census codes for industries and occupations. Codes for exposure to formaldehyde and five other agents were assigned to each classified job held by the study subjects, using a job–exposure matrix assessing both likelihood and degree of exposure and created *ad hoc*, and combined with information on duration of exposure; finally, three summary variables for presumed exposure were derived. The risk for laryngeal cancer, adjusted for age, smoking and drinking habits and length of education, was not associated with exposure to formaldehyde to a significant degree. The odds ratios were 1.0 (95% CI, 0.6–1.7) for patients with any 'low' exposure, 1.0 (0.4–2.1) for any 'medium' exposure and 2.0 (0.2–20; two exposed cases) for any 'high' exposure. Odds ratios of 0.8 (0.4–1.3) and 1.3 (0.6–3.1) were seen for exposure < 10 years and ≥ 10 years and of 1.0 (0.5–2.0) and 1.3 (0.5–3.3) for a medium and high formaldehyde score, respectively; the latter were calculated as the sum of years exposed weighted by the level of exposure in each of the years.

2.4.3 Cancers at other sites

In a study of 578 male leukaemia cases, 622 male non-Hodgkin's lymphoma cases and 1245 population-based controls in Iowa and Minnesota (United States), Linos *et al.* (1990) observed elevated risks for both leukaemia (odds ratio, 2.1 [95% CI, 0.4–10]) and non-Hodgkin's lymphoma (3.2 [0.8–13]) among subjects who had been employed in funeral homes and crematoria, indicating some degree of professional exposure to formaldehyde and other compounds. The risks were particularly high for the acute myeloid subtype of leukaemia (6.7 [1.2–36]) and the follicular subtype of non-Hodgkin's lymphoma (6.7 [1.2–37]), however, these estimates were each based on only three exposed cases.

In the study of Gérin et al. (1989) in Montréal, Canada (see p. 274), 206 cases of non-Hodgkin's lymphoma were compared with cases of other cancer. No association was found with estimated exposure to formaldehyde.

Merletti et al. (1991) reported a case–control study of 103 male residents of Turin, Italy, with a diagnosis of cancer of the oral cavity or oropharynx notified to the population-based cancer registry of the city, and a random sample of 679 males, stratified by age, chosen from files of residents of Turin. Detailed occupational (since 1945) and lifelong smoking and drinking histories were obtained by personal interview, with response rates of 84% for cases and 57% for controls. Each job held for at least six months was coded according to the International Standard Classification of Occupations and the International Standard Industrial Classification, and a job–exposure matrix for 13 agents (including formaldehyde) which are known or suspected respiratory tract carcinogens and three non-specific exposures (including dust) were applied to the occupation–industry code combination of study subjects; the matrix was developed at the International Agency for Research on Cancer for use in a similar study of laryngeal cancer. Study subjects were grouped into three categories of presumed frequency and intensity of exposure to formaldehyde, with a 'no exposure' group (exposure not higher than that of the general population) as the reference level. An association was suggested between cancer of the oral cavity or oropharynx and exposure to formaldehyde, with odds ratios of 1.6 (95% CI, 0.9–2.8) for any exposure and 1.8 (0.6–5.5) for 'probable or definite' exposure; however, only 25 and six cases were exposed, respectively. No relationship was seen with duration of exposure to formaldehyde: non-significantly raised odds ratios were estimated of 1.7 for 1–15 years of exposure and 1.5 for ≥ 16 years within the 'any exposure' category, and 2.1 and 1.4, respectively, within the 'probable or definite' exposure category. Separate results in association with exposure to formaldehyde were not reported for the 12 men with oropharyngeal cancer.

Goldoft et al. (1993) interviewed nine of 14 patients with nasal or nasopharyngeal melanoma as part of a population-based case–control study of sinonasal cancer described above (Vaughan et al., 1986a,b; see pp. 269–270). The frequency of exposure to formaldehyde was compared with the frequency of exposure of the control subjects included in the study of Vaughan et al. One subject had lived in a residence insulated with formaldehyde-based foam ([0.3 expected] not significant). None of the melanoma patients reported specific occupational exposure to formaldehyde (0.3 expected), and none reported having been employed in industries likely to result in exposure to formaldehyde (0.8 expected).

The case–control studies of cancer and exposure to formaldehyde are summarized in Table 13.

Table 13. Case–control studies of formaldehyde by cancer site

Authors and country	Subjects	Exposure estimates	Odds ratio (95% CI)	Comments
Sinonasal cancer				
Hemberg et al. (1983) Denmark, Finland, Sweden	167 patients (distribution by sex not given) 167 controls	Employment in particle-board or plywood industry	—	No exposed subjects
Brinton et al. (1984) United States	160 patients (93 men, 67 women) 290 controls	Ever	0.4 (0.1–1.8)	Control for tobacco use did not change results
Olsen et al. (1984) Denmark	488 patients (distribution by sex not given) 2465 controls	Probably exposed ≥ 10 years previously (men)	3.1 (1.8–5.3)	Unadjusted
		Probably exposed ≥ 10 years previously (men)	1.6 (0.7–3.6)	Adjusted for exposure to wood dust
Olsen & Asnaes (1986) Denmark	215 men with squamous-cell carcinoma 2465 controls	Probably exposed ≥ 10 years previously	2.4 (0.8–7.4)	Adjusted for exposure to wood dust
	39 men with adenocarcinoma 2465 controls	Probably exposed ≥ 10 years previously	1.8 (0.5–6.0)	Adjusted for exposure to wood dust
Hayes et al. (1986) Netherlands	63 male patients 161 controls	Any, with no or little exposure to wood dust Industrial hygienist A Industrial hygienist B	2.5 [1.0–5.9] 1.6 [0.8–3.1]	
	28 male patients 34 controls	Any, with moderate or high exposure to wood dust Industrial hygienist A Industrial hygienist B	1.9 [0.6–6.5] NR	
	45 male patients with squamous-cell carcinoma 161 controls	Any, with no or little exposure to wood dust Industrial hygienist A Industrial hygienist B	3.0 [1.2–7.8] 1.9 [0.9–4.1]	Controlling for cigarette use did not change the result
Vaughan et al. (1986a) United States	53 patients (distribution by sex not given) 552 controls	Low Medium or high High exposure score (exposure level weighted by period of exposure)	0.8 (0.4–1.7) 0.3 (0.0–1.3) 0.3 (0.0–2.3)	Occupational exposure; adjusted for sex, age, cigarette smoking and alcohol consumption

Table 13 (contd)

Authors and country	Subjects	Exposure estimates	Odds ratio (95% CI)	Comments
Sinonasal cancer (contd)				
Vaughan et al. (1986b) United States	53 patients (distribution by sex not given) 552 controls	Any, from 'mobile home' Any from particle-board or plywood < 10 years ≥ 10 years	0.6 (0.2–1.7) 1.8 (0–9–3.8) 1.5 (0.7–3.2)	Residential exposure; adjusted for sex, age, cigarette smoking and alcohol consumption
Roush et al. (1987) United States	169 male patients 509 male controls	Probably exposed for most of working life Plus exposure ≥ 20 years before death Plus exposure to high level in some year Plus exposure to high level ≥ 20 years before death	0.8 (0.5–1.3) 1.0 (0.5–1.8) 1.0 (0.5–2.2) 1.5 (0.6–3.9)	Information on histological subtypes not available; information on occupation obtained from death certificates and city directories; adjusted for age and calender period
Luce et al. (1993) France	77 patients (59 men, 18 women) with squamous-cell carcinoma 407 controls	Possible (men) Probable or definite (men) < 20 years ≥ 20 years	1.0 (0.4–2.4) 1.1 (0.5–2.5) 0.8 (0.3–2.0)	Adjusted for age, exposure to wood dust, glues and adhesives
Nasopharyngeal cancer				
Olsen et al. (1984) Denmark	266 patients (distribution by sex not given) 2465 controls	Possible or probable exposure Men Women	0.7 (0.3–1.7) 2.6 (0.3–22)	Unadjusted
Vaughan et al. (1986a) United States	27 patients (distribution by sex not given) 552 controls	Low Medium or high High exposure score (exposure level weighted by period of exposure)	1.2 (0.5–3.3) 1.4 (0.4–4.7) 2.1 (0.6–7.8)	Occupational exposure; adjusted for cigarette smoking and race

Table 13 (contd)

Authors and country	Subjects	Exposure estimates	Odds ratio (95% CI)	Comments
Nasopharyngeal cancer (contd)				
Vaughan et al. (1986b) United States	27 patients (distribution on sex not given) 552 controls	Any, from 'mobile home' < 10 years ≥ 10 years Any, from particle-board or plywood < 10 years ≥ 10 years	 2.1 (0.7–6.6) 5.5 (1.6–19) 1.4 (0.5–3.4) 0.6 (0.2–2.3)	Residential exposure; adjusted for cigarette smoking and ethnic origin
Roush et al. (1987) United States	147 male patients 509 male controls	Probably exposed for most of working life Plus exposure ≥ 20 years before death Plus exposure to high level in some year Plus exposure to high level ≥ 20 years before death	1.0 (0.6–1.7) 1.3 (0.7–2.4) 1.4 (0.6–3.1) 2.3 (0.9–6.0)	Information on occupation obtained from death certificates and city directories; adjusted for age and calender period
West et al. (1993) Philippines	104 patients (76 male, 28 female) 193 controls	< 15 years ≥ 15 years First exposure < 25 years before diagnosis First exposure ≥ 25 years before diagnosis	2.7 (1.1–6.6) 1.2 (0.5–3.2) 1.3 (0.6–3.2) 2.9 (1.1–7.6)	Adjusted for other occupational exposure
Cancer of the oral cavity, oro- and hypopharynx				
Vaughan et al. (1986a)	205 patients with oro- or pharyngeal cancer (distribution by sex not given) 552 controls	Low Medium High High exposure score (exposure level weighted by period of exposure)	0.8 (0.5–1.4) 0.8 (0.4–1.7) 0.6 (0.1–2.7) 1.5 (0.7–3.0)	Occupational exposure; adjusted for sex, age, cigarette smoking and alcohol consumption
Vaughan et al. (1986b)	205 patients with oro- or pharyngeal cancer (distribution by sex not given) 552 controls	Any, from 'mobile home' < 10 years ≥ 10 years Any, from particle-board or plywood < 10 years ≥ 10 years	 0.9 (0.5–1.8) 0.8 (0.2–2.7) 1.1 (0.7–1.9) 0.8 (0.5–1.4)	Residential exposure; adjusted for sex, age, cigarette smoking and alcohol consumption

Table 13 (contd)

Authors and country	Subjects	Exposure estimates	Odds ratio (95% CI)	Comments
Cancer of the oral cavity, oro- and hypopharynx (contd)				
Merletti et al. (1991) Italy	86 male patients with oral or oropharyngeal cancer 373 controls	Any Probable or definite	1.6 (0.9–2.8) 1.8 (0.6–5.5)	Adjusted for age, level of education, area of birth, tobacco smoking and alcohol drinking
Lung cancer				
Andersen et al. (1982) Denmark	84 cases among doctors (79 men and five women) 252 controls	Ever employed in exposed speciality	1.0 (0.4–2.4)	Both cases and controls were medical doctors
Fayerweather et al. (1983) United States	181 male cases 181 controls	Any, < 5 years Any, ≥ 5 years	1.2 [0.6–2.8] 0.8 [0.4–1.6]	Estimates but not CIs adjusted for smoking habits
Coggon et al. (1984) England and Wales	598 male patients 1180 controls	Any High	1.5 (1.2–1.8) 0.9 (0.6–1.4)	Matched analysis, but unadjusted for smoking habits; all subjects < 40 years at death
Bond et al. (1986) United States	308 male patients 588 controls	Any Any, ≥ 15 years before death	0.6 (0.3–1.3) 0.3 (0.1–0.9)	Unadjusted
Gérin et al. (1989) Canada	857 male patients 1523 male cancer controls (a) 533 male population controls (b)	Any, < 10 years ≥ 10 years Low Medium High	0.8 (0.6–1.2) (a) 1.0 (0.6–1.8) (b) 0.5 (0.3–0.8) (a) 0.5 (0.3–0.8) (b) 1.0 (0.7–1.4) (a) 0.9 (0.5–1.6) (b) 1.5 (0.8–2.8) (a) 1.0 (0.4–2.4) (b)	Adjusted for age, ethnic group, socioeconomic status, cigarette smoking, dirtiness of job and other occupational risk factors

Table 13 (contd)

Authors and country	Subjects	Exposure estimates	Odds ratio (95% CI)	Comments
Lung cancer (contd)				
Gérin et al. (1989) Canada (contd)	Adenocarcinoma subtype 162 male patients 1523 male cancer controls (a) 533 male population controls (b)	Any, < 10 years ≥ 10 years Low Medium High	0.6 (0.3–1.3) (a) 0.8 (0.3–2.0) (b) 0.5 (0.2–1.2) (a) 0.5 (0.2–1.3) (b) 0.8 (0.4–1.6) (a) 1.0 (0.4–2.5) (b) 2.3 (0.9–6.0) (a) 2.2 (0.7–7.6) (b)	
Partanen et al. (1990) Finland	118 male woodworkers 354 controls	Any Any, ≥ 10 years since first exposure	0.7 [0.2–2.8] 0.9 [0.2–3.8]	Adjusted for vital status and smoking
Laryngeal cancer				
Wortley et al. (1992) United States	235 patients (185 men, 50 women) 547 controls	Any Low Medium High ≥ 10 years previously	1.0 (0.6–1.7) 1.0 (0.4–2.1) 2.0 (0.2–20) 1.0 (0.3–3.0)	Adjusted for age, smoking, drinking and level of education

CI, confidence interval

2.5 Meta-analyses

Recent reviews (Purchase & Paddle, 1989; McLaughlin, 1994) and meta-analyses (Blair et al., 1990a; Partanen, 1993) summarize most of the available data. Some differences exist between the analyses of Blair et al. (1990a) and Partanen (1993). Partanen (1993) used lagged and confounder-adjusted inputs, whenever available, and developed summary relative risks using a log-Gaussian, fixed-effects model. Blair et al. (1990a) simply summed observed and expected numbers. Partanen (1993) also included three studies not incorporated by Blair et al. (1990a) and used only the values for men; there were also differences in exposure contrasts. The two meta-analyses were in overall agreement with regard to the risks for lung cancer, nasopharyngeal carcinoma and miscellaneous cancers of the upper respiratory tract but differed with regard to the risk for cancer of the nasal cavities and paranasal sinuses: Blair et al. (1990a) found a relative risk of 1.1 (95% CI, 0.7–1.5) for the more highly exposed category, while Partanen (1993) found a risk of 1.7 (1.0–2.8). For the mixed category of cancers of the oropharynx, lip, tongue, salivary glands and mouth, the aggregated data did not suggest associations with exposure to formaldehyde.

The results of the two meta-analyses are summarized in Table 14.

3. Studies of Cancer in Experimental Animals

3.1 Inhalation

3.1.1 Mouse

Groups of 42–60 C3H mice [sex and age unspecified] were started on a regimen of exposure to formaldehyde (USP grade) vapour at concentrations of 0, 0.05, 0.1 or 0.20 mg/L [0, 50, 100 or 200 mg/m^3] for 1 h per day, three times a week, ostensibly for 35 weeks. Treatment of mice with the highest concentration was discontinued after the eleventh exposure because of severe toxicity, and 36 of the mice exposed to 0.05 mg/L for 35 weeks were subsequently exposed to 0.15 mg/L [150 mg/m^3] for a further 29 weeks. Surviving animals in the initial groups were killed at 35 weeks and those on extended treatment at 68 weeks. The nasal epithelium was not examined, either grossly or microscopically. There was no evidence of induction of pulmonary tumours at any dose. Basal-cell hyperplasia, squamous-cell metaplasia and atypical metaplasia were seen in the trachea and bronchi of most of the exposed mice but not in untreated controls (Horton et al., 1963). [The Working Group noted the high doses used, the short intervals of exposure, the short duration of the experiment and the lack of pathological examination of the nose.]

Groups of 119–120 male and 120–121 female B6C3F1 mice, six weeks of age, were exposed to 0, 2.0, 5.6 or 14.3 ppm [0, 2.5, 6.9, 17.6 mg/m^3] formaldehyde (> 97.5% pure) vapour by whole-body exposure for 6 h per day on five days per week, for up to 24 months, followed by a six-month observation period with no further exposure. Ten males and 10 females

Table 14. Aggregated risk ratios (RR), 95% confidence intervals (95% CI) and observed (O) and expected (E) frequencies of respiratory cancers in the meta-analyses of Blair et al. (1990a) and Partanen (1993)

Site	Level or duration of exposure to formaldehyde											
	Any				Low/medium				Substantial			
	Blair et al.		Partanen		Blair et al.		Partanen		Blair et al.		Partanen	
	O/E	RR (95% CI)	O/E	RR (95% CI)	O/E	RR (95% CI)	O/E	RR (95% CI)	O/E	RR (95% CI)	O/E	RR (95% CI)
Lung												
Medical professions[a]	29/89	0.3 (0.2–0.5)	54/160	0.3 (0.3–0.4)								
Nonmedical professions[b]	490/520	0.9 (0.9–1.0)	474/486	1.0 (0.9–1.1)								
Industrial workers	1181/1097	1.1 (1.1–1.1)	833/752	1.1 (1.0–1.2)	514/422	1.2 [1.1–1.3]	518/425	1.2 (1.1–1.3)	250/240	1.0 [0.9–1.2]	233/216	1.1 (1.0–1.2)
Nose and nasal sinuses	60/56	1.1 (0.8–1.4)	93/78	1.1 (0.8–1.5)	38/46	0.8 (0.6–1.1)	33/30	1.1 (0.7–1.8)	30/28	1.1 (0.7–1.5)	36/21	1.7 (1.0–2.8)
Nasopharynx	31/25	1.2 (0.8–1.7)	36/21	2.0 (1.4–2.9)	30/27	1.1 (0.7–1.6)	23/16	1.6 (1.0–2.7)	13/6	2.1 (1.1–3.5)	11/4	2.7 (1.4–5.6)
Other respiratory			69/57	1.2 (0.9–1.6)			52/48	1.1 (0.7–1.5)			23/20	1.2 (0.6–2.1)

Blair et al. (1990a) included the following studies in their analysis: Harrington & Shannon (1975), Petersen & Milham (1980), Jensen & Andersen (1982), Fayerweather et al. (1983), Friedman & Ury (1983), Marsh (1983), Milham (1983), Walrath & Fraumeni (1983), Wong (1983), Acheson et al. (1984a,c), Coggon et al. (1984), Harrington & Oakes (1984), Levine et al. (1984), Liebling et al. (1984), Malker & Weiner (1984), Olsen et al. (1984), Walrath & Fraumeni (1984), Stayner et al. (1985), Partanen et al. (1985), Walrath et al. (1985), Bertazzi et al. (1986), Blair et al. (1986), Bond et al. (1986), Gallagher et al. (1986), Hayes et al. (1986), Logue et al. (1986), Stroup et al. (1986), Vaughan et al. (1986a,b), Blair et al. (1987), Roush et al. (1987), Stayner et al. (1988), Bertazzi et al. (1989), Gérin et al. (1989), Blair et al. (1990b), Hayes et al. (1990)

Partanen (1993) included in his analysis both the above studies and: Brinton et al. (1984), Partanen et al. (1990), Merletti et al. (1991)

[a] Anatomists, pathologists, forensic medicine specialists
[b] Funeral directors, embalmers, undertakers, medicinal drug users

from each group were killed at 6 and 12 months, 0–20 of each sex at 18 months, 17–41 at 24 months and 0–16 at 27 months. Between 0 and 24 months, 78 male and 30 female controls, 77 and 34 exposed to 2 ppm formaldehyde vapour, 81 and 19 exposed to 5.6 ppm and 82 and 34 exposed to 14.3 ppm died; all animals that died or were killed were examined grossly. Thorough histopathological examinations were performed on control and high-dose mice, on multiple sections of the nasal cavity and on all lesions identified grossly in the other two groups. Squamous-cell carcinomas occurred in the nasal cavities of 2/17 male mice at the high dose killed at 24 months. There were no nasal cavity tumours in males mice treated with the lower doses of formaldehyde, in females at any dose or among 21 male or 31 female control mice killed at 24 months ($p > 0.05$). A variety of non-neoplastic lesions (such as squamous-cell hyperplasia, squamous-cell metaplasia and dysplasia) were commonly found in the nasal cavities of mice exposed to formaldehyde, particularly at 14.3 ppm (Kerns et al., 1983a,b; Gibson, 1984).

3.1.2 Rat

Groups of 119–120 male and 120 female Fischer 344 rats, seven weeks of age, were exposed to 0, 2.0, 5.6 or 14.3 ppm [0, 2.5, 6.9, 17.6 mg/m^3] formaldehyde (> 97.5% pure) vapour by whole-body exposure for 6 h per day on five days per week for up to 24 months and were then observed for six months with no further exposure. Ten males and 10 females from each group were killed at 6 and 12 months, 19–20 of each sex at 18 months, 13–54 at 24 months, 0–10 at 27 months and 0–6 at 30 months. Between 0 and 24 months, 6 males and 13 females in the control group, 10 and 16 exposed to 2 ppm, 19 of each sex exposed to 5.6 ppm and 57 and 67 exposed to 14.3 ppm died; all animals that died or were killed were examined grossly. Histopathological examinations were performed on multiple sections of the nasal cavity, on all lesions identified grossly and on all major tissues of each organ system (approximately 40/animal) from control and high-dose rats. The findings for the nasal cavity are summarized in Table 15. While no nasal cavity malignancies were found in rats exposed to 0 or 2.0 ppm formaldehyde, two squamous-cell carcinomas (one among 119 males and one among 116 females examined) occurred in the group exposed to 5.6 ppm and 107 (51 among 117 males and 52 among 115 females examined) in those exposed to 14.3 ppm ($p < 0.001$). Five additional nasal cavity tumours (classified as carcinoma, undifferentiated carcinoma/sarcoma and carcinosarcoma) were identified in rats exposed to 14.3 ppm; two of these tumours were found in rats that also had squamous-cell carcinomas of the nasal cavity. There was a significant overall increase in the incidence of polypoid adenomas in treated animals (males and females combined) when compared with controls [$p = 0.02$, Fisher's exact test]. The incidences of polypoid adenomas were marginally significantly elevated in females at the low dose and in males at the middle dose (see also Table 15). A variety of non-neoplastic lesions were commonly found in the nasal cavities of rats exposed to formaldehyde, particularly at 14.3 ppm (Kerns et al., 1983a,b; Gibson, 1984). More than half (57%) of the squamous-cell carcinomas in rats exposed to 14.3 ppm formaldehyde were observed on the anterior portion of the lateral side of the nasoturbinate and the adjacent lateral wall, 26% were located on the midventral nasal septum, 10% on the dorsal septum and roof of the dorsal meatus and a small number (3%) on the maxilloturbinate (Morgan et al., 1986a).

Table 15. Neoplastic lesions in the nasal cavities of Fischer 344 rats exposed to formaldehyde vapour

Lesion	Exposure (ppm)							
	0		2.0		5.6		14.3	
	M	F	M	F	M	F	M	F
(No. of nasal cavities examined	118	114	118	118	119	116	117	115)
Squamous-cell carcinoma	0	0	0	0	1	1	51^a	52^a
Nasal carcinoma	0	0	0	0	0	0	1^b	1
Undifferentiated carcinoma or sarcoma	0	0	0	0	0	0	2^b	0
Carcinosarcoma	0	0	0	0	0	0	1	0
Osteochondroma	1	0	0	0	0	0	0	0
Polypoid adenoma^e	1	0	{ 4	4^c	6^d	0	4	1 }^{e}

From Morgan et al. (1986a)
^{a} $p < 0.001$, pair-wise comparisons
^{b} One animal also had a squamous-cell carcinoma.
^{c} [$p = 0.07$, Fisher's exact test in comparison with female controls]
^{d} [$p = 0.06$, Fisher's exact test in comparison with male controls]
^{e} [$p = 0.02$, Fisher's exact test in comparison of all treated rats with controls]

In a study to investigate the carcinogenicity of bis(chloromethyl)ether formed *in situ* in inhalation chambers, by mixing formaldehyde and hydrogen chloride gas at high concentrations before introduction into the chamber in order to maximize formation of bis(chloromethyl)ether, 99 male Sprague–Dawley rats, eight weeks of age, were exposed to a mixture of 14.7 ppm [18.1 mg/m^3] formaldehyde vapour [purity unspecified] and 10.6 ppm [15.8 mg/m^3] hydrogen chloride gas for 6 h per day on five days per week for life. The average level of bis(chloromethyl)ether was 1 ppb [4.7 μg/m^3]. Groups of 50 rats were sham-exposed to air or were untreated. The animals were allowed to die naturally and were then necropsied. Histological sections of nasal cavities, respiratory tract, major organs and gross lesions were prepared. No nasal cancers were found in the controls, but 28 of the treated rats developed tumours of the nasal cavity, 25 of which were squamous-cell carcinomas [$p < 0.001$, Fisher's exact test] and three of which were papillomas. Mortality was greater in the treated group than in controls throughout the experiment; about 50% of the exposed rats were still alive at 223 days, when the first nasal carcinoma was observed. About two-thirds of the exposed rats showed squamous-cell metaplasia of the nasal mucosa; these lesions were not seen in controls (Albert et al., 1982).

In the sames series of experiments, groups of 99–100 male Sprague-Dawley rats, nine weeks of age, were exposed for 6 h per day on five days per week for life to: (1) 14.3 ppm [17.6 mg/m^3] formaldehyde [purity unspecified] and 10 ppm [14.9 mg/m^3] hydrogen chloride gas mixed before dilution in the exposure chamber to maximize formation of bis(chloromethyl)ether; (2) 14.1 ppm [17.3 mg/m^3] formaldehyde and 9.5 ppm [14.2 mg/m^3] hydrogen chloride gas not mixed before introduction into the exposure chamber; (3) 14.2 ppm [17.5 mg/m^3] formaldehyde vapour alone; (4) 10.2 ppm [15.2 mg/m^3] hydrogen chloride gas alone; or (5) air (sham-exposed controls). A control group of 99 rats was also available. The findings in the nasal cavity are summarized in Table 16. At the end of the experiment, 38 squamous-cell carcinomas of the nasal cavities and 10 polyps or papillomas were observed in rats exposed to formaldehyde alone; none were seen in the controls (p = 0.001, Fisher's exact test). No differences were reported between groups in the incidences of tumours outside the nasal cavity (Albert et al., 1982; Sellakumar et al., 1985.)

Table 16. Neoplastic lesions in the nasal cavities of rats exposed to formaldehyde (HCHO) and/or hydrogen chloride (HCl) vapour

Lesion	Group 1: Premixed HCl (10 ppm) and HCHO (14.3 ppm)	Group 2: Non-premixed HCl (9.5 ppm) and HCHO (14.1 ppm)	Group 3: HCHO (14.2 ppm)	Group 4: HCl (10.2 ppm)	Group 5: Air controls	Colony controls
(No. of rats examined	100	100	100	99	99	99)
Squamous-cell carcinoma	45	27	38	0	0	0
Adenocarcinoma	1	2	0	0	0	0
Mixed carcinoma	0	0	1	0	0	0
Fibrosarcoma	1	0	1	0	0	0
Aesthesioneuroepithelioma	1	0	0	0	0	0
Polyps or papillomas	13	11	10	0	0	0
Tumours in organs outside the respiratory tract	22	12	10	19	25	24

From Sellakumar et al. (1985)

Nine groups of 45 male Wistar rats [age unspecified], initially weighing 80 g, were exposed to 0, 10 or 20 ppm [0, 12.3 or 25 mg/m^3] of formaldehyde vapour [purity unspecified] starting one week after acclimatization. Whole-body exposures for 6 h per day on five days per week were continued for four, eight or 13 weeks; thereafter, the rats were observed during recovery periods of 126, 122 or 117 weeks, respectively, when all survivors were killed. All rats were autopsied and examined by gross pathology; histological examination was limited to six cross-sections of the nose of each rat. Hyperplasia and metaplasia of the nasal epithelium were found to persist in rats exposed to formaldehyde. Significant tumour incidences are presented in Table 17. In control rats, the only nasal tumours reported were two squamous-cell carcinomas among 45 rats that were exposed to air for eight weeks: One was a small tumour found at 130 weeks which appeared to involve a nasolachrymal duct; the second was a large squamous-cell carcinoma in a rat killed at week 94, which formed a large mass outside the nasal cavity and was thought to have arisen in a nasolachrymal duct or maxillary sinus. The tumours were considered by the authors not to resemble those observed in the rats exposed to formaldehyde. Rats exposed to 10 ppm formaldehyde also had two squamous-cell carcinomas: One was reported to be a small nasolachrymal-duct tumour in a survivor at 130 weeks, and the second occurred largely outside the nasal cavity in association with an abnormal incisor tooth in a rat killed at week 82. Rats exposed to 20 ppm formaldehyde had 10 tumours: Polypoid adenomas of the nasal cavity were found in one rat exposed for four weeks and killed at 100 weeks and in another rat exposed for eight weeks and killed at 110 weeks; and there were six squamous-cell carcinomas, two of which were thought to originate in the nasolachrymal ducts, one of which appeared to be derived from the palate, and the three others, all in the group exposed for 13 weeks, appeared to arise from the naso- or maxillo-turbinates and formed large tumours that invaded the bone and subcutaneous tissues. The other two neoplasms observed in treated animals were an ameloblastoma found at week 73 and an exophytic tumour of the nasal septum of doubtful malignancy, which was designated a carcinoma *in situ*, in a rat that died at 81 weeks. The authors concluded that the nasal tumours were induced by formaldehyde only at 20 ppm, at an incidence of 4.5% (6 tumors/132 rats) [$p = 0.01$, Fisher's exact test] (Feron *et al.*, 1988). [The Working Group noted that positive findings were made in spite of the short duration of exposure.]

Table 17. Nasal tumours in rats exposed to formaldehyde for various periods followed by observation up to 126 weeks

Exposure time; no. of rats	Tumour	Dose (ppm [mg/m^2])		
		0	10 [12.3]	20 [25]
4 weeks				
No. of rats		44	44	45
	Polypoid adenoma	0	0	1[a]
	Squamous-cell carcinoma	0	0	1
8 weeks				
No. of rats		45	44	43
	Polypoid adenoma	0	0	1[a]
	Squamous-cell carcinoma	2	1	1
13 weeks				
No. of rats		45	44	44
	Squamous-cell carcinoma	0	1	3[a]
	Cystic squamous-cell carcinoma	0	0	1
	Carcinoma *in situ*	0	0	1[a]
	Ameloblastoma	0	0	1

From Feron *et al.* (1988)
[a]Considered by the authors to be causally related to exposure to formaldehyde

A total of 720 male specific pathogen-free Wistar rats initially weighing 30–50 g were acclimatized for one week, and then the nasal mucosa of 480 of the rats was severely injured bilaterally by electrocoagulation. One week later, groups of 180 rats were exposed to 0, 0.1, 1.0 or 10 ppm [0, 0.123, 1.23 or 12.3 mg/m^3] of formaldehyde [purity unspecified] vapour by whole-body exposure for 6 h per day on five days per week. One-half of the animals (30 undamaged, 60 damaged rats) were exposed for 28 months, and the other half (30 undamaged, 60 damaged) were exposed for only three months and then allowed to recover for 25 months with no further treatment. All surviving rats were killed at 29 months, autopsied and examined grossly; histological examination was restricted to six cross-sections of the nose of each rat. The neoplastic lesions found in the nasal cavity are summarized in Table 18. A high incidence of nasal tumours (17/58) was found in rats with damaged noses and exposed to 10 ppm formaldehyde for 28 months; only one was found in 54 controls [$p < 0.001$; Fisher's exact test], and only one of the 26 rats with undamaged noses that were exposed to 10 ppm formaldehyde for 28 months developed a nasal tumour. The tumour incidences in the other groups were low (0–4%). Eight additional squamous-cell carcinomas found in this study that appeared to be derived from the nasolachrymal ducts were excluded from the analysis (Woutersen et al., 1989).

Table 18. Nasal tumours in male Wistar rats with damaged or undamaged noses and exposed to formaldehyde vapour for 28 or three months, followed by a 25-month recovery period

Exposure time; no. of rats	Tumour	Exposure (ppm [mg/m^3])							
		0		0.1 [0.123]		1.0 [1.23]		10.0 [12.3]	
		U	D	U	D	U	D	U	D
28 months Effective number		26	54	26	58	28	56	26	58
	Squamous-cell carcinoma	0	1	1	1	1	0	1	15
	Adenosquamous carcinoma	0	0	0	0	0	0	0	1
	Adenocarcinoma	0	0	0	0	0	0	0	1
3 months Effective number		26	57	30	57	29	53	26	54
	Squamous-cell carcinoma	0	0	0	2	0	2	1	1
	Carcinoma in situ	0	0	0	0	0	0	0	1
	Polypoid adenoma	0	0	0	0	0	0	1	0

From Woutersen et al. (1989)
U, undamaged; D, damaged nose

In a study to explore the interaction between formaldehyde and wood dust (see also p. 165 of the monograph on wood dust), two groups of 16 female Sprague-Dawley rats, 11 weeks of age, were exposed either to air or to formaldehyde [purity unspecified] at an average concentration of 12.4 ppm [15.3 mg/m^3]. Exposures were for 6 h per day for five days a week for a total of 104 weeks. At the end of the experiment, surviving animals were killed, and histological sections were prepared from five cross-sections of the nose of each rat. Pronounced squamous-cell metaplasia or metaplasia with dysplasia was observed in 10/16 rats exposed to formaldehyde and in 0/15 controls. One exposed rat developed a squamous-cell carcinoma [not significant]. Neither the frequency nor the latent periods of induction of tumours outside the nasal cavity differed from those in controls (Holmström et al., 1989a). [The Working Group noted the small numbers of animals used in the study.]

3.1.3 Hamster

A group of 88 male Syrian golden hamsters [age unspecified] were exposed to 10 ppm [12.3 mg/m^3] formaldehyde [purity unspecified] for 5 h a day on five days a week for life; 132 untreated controls were available. At necropsy, all major tissues were preserved, and histological sections were prepared from two transverse sections of the nasal turbinates of each animal, longitudinal sections were taken of the larynx and trachea, and all lung lobes were cut through the major bronchus. No tumours of the nasal cavities or respiratory tract were found in either the controls or the animals exposed to formaldehyde (Dalbey, 1982).

In a second study in the same report, 50 male Syrian golden hamsters [age unspecified] were exposed to 30 ppm [36.9 mg/m^3] formaldehyde [purity unspecified] for 5 h once per week for life. A group of 50 untreated hamster served as controls. When the animals died, their respiratory tract tissues were preserved, stained with Wright's stain, rendered semitransparent and evaluated for 'subgross' evidence of tumours. Areas of dense staining of 1 mm or more were scored as tumours. Multiple transverse sections of the nasal turbinates were evaluated similarly. No nasal tumours were observed in control or treated hamsters (Dalbey, 1982).

3.2 Oral administration

Rat: In a lifetime study, formaldehyde was administered in drinking-water to male and female Sprague-Dawley rats beginning at various ages. Groups of 50 male and 50 female rats received 10, 50, 100, 500, 1000 or 1500 ppm [mg/L] formaldehyde from seven weeks of age for life; two control groups of 50 males and 50 females and 100 males and 100 females received 15 mg/L (ppm) methanol or nothing, respectively, in their drinking water. Two groups of 18–20 male and female breeder rats, 25 weeks old, were given formaldehyde at 0 or 2500 ppm for life. The offspring of these breeders, 36–59 males and 37–49 females, were initially exposed to 0 or 2500 ppm formaldehyde via their mothers starting on day 13 of gestation and then received these levels in the drinking-water for life. The survival rates in the treated groups were similar to those of controls. All animals were necropsied, and extensive histological examinations were performed. The authors reported an increased, dose-related incidence of leukaemias in the treated groups (see Table 19). They also observed a variety of malignant and benign tumours of the stomach and intestines in the treated animals. Although the incidences of intestinal tract tumours were low, there were no comparable tumours in the control groups in this study, and some of these tumours were reported to be uncommon among historical controls (Soffritti et al., 1989).

Table 19. Incidences of leukaemia and gastrointestinal tract tumours after administration of formaldehyde to rats in drinking-water (males and females combined)

Treatment	No. of tumours (benign and malignant)		
	Leukaemia	Gastrointestinal tract	
		Stomach	Intestine
7 weeks old			
(0 ppm + 15 ppm methanol)	8/100	0	0
0 ppm	7/200	0	0
10 ppm	3/100	2/100	1/100
50 ppm	9/100	0	2/100
100 ppm	9/100	0	0
500 ppm	12/100 [a]	0	0 [a]
1000 ppm	13/100	1/100	1/100
1500 ppm	18/100	2/100	6/100
25 weeks old (breeders)			
0 ppm	1/40	0/40	0/40
2500 ppm	4/36	2/36	0/36
Offspring			
0 ppm	6/108 [b]	0/108 [b]	0/108 [b]
2500 ppm	4/73	5/73	8/73

From Soffritti et al. (1989)
[a][Significant linear dose–response relationship when formaldehyde-treated groups are compared with water controls, $p < 0.001$, or water-methanol controls, $p < 0.01$, Cochran–Mantel-Haenszel test]
[b][$p = 0.01$; Fisher's exact test]

Concerns about the results in this study and their interpretation have been published by Feron et al. (1990). They noted that leukaemia incidences in untreated Sprague-Dawley rats vary widely and that incidences similar to those seen in the group receiving the highest dose of formaldehyde have been reported previously among controls in the same laboratory and others. The Working Group, however, noted the absence of gastric or intestinal tumours among the 300 control animals, while, on the basis of the authors' report of historical incidences, three gastric and one intestinal tumour would have been expected. Furthermore, the reporting of the data was limited. The Group subjected the available data from this study to statistical analysis, despite the above reservations. The groups treated at seven weeks were found to differ significantly with regard to both leukaemia and intestinal tumour incidence from the 300 combined controls [$p < 0.05$, Fisher's exact test]. The incidence of leukaemia in the treated groups also differed significantly [$p < 0.001$, Fisher's exact test] from that in the untreated controls; however, the difference in intestinal tumours was only marginally significant

[$p = 0.055$, Fisher's exact test]. When the groups treated at seven weeks were compared with the controls given methanol, the differences were not significant. A significant, linear dose–response relationship was found for the incidences of both leukaemia and intestinal tumours, in comparison with either the untreated [$p < 0.01$] or the methanol controls [$p < 0.01$] [Cochran–Mantel Haenszel test].

Wistar rats, obtained at five weeks of age and acclimatized for nine days, were divided into four groups of 70 males and 70 females and were treated for up to 24 months with drinking-water containing formaldehyde generated from 95% pure paraformaldehyde and 5% water. The mean doses of formaldehyde were 0, 1.2, 15 or 82 mg/kg bw per day for males and 0, 1.8, 21 or 109 mg/kg bw per day for females. Selected animals were killed at 53 and 79 weeks, and all surviving animals were killed at 102 weeks. Thorough necropsies were done on all animals. Extensive histological examinations were made of animals in the control and high-dose groups; somewhat less extensive examinations were made of animals receiving the low and middle doses, but the liver, lung, stomach and nose were examined in each case. Treatment-related hyperplastic lesions, ulceration and atrophy were found in the stomachs, but the incidence of tumours did not vary notably between groups. Two benign gastric papillomas were observed—one in a male at the low dose and the other in a female control. The authors noted that the other tumours observed were common in this strain of rat and that there was no indication of a treatment-related response (Til *et al.*, 1989).

Four groups of 20 male and 20 female Wistar rats, four weeks of age, were given formaldehyde (prepared from 80% pure paraformaldehyde) in their drinking-water at concentrations of 0, 0.02, 0.1 or 0.5% for up to 24 months. Six rats were chosen at random from each group and killed after 12 and 18 months of treatment; surviving animals were killed at 24 months and necropsied, and histological examinations were performed on major organs. [The Working Group noted that gross or microscopic examination of the nasal cavities was not mentioned specifically.] Rats given the high dose had reduced body weight gain and high mortality. Non-neoplastic lesions, such as squamous- and basal-cell hyperplasia, erosion and ulceration, were seen in the stomachs and forestomachs of rats given 0.5% at 12 months. The incidences of tumours in all groups were similar to those occurring spontaneously in this strain of rat. The authors reported that there were no significant differences in the incidences of any tumours from those in the control groups (Tobe *et al.*, 1989). [The Working Group noted the lack of detailed reporting of tumours and the small numbers of animals used.]

In a study to evaluate the effects of formaldehyde on gastric carcinogenesis induced by oral administration of N-methyl-N'-nitro-N-nitrosoguanidine (MNNG) (see below), two groups of 10 male Wistar rats, seven weeks of age, received tap water for the first eight weeks of the study. During weeks 8–40, one group then received pure water and the other group received 0.5% formaldehyde in the drinking-water. Animals still alive at 40 weeks were killed, rats surviving beyond 30 weeks being considered as effective animals for the study. Necropsy was performed on most animals that died and all animals that were killed, and the stomach and other abdominal organs were examined grossly and histologically. Eight of 10 animals that had received formaldehyde in drinking-water and none of the controls developed forestomach papillomas ($p < 0.01$, Fisher's exact test) (Takahashi *et al.*, 1986).

3.3 Skin application

Mouse: Two groups of 16 male and 16 female Oslo hairless mice [age unspecified] received topical applications of 200 µl of 1 or 10% formaldehyde in water on the skin of the back twice a week for 60 weeks. All of the animals treated with 10% formaldehyde were necropsied and the brain, lungs, nasal cavities and all tumours of the skin and other organs were examined histologically. Virtually no changes were found in the mice treated with 1% formaldehyde. The higher dose induced slight epidermal hyperplasia and a few skin ulcers. There were no benign or malignant skin tumours or tumours in other organs in either group (Iversen, 1986). [The Working Group noted the incomplete reporting of the data.]

3.4 Subcutaneous injection

Rat: In a study reported as an abstract, 10 rats [strain, age and sex unspecified] were injected subcutaneously once a week for 15 months with 1 ml of a 0.4% aqueous solution of formaldehyde and then observed for life. Spindle-cell sarcomas were found in three rats: two in the skin at the injection site and one in the peritoneal cavity (Watanabe *et al.*, 1954). [The Working Group noted the lack of controls.]

3.5 Administration with known carcinogens and other modifying factors

3.5.1 Mouse

Two groups of 50 female CBA × C57Bl6 mice, weighing 10–12 g, received drinking-water containing *N*-nitrosodimethylamine (NDMA) at a concentration of 10 mg/L and formaldehyde at a concentration of 0.5 mg/L for 26 or 39 weeks. Other groups of mice were treated either with NDMA alone for 26 and 39 weeks or formaldehyde alone for 39 weeks. Animals were killed after completion of treatment and necropsied, and the liver, kidney, lung, spleen and all gross lesions were examined histologically. No tumours were observed in the group receiving formaldehyde alone for 39 weeks. Combined administration of NDMA and formaldehyde increased the proportions of surviving mice bearing tumours in the liver, kidney and/or lung in the groups treated for 26 weeks and for 39 weeks, as compared with mice treated with NDMA alone (11/15 versus 17/30 and 19/19 versus 20/25) ($p = 0.049$, Fisher's exact test). The effect was not associated with obvious changes in the relative incidence of tumours at any site (Litvinov *et al.*, 1984).

Oslo hairless mice [age unspecified] each received a single application of 51.2 µg 7,12-dimethylbenz[*a*]anthracene (DMBA) in 100 µl of reagent-grade acetone on the skin of the back. Nine days later, the first group of 16 male and 16 females mice received twice weekly applications of 200 µl 10% formaldehyde in water (technical-grade formalin) on the skin of the back. A second group of 16 males and 16 females received 17 nmol 12-*O*-tetradecanoylphorbol-13-acetate (TPA) on the skin of the back twice a week. A third group, of 176 mice [sex unspecified], was given no further treatment. Animals were observed weekly for 60 weeks (groups 1 and 2) or 80 weeks (group 3). All of the animals treated with 10% formaldehyde were necropsied, and the brain, lungs, nasal cavities and all tumours of the skin and other organs were examined histologically. In group 1, 3/32 mice had lung adenomas and 11/32 (34%) had 25 neoplasms of the skin, including three squamous-cell carcinomas and 22 papillomas. In mice receiving DMBA alone, 225 skin tumours (including six squamous-cell carcinomas) occurred in 85/176 (48%) animals. Statistical analysis of the results for these two groups was reported by the authors to show no significant effect of formaldehyde on the skin tumour yield initiated by DMBA ($p > 0.30$, Gail test), but formaldehyde significantly enhanced the rate of skin tumour induction ($p = 0.01$, Peto's test), thus reducing the latent period for the tumours (Iversen, 1986). [The Working Group noted the incomplete reporting of the tumours.]

3.5.2 Rat

Two groups of 30 and 21 male Wistar rats, seven weeks of age, received MNNG in the drinking-water at a concentration of 100 mg/L and a standard diet containing 10% sodium chloride for eight weeks. Thereafter, the rats received the standard diet with 0 or 0.5% formaldehyde in the drinking-water for a further 32 weeks. Animals still alive at 40 weeks were killed, rats surviving 30 weeks or more being considered effective animals for the study. Necropsies were performed on most animals that died and on all animals killed at week 40. Malignant tumours of the stomach and duodenum were found in 5/17 (29%) rats that received both MNNG and formaldehyde and in 4/30 (13%) rats that received MNNG [not significant]. Adenocarcinomas of the glandular stomach were found in 4/17 (23.5%) rats given the combined treatment and in 1/30 rats given MNNG alone ($p < 0.05$, Fisher's exact test). Papillomas of the forestomach were found in 15/17 rats given the combined treatment, in 0/30 given MNNG alone ($p < 0.01$, Fisher's exact test) and in 8/10 given formaldehyde alone (not significant; see section 3.2). The incidence of adenomatous hyperplasia of the fundus of the glandular stomach was significantly greater in the group given the combined treatment (15/17) than in those given MNNG alone (0/3) ($p < 0.01$, Fisher's exact test) (Takahashi *et al.*, 1986).

3.5.3 Hamster

Groups of male Syrian golden hamsters [age unspecified] were treated in various ways: 50 were exposed by inhalation to 30 ppm [36.9 mg/m^3] formaldehyde [purity unspecified] for 5 h per day once a week for life; 100 hamsters were injected subcutaneously with 0.5 mg *N*-nitrosodiethylamine (NDEA) once a week for 10 weeks and then given no further treatment; 50 hamsters were injected with NDEA once a week for 10 weeks, exposed to 30 ppm formaldehyde for 5 h 48 h before each injection of NDEA and then received weekly exposure to 30 ppm formaldehyde for life; and the fifth group of [presumably 50] hamsters was injected with NDEA once a week for 10 weeks and then exposed to 30 ppm formaldehyde for 5 h per day once a week for life, beginning two weeks after the last NDEA injection. A group of 50 animals were untreated. After the animals had died, the respiratory tract tissues were removed, stained with Wright's stain, rendered semitransparent and evaluated for 'subgross' evidence of tumours. Areas of dense staining greater than 1 mm in 2–3-mm transverse-step sections of nasal turbinates were scored as tumours. No tumours were observed in untreated hamsters or those exposed only to formaldehyde, but 77% of hamsters treated with NDEA alone had tumours at one or more sites in the respiratory tract. Ten or more such lesions from each tissue were examined histologically, and all were found to be adenomas. Lifetime exposure of NDEA-treated hamsters to formaldehyde did not increase the number of tumour-bearing animals. The incidences of nasal tumours in NDEA-treated groups were low (0–2%). The only significant increase was in the multiplicity of tracheal tumours in the group receiving formaldehyde concurrently with and subsequent to NDEA injection as compared with that in animals receiving NDEA alone ($p < 0.05$, Kolmogorov–Smirnoff test) (Dalbey, 1982).

4. Other Data Relevant to an Evaluation of Carcinogenicity and its Mechanisms

4.1 Absorption, distribution, metabolism and excretion

4.1.1 Humans

In humans, as in other animals, formaldehyde is an essential metabolic intermediate in all cells. It is produced endogenously from serine, glycine, methionine and choline, and it is generated in the demethylation of N-, O- and S-methyl compounds. It is an essential intermediate in the biosynthesis of purines, thymidine and certain amino acids.

The endogenous concentration of formaldehyde, determined by gas chromatography–mass spectrometry (Heck et al., 1982) in the blood of human subjects not exposed to formaldehyde, was 2.61 ± 0.14 μg/g of blood (mean ± SE; range, 2.05–3.09 μg/g) (Heck et al., 1985), i.e. about 0.1 mmol/L (assuming that 90% of the blood volume is water and the density of human blood is 1.06 g/cm^3 (Smith et al., 1983)). This concentration represents the total concentration of endogenous formaldehyde in the blood, both free and reversibly bound.

The possibility that gaseous formaldehyde may be adsorbed to respirable particles, inhaled and subsequently released into the lung has been examined. Risby et al. (1990) developed and validated a model to describe the adsorption of formaldehyde to and release from respirable carbon black particles. They concluded that of an airborne concentration of 6 ppm [7.4 mg/m^3], only 2 ppb [0.0025 mg/m^3] would be adsorbed to carbon black. Rothenberg et al. (1989) investigated the adsorption of formaldehyde to dust particles in homes and offices and concluded that, even with a concentration of 1 ppm formaldehyde (1.2 mg/m^3), the particle-associated dose to the pulmonary compartment of an adult human would be approximately 0.05 μg/h, whereas the dose of vapour-phase formaldehyde delivered to the upper respiratory tract would be 500 μg/h, i.e. four orders of magnitude larger.

Since formaldehyde can induce allergic contact dermatitis in humans (section 4.2.1), it can be concluded that formaldehyde or its metabolites penetrate human skin (Maibach, 1983). The kinetics of this penetration were determined in vitro using a full-thickness skin sample mounted in a diffusion cell at 30 °C (Lodén, 1986). The rate of 'resorption' of ^{14}C-formaldehyde (defined as the uptake of ^{14}C into phosphate-buffered saline, pH 7.4, flowing unidirectionally beneath the sample) was 16.7 μg/cm^2 per h when a 3.7% solution of formaldehyde was used, and increased to 319 μg/cm² per h when a 37% solution was used. The presence of methanol in both of these solutions (at 3.3–4.9% and 10–15%, respectively) may have affected the uptake rate, and it is unclear whether the resorbed ^{14}C was due only to formaldehyde. Skin retention of formaldehyde represented a significant fraction of the total amount of formaldehyde absorbed.

The concentration of formaldehyde was measured in the blood of six human volunteers immediately after exposure by inhalation to 1.9 ppm [2.3 mg/m³] for 40 min. The measured value was 2.77 ± 0.28 μg/g, which was not different from the pre-exposure concentration due to metabolically formed formaldehyde (see above). The absence of an increase is understandable, since formaldehyde is rapidly metabolized by human erythrocytes (Malorny et al., 1965), which contain formaldehyde dehydrogenase (Uotila & Koivusalo, 1987) and aldehyde dehydrogenase (Inoue et al., 1979).

A gas chromatographic method was used to examine the urinary excretion of formate by veterinary medical students exposed to low concentrations of formaldehyde, in order to determine whether monitoring of formate is a useful biomarker for human exposure to formaldehyde (Gottschling et al., 1984). The average baseline level of formate in the urine of 35 unexposed subjects was 12.5 mg/L, but the level varied considerably both within and among subjects (range, 2.4–28.4 mg/L). No significant changes in concentration were detected over a three-week period of exposure to formaldehyde at a concentration in air of less than 0.4 ppm [0.5 mg/m³]. The authors concluded that biological monitoring of formic acid in the urine to determine exposure to formaldehyde is not a feasible technique at this concentration.

4.1.2 Experimental systems

The steady-state concentrations of endogenous formaldehyde have been determined by gas chromatography–mass spectrometry (Heck et al., 1982) in the blood of Fischer 344 rats (2.24 ± 0.07 μg/g of blood (mean ± SE)) (Heck et al., 1985) and three rhesus monkeys (2.04 ± 0.40 μg/g of blood; range, 1.24–2.45 μg/g) (Casanova et al., 1988). These concentrations are similar to those measured in humans by the same method (see section 4.1.1). The blood concentrations of formaldehyde immediately after exposure of rats once to 14.4 ppm [17.6 mg/m³] (2 h) or exposure of monkeys subacutely to 6 ppm [7.3 mg/m³] (6 h/day, five days/week, four weeks) were indistinguishable from those before exposure.

As reported in an abstract, more than 93% of a dose of inhaled formaldehyde was absorbed readily by the tissues of the respiratory tract (Patterson et al., 1986). In rats, formaldehyde is absorbed almost entirely in the nasal passages (Chang et al., 1983; Heck et al., 1983). In rhesus monkeys, absorption occurs primarily in the nasal passages but also in the trachea and proximal regions of the major bronchi (Monticello et al., 1989; Casanova et al., 1991). The efficiency and sites of formaldehyde uptake are determined by nasal anatomy, which differs greatly among species (Schreider, 1986). The structure of the nose gives rise to complex airflow patterns, which have been correlated with the location of formaldehyde-induced nasal lesions in both rats and monkeys (Morgan et al., 1991).

After exposure by inhalation, absorbed formaldehyde can be oxidized to formate and carbon dioxide or can be incorporated into biological macromolecules via tetrahydrofolate-dependent one-carbon biosynthetic pathways (see Figure 1). The fate of inhaled formaldehyde was studied in Fischer 344 rats exposed to ^{14}C-formaldehyde (at 0.63 or 13.1 ppm [0.8 or 16.0 mg/m^3]) for 6 h. About 40% of the inhaled ^{14}C was eliminated as expired ^{14}C-carbon dioxide over a 70-h period; 17% was excreted in the urine, 5% was eliminated in the faeces and 35–39% remained in the tissues and carcass. Elimination of radioactivity from the blood of rats after exposure by inhalation to 0.63 ppm or 13.1 ppm ^{14}C-formaldehyde is multiphasic. The terminal half-time of the radioactivity was approximately 55 h (Heck et al., 1983), but the half-time of formaldehyde in rat plasma after intraperitoneal administration is reported to be approximately 1 min (Rietbrock, 1965). Analysis of the time course of residual radioactivity in plasma and erythrocytes after inhalation or intravenous injection of ^{14}C-formaldehyde or intravenous injection of ^{14}C-formate showed that the radioactivity is due to incorporation of ^{14}C (as ^{14}C-formate) into serum proteins and erythrocytes and subsequent release of labelled proteins and cells into the circulation (Heck et al., 1983).

The fate of ^{14}C-formaldehyde after topical application to Fischer 344 rats, Dunkin–Hartley guinea-pigs and cynomolgus monkeys was described by Jeffcoat et al. (1983). Aqueous formaldehyde was applied to a shaven area of the lower back, and the rodents were placed in metabolism cages for collection of urine, faeces, expired air and ^{14}C-formaldehyde evaporated from the skin. Monkeys were seated in a restraining chair and were fitted with a plexiglass helmet for collection of exhaled ^{14}C-carbon dioxide. The concentrations of ^{14}C in tissues, blood and carcass of rodents were determined at the end of the experiment. Rodents excreted about 6.6% of the dermally applied dose in the urine over 72 h, while 21–28% was collected in the air traps. It was deduced that almost all of the air-trapped radioactivity was due to evaporation of formaldehyde from the skin, since less than 3% of the radioactivity (i.e. 0.6–0.8% of the applied ^{14}C) was due to ^{14}C-carbon dioxide. Rodent carcass contained 22–28% of the ^{14}C and total blood about 0.1%; a substantial fraction of ^{14}C (3.6–16%) remained in the skin at the site of application. In monkeys, only 0.24% of the dermally applied ^{14}C-formaldehyde was excreted in the urine, and 0.37% was accounted for as ^{14}C-carbon dioxide in the air traps; about 0.015% of the radioactivity was found in total blood and 9.5% in the skin at the site of application. Less than 1% of the applied dose was excreted or exhaled, in contrast to rodents in which nearly 10% was eliminated by these routes. Coupled with the observation of lower blood levels of ^{14}C in monkeys than in rodents, the results suggest that the skin of monkeys may be less permeable to aqueous formaldehyde than that of rodents.

Formaldehyde is absorbed rapidly and almost completely from the rodent intestinal tract (Buss et al., 1964). In rats, about 40% of an oral dose of ^{14}C-formaldehyde (7 mg/kg) was eliminated as ^{14}C-carbon dioxide within 12 h, while 10% was excreted in the urine and 1% in the faeces. A substantial portion of the radioactivity remained in the carcass as products of metabolic incorporation.

Formaldehyde reacts rapidly with glutathione, forming a hemithioacetal, S-hydroxymethylglutathione, which is a substrate for the cytosolic enzyme, formaldehyde dehydrogenase [formaldehyde:NAD$^+$ oxidoreductase (glutathione-formylating), EC 1.2.1.1] (Uotila & Koivusalo, 1974a). With NAD$^+$ as a cofactor, this enzyme catalyses the oxidation of S-hydroxymethylglutathione to S-formylglutathione. The latter compound is hydrolysed to formate by

S-formylglutathione hydrolase [EC 3.1.2.12], regenerating free glutathione (Uotila & Koivusalo, 1974b).

Figure 1. Metabolism and fate of formaldehyde

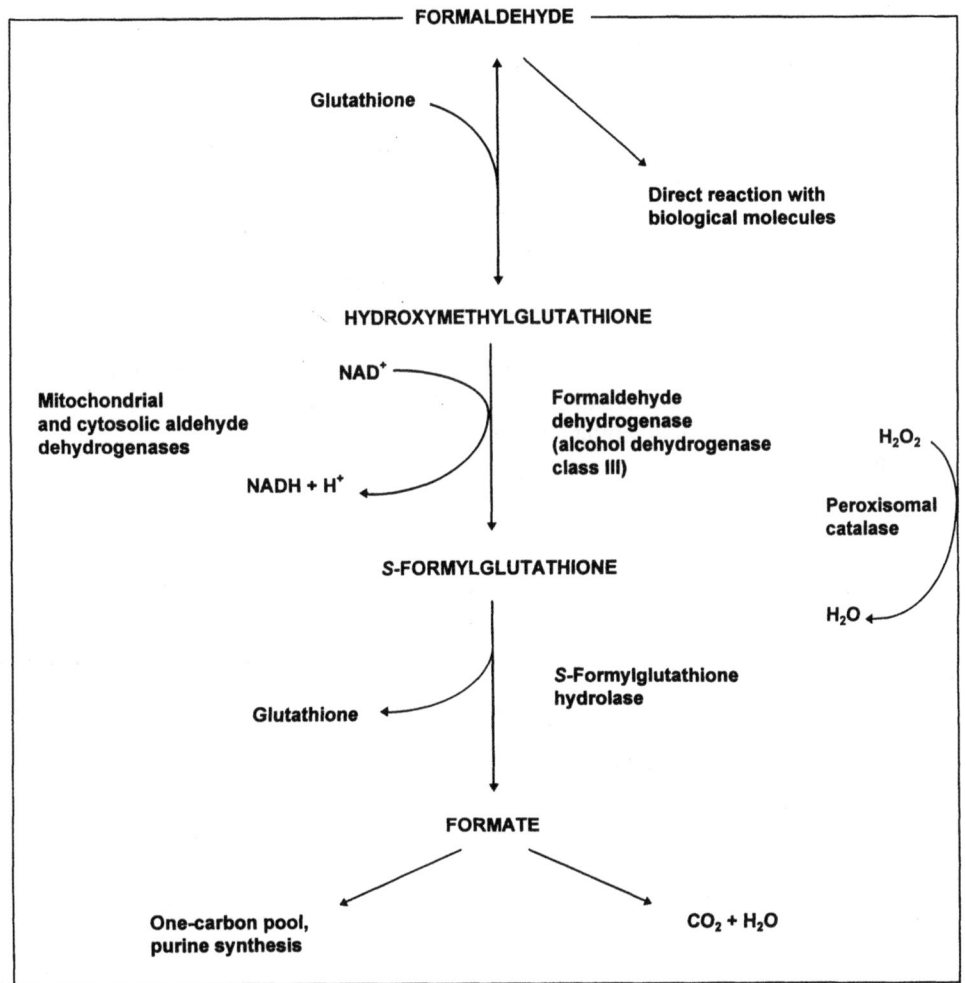

Formaldehyde dehydrogenase has been identified in a number of tissues in several species (Koivusalo et al., 1982). The activity of formaldehyde dehydrogenase is similar in the respiratory and olfactory mucosa of rats (Casanova-Schmitz et al., 1984a; Bogdanffy et al., 1986; Keller et al., 1990). This enzyme is structurally identical to another well-characterized enzyme, class III alcohol dehydrogenase [alcohol:NAD^+ oxidoreductase, EC 1.1.1.1] (Kaiser et al., 1991; Danielsson & Jörnvall, 1992), which catalyses the oxidation of long-chain primary alcohols to aldehydes; in contrast to the well-characterized class I alcohol dehydrogenase, however, it has low affinity for ethanol and is not inhibited by 4-methylpyrazole. Class III alcohol dehydrogenase does not require glutathione when catalysing the oxidation of primary alcohols, but a thiol group is essential for the oxidation of formaldehyde, presumably because a hemithioacetal is formed which is structurally similar to a primary alcohol (Holmquist & Vallee, 1991). Numerous other thiols perform this function at nearly the same rate as glutathione (Holmquist & Vallee, 1991). Aldehydes other than formaldehyde are not oxidized by the enzyme.

Because formaldehyde dehydrogenase and class III alcohol dehydrogenase are identical, it cannot be concluded that the normal function of 'formaldehyde dehydrogenase' *in vivo* is solely to catalyse the oxidation of formaldehyde. Lam *et al.* (1985) and Casanova and Heck (1987) found that depletion of glutathione, either by inhalation of acrolein or by intraperitoneal injection of phorone, increased the amount of DNA–protein cross-links in the nasal mucosa of rats exposed to formaldehyde, implying that formaldehyde oxidation (detoxification) was partially inhibited. The authors postulated that depletion of glutathione had decreased the concentration of *S*-hydroxymethylglutathione, resulting in an increase in the tissue concentration of formaldehyde. Dicker and Cederbaum (1985, 1986) showed, however, that phorone not only depletes glutathione but can also inhibit a mitochondrial low-K_m aldehyde dehydrogenase, which may also be important for the oxidation of formaldehyde. The low-K_m mitochondrial aldehyde dehydrogenase [aldehyde:NAD^+ oxidoreductase, EC 1.2.1.3] catalyses the oxidation of both formaldehyde and acetaldehyde, although acetaldehyde is the preferred substrate of both. This enzyme is strongly inhibited by cyanamide, which acts by inhibiting the uptake and oxidation of formaldehyde by mitochondria and isolated rat hepatocytes (Dicker & Cederbaum, 1984). Inhibition of formaldehyde oxidation in hepatocytes was incomplete, however, presumably because formaldehyde was also being oxidized by the cytosolic formaldehyde dehydrogenase. The authors concluded that both formaldehyde dehydrogenase and the low-K_m mitochondrial aldehyde dehydrogenase contribute to the overall metabolism of formaldehyde in isolated rat hepatocytes, but, as the two enzymes have different K_m values, the importance of each is dependent on the formaldehyde concentration (Dicker & Cederbaum, 1986).

The experiments of Dicker and Cederbaum (1984, 1985, 1986) are useful for understanding the metabolism of formaldehyde in general and in hepatocytes in particular, but their relevance to the toxicology of inhaled formaldehyde is uncertain. Although aldehyde dehydrogenase activity was identified in rat nasal mucosa (Casanova-Schmitz *et al.*, 1984a; Bogdanffy *et al.*, 1986), it is not known whether this activity is due to the low-K_m mitochondrial aldehyde dehydrogenase. Moreover, the subcellular location of the low-K_m enzyme within the mitochondria might restrict its accessibility to exogenous formaldehyde and, therefore, impair its ability to metabolize the compound. Thus, the role of this dehydrogenase in the detoxification of inhaled formaldehyde is presently unknown.

Oxidation of formaldehyde to formate may also be mediated by catalase, which is located in peroxisomes. In this reaction, formaldehyde acts as a hydrogen donor for the peroxidative decomposition of the catalase–hydrogen peroxide complex. This reaction contributes less to the overall metabolism of formaldehyde in isolated, perfused rat liver than other pathways, owing to the rate-limiting generation of hydrogen peroxide (Waydhas *et al.*, 1978). The latter compound is also decomposed by the glutathione peroxidase system, resulting in depletion of glutathione and the production of oxidized glutathione. In hepatocytes in which glutathione has been depleted, hydrogen peroxide production is increased, which may result in increased metabolism of formaldehyde via catalase (Jones *et al.*, 1978).

Incubation of formaldehyde with human nasal mucus *in vitro* resulted in the reversible formation of protein adducts, primarily with albumin, suggesting that a portion of the inhaled formaldehyde is retained in the mucous blanket (Bogdanffy *et al.*, 1987). No adducts were found in high relative-molecular-mass glycoproteins. Absorbed formaldehyde may react with nucleophiles (e.g. amino and sulfhydryl groups) at or near the absorption site, or it can be oxidized to formate and exhaled as carbon dioxide or incorporated into biological macromolecules via tetrahydrofolate-dependent one-carbon biosynthetic pathways.

Several of the urinary excretion products of formaldehyde in rats have been identified after intraperitoneal administration of ^{14}C-formaldehyde. After injecting Wistar rats with 0.26 mg/kg bw, Hemminki (1984) detected formate and a sulfur-containing metabolite (thought to be a derivative of thiazolidine-4-carboxylic acid) and products presumed to result from one-carbon metabolism. Thiazolidine-4-carboxylate, which is formed via the nonenzymatic condensation of formaldehyde with cysteine, was not detected in urine.

After Sprague–Dawley rats were injected intraperitoneally with 4 or 40 mg/kg bw of ^{14}C-formaldehyde, formate (80% of the total radioactivity in urine), N-(hydroxymethyl)- and N,N'-bis(hydroxymethyl)urea (15% of urinary radioactivity) (which appeared to have resulted from the condensation of formaldehyde with urea) and an unidentified product (5% of the total) were identified (Mashford & Jones, 1982). As the urine of the Sprague–Dawley rat contains little, if any, cysteine, formation of thiazolidine-4-carboxylate is precluded and urea-containing adducts can be formed. The existence of these adducts suggests that, at least in Sprague–Dawley rats administered large doses of formaldehyde, a portion of the injected material (about 3–5% at a dose of 40 mg/kg bw) is excreted unchanged in the urine. After exposure by inhalation, however, it is questionable whether a significant amount of formaldehyde is excreted unchanged in the urine, since such high dose levels are not attainable by this route.

The formation of DNA–protein cross-links by formaldehyde in the nasal respiratory mucosa of rats after exposure to 6 ppm [7.3 mg/m^3] and more has been demonstrated by a variety of techniques, including decreased extractability of DNA from proteins (Casanova-Schmitz & Heck, 1983), double-labelling studies with ^3H- and ^{14}C-formaldehyde (Casanova-Schmitz *et al.*, 1984b; Casanova & Heck, 1987; Heck & Casanova, 1987) and isolation of DNA from respiratory mucosal tissue and quantification of cross-links by high-performance liquid chromatography after exposure to ^{14}C-formaldehyde (Casanova *et al.*, 1989, 1995). The formation of DNA–protein cross-links is a nonlinear function of concentration (Casanova & Heck, 1987; Casanova *et al.*, 1989, 1995; Heck & Casanova, 1995; see Figure 2). Cross-links were not detected in the olfactory mucosa or in the bone marrow of rats (Casanova-Schmitz *et al.*, 1984b; Casanova & Heck, 1987).

Figure 2. Concentration of DNA–protein cross-links formed per unit time in the turbinates and lateral wall/septum of Fischer 344 rats and rhesus monkeys in relation to airborne formaldehyde concentration

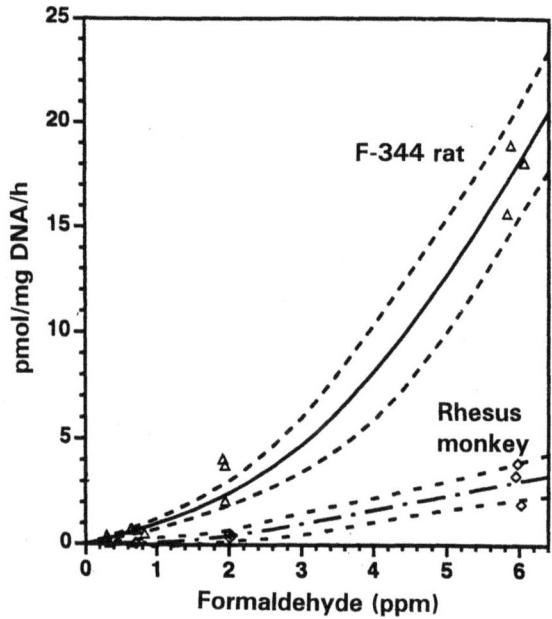

Reproduced, with permission, from Casanova et al. (1991)
All animals were exposed for 6 h. Dashed lines are the 95% confidence limits around the mean for each species.

DNA–protein cross-links were also measured in the respiratory tracts of groups of three rhesus monkeys immediately after single, 6-h exposures to airborne ^{14}C-formaldehyde (0.7, 2 or 6 ppm [0.9, 2.4 or 7.3 mg/m^3]) (Casanova et al., 1991). The concentrations of cross-links in the nose of monkeys decreased in the order: middle turbinates > lateral wall–septum > nasopharynx, and this order is consistent with the location and severity of lesions in monkeys exposed to 6 ppm (Monticello et al., 1989). Very low levels of cross-links were also found in the trachea and carina of some monkeys, but none were detected in the maxillary sinus. The yield of cross-links in the nose of monkeys was approximately an order of magnitude lower than that in the nose of rats, due largely to species differences in minute volume and quantity of exposed tissue (Casanova et al., 1991; Figure 2). A pharmacokinetic model based on these results indicated that the concentrations of DNA–protein cross-links in the human nose would be lower than those in the noses of monkeys and rats (Casanova et al., 1991).

The yields of DNA–protein cross-links produced in rats exposed to formaldehyde (at 0.7, 2, 6 or 15 ppm [0.9, 2.4, 7.3 or 18.3 mg/m^3] for 6 h/day, five days/week for 11 weeks and four days) were compared with those produced in naive (previously unexposed) rats (Casanova et al.,

1995). The acute yields of cross-links (pmol/mg DNA) were determined in the lateral meatus (susceptible tumour site; see section 3.1 (Morgan et al., 1986a)) and in the medial and posterior meatuses (low susceptibility site (Morgan et al., 1986a)) after a single 3-h exposure of pre-exposed and naive rats to the same concentration of ^{14}C-formaldehyde. At 0.7 and 2 ppm, the acute yields of cross-links in the lateral meatus of pre-exposed rats were indistinguishable from those of naive rats; at 6 and 15 ppm, the acute yields in pre-exposed rats were approximately half those of naive rats, and the difference was significant (Figure 3). Pre-exposed animals had lower concentrations of cross-links than naive rats at 6 and 15 ppm partly because of an increase in total DNA in the target tissue caused by cell proliferation (Heck & Casanova, 1995; see section 4.2.2). The acute yields of DNA–protein cross-links in the medial and posterior meatuses were similar in pre-exposed and naive rats at all concentrations and were lower than the acute yields in the lateral meatus. This result is consistent with the location and severity of lesions in the rat nose (Morgan et al., 1986a).

In order to determine whether DNA–protein cross-links accumulate with repeated exposure, the cumulative yield was investigated using reduced DNA extractability as a measure of cross-linking. Rats were exposed subchronically to unlabelled formaldehyde (6 or 10 ppm [7.3 or 12.2 mg/m^3]; 6 h/day, five days/week, 11 weeks and four days) (Casanova et al., 1995), and the cumulative yields of DNA–protein cross-links in the nasal mucosa of pre-exposed rats were compared with those in naive rats after a single 3-h exposure to the same concentration of unlabelled formaldehyde. A concentration-dependent increase in the yield of DNA–protein cross-links over that in unexposed controls was seen in both pre-exposed and naive rats. The yield was not higher in pre-exposed than in naive rats, suggesting that no accumulation had occurred in pre-exposed rats. The results suggest that DNA–protein cross-links in the rat nasal mucosa are rapidly repaired.

An anatomically based pharmacokinetic model was developed for determining the site-specificity of cross-link formation in the nasal mucosa of Fischer 344 rats (Heck & Casanova, 1995) and rhesus monkeys (Casanova et al., 1991). The model is based on the assumption that the site-specificity of cross-links is due to nasal airflow and absorption patterns, rather than to site-specific differences in metabolism (Casanova et al., 1991; Heck & Casanova, 1995). Parameter estimation indicates that at concentrations of less than about 3 ppm [3.7 mg/m^3], about 90% of a dose of inhaled formaldehyde is eliminated by saturable metabolism, 10% is eliminated by nonsaturable pathways and only $1/10^6$ (i.e. 10^{-4} %) exists as DNA–protein cross-links immediately after exposure. The amount bound to DNA increases sublinearly with respect to concentration but linearly with respect to time during exposure (Heck & Casanova, 1995). Computer simulations of nasal airflow and formaldehyde absorption patterns at specific sites in the nose of rats are generally consistent with the experimental results on the site-specificity of DNA–protein cross-links (Kimbell et al., 1993).

4.2 Toxic effects

The toxicity of formaldehyde in humans and experimental systems has been reviewed (IARC, 1982; Heck & Casanova-Schmitz, 1984; Feinman, 1988; WHO, 1989; Heck et al., 1990;

American Conference of Governmental Industrial Hygienists, 1991; Bardana & Montanaro, 1991; Restani & Galli, 1991; Vaught, 1991; Leikauf, 1992).

Figure 3. Acute yields of DNA–protein cross-links (mean ± SE) in the lateral meatus (LM) and medial and posterior meatuses (M:PM) of pre-exposed and naive (previously unexposed) Fischer 344 rats immediately after a single 3-h exposure to ^{14}C-formaldehyde

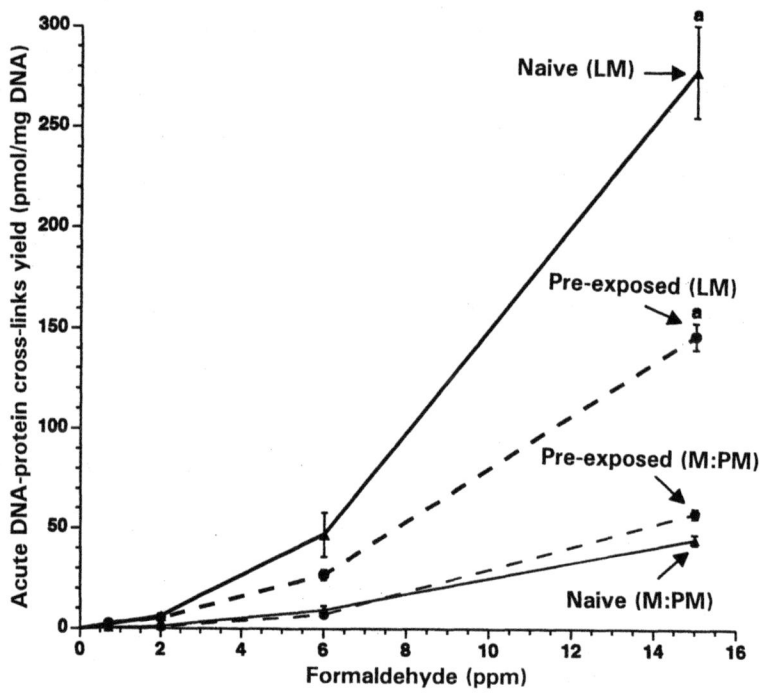

Adapted, with permission, from Casanova *et al.* (1995)
Pre-exposed rats were exposed subchronically to the same concentrations of unlabelled formaldehyde (6 h/day, five days/week, for 11 weeks and four days), while naive rats were exposed to room air. Exposure to ^{14}C-formaldehyde occurred on the fifth day of the twelfth week, and the acute yields pertain to the DNA–protein cross-links produced at that time.

4.2.1 Humans

(a) Acute effects

(i) Odour detection

The threshold for detection of formaldehyde odour was determined among 22 nonsmokers and 22 aged-matched, heavy smokers (all female) (Berglund & Nordin, 1992). Odour was detected at 25–144 ppb (31–177 μg/m^3) by nonsmokers and at 20–472 ppb (25–581 μg/m^3) by smokers ($p < 0.01$).

(ii) Irritation

The following studies of healthy humans given short-term exposures to formaldehyde under controlled conditions indicate that the irritation threshold for eyes, nose and throat is 0.5–1 ppm (0.6–1.2 mg/m^3).

Irritation thresholds were determined in subjects exposed to steadily increasing (0–3.2 ppm [0–3.9 mg/m^3] over 37 min) or to constant formaldehyde concentrations (0, 1, 2, 3 or 4 ppm [0, 1.2, 2.4, 3.7 or 4.9 mg/m^3], 1.5 min per exposure). The thresholds for eye and nose irritation were between 1 and 2 ppm (1.2–2.5 mg/m^3); the threshold for throat irritation was > 2 ppm (Weber-Tschopp *et al.*, 1977)..

Workers exposed to 0.35–1.0 ppm [0.43–1.2 mg/m^3] for 6 min had a significant irritation response at 1.0 ppm; nonsignificant responses were reported at 0.7 and 0.9 ppm [0.9 and 1.1 mg/m^3] (Bender *et al.*, 1983).

Among nonsmokers exposed to 0.5–3.0 ppm [0.6–3.7 mg/m^3], some subjects reported eye irritation at 1.0 ppm, and one reported nose and throat irritation at 0.5 ppm (Kulle *et al.*, 1987). Tolerance to the irritating effects of formaldehyde developed during prolonged exposure to concentrations above 1 ppm (Andersen & Mølhave, 1983).

Respiratory and ocular irritation has been reported by occupants of mobile homes (see section 1) and offices where there are low levels of formaldehyde (Hanrahan *et al.*, 1984; Bracken *et al.*, 1985; Ritchie & Lehnen, 1987; Broder *et al.*, 1988a,b,c; Liu *et al.*, 1991) and by medical students, histology technicians and embalmers, who may be exposed briefly to higher concentrations (Kilburn *et al.*, 1985; Holness & Nethercott, 1989; Uba *et al.*, 1989). In general, the reported thresholds for irritation in uncontrolled environments are lower than those in controlled exposures. The answers to a questionnaire indicated that a few individuals experienced sensory irritation at concentrations as low as 0.1 ppm [0.12 mg/m^3]; however, the contribution of other substances is unknown.

(iii) Pulmonary function

Fifteen healthy nonsmokers and 15 asthmatic subjects were exposed to 2 ppm [2.4 mg/m^3] formaldehyde for 40 min to determine whether acute exposures could induce asthmatic symptoms (Schachter *et al.*, 1986; Witek *et al.*, 1987). On separate days, the subjects either remained at rest or engaged in moderate exercise, and pulmonary function was measured before, during, immediately after or 24 h after exposure. No significant airway obstruction or changes in pulmonary function were noted. Neither healthy nor asthmatic subjects had bronchial hyperreactivity, as shown by responsiveness to methacholine.

Similar observations were made on a group of 15 hospital laboratory workers who had been exposed to formaldehyde (Schachter *et al.*, 1987). The subjects were exposed in an environmental chamber to 2.0 ppm [2.4 mg/m^3] for 40 min on four occasions, during two of which the subjects were at rest and during two of which they performed moderate exercise. Lung function was unaltered on all four days, and there were no delayed obstructive changes or increased reactivity to methacholine.

Healthy nonsmokers (nine subjects for 3 h, 22 for 1 h) and asthmatic subjects (nine subjects for 3 h, 16 for 1 h) were exposed to 3.0 ppm [3.7 mg/m^3] formaldehyde, either at rest or when engaged in intermittent heavy exercise. Pulmonary function and nonspecific airway reactivity were assessed before, during and up to 24 h after exposure. No significant changes were observed among asthmatic subjects. Small decreases (< 5%) in pulmonary function (forced expiratory volume at one second, forced vital capacity) were observed in healthy nonsmokers exposed to formaldehyde while engaging in heavy exercise. Two normal and two asthmatic subjects had decrements greater than 10% at two times. There were no changes in nonspecific airway reactivity (as judged by the methacholine challenge test) (Sauder et al., 1986; Green et al., 1987; Sauder et al., 1987).

Healthy nonsmokers were exposed for 3 h at rest to 0, 0.5, 1.0, 2.0 or 3.0 ppm [0, 0.6, 1.2, 2.4 or 3.7 mg/m^3] formaldehyde; they were also exposed to 2.0 ppm while exercising. Nasal flow resistance was increased at 3.0 ppm but not at 2.0 ppm. There was no significant decrement in pulmonary function or increase in bronchial reactivity to methacholine with exposure to 3.0 ppm at rest or to 2.0 ppm with exercise (Kulle et al., 1987).

A group of 24 healthy nonsmokers were exposed while engaged in intermittent heavy exercise for 2 h to formaldehyde at 3 ppm [3.7 mg/m^3] or to a mixture of formaldehyde and 0.5 mg/m^3 of respirable carbon aerosol, in order to determine whether adsorption of formaldehyde on respirable particles elicits a pulmonary response. Small (< 5%) decreases were seen in forced vital capacity and forced expiratory volume, but these effects were not considered to be clinically significant (Green et al., 1989). As noted previously, Risby et al. (1990) and Rothenberg et al. (1989) estimated that the amount of formaldehyde adsorbed onto carbon black or dust particles and delivered to the deep lung by particle inhalation is minuscule in relation to the amount that remains in the vapour phase and is adsorbed in the upper respiratory tract.

In a study of controlled exposure to formaldehyde, 18 subjects, nine of whom had complained of adverse effects from urea–formaldehyde foam insulation installed in their homes, were exposed to 1 ppm [1.2 mg/m^3] formaldehyde or to off-gas products of urea–formaldehyde foam insulation containing 1.2 ppm [1.5 mg/m^3] formaldehyde, for 90 min (Day et al., 1984). No statistically or clinically significant change in pulmonary function was seen either during or 8 h after exposure, and no evidence was obtained that urea–formaldehyde foam insulation off-gas acts as a lower airway allergen. When 15 asthmatic subjects were exposed for 90 min to concentrations of 0.008–0.85 mg/m^3 formaldehyde, no change in pulmonary function was seen, and there was no evidence of an increase in bronchial reactivity (Harving et al., 1990).

(b) Chronic effects

(i) Effects on the nasal mucosa

The possibility that formaldehyde may induce pathological or cytogenetic changes in the nasal mucosa has been examined in subjects exposed either in residential environments or in occupational settings. Samples of cells were collected with a swab inserted 2–3 cm into the nostrils of subjects living in urea–formaldehyde foam-insulated homes and of subjects living in homes without this type of insulation and were examined cytologically. Small but significant increases were observed in the prevalence of squamous metaplastic cells in the samples from the occupants of urea–formaldehyde foam-insulated homes (Broder et al., 1988a,b,c). A follow-up study one year later (Broder et al., 1991) showed a decrease in nasal signs that was unrelated to any decrease in formaldehyde levels.

Cell smears were collected with a swab inserted 6–8 cm into the nose from 42 workers employed in two phenol–formaldehyde plants and 38 controls with no known exposure to formaldehyde. The formaldehyde concentrations in the plants were 0.02–2.0 ppm [0.02–2.4 mg/m^3], with occasional peaks as high as 9 ppm [11.0 mg/m^3], and the average length of employment in the plants was about 17 years. Atypical squamous metaplasia was detected as a function of age > 50, but there was no association with exposure to formaldehyde (Berke, 1987).

Biopsy samples were taken from the anterior edge of the inferior turbinate of the nose of 37 workers in two particle-board plants, 38 workers in a laminate plant and 25 controls of similar ages. The formaldehyde concentrations in the three plants were 0.1–1.1 mg/m^3, with peak concentrations up to 5 mg/m^3. Simultaneous exposure to wood dust occurred in the particle-board plants but not in the laminate plant. The average length of employment was 10.5 years. Exposure to formaldehyde appeared to be associated with squamous metaplasia and mild dysplasia, but no concentration–response relationship was observed, and the histological score was not related to years of employment. There was no detectable difference in the nasal histology of workers exposed to formaldehyde alone and to formaldehyde and wood dust (Edling et al., 1987b, 1988).

Biopsy samples were collected from the medial or inferior aspect of the middle turbinate, 1 cm behind the anterior border, from 62 workers engaged in the manufacture of resins for laminate production, 89 workers employed in furniture factories who were exposed to particle-board and glue, and 32 controls, who were mainly clerks in a local government office. The formaldehyde concentrations in the resin manufacturing plant were 0.05–0.5 mg/m^3, with frequent peaks over 1 mg/m^3. The concentrations in the furniture factories were 0.2–0.3 mg/m^3, with rare peaks to 0.5 mg/m^3; these workers were also exposed to wood dust (1–2 mg/m^3). The control group was exposed to concentrations of formaldehyde of 0.09–0.17 mg/m^3. The average length of employment was about 10 years. The histological scores of workers exposed to formaldehyde alone were slightly but significantly higher than those of controls, but the histological scores of workers exposed to formaldehyde and wood dust together did not differ from those of controls. No correlation was found between histological score and either duration or concentration of exposure (Holmström et al. (1989b). [The possible effect of age on nasal cytology, as noted by Berke (1987), was not determined.]

A nasal biopsy sample was taken from the anterior curvature of the middle turbinate from 37 workers exposed at a chemical company where formaldehyde resins were produced and from 37 age-matched controls. The formaldehyde concentrations in the company ranged from 0.5 to > 2 ppm [0.6–> 2.4 mg/m^3], and the average length of employment was 20 years. Hyperplasia and squamous metaplasia were commoner among the exposed workers than the controls, but the difference was not significant. The histological scores increased with age and with exposure concentration and duration, but the changes were not significant (Boysen et al., 1990).

Histopathological abnormalities of respiratory nasal mucosa cells were determined in 15 nonsmokers (seven women, eight men) who were exposed to formaldehyde released from a urea–formaldehyde glue in a plywood factory. Each subject was paired with a control matched for age and sex. The mean age of the controls was 30.6 ± 8.7 years and that of exposed workers was 31.0 ± 8.0 years. The mean levels of exposure to formaldehyde (8-h time-weighted) were about 0.1 mg/m³ in the sawmill and shearing-press department and 0.39 mg/m³ in the warehouse area. Peak exposure levels were not given. There was concurrent exposure to low levels of wood dust (respirable mass, 0.23 mg/m³ in the warehouse, 0.73 mg/m³ during sawing). Nasal respiratory cell samples were collected from near the inner turbinate with an endocervical cytology brush. The exposed group had chronic inflammation of the nasal respiratory mucosa and a higher frequency of squamous metaplasia than the controls (mean scores, 2.3 ± 0.5 in the exposed group, 1.6 ± 0.5 in the control group; $p < 0.01$, Mann–Whitney U test) (Ballarin et al., 1992).

The effects of formaldehyde, other than cancer, on the nasal mucosa are summarized in Table 20.

(ii) Pulmonary function

Pulmonary function has been assessed in residents of mobile and conventional homes (Broder et al., 1988a,b,c) and mobile offices (Main & Hogan, 1983) exposed to concentrations of 0.006–1.6 ppm [0.007–2.0 mg/m³]. No changes were seen in pulmonary function or airway resistance.

Lung function tests were performed on particle-board and plywood workers (Holmström & Wilhelmsson, 1988; Horvath et al., 1988; Imbus & Tochilin, 1988; Malaka & Kodama, 1990), workers using acid-hardening paints (Alexandersson & Hedenstierna, 1988, 1989), embalmers (Levine et al., 1984b; Holness & Nethercott, 1989), urea–formaldehyde resin producers (Holmström & Wilhelmsson, 1988; Nunn et al., 1990), medical students (Uba et al., 1989) and anatomy and histology workers (Khamgaonkar & Fulare, 1991). These groups were often exposed to formaldehyde in combination with other substances. The formaldehyde concentrations were < 0.02–> 5 ppm [< 0.02–> 6.0 mg/m³]. In most of the studies, formaldehyde alone or in combination with other agents caused transient, reversible declines in lung function, but there was no evidence that formaldehyde induces a chronic decrement in lung function.

(iii) Effects on the skin

Formaldehyde is a skin irritant and can cause allergic contact dermatitis. It is difficult to distinguish between these two effects (Maibach, 1983). The estimated percentages of people with positive reactions in patch tests were 8.4% in the United States, 7.4% in Saskatoon, Canada, 9.2% in Cologne, Germany, and 5.5% of men and 12.4% of women in Hamburg, Germany (Cronin, 1991). Maibach (1983), however, indicated that these estimates may be considerably inflated, as they are usually uncorrected for the 'excited skin state' and are often unconfirmed. He estimated that the results of more than 40% of patch tests are unreproducible, especially for substances such as formaldehyde, as the concentrations that evoke an allergic response and an irritant response are similar.

In order to determine whether specific immunoglobulin (Ig) E antibodies are involved in contact dermatitis after exposure to formaldehyde, 23 patients with a history of a positive epicutaneous test to formaldehyde were studied. Fifteen (65%) showed a positive reaction on retesting. The findings do not support the hypothesis that specific IgE antibodies are active in the

Table 20. Findings in nasal mucosa of people with occupational exposure to formaldehyde

Reference	Industry	Concentration of formaldehyde (mg/m^3)	No. of exposed	No. of controls	Method	Findings
Edling et al. (1987b)	Formaldehyde (laminate plant)	0.5-1.1	38	25	Nasal biopsy	Histological score: exposed 2.8, controls 1.8 ($p < 0.05$) Four exposed men had mild dysplasia
Edling et al. (1988)	Formaldehyde Wood dust (laminated particle-board)	0.1–1.1 (peaks to 5) 0.6–1.1	75	25	Nasal biopsy	Histological score: exposed 2.9, controls 1.8 ($p < 0.05$) Six men had mild dysplasia
Berke (1987)	Formaldehyde (phenol?) (laminate)	0.02–2.4 (peaks to 11–18.5)	42	38	Swab smears Clinical examination	No positive correlation between exposure to formaldehyde and abnormal cytology More mucosal abnormalities in non-smoking exposed workers ($p = 0.004$)
Boysen et al. (1990)	Formaldehyde (production of formaldehyde and formaldehyde resins)	0.6–> 2.4	37	37	Nasal biopsy	Histological score: exposed 1.9, controls, 1.7 ($p > 0.05$) Three exposed and none of the controls had dysplasia
Holmström et al. (1989b)	Formaldehyde (resins for laminate production)	0.05–0.5 (peaks to > 1)	62	32	Nasal biopsy	Histological score: exposed 2.16, controls 1.56 ($p < 0.05$) No case of dysplasia
Ballarin et al. (1992)	Formaldehyde Wood dust (plywood factory)	0.1–0.39 0.23–0.73	15	15	Nasal scrapes	Micronuclei in nasal mucosal cells: exposed 0.90, controls 0.25 ($p < 0.010$) Cytological score: exposed 2.3, controls 1.6 ($p < 0.01$) One exposed had mild dysplasia

pathogenesis of contact sensitivity to formaldehyde, in either atopic or nonatopic patients (Lidén et al., 1993).

Contact urticaria has also, but rarely, been associated with exposure to formaldehyde. Cases have been reported in a nonatopic histology technician (Rappaport & Hoffman, 1941), a worker exposed through contact with formaldehyde-treated leather (Helander, 1977) and a worker in a pathology laboratory (Lindskov, 1982). Information about the mechanisms of contact urticaria is limited (Maibach, 1983).

(c) Allergy

Immunological tests were performed on 23 asthmatic subjects who lived in urea–formaldehyde foam-insulated homes and on four asthmatic subjects living in conventionally insulated homes. The authors concluded that long-term exposure to formaldehyde had not affected the six immune parameters measured, but that short-term acute exposure resulted in minor immunological changes (Pross et al., 1987).

No IgE-mediated sensitization could be attributed to formaldehyde in 86 individuals at risk of exposure to formaldehyde (Kramps et al., 1989), and none of 63 practising pathologists had allergen-specific IgE directed against formaldehyde, although 29 subjects complained of sensitivityy to formaldehyde (Salkie, 1991).

The immune responses of a large number of people exposed to formaldehyde were investigated, including people living in mobile homes or working in buildings insulated with urea–formaldehyde foam, patients undergoing haemodialysis with formaldehyde-sterilized dialysers, physicians and dialysis nurses exposed to formaldehyde, histology technicians, medical and pathology students, and workers in an aircraft factory who were exposed to formaldehyde and other substances (including phenol and solvents) (Patterson et al., 1989; Grammer et al., 1990; Dykewicz et al., 1991). The authors of the last paper stated that none of their studies indicated an immunological basis for respiratory or conjunctival symptoms (conjunctivitis, rhinitis, coughing, wheezing, shortness of breath) seen after exposure to gaseous formaldehyde.

Elevated serum levels of IgE, IgG or IgM antibodies were observed in several individuals exposed to formaldehyde (Thrasher et al., 1987, 1988, 1990). The experimental design and methods used have been criticized, however, for lack of adequate controls, lack of a correlation between disease and immunological abnormalities, lack of information about the diseased and comparison populations and use of unproven diagnostic tests (Beavers, 1989; Greenberg & Stave, 1989).

4.2.2 Experimental systems

Formaldehyde has been shown to be toxic *in vitro* in a variety of experimental systems, including human cells. It decreased growth rate, cloning efficiency and the ability of cells to exclude trypan blue while inducing squamous differentiation of cultured human bronchial epithelial cells (Grafström, 1990). These effects occurred simultaneously with elevated levels of intracellular calcium ion, decreased levels of free low-relative-molecular-mass thiols, including glutathione, and the appearance of genotoxicity (see section 4.4).

(a) Acute effects

(i) Irritation

A quantitative measure of sensory irritation in rodents is provided by the reflex decrease in respiratory rate of mice or rats caused by stimulation of trigeminal nerve receptors in the nasal passages. In comparison with other aldehydes (Steinhagen & Barrow, 1984), formaldehyde is a potent respiratory tract irritant, eliciting a 50% decrease in respiratory frequency in B6C3F1 mice at 4.9 ppm [6.0 mg/m^3] and Fischer 344 rats at 31.7 ppm [38.7 mg/m^3] (Chang *et al.*, 1981). Swiss–Webster mice exposed to the concentration that elicits a 50% decrease in respiratory frequency (3.1 ppm [3.8 mg/m^3]) for five days (6 h/day) developed mild histopathological lesions in the anterior nasal cavity, but no lesions were found in the posterior nasal cavity or in the lung (Buckley *et al.*, 1984).

In addition to decreasing the respiratory rate, formaldehyde may also alter the tidal volume, resulting in a decrease in minute ventilation. Exposure to formaldehyde over a 10-min test period induced prompt reductions in both respiratory rates and minute volumes of mice and rats, whether or not they were exposed before testing to 6 ppm [7.4 mg/m^3] formaldehyde for 6 h per day for four days (Fig. 4). These effects were observed at lower concentrations of formaldehyde in mice than in rats (Chang *et al.*, 1983). A similar effect has been demonstrated in C57Bl6/F1 mice and CD rats (Jaeger & Gearhart, 1982).

Rats exposed to 28 ppm [34.1 mg/m^3] formaldehyde for four days developed tolerance to its sensory irritancy, but rats exposed to 15 ppm [18.3 mg/m^3] for one, four or 10 days did not (Chang & Barrow, 1984).

(ii) Pulmonary hyperreactivity

Formaldehyde induced pulmonary hyperreactivity in guinea-pigs: exposure to 0.03 ppm [0.04 mg/m^3] caused transient bronchoconstriction and hyperreactivity to infused acetylcholine when the duration of exposures was 8 h, but higher concentrations (10 ppm [12.2 mg/m^3]) were required to induce bronchoconstriction when the duration was 2 h. These effects occurred with no evidence of tracheal epithelial damage after exposure to 3.4 ppm [4.1 mg/m^3] for 8 h. The mechanism by which they occur is unknown (Swiecichowski *et al.*, 1993).

The effects of formaldehyde (vaporized formalin) on pulmonary flow were determined in cynomolgus monkeys, which were tranquilized before exposure and received an endotracheal tube transorally. Pulmonary flow resistance was increased at a concentration of 2.5 ppm [3.0 mg/m^3]. Airway narrowing was not correlated with methacholine reactivity (Biagini *et al.*, 1989). [The Working Group questioned the relevance of these findings, in view of the method of administration.]

(iii) Cytotoxicity and cell proliferation in the respiratory tract

The acute and subacute effects of formaldehyde in experimental animals are summarized in Table 21. A critical issue for the mechanism of carcinogenesis is whether low concentrations of formaldehyde increase the rate of cell turnover in the nasal epithelium. Subacute exposure to a low concentration of formaldehyde (1 ppm [1.2 mg/m^3], 6 h/day, three days) has been reported to induce a small, transient increase in nasal epithelial cell turnover in Wistar rats (Zwart *et al.*,

310 IARC MONOGRAPHS VOLUME 62

Figure 4. Representative time–response curves for the minute volume of naive and formaldehyde-treated mice and rats during 10-min exposures to various concentrations of formaldehyde

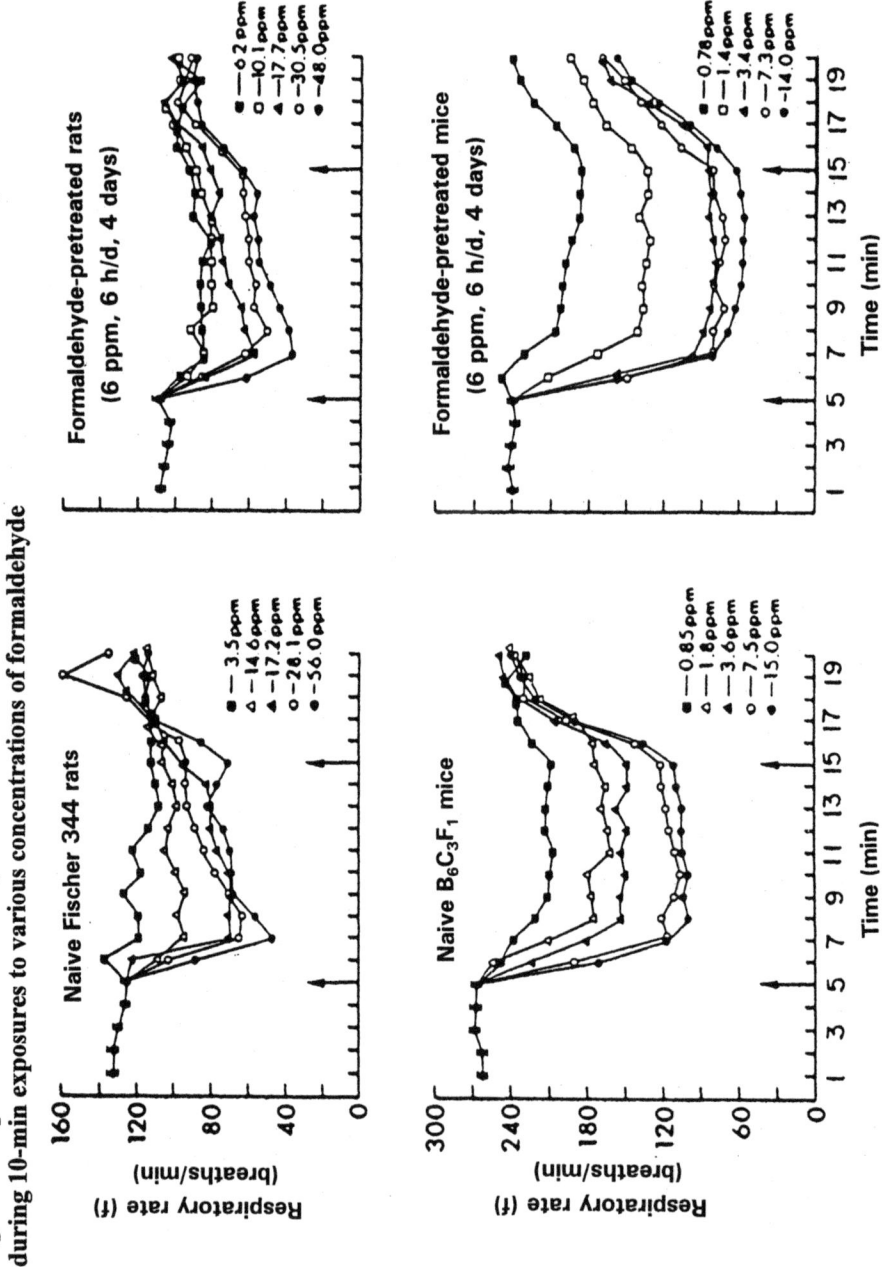

From Chang *et al.*(1981)
Data for the pre-exposure period are means ± SE of 19 or (A) 22 or (B) 28 animals, and the points for each concentration are means for four animals. Arrows indicate beginning and end of exposure.

Figure 4 (contd)

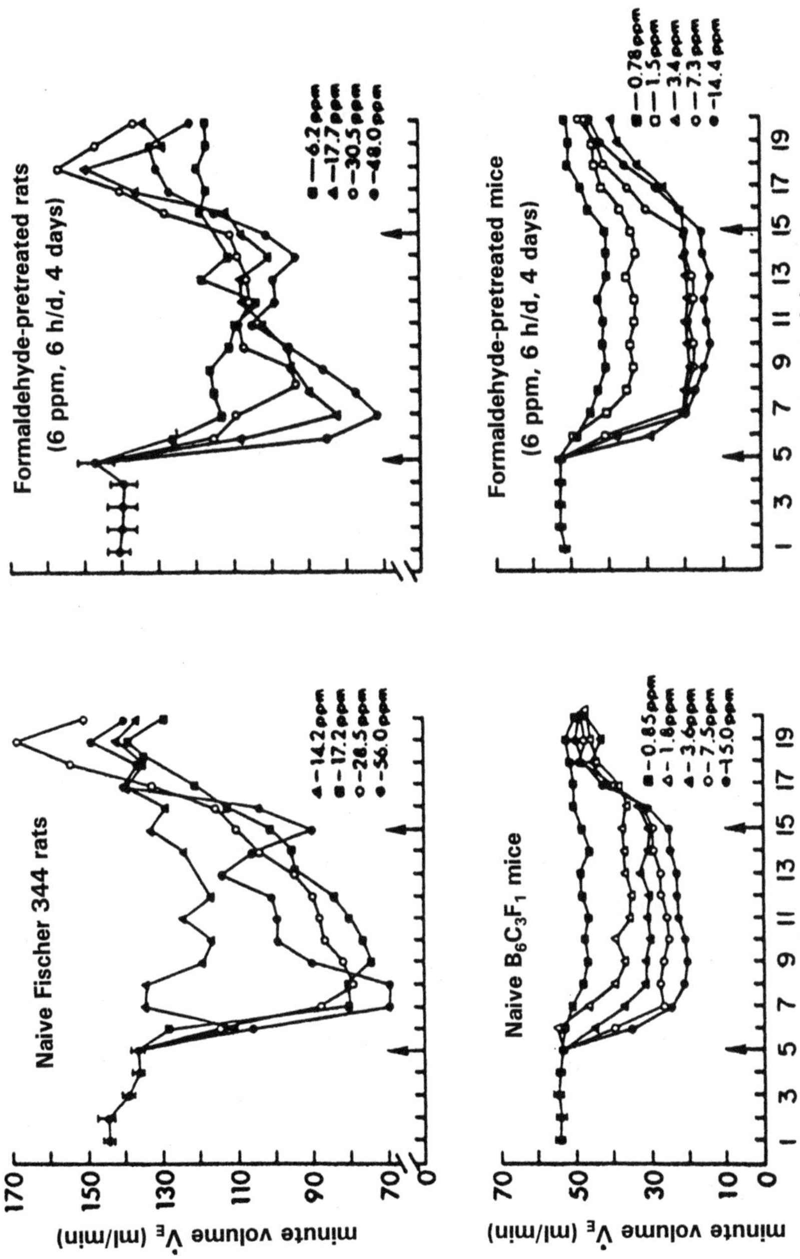

1988), but the apparent increase was not shown to be significant, and it was not confirmed in later studies (Reuzel et al., 1990). Other investigators did not detect an increase in cell turnover in the nasal epithelium of Fischer 344 rats exposed to 0.7 or 2 ppm [0.9 or 2.4 mg/m^3] (6 h/day, one, four or nine days) (Monticello et al., 1991) or to 0.5 or 2 ppm [0.6 or 2.4 mg/m^3] (6 h/day, three days) (Swenberg et al., 1983). Low concentrations of formaldehyde (0.5 or 2 ppm; 6 h/day, one, two, four, nine or 14 days) also did not inhibit mucociliary function in the nasal passages of Fischer 344 rats (Morgan et al., 1986b,c), and no injury to the nasal epithelium of rats of this strain was detected ultrastructurally after exposure to 0.5 or 2 ppm (6 h/day, one or four days) (Monteiro-Riviere & Popp, 1986).

Wistar rats exposed to 3 ppm [3.7 mg/m^3] (6 h/day, three days (Zwart et al., 1988) or 22 h/day, three days (Reuzel et al., 1990)) had a transient increase in cell replication. Higher formaldehyde concentrations ($\geq$ 6 ppm [7.3 mg/m^3]) induced erosion, epithelial hyperplasia, squamous metaplasia and inflammation in a site-specific manner in the nasal mucosa (Monticello et al., 1991). Mice are less responsive than rats, probably because they are better able than rats to reduce their minute ventilation when exposed to high concentrations of formaldehyde (Chang et al., 1983; Swenberg et al., 1983). Fischer 344 rats exposed to 6, 10 or 15 ppm [7.3, 12.2 or 18.3 mg/m^3] (6 h/day, one, four or nine days, or 6 h/day, five days/week, six weeks) had an enhanced rate of cell turnover (Monticello et al., 1991). The severity of nasal epithelial responses at 15 ppm was much greater than at 6 ppm (Monteiro-Riviere & Popp, 1986). Rhesus monkeys exposed to 6 ppm (6 h/day, five days) developed similar nasal lesions to rats. Mild lesions, characterized as multifocal loss of cilia, were also detected in the larynx, trachea and carina (Monticello et al., 1989).

The relative importance of concentration and total dose on cell proliferation was examined in Fischer 344 and Wistar rats exposed to a range of concentrations for various lengths of time, such that the total inhaled dose was constant. Exposures were for three or 10 days (Swenberg et al., 1983) or four weeks (Wilmer et al., 1987). All of the investigators concluded that concentration, not total dose, is the primary determinant of the cytotoxicity of formaldehyde. A similar conclusion was reached when rats were exposed for 13 weeks (Wilmer et al., 1989).

The effects of simultaneous exposure to formaldehyde and ozone were investigated in Wistar rats exposed to 0.3, 1 or 3 ppm [0.4, 1.2 and 3.7 mg/m^3] formaldehyde, 0.2, 0.4 or 0.8 ppm [0.4, 0.8 or 1.6 mg/m^3] ozone or mixtures of 0.4 ppm ozone with 0.3, 1 or 3 ppm formaldehyde or 1 ppm formaldehyde with 0.2, 0.4 or 0.8 ppm ozone (22 h/day, three days). Both formaldehyde (3 ppm) and ozone (0.4 or 0.8 ppm) induced cell proliferation in the most anterior region of the respiratory epithelium. In a slightly more posterior region, ozone had no effect on cell replication, but formaldehyde either enhanced cell proliferation (3 ppm) or appeared to inhibit it slightly (0.3 or 1 ppm). Combined exposures to low concentrations (0.4 ppm ozone and 0.3 ppm formaldehyde, 0.4 or 0.8 ppm ozone and 1 ppm formaldehyde) induced less cell proliferation than ozone alone; however, more than additive increases in cell proliferation were detected in the anterior nose after exposure to 0.4 ppm ozone in combination with 3 ppm formaldehyde, and in a slightly more posterior region after exposure to 0.4 ppm ozone with 1 or 3 ppm formaldehyde. The results suggested to the authors a complex response of the nasal epithelium to low (just nonirritating) concentrations of these irritants but a

Table 21. Cytotoxicity and cell proliferation induced by acute and subacute exposure to formaldehyde

Species	Exposure	Effects	Reference
Fischer 344 rat, male; B6C3F1 mouse, male	0, 0.6, 2.4, 7.4, 18.5 mg/m^3, 6 h/day, 3 days	0.6, 2.4: No increase in cell replication rate in nasal mucosa 7.4: Increased cell turnover (rats only) 18.5: Cell proliferation (rats and mice)	Swenberg et al. (1983)
Fischer 334 rat, male; B6C3F1 mouse, male	0, 18.5 mg/m^3, 6 h/day, 1 or 5 days	18.5: Cell proliferation induced in nasal mucosa of both species; rat responses exceeded mouse responses	Chang et al. (1983)
Fischer 344 rat, male	3.7 mg/m^3 × 12 h/day, 7.4 mg/m^3 × 6 h/day, 14.4 mg/m^3 × 3 h/day ($C \times t = 44$ mg/m^3-h/day), 3 or 10 days	Cell proliferation related more closely to concentration than to time; proliferation less after 10 than after 3 days of exposure, indicating adaptation	Swenberg et al. (1983)
Fischer 344 rat, male	0, 0.6, 2.4, 7.4, 18.5 mg/m^3, 6 h/day, 1, 2, 4, 9 or 14 days	0.6: No effects on mucociliary function 2.4: Minimal effects 7.4: Moderate inhibition 18.5: Marked inhibition	Morgan et al. (1986c)
Fischer 344 rat, male	0, 2.4, 18.5 mg/m^3, 10, 20, 45 or 90 min or 6 h	2.4: No effect on mucociliary function 18.5: Inhibition of mucociliary function, marked recovery 1 h after exposure	Morgan et al. (1986b)
Fischer 344 rat, male	0, 0.6, 2.4 mg/m^3, 6 h/day, 1 or 4 days; 7.4 mg/m^3, 6 h/day, 1, 2 or 4 days; 18.5 mg/m^3, 6 h/day, 1 or 2 days	0.6, 2.4: No lesions 7.4, 18.5: Non-cell-specific, dose-related injury, including hypertrophy, nonkeratinized squamous cells, nucleolar segregation	Monteiro-Riviere & Popp (1986)
Wistar rat, male	0, 6.2 mg/m^3 × 8 h/day, 12.3 mg/m^3 × 8 h/day ($C \times t = 49$ or 98 mg/m^3-h/day); 2.13 mg/m^3 × 8 × 0.5 h/day, 25 mg/m^3 × 8 × 0.5 h/day ($C \times t = 49$ or 98 mg/m^3-h/day), 5 days/week, 4 weeks	Labelling index increased at all concentrations; cell proliferation more closely related to concentration than to total dose	Wilmer et al. (1987)
Wistar rat, male and female	0, 0.37, 1.2, 3.7 mg/m^3, 6 h/day, 3 days	0.37, 1.2: Small increase in cell turnover at 1.2 ppm, but significance not shown and not confirmed in later studies (Reuzel et al., 1990); 3.7: significant, transient increase in cell turnover	Zwart et al. (1988)

Table 21 (contd)

Species	Exposure	Effects	Reference
Rhesus monkey, male	0, 7.4 mg/m^3, 6 h/day, 5 days/week, 1 or 6 weeks	Lesions similar to those in rats (Monticello et al., 1991) but more widespread, extending to trachea and major bronchi; increased cell replication in nasal passages, trachea and carina; percentage of nasal surface area affected increased between 1 and 6 weeks	Monticello et al. (1989)
Wistar rat, male	0, 0.37, 1.2, 3.7 mg/m^3, 22 h/day, 3 days Also investigated effect of simultaneous exposure to 0.4, 0.8 or 1.6 mg/m^3 ozone	0.37, 1.2: Either no increase or inhibition of cell proliferation 3.7: Increased cell replication 0.8 mg/m^3 ozone + 1.2 or 3.7 mg/m^3 formaldehyde: Synergistic increase in cell turnover 1.6 mg/m^3 ozone + 1.2 mg/m^3 formaldehyde: Inhibition of cell turnover	Reuzel et al. (1990)
Fischer 344 rat, male	0, 0.86, 2.4, 7.4, 12.3, 18.5 mg/m^3, 6 h/day, 1, 4, or 9 days or 6 weeks	0.86, 2.4: No effect on cell turnover 7.4, 12.3, 18.5: Concentration- and site-dependent cell proliferation induced at all exposure times	Monticello et al. (1991)

C, concentration; t, time

synergistic increase in cell proliferation at irritating concentrations. To induce a synergistic effect on cell proliferation, at least one of the compounds must be present at a cytotoxic concentration (Reuzel *et al.*, 1990).

(iv) Enzyme induction

No increase in the activity of formaldehyde or aldehyde dehydrogenase was seen in the nose of Fischer 344 rats exposed to 15 ppm [18.3 mg/m^3] (6 h/day, five days/week, two weeks) (Casanova-Schmitz *et al.*, 1984a). A large increase in the activity of rat pulmonary cytochrome P450 was seen, however, after exposure to 0.5, 3 or 15 ppm formaldehyde [0.6, 3.7 or 18.3 mg/m^3] (6 h/day, four days) (Dallas *et al.*, 1986), although Dinsdale *et al.* (1993), using the same rat strain, could not confirm these results and found no increase in pulmonary cytochrome P450 activity after exposure to 10 ppm [12.2 mg/m^3] formaldehyde (6 h/day, four days).

(b) Chronic effects

(i) Cytotoxicity and cell proliferation in the respiratory tract

The subchronic and chronic effects of formaldehyde in different animal species exposed by inhalation are summarized in Table 22. No increases in cell turnover or DNA synthesis were found in the nasal mucosa after subchronic or chronic exposure to concentrations ≤ 2 ppm [≤ 2.4 mg/m^3] (Rusch *et al.*, 1983; Zwart *et al.*, 1988; Monticello *et al.*, 1993; Casanova *et al.*, 1995). Small, site-specific increases in the rate of cell turnover were noted at 3 ppm [3.7 mg/m^3] (6 h/day, 5 days/week, 13 weeks) in Wistar rats (Zwart *et al.*, 1988) and in the rate of DNA synthesis at 6 ppm [7.3 mg/m^3] (6 h/day, 5 days/week, 12 weeks) in Fischer 344 rats (Casanova *et al.*, 1995). At these concentrations, however, an adaptive response occurs in rat nasal epithelium, as cell turnover rates after six weeks (Monticello *et al.*, 1991) or 13 weeks (Zwart *et al.*, 1988) are lower than those after one to four days of exposure. Monticello *et al.* (1993) detected no increase in cell turnover in the nasal passages of Fischer 344 rats exposed to 6 ppm [7.3 mg/m^3] formaldehyde for three months (6 h/day, 5 days/week), but, as already noted, Casanova *et al.* (1995) detected a small increase in DNA synthesis under these conditions. Large, sustained increases in cell turnover were observed at 10 and 15 ppm [12.2 and 18.3 mg/m^3] (6 h/day, 5 days/week, 3, 6, 12 or 18 months) (Monticello *et al.*, 1993). The effects of subchronic exposure to various concentrations of formaldehyde on DNA synthesis in the rat nose are illustrated in Figure 5.

Additional studies have shown the importance of increased cell turnover in the induction of rat nasal tumours (Appelman *et al.*, 1988; Woutersen *et al.*, 1989). The investigators damaged the nasal mucosa of Wistar rats by bilateral intranasal electrocoagulation and evaluated the susceptibility of the rats to formaldehyde at concentrations of 0.1, 1 or 10 ppm [0.1, 1.2 or 12.2 mg/m^3] (6 h/day, 5 days/week, 13 or 52 weeks, 28 months, or three months of exposure followed by a 25-month observation period). In rats with undamaged mucosa, the effects of exposure were seen only at 10 ppm; these effects were limited to degenerative, inflammatory and hyperplastic changes. These noncancerous effects were increased by electrocoagulation. In the group exposed to 10 ppm for 28 months, nasal tumours were induced in 17/58 rats. No compound-related tumours were induced at 0.1 or 1 ppm. It was concluded that the damaged

Table 22. Cytotoxicity and cell proliferation induced by subchronic and chronic exposures to formaldehyde

Species	Exposure	Effects	Reference
Fischer 344 rat, Syrian hamster, male and female; cynomolgus monkey, male	0, 0.25, 1.2, 3.7 mg/m^3, 22 h/day, 7 days/week, 26 weeks	Rats: Squamous metaplasia in nasal turbinates at 3.7 mg/m^3 only Hamsters: No significant toxic response Monkey: Squamous metaplasia in nasal turbinates at 3.7 mg/m^3 only	Rusch et al. (1983)
B6C3F1 mouse, male	0, 2.5, 4.9, 12.3, 24.7, 49.2 mg/m^3, 6 h/day, 5 days/week, 13 weeks	2.5, 4.9: No lesion induced 12.3, 24.7, 49.2: Squamous metaplasia, inflammation of nasal passages, trachea and larynx; 80% mortality at 49.2 mg/m^3	Maronpot et al. (1986)
Wistar rat, male and female	0, 0.37, 1.2, 3.7 mg/m^3, 6 h/day, 5 days/week, 13 weeks	0.37, 1.2: No increase in cell replication 3.7: Increased cell turnover in nasal epithelium but cell proliferation lower than after 3 days	Zwart et al. (1988)
Wistar rat, male and female	0, 1.2, 12.3, 24.7 mg/m^3, 6 h/day, 5 days/week, 13 weeks	1.2: Results inconclusive 12.3, 24.7: Squamous metaplasia, epithelial erosion, cell proliferation in nasal passages and larynx; no hepatotoxicity	Woutersen et al. (1987)
Wistar rat, male	0, 0.12, 1.2, 12.3 mg/m^3, 6 h/day, 5 days/week, 13 or 52 weeks Nasal mucosa of some rats injured by bilateral intranasal electrocoagulation to induce cell proliferation	0: Electrocoagulation induced hyperplasia and squamous metaplasia, still visible after 13 weeks but slight after 52 weeks 0.12, 1.2: Focal squamous metaplasia after 13 or 52 weeks; no adverse effects in animals with undamaged nasal mucosa 12.3: Squamous metaplasia and degeneration in respiratory epithelium (both intact and damaged nose) and olfactory epithelium (damaged nose only)	Appelman et al. (1988)
Wistar rat, male	0, 1.2 mg/m^3 × 8 h/day, 2.4 mg/m^3 × 8 h/day ($C × t$ = 9.8 or 19.7 mg/m^3-h/day), 5 days/week, 13 weeks; 2.4 mg/m^3 × 8 × 0.5 h/day, 4.9 mg/m^3 × 8 × 0.5 h/day ($C × t$ = 9.8 or 19.7 mg/m^3 h/day), 5 days/week, 13 weeks	1.2, 2.5: No observed toxic effect 4.9: Epithelial damage, squamous metaplasia, occasional keratinization; concentration, not total dose, determines severity of toxic effect	Wilmer et al. (1989)
Fischer 344 rat, male	0, 0.86, 2.5, 7.4, 12.3, 18.5 mg/m^3, 6 h/day, 5 days/week, 3 months	0.86, 2.5, 7.4: No increase in cell replication detected 12.3, 18.5: Sustained cell proliferation	Monticello et al. (1993)
Fischer 334 rat, male	0, 0.86, 2.5, 7.4, 18.5 ppm, 6 h/day, 5 days/week, 12 weeks	0.86, 2.5: DNA synthesis rates in nasal mucosa similar in naive (previously unexposed) and subchronically exposed rats 7.4, 18.5: DNA synthesis rates higher in subchronically exposed than in naive rats, especially at 18.5 mg/m^3	Casanova et al. (1995)

C, concentration; t, time

Figure 5. Cell turnover in the lateral meatus (LM) and medial and posterior meatuses (M:PM) of pre-exposed and naive (previously unexposed) Fischer 344 rats, as measured by incorporation of ^{14}C derived from inhaled ^{14}C-formaldehyde into nucleic acid bases (deoxyadenosine, deoxyguanosine and thymidine) and thence into DNA, during a single 3-h exposure to 0.7, 2, 6 or 15 ppm [0.86, 2.5, 7.4 or 18.5 mg/m^3]

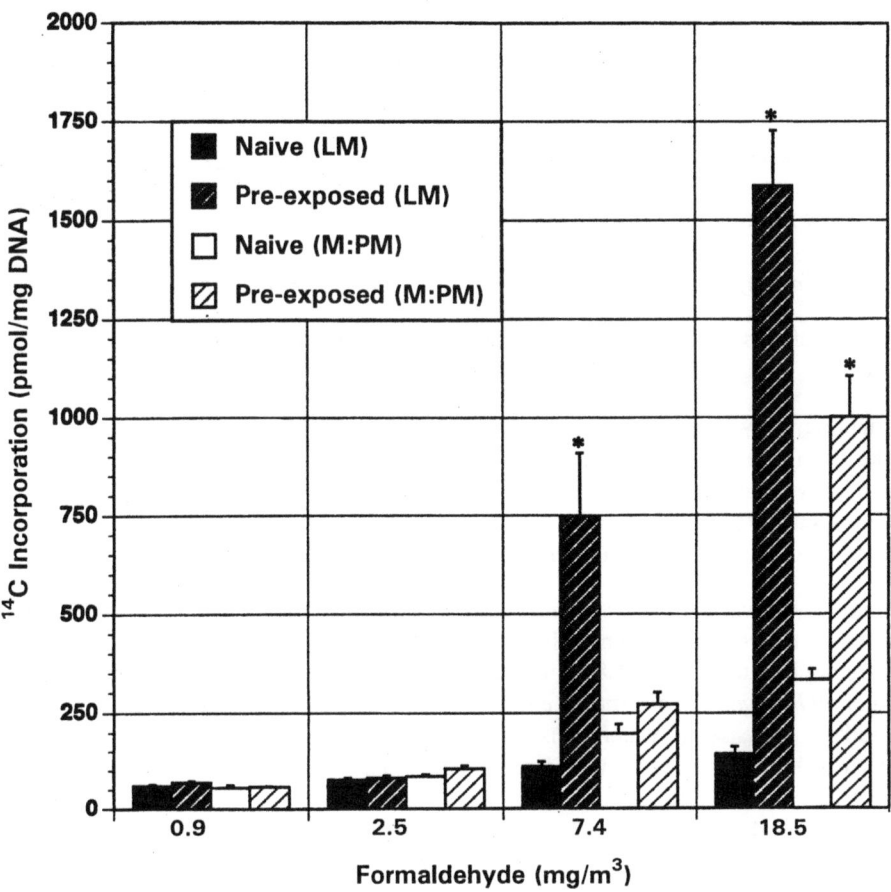

Reproduced, with permission, from Casanova et al. (1995)

Pre-exposed rats were exposed subchronically to the same concentrations of unlabelled formaldehyde (6 h/day, 5 days/week, 11 weeks and four days), while naive rats were exposed to room air. The exposure to ^{14}C-formaldehyde occurred on the fifth day of the twelfth week. The asterisk denotes a significant difference between pre-exposed and naive rats.

mucosa was more susceptible to the cytotoxic effects of formaldehyde and that severe damage contributes to the induction of nasal tumours.

Rhesus monkeys exposed to 6 ppm [7.3 mg/m^3] formaldehyde (6 h/day, 5 days/week) had a larger percentage of the nasal mucosal surface area affected after six weeks than after five days. Cell proliferation was detected in the nasal passages, larynx, trachea and carina, but the effects in the lower airways were minimal in comparison with the effects in the nasal passages (Monticello *et al.*, 1989). Other studies showed that Fischer 344 rats exposed to 1 ppm [1.2 mg/m^3] (22 h/day, 7 days/week, 26 weeks) developed no detectable nasal lesions (Rusch *et al.*, 1983), but Fischer 344 rats exposed to 2 ppm [2.4 mg/m^3] (6 h/day, 5 days/week, 24 months) developed mild squamous metaplasia in the nasal turbinates (Kerns *et al.*, 1983b). Although the total dose received by the former group was 2.5 times higher than that received by the latter, the incidence and severity of lesions was less, again demonstrating the greater importance of concentration than total dose (Rusch *et al.*, 1983).

(ii) Toxicity in the gastrointestinal tract after oral administration

The toxic effects of formaldehyde given by oral administration have been reviewed (Restani & Galli, 1991).

Formaldehyde was administered orally to rats and dogs at daily doses of 50, 100 or 150 mg/kg bw (rats) or 50, 75 or 100 mg/kg bw (dogs) for 91 consecutive days. Significant changes in body weight were observed at the higher doses, but clinical and pathological studies revealed no specific treatment-related effects on the kidney, liver or lung, which were considered possible target organs, or on the gastrointestinal mucosa (Johannsen *et al.*, 1986).

Formaldehyde was administered in the drinking-water to male and female Wistar rats for up to two years. In the chronic portion of the study, the mean daily doses of formaldehyde were 1.2, 15 or 82 mg/kg bw (males) and 1.8, 21 or 109 mg/kg bw (females). Controls received drinking-water either *ad libitum* or in an amount equal to that consumed by the highest-dose group, which had a marked decrease in water consumption. Pathological changes after two years were essentially restricted to the highest-dose group and consisted of a thickened and raised limiting ridge of the forestomach and gastritis and hyperplasia of the glandular stomach. The no-adverse-effect level was estimated to be 82 mg/kg bw per day (males) or 109 mg/kg bw per day (females) (Til *et al.*, 1988, 1989).

In another experiment in which formaldehyde was administered in the drinking-water to male and female Wistar rats, fixed concentrations (0, 0.02, 0.1 and 0.5%) were given for up to two years. Estimated from the water intake, these concentrations corresponded, on average, to 0, 10, 50 and 300 mg/kg bw per day. All rats that received the highest dose died during the study. The lesions induced in the stomach were similar to those reported by Til *et al.* (1988, 1989). No treatment-related tumour was found. The no-effect level was estimated to be 0.02% (10 mg/kg bw per day), as forestomach hyperkeratosis was observed in a small number of rats (2/14) receiving 0.1% formaldehyde (50 mg/kg bw per day) (Tobe *et al.*, 1989).

(c) Immunotoxicity

The possibility that formaldehyde may induce changes in the immune response was examined in B6C3F1 mice exposed to 15 ppm [18.3 mg/m^3] formaldehyde (6 h/day, 5 days/week, 3 weeks). A variety of immune function tests revealed no significant changes, except for an increase in host resistance to challenge with the bacterium, *Listeria monocytogenes*, implying an increased resistance to infection. Exposure did not alter the number or impair the function of resident peritoneal macrophages, but it increased the competence for release of hydrogen peroxide from peritoneal macrophages (Dean *et al.*, 1984; Adams *et al.*, 1987).

Sprague–Dawley rats were exposed to 12.6 ppm [15.4 mg/m^3] formaldehyde (6 h/day, 5 days/week, 22 months) and then vaccinated with pneumococcal polysaccharide antigens and tetanus toxoid. They were tested three to four weeks later for the development of antibodies. An IgG response to pneumococcal polysaccharides and to tetanus toxoid and an IgM response to tetanus toxoid were found in both exposed and control groups. No evidence was obtained that long-term exposure to a high concentration of formaldehyde impairs B-cell function, as measured by antibody production (Holmström *et al.*, 1989c).

In order to investigate the induction of sensitivity to formaldehyde, undiluted formalin was painted on shaven and epilated dorsal sites on guinea-pigs; a second application was administered two days later at naive sites, to give a total dose of 74 mg/animal. Other animals received diluted formalin at doses of 12–9.3 mg/animal. All animals receiving 74 mg developed skin sensitivity when tested seven days after exposure. A significant dose–response relationship was observed for degree of sensitization and for percentage of animals sensitized; however, pulmonary sensitivity was not induced when formaldehyde was administered dermally, by injection or by inhalation, and no cytophilic antibodies were detected in blood (Lee *et al.*, 1984).

4.3 Reproductive and developmental effects

4.3.1 Humans

The incidence of spontaneous abortion was studied among hospital staff in Finland who used ethylene oxide (see IARC, 1994b), glutaraldehyde and formaldehyde for sterilizing instruments. Potentially exposed women were identified in 1980 with the help of supervising nurses at all of the approximately 80 general hospitals of the country, and an equal number of control women were selected by the supervising nurse from among nursing auxiliaries in the same hospitals who had no exposure to sterilizing agents, anaesthetic gases or X-rays. Study subjects were administered a postal questionnaire which requested personal data and information on smoking habits, intake of alcohol, reproductive history, including number of pregnancies and their outcome, and occupation at the time of each pregnancy. Information about exposure to chemical sterilizing agents was obtained from the supervising nurses. The crude rates of spontaneous abortions were 16.7% for sterilizing staff who were considered to have been exposed during the first trimester of pregnancy, 6.0% for sterilizing staff who left employment when they learnt they were pregnant (the difference being significant) and 10.6% among controls. When adjusted for age, parity, decade of pregnancy, smoking habits and alcohol and coffee consumption, the rate associated with exposure to ethylene oxide, with or without other agents, was 12.7%, which was significantly increased ($p < 0.05$), and that associated with formaldehyde, with or without other agents, was 8.4%, which was comparable to the reference level of 10.5% (Hemminki et al., 1982).

In a nationwide record linkage study in Finland, all nurses who had been pregnant between the years 1973 and 1979 and who had worked in anaesthesia, surgery, intensive care, operating rooms or internal departments of a general hospital (and in paediatric, gynaecological, cancer and lung departments for the part of the study concerned with malformations) were identified. Each of the 217 women treated for spontaneous abortion according to the files of the Finnish hospital discharge register and the 46 women notified to the Register of Congenital Malformations was individually matched on age and hospital with three control women, who were selected at random from the same population of nurses and matched for age and hospital where they were employed. Information was obtained from supervising nurses by postal questionnaires on the exposure of cases and controls to sterilizing agents (ethylene oxide, glutaraldehyde and formaldehyde), anaesthetic gases, disinfectant soaps, cytostatic drugs and X-radiation. Exposure to formaldehyde during pregnancy was reported for 3.7% of the nurses who were later treated for spontaneous abortion and for 5.2% of their controls, yielding a crude odds ratio of 0.7 [95% CI, 0.28–1.7]. Exposure to formaldehyde was also reported for 8.8% of nurses who gave birth to a malformed child and to 5.3% matched controls, to give an odds ratio of 1.7 [95% CI, 0.39–7.7]; the latter analysis was based on eight exposed subjects (Hemminki et al., 1985).

The occurrence of spontaneous abortions among women working in laboratories in Finland and congenital malformations and birth weights of the children were investigated in a matched retrospective case–control study. The final population in the study of spontaneous abortion was 206 cases and 329 controls; that in the study of congenital malformations was 36 cases and 105 controls. Information on occupational exposure, health status, medication, contraception, smoking and alcohol consumption during the first trimester of the pregnancy was collected by postal questionnaire. The odds ratio for spontaneous abortion was increased among women who had been exposed to formalin for at least three days per week (odds ratio, 3.5; 95% CI, 1.1–11). A greater proportion of the cases (8/10) than the controls (4/7) who had been exposed to formalin had been employed in pathology and histology laboratories. Most of the cases (8/10) and controls (5/7) who were exposed to formalin were also exposed to xylene (see IARC, 1989c). The authors stated that the results for individual chemicals should be interpreted cautiously because laboratory personnel are often exposed to several solvents and other chemicals simultaneously. No association was observed between exposure to formalin and congenital malformations [data not shown] (Taskinen et al., 1994).

4.3.2 Experimental systems

The reproductive and developmental toxicity of formaldehyde has been reviewed (Feinman, 1988; WHO, 1989).

Whether administered by inhalation, ingestion or the skin to various rodent species, formaldehyde did not exert adverse effects on reproductive parameters or fetal development (Marks et al., 1980; Feinman, 1988). Additional studies have confirmed this assessment. Groups of 25 pregnant Sprague–Dawley rats were exposed to formaldehyde (0, 5, 10, 20 or 40 ppm [0, 6, 12, 24 or 49 mg/m^3]; 6 h/day, days 6–20 of gestation). On day 21, the rats were killed and maternal and fetal parameters were evaluated. The authors concluded that formaldehyde was neither embryolethal nor teratogenic when given under these conditions. The mean fetal body weight at 20 ppm was 5% less than that of controls ($p < 0.05$) in males but was not reduced in females; at 40 ppm, mean fetal body weight was about 20% less than that in controls ($p < 0.01$) in both males and females. The decrease in fetal weight in the group given the high dose was attributable to maternal toxicity (Saillenfait *et al.*, 1989).

Groups of 25 mated female Sprague–Dawley rats were exposed to formaldehyde at 2, 5 or 10 ppm [2.5, 6 or 12 mg/m^3] (6 h/day) on days 6–15 of gestation. At 10 ppm, there was a significant decrease in maternal food consumption and weight gain. None of the parameters of pregnancy, including numbers of corpora lutea, implantation sites, live fetuses, dead fetuses and resorptions, or fetal weights were affected by treatment (Martin, 1990).

Formaldehyde was applied topically to pregnant Syrian hamsters on day 8, 9, 10 or 11 of gestation by clipping the hair on the dorsal body and applying 0.5 ml formalin (37% formaldehyde) with a syringe directly onto the skin. In order to prevent grooming, the animals were anaesthetized with nembutal (13 mg intraperitoneally) during the 2-h treatment. On day 15, fetuses were removed from four to six hamsters per group and examined. The number of resorptions was increased, but no teratogenic effects or effects on fetal weight or length were detected. The authors suggested that the increase in resorptions may have been caused by stress (Overman, 1985).

4.4 Genetic and related effects

The mutagenicity of formaldehyde has been reviewed (IARC, 1982, 1987d; Ma & Harris, 1988; WHO, 1989; Feron *et al.*, 1991).

4.4.1 *Humans*

(a) DNA–protein cross-links

No data were available to the Working Group.

(b) Mutation and allied effects

The effects of formaldehyde on the frequencies of chromosomal aberrations and sister chromatid exchange in peripheral lymphocytes of people occupationally exposed to formaldehyde were reviewed previously (IARC, 1987d). Both positive and negative results were obtained, but their interpretation was difficult because of the small number of subjects studied and inconsistencies in the findings. Since then, further data on the cytogenetic effects of formaldehyde in humans have been published.

In a study of workers exposed to formaldehyde in a factory manufacturing wood-splinter materials, short-term cultures of peripheral lymphocytes were examined from a group of 20 workers aged 27–57 (mean, 42.3 years), of whom 10 were men and 10 were women. They had been exposed to formaldehyde at 8-h time-weighted concentrations of 0.55–10.36 mg/m^3 for periods of 5–≥ 16 years. The control group consisted of 19 people [sex and age unspecified] employed in the same plant whose habits and social status were similar to those of the exposed group but who had unknown occupational contact with chemicals. No significant difference was observed between control and exposed groups with respect to any of the chromosomal anomalies (including chromatid and chromosome gaps, breaks, exchanges, breaks per cell, percentage of cells with aberrations) scored in the study (controls: 3.6% aberrant cells, 0.08 breaks per cell; exposed: 3.08% aberrant cells, 0.045 breaks per cell). The authors noted that the frequency of aberrations in the control group was higher than that seen in the general population (1.2–2% aberrant cells) (Vargová et al., 1992). [The Working Group noted that, although the text states that there were 20 people in the exposed group, Table II of the paper gives a figure of 25. The Group also noted the lack of detail on the smoking habits of the subjects.]

In the study of Ballarin et al. (1992), described on p. 306, the frequency of micronuclei in respiratory nasal mucosa cells was also investigated. At least 6000 cells from each individual were scored for micronuclei. A significant excess of micronucleated cells was seen in the exposed group (mean percentage of micronucleated cells, 0.90 ± 0.47; range, 0.17–1.83 in exposed group; 0.25 ± 0.22; range, 0.0–0.66 in controls; Mann–Whitney U test: $p < 0.01$). The authors noted the absence of a dose–response relationship between exposure to formaldehyde and the frequency of micronuclei and that concurrent exposure to wood dust could have contributed to the excess of micronucleated cells seen in the exposed group.

In a prospective study of the effect of formaldehyde on the frequency of micronuclei in oral and nasal mucosal cells and peripheral lymphocytes from a group of 29 student morticians, samples of blood and epithelial cells were taken before the students started the course (baseline samples) and again after the first nine weeks in an embalming laboratory. During the 85-day study period, the subjects had average cumulative formaldehyde exposures of 14.8 ppm-h [17.8 mg/m^3-h], with an average air concentration of 1.4 ppm [1.7 mg/m^3]. Epithelial cells were taken with a cytopathology brush from each inner cheek and from the inferior turbinate of each nostril. Weakly positive results were found in lymphocytes, positive results in buccal epithelium and negative results in nasal epithelium (Suruda et al., 1993). [The Working Group noted the inadequate reporting of the data in this study and was unable to evaluate it.]

(c) Sperm abnormalities

Eleven hospital autopsy service workers and 11 matched controls were evaluated for sperm count, abnormal sperm morphology and the frequency of one or two fluorescent bodies. Subjects were matched for sex, age and use of alcohol, tobacco and marijuana; additional information was collected on health, medications and other exposure to toxins. Exposed and control subjects were sampled three times at two- to three-month intervals. Ten exposed subjects had been employed for 4.3 months (range, 1–11 months) before the first sample was taken, and one had been employed for several years. Exposure to formaldehyde was intermittent, with a time-weighted average of 0.61–1.32 ppm [0.73–1.58 mg/m^3] (weekly exposure, 3–40 ppm-h [3.6–48 mg/m^3-h]). No significant difference was observed between the exposed and control groups with regard to sperm parameters (Ward et al., 1984).

(d) Urinary mutagenicity

Hospital autopsy service workers in Galveston, TX (United States), consisting of 15 men and four women aged < 30–> 50, and a control group from the local medical school, consisting of 15 men and five women in the same age range and matched for consumption of tobacco, marijuana, alcohol and coffee, were studied for urinary mutagenicity (Connor et al., 1985). Individuals were sampled three times at approximately two-month intervals. The time-weighted average exposures to formaldehyde in the work areas were estimated to be 0.61–1.32 ppm [0.73–1.58 mg/m^3]. Urine (150–200 ml from each subject) was treated with β-glucuronidase and passed through an XAD-2 column, which was then washed with water. The fraction that eluted with acetone was assayed for mutagenicity in *Salmonella typhimurium* TA98 and TA100 in the presence and absence of an exogenous metabolic activation system from livers of Aroclor-1254-induced rats. No increase in mutagenicity was seen in the autopsy workers as compared with the control group.

4.4.2 Experimental systems

(a) DNA–protein cross-links

Formaldehyde induces DNA–protein cross-links in mammalian cells *in vitro* and *in vivo* (see Table 23). The precise nature of these cross-links is unknown. Studies of the repair of DNA–protein cross-links caused by formaldehyde *in vitro* showed that they are removed from several types of normal cells and xeroderma pigmentosum cells, with a half-time of 2–3 h. These removal rates were similar at non-toxic and toxic concentrations of formaldehyde. In formaldehyde-exposed normal cells, active removal of DNA adducts by DNA excision repair was indicated by formation of DNA single-strand breaks, which could be accumulated in the presence of DNA repair synthesis inhibitors (Grafström et al., 1984).

Groups of four male Fischer 344 rats were exposed for 6 h to 0.3, 0.7, 2, 6 or 10 ppm [0.4, 0.9, 2.4, 7.3 or 12.2 mg/m^3] ^{14}C-formaldehyde in a nose-only inhalation chamber. Individual male rhesus monkeys (*Macaca mulatta*) were exposed for 6 h to 0.7, 2 or 6 ppm ^{14}C-formaldehyde in a mouth-only inhalation chamber. DNA–protein cross-links induced by exposure to formaldehyde were measured in the nasal mucosa of several regions of the upper respiratory tract of exposed animals. The concentration of cross-links increased non-linearly with the airborne concentration in both species. The concentrations of cross-links in the turbinates and anterior nasal mucosa were significantly lower in monkeys than in rats. Cross-links were also formed in the nasopharynx and trachea of monkeys, but they were not detected in the sinus, proximal lung or bone marrow. The authors suggested that the differences between the species with respect to DNA–protein cross-link formation may be due to differences in nasal cavity deposition and in the elimination of absorbed formaldehyde (Heck et al., 1989; Casanova et al., 1991).

(b) Mutation and allied effects (see also Table 23 and Appendices 1 and 2)

Formaldehyde induced mutation and DNA damage in bacteria and mutation, gene conversion, DNA strand breaks and DNA–protein cross-links in fungi. In *Drosophila*

Table 23. Genetic and related effects of formaldehyde

Test system		Result[a]		Dose[b] (LED/HID)	Reference
		Without exogenous metabolic system	With exogenous metabolic system		
*	Misincorporation of DNA bases into synthetic polynucleotides in vitro	+	0	30	Snyder & Van Houten (1986)
PRB	Prophage induction, SOS repair test, DNA strand breaks, cross-links or related damage	+	0	0.0075	Kuykendall & Bogdanffy (1992)
PRB	Prophage induction, SOS repair test, DNA strand breaks, cross-links or related damage	+	0	20	Le Curieux et al. (1993)
ECB	Escherichia coli (or E. coli DNA) strand breaks, cross-links or related damage; DNA repair	+	0	600	Wilkins & MacLeod (1976)
ECB	Escherichia coli (or E. coli DNA) strand breaks, cross-links or related damage; DNA repair	+	0	60	Poverenny et al. (1975)
ECD	Escherichia coli polA/W31110-P3478, differential toxicity (spot test)	+	0	10	Leifer et al. (1981)
ECL	Escherichia coli K12 KS160-KS66 polAI, differential toxicity	+	0	60	Poverenny et al. (1975)
ECK	Escherichia coli K12, forward or reverse mutation	+	0	60	Zijlstra (1989)
ECK	Escherichia coli K12, forward or reverse mutation	+	0	18.8	Graves et al. (1994)
ECK	Escherichia coli K12, forward or reverse mutation	+	0	120	Crosby et al. (1988)
SAF	Salmonella typhimurium, forward mutation	+	+	10	Temcharoen & Thilly (1983)
SA0	Salmonella typhimurium TA100, reverse mutation	(+)	0	25	Marnett et al. (1985)
SA0	Salmonella typhimurium TA100, reverse mutation	–	–	30	Gocke et al. (1981)
SA0	Salmonella typhimurium TA100, reverse mutation	–	+	16.6	Haworth et al. (1983)
SA0	Salmonella typhimurium TA100, reverse mutation	(+)	+	30 (toxic above 125 µg/plate)	Connor et al. (1983)
SA0	Salmonella typhimurium TA100, reverse mutation	+	0	7.5	Takahashi et al. (1985)
SA0	Salmonella typhimurium TA100, reverse mutation	+	+[c]	4.5	Pool et al. (1984)
SA0	Salmonella typhimurium TA100, reverse mutation	+	0	9.3	O'Donovan & Mee (1993)
SA0	Salmonella typhimurium TA100, reverse mutation	(+)	+	3	Schmid et al. (1986)
SA2	Salmonella typhimurium TA102, reverse mutation	+	0	10	Marnett et al. (1985)
SA2	Salmonella typhimurium TA102, reverse mutation	+	0	10	Le Curieux et al. (1993)
SA2	Salmonella typhimurium TA102, reverse mutation	+	0	35.7	O'Donovan & Mee (1993)
SA4	Salmonella typhimurium TA104, reverse mutation	+	0	10	Marnett et al. (1985)
SA5	Salmonella typhimurium TA1535, reverse mutation	–	–	30	Gocke et al. (1981)
SA5	Salmonella typhimurium TA1535, reverse mutation	–	–	50	Haworth et al. (1983)
SA5	Salmonella typhimurium TA1535, reverse mutation	0	–[c]	9	Pool et al. (1984)

Table 23 (contd)

Test system		Result[a]		Dose[b] (LED/HID)	Reference
		Without exogenous metabolic system	With exogenous metabolic system		
SA5	*Salmonella typhimurium* TA1535, reverse mutation	-	0	143	O'Donovan & Mee (1993)
SA7	*Salmonella typhimurium* TA1537, reverse mutation	-	-	30	Gocke *et al.* (1981)
SA7	*Salmonella typhimurium* TA1537, reverse mutation	-	-	50	Haworth *et al.* (1983)
SA7	*Salmonella typhimurium* TA1537, reverse mutation	-	0	143	O'Donovan & Mee (1993)
SA8	*Salmonella typhimurium* TA1538, reverse mutation	-	-	30	Gocke *et al.* (1981)
SA8	*Salmonella typhimurium* TA1538, reverse mutation	-	0	143	O'Donovan & Mee (1993)
SA9	*Salmonella typhimurium* TA98, reverse mutation	+	0	5	Marnett *et al.* (1985)
SA9	*Salmonella typhimurium* TA98, reverse mutation	-	-	30	Gocke *et al.* (1981)
SA9	*Salmonella typhimurium* TA98, reverse mutation	-	(+)	16.6	Haworth *et al.* (1983)
SA9	*Salmonella typhimurium* TA98, reverse mutation	-	(+)	30 (toxic above 100 μg/plate)	Connor *et al.* (1983)
SA9	*Salmonella typhimurium* TA98, reverse mutation	0	(+)[c]	3	Pool *et al.* (1984)
SA9	*Salmonella typhimurium* TA98, reverse mutation	+	0	17.9	O'Donovan & Mee (1993)
SAS	*Salmonella typhimurium* TA97, reverse mutation	+	0	5	Marnett *et al.* (1985)
SAS	*Salmonella typhimurium* (other miscellaneous strains), reverse mutation	-	-	100 (toxic at 250 μg/ml)	Connor *et al.* (1983)
ECW	*Escherichia coli* WP2 *uvrA*, reverse mutation	+	0	15	Takahashi *et al.* (1985)
ECW	*Escherichia coli* WP2 *uvrA*(pKM101), reverse mutation	+	0	17.9	O'Donovan & Mee (1993)
EC2	*Escherichia coli* WP2, reverse mutation	+	0	1.2	Nishioka (1973)
EC2	*Escherichia coli* WP2(pKM101), reverse mutation	+	0	35.7	O'Donovan & Mee (1993)
EC2	*Escherichia coli* WP2, reverse mutation	+	0	60	Takahashi *et al.* (1985)
ECR	*Escherichia coli* (other miscellaneous strains), reverse mutation	+	0	900	Panfilova *et al.* (1966)
ECR	*Escherichia coli* (other miscellaneous strains), reverse mutation	+	0	80	Demerec *et al.* (1951)
ECR	*Escherichia coli* (other miscellaneous strains), reverse mutation	+	0	30	Takahashi *et al.* (1985)
SSB	*Saccharomyces* species, DNA strand breaks, cross-links or related damage	+	0	990	Magaña-Schwencke *et al.* (1978)
SSB	*Saccharomyces* species, DNA strand breaks, cross-links or related damage	+	0	500	Magaña-Schwencke & Ekert (1978)
SSB	*Saccharomyces* species, DNA strand breaks, cross-links or related damage	+	0	500	Magaña-Schwencke & Moustacchi (1980)
SCH	*Saccharomyces cerevisiae*, gene conversion	+	0	540	Chanet *et al.* (1975)
SCH	*Saccharomyces cerevisiae*, homozygosis by mitotic recombination or gene conversion	+	0	18.5	Zimmermann & Mohr (1992)

Table 23 (contd)

Test system		Result[a]		Dose[b] (LED/HID)	Reference
		Without exogenous metabolic system	With exogenous metabolic system		
NCF	*Neurospora crassa*, forward mutation	+		100	de Serres et al. (1988)
NCR	*Neurospora crassa*, reverse mutation	-		732	Dickey et al. (1949)
NCR	*Neurospora crassa*, reverse mutation	+		37 500	Jensen et al. (1952)
NCR	*Neurospora crassa*, reverse mutation	-		300	Kølmark & Westergaard (1953)
PLM	Plants (other), mutation	+		0.0	Auerbach et al. (1977)
DMG	*Drosophila melanogaster*, genetic crossing over or recombination	+		2700	Ratnayake (1970)
DMG	*Drosophila melanogaster*, genetic crossing over or recombination	+		420	Alderson (1967)
DMG	*Drosophila melanogaster*, genetic crossing over or recombination	+		1260	Sobels & van Steenis (1957)
DMX	*Drosophila melanogaster*, sex-linked recessive lethal mutations	+		420	Alderson (1967)
DMX	*Drosophila melanogaster*, sex-linked recessive lethal mutations	(+)		1940	Ratnayake (1968)
DMX	*Drosophila melanogaster*, sex-linked recessive lethal mutations	+		2380	Ratnayake (1970)
DMX	*Drosophila melanogaster*, sex-linked recessive lethal mutations	+		1940	Auerbach & Moser (1953)
DMX	*Drosophila melanogaster*, sex-linked recessive lethal mutations	+		1080	Kaplan (1948)
DMX	*Drosophila melanogaster*, sex-linked recessive lethal mutations	+		420	Khan (1967)
DMX	*Drosophila melanogaster*, sex-linked recessive lethal mutations	+		270	Stumm-Tegethoff (1969)
DMX	*Drosophila melanogaster*, sex-linked recessive lethal mutations	+		1260	Sobels & van Steenis (1957)
DMH	*Drosophila melanogaster*, heritable translocation	+		2700	Ratnayake (1970)
DMH	*Drosophila melanogaster*, heritable translocation	+		420	Khan (1967)
DML	*Drosophila melanogaster*, dominant lethal mutation	+		1940	Auerbach & Moser (1953)
DML	*Drosophila melanogaster*, dominant lethal mutation	+		1400	Šrám (1970)
*	*Caenorhabditis elegans*, recessive lethal mutations	+		700	Johnsen & Baillie (1988)
DIA	DNA strand breaks, cross-links or related damage, animal cells *in vitro*	+	0	6	Ross & Shipley (1980)
DIA	DNA strand breaks, cross-links or related damage, animal cells *in vitro*	+	0	3.75	Ross et al. (1981)
DIA	DNA strand breaks, cross-links or related damage, animal cells *in vitro*	+	0	22.5	Demkowicz-Dobrzanski & Castonguay (1992)
DIA	DNA strand breaks, cross-links or related damage, animal cells *in vitro*	+	0	7.5	O'Connor & Fox (1987)
G9H	Gene mutation, Chinese hamster V79 cells, *hprt* locus	+	0	9	Grafström et al. (1993)
SIC	Sister chromatid exchange, Chinese hamster cells *in vitro*	+	0	1	Obe & Beek (1979)
SIC	Sister chromatid exchange, Chinese hamster cells *in vitro*	+	+	3.2	Natarajan et al. (1983)
SIC	Sister chromatid exchange, Chinese hamster cells *in vitro*	+	+	1.8	Basler et al. (1985)
CIC	Chromosomal aberrations, Chinese hamster cells *in vitro*	+	+	6.5	Natarajan et al. (1983)
CIC	Chromosomal aberrations, Chinese hamster cells *in vitro*	+	0	18	Ishidate et al. (1981)

FORMALDEHYDE

Table 23 (contd)

Test system		Result[a]		Dose[b] (LED/HID)	Reference
		Without exogenous metabolic system	With exogenous metabolic system		
TCM	Cell transformation, C3H10T1/2 mouse cells	+[d]	0	0.5	Ragan & Boreiko (1981)
DIH	DNA strand breaks, cross-links or related damage, human cells *in vitro*	+	0	24	Fornace et al. (1982)
DIH	DNA strand breaks, cross-links or related damage, human cells *in vitro*	+	0	1.5	Craft et al. (1987)
DIH	DNA strand breaks, cross-links or related damage, human cells *in vitro*	+	0	3	Grafström et al. (1986)
DIH	DNA strand breaks, cross-links or related damage, human cells *in vitro*	+	0	3	Snyder & Van Houten (1986)
DIH	DNA strand breaks, cross-links or related damage, human cells *in vitro*	+	0	3	Saladino et al. (1985)
DIH	DNA strand breaks, cross-links or related damage, human cells *in vitro*	+	0	3	Grafström et al. (1984)
DIH	DNA strand breaks, cross-links or related damage, human cells *in vitro*	+	0	12	Grafström (1990)
UIH	Unscheduled DNA synthesis, human bronchial epithelial cells *in vitro*	–	0	3 (> 0.1 mmol/L was lethal)	Doolittle et al. (1985)
GIH	Gene mutation, human cells *in vitro*	+	0	3	Grafström et al. (1985)
GIH	Gene mutation, human cells *in vitro*	+	0	3.9	Goldmacher & Thilly (1983)
GIH	Gene mutation, human cells *in vitro*	+	0	0.9	Craft et al. (1987)
GIH	Gene mutation, human cells *in vitro*	+	0	4.5	Crosby et al. (1988)
GIH	Gene mutation, human cells *in vitro*	+	0	4.5	Liber et al. (1989)
GIH	Gene mutation, human cells *in vitro*	+	0	3	Grafström (1990)
RIH	DNA repair exclusive of unscheduled DNA synthesis, human cells *in vitro*	+	0	6	Grafström et al. (1984)
SHL	Sister chromatid exchange, human lymphocytes *in vitro*	+	0	5.4	Obe & Beck (1979)
SHL	Sister chromatid exchange, human lymphocytes *in vitro*	+	0	5	Kreiger & Garry (1983)
SHL	Sister chromatid exchange, human lymphocytes *in vitro*	+	+	3.75	Schmid et al. (1986)
CHF	Chromosomal aberrations, human fibroblasts *in vitro*	+	0	60	Levy et al. (1983)
CHL	Chromosomal aberrations, human lymphocytes *in vitro*	+	0	10	Miretskaya & Shvartsman (1982)
CHL	Chromosomal aberrations, human lymphocytes *in vitro*	+	+	7.5	Schmid et al. (1986)
CHL	Chromosomal aberrations, human lymphocytes *in vitro*	+	0	3.75	Dresp & Bauchinger (1988)
DVA	DNA–protein cross-links, rat cells *in vivo*	(+)		1.5 inhal. 6 h	Casanova-Schmitz et al. (1984b)
DVA	DNA–protein cross-links, rat cells *in vivo*	+		1.5 inhal. 6 h	Lam et al. (1985)
DVA	DNA–protein cross-links, rat cells *in vivo*	+		0.25 inhal. 3 h	Heck et al. (1986)
DVA	DNA–protein cross-links, rat cells *in vivo*	+		0.25 inhal. 3 h	Casanova & Heck (1987)
DVA	DNA–protein cross-links, rat cells *in vivo*	+		0.08 inhal. 6 h	Casanova et al. (1989)
DVA	DNA–protein cross-links, rhesus monkey nasal turbinate cells *in vivo*	+		0.05 inhal. 6 h	Heck et al. (1989)

Table 23 (contd)

Test system		Result[a]		Dose[b] (LED/HID)	Reference
		Without exogenous metabolic system	With exogenous metabolic system		
DVA	DNA–protein cross-links, rhesus monkey nasal turbinate cells in vivo	+		0.05 inhal. 6 h	Casanova et al. (1991)
*	DNA–protein cross-links, rat tracheal implant cells in vivo	+		2 mg/ml instil.	Cosma et al. (1988)
SVA	Sister chromatid exchange, rat cells in vivo	–		3.9 inhal. 6 h/d × 5	Kligerman et al. (1984)
*	Micronucleus induction, newt (Pleurodeles waltl) in vivo	–		5 μg/ml, 12 d	Siboulet et al. (1984)
MVM	Micronucleus induction, mouse in vivo	–		25 ip × 1	Natarajan et al. (1983)
MVM	Micronucleus induction, mouse in vivo	–		30 ip × 1	Gocke et al. (1981)
MVR	Micronucleus induction, rat (gastrointestinal tract) in vivo	+		200 po × 1	Migliore et al. (1989)
CBA	Chromosomal aberrations, mouse bone-marrow cells in vivo	–		25 ip × 1	Natarajan et al. (1983)
CBA	Chromosomal aberrations, rat bone-marrow cells in vivo	+		0.07 inhal. 4 h/d, 4 months	Kitaeva et al. (1990)
CBA	Chromosomal aberrations, rat bone-marrow cells in vivo	–		3.9 inhal. 6 h/d × 5, 8 weeks	Dallas et al. (1992)
CLA	Chromosomal aberrations, rat leukocytes in vivo	–		3.9 inhal. 6 h/d × 5	Kligerman et al. (1984)
CCC	Chromosomal aberrations, mouse spermatocytes treated in vivo, spermatocytes observed	–		50 ip × 1	Fontignie-Houbrechts (1981)
CVA	Chromosomal aberrations, mouse spleen cells in vivo	–		25 ip × 1	Natarajan et al. (1983)
CVA	Chromosomal aberrations, rat pulmonary lavage cells in vivo	+		3.9 inhal. 6 h/d × 5	Dallas et al. (1992)
GVA	Gene mutation, rat cells in vivo (p53 point mutations in nasal carcinomas)	+		3.9 inhal. 6 h/d, 2 years	Recio et al. (1992)
MST	Mouse spot test	–		3.9 inhal. 6 h/d × 3	Jensen & Cohr (1983) [Abstract]
DLM	Dominant lethal mutation, mouse	(+)		50 ip × 1	Fontignie-Houbrechts (1981)
DLM	Dominant lethal mutation, mouse	–		20 ip × 1	Epstein et al. (1972)
DLR	Dominant lethal mutation, rat	(+)		0.2 inhal. 4 h/d × 120	Kitaeva et al. (1990)
DLM	Dominant lethal mutation, mouse	–		20 ip × 1	Epstein & Shafner (1968)
MVH	Micronucleus formation, human lymphocytes in vivo	(+)		0.06[c] inhal. 8-h TWA	Suruda et al. (1993)
MVH	Micronucleus formation, human cells (buccal epithelium) in vivo	+		0.06[c] inhal. 8-h TWA	Suruda et al. (1993)
MVH	Micronucleus formation, human cells (nasal epithelium) in vivo	–		0.06[c] inhal. 8-h TWA	Suruda et al. (1993)
MVH	Micronucleus formation, human cells (nasal epithelium) in vivo	+		0.06[c] inhal. 8-h TWA	Ballarin et al. (1992)
SLH	Sister chromatid exchange, human lymphocytes in vivo	–		0.5 inhal. 8-h TWA	Thomson et al. (1984)
SLH	Sister chromatid exchange, human lymphocytes in vivo	–		0.5 inhal. 8-h TWA	Bauchinger & Schmid (1985)
SLH	Sister chromatid exchange, human lymphocytes in vivo	+		0.2 inhal. 8-h TWA	Yager et al. (1986)
SLH	Sister chromatid exchange, human lymphocytes in vivo	–		0.06[c] inhal. 8-h TWA	Suruda et al. (1993)
CLH	Chromosomal aberrations, human lymphocytes in vivo	–		0.5 inhal. 8-h TWA	Thomson et al. (1984)

Table 23 (contd)

Test system		Result[a]		Dose[b] (LED/HID)	Reference
		Without exogenous metabolic system	With exogenous metabolic system		
CLH	Chromosomal aberrations, human lymphocytes *in vivo*	−		0.8 inhal. 8-h TWA	Fleig *et al.* (1982)
CLH	Chromosomal aberrations, human lymphocytes *in vivo*	+		0.5 inhal. 8-h TWA	Bauchinger & Schmid (1985)
CLH	Chromosomal aberrations, human lymphocytes *in vivo*	−		0.4 inhal.	Vargová *et al.* (1992)
SPR	Sperm morphology, rats *in vivo*	+		200 po × 1	Cassidy *et al.* (1983)
SPM	Sperm morphology, mice *in vivo*	−		100 po × 5	Ward *et al.* (1984)
SPH	Sperm morphology, humans *in vivo*	−		0.2 inhal. 8-h TWA	Ward *et al.* (1984)

*Not on profile

[a] +, positive; (+) weak positive; −, negative; 0, not tested; ?, inconclusive (variable response in several experiments within an adequate study)

[b] In-vitro tests, μg/ml; in-vivo tests, mg/kg bw

[c] Tested with S9 without co-factors

[d] Positive only in presence of 12-*O*-tetradecanoylphorbol 13-acetate (TPA)

[e] Based on a mean 8-h time-weighted average of 0.33 ppm (range, 0.1–0.96 ppm); peak exposures up to 6.6 ppm

melanogaster, administration of formaldehyde in the diet induced sex-linked recessive lethal mutations, dominant lethal effects, heritable translocations and crossing-over in spermatogonia. In a single study, it induced recessive lethal mutations in a nematode. It induced chromosomal aberrations, sister chromatid exchange, DNA strand breaks and DNA–protein cross-links in animal cells and, in single studies, gene mutation in Chinese hamster V79 cells and transformation of mouse C3H10T1/2 cells *in vitro*. It induced DNA–protein cross-links, chromosomal aberrations, sister chromatid exchange and gene mutation in human cells *in vitro*. Experiments in human and Chinese hamster lung cells indicate that formaldehyde can inhibit repair of DNA lesions caused by the agent itself or by other mutagens (Grafström, 1990; Grafström *et al.*, 1993).

While there is conflicting evidence that formaldehyde can induce chromosomal anomalies in the bone marrow of rodents exposed by inhalation *in vivo*, recent studies have shown that it induces cytogenetic damage in the cells of tissues that are more locally exposed, either by gavage or by inhalation. Thus, groups of five male Sprague–Dawley rats were given 200 mg/kg bw formaldehyde orally, were killed 16, 24 or 30 h after treatment and were examined for the induction of micronuclei and nuclear anomalies in cells of the gastrointestinal epithelium. The frequency of mitotic figures was used as an index of cell proliferation. Treated rats had significant (greater than five fold) increases in the frequency of micronucleated cells in the stomach, duodenum, ileum and colon; the stomach was the most sensitive, with a 20-fold increase in the frequency of micronucleated cells 30 h after treatment, and the colon the least sensitive. The frequency of nuclear anomalies was also significantly increased at these sites. These effects were observed in conjunction with signs of severe local irritation (Migliore *et al.*, 1989).

In the second experiment, male Sprague–Dawley rats were exposed to 0, 0.5, 3 or 15 ppm [0, 0.62, 3.7 or 18.5 mg/m^3] formaldehyde for 6 h per day on five days per week, for one and eight weeks. There was no significant increase in chromosomal abnormalities in the bone-marrow cells of formaldehyde-exposed rats relative to controls, but there was a significant increase in the frequency of chromosomal aberrations in pulmonary lavage cells (lung alveolar macrophages) from rats that inhaled 15 ppm formaldehyde. Aberrations, which were predominantly chromatid breaks, were seen in 7.6 and 9.2% of the scored pulmonary lavage cells from treated animals and in 3.5 and 4.8% of cells from controls, after one and eight weeks, respectively (Dallas *et al.*, 1992).

Assays for dominant lethal mutations in rodents *in vivo* gave inconclusive results. In single studies, formaldehyde induced sperm-head anomalies in rats but not in mice.

(c) Mutational spectra

The spectrum of mutations induced by formaldehyde was studied in human lymphoblasts *in vitro*, in *Escherichia coli* and in naked pSV2gpt plasmid DNA (Crosby *et al.*, 1988). Thirty TK6 X-linked *hprt*⁻ human lymphoblast colonies induced by eight treatments with 150 μmol/L formaldehyde were characterized by Southern blot analysis. Fourteen (47%) of these mutants had visible deletions of some or all of the X-linked *hprt* bands, indicating that formaldehyde can induce large losses of DNA in human TK6 lymphoblasts. The remainder of the mutants showed normal restriction patterns, which, according to the authors, probably consisted of point mutations or smaller insertions or deletions that were too small to detect by Southern blot analysis. In *E. coli*, the mutations induced by formaldehyde were characterized in the xanthine guanine phosphoribosyl transferase (*gpr*) gene. Exposure of *E. coli* to 4 mmol/L formaldehyde for 1 h induced large insertions (41%), large deletions (18%) and point mutations (41%). DNA sequencing revealed that most of the point mutations were transversions at GC base-pairs. In contrast, exposure of *E. coli* to 40 mmol/L formaldehyde for 1 h produced 92% point mutations, 62% of which were transitions at a single AT base-pair in the gene. Therefore, formaldehyde produced different genetic alterations in *E. coli* at different concentrations. When naked pSV2gpt plasmid DNA was exposed to 3.3 or 10 mmol/L formaldehyde and transformed into *E. coli*, most of the resulting mutations were frameshifts, again suggesting a different mechanism of mutation.

Sixteen of the 30 formaldehyde-induced human lymphoblast TK6 X-linked *hprt* mutants referred to above which were not attributable to deletion were examined by Southern blot, northern blot and DNA sequence analysis (Liber *et al.*, 1989). Of these, nine produced mRNA of normal size and amount, three produced mRNA of normal size but in reduced amounts and three produced no detectable mRNA. Sequence analyses of cDNA prepared from *hprt* mRNA were performed on one spontaneous and seven formaldehyde-induced mutants by normal northern blotting. The spontaneous mutant was caused by an AT→GC transition. Six of the formaldehyde-induced mutants were base substitutions, all of which occurred at AT base-pairs. There was an apparent hot spot, in that four of six independent mutants were AT→CG transversions at a specific site. The remaining mutant had lost exon 8.

Table 24. DNA sequence analysis of *p53* cDNA (polymerase chain reaction fragment D) from squamous-cell carcinomas of nasal passages induced in rats by formaldehyde

DNA sequence[a]	Mutation (codon)[b]	Equivalent human *p53* codon no.	Location in conserved region
$_{396}$C→A	TTC→TTA (132) phe→leu	134	II
$_{398}$G→T	TGC→TTC (133) cys→phe	135	II
$_{638}$G→T	AGC→ATC (213) ser→ile	215	
$_{812}$G→A	CGT→CAT (271) arg→his	273	V
$_{842}$G→C	CGG→CCG (281) arg→pro	283	V

From Recio *et al.* (1992)
[a]The A in the start codon is designated as base position 1.
[b]The start codon ATG is designated as codon 1.

DNA sequence analysis of polymerase chain reaction-amplified cDNA fragments containing the evolutionarily conserved regions II–V of the rat *p53* gene was used to examine *p53* mutations in 11 primary nasal squamous-cell carcinomas induced in rats by formaldehyde. The rats has been exposed by inhalation to 15 ppm [18.5 mg/m^3] formaldehyde for up to two years. Point mutations at GC base-pairs in the *p53* complementary DNA sequence were found in five of the tumours (Table 24). The authors pointed out that all five human counterparts of the mutated *p53* codons listed in the Table have been identified as mutants in a variety of human cancers; the CpG dinucleotide at codon 273 (codon 271 in the rat) is a mutational hot spot occurring in many human cancers (Recio *et al.*, 1992).

5. Summary of Data Reported and Evaluation

5.1 Exposure data

Formaldehyde is produced worldwide on a large scale by catalytic, vapour phase oxidation of methanol. Annual world production is about 12 million tonnes. It is used mainly in the production of phenolic, urea, melamine and acetal resins, which have wide use in the production of adhesives and binders for the wood, plastics, textiles, leather and related industries. Formaldehyde is also used extensively as an intermediate in the manufacture of industrial chemicals, such as 1,4-butanediol and 4,4′-diphenylmethane diisocyanate (for polyurethanes and particle-board), pentaerythritol (for surface coatings and explosives) and hexamethylene tetramine (for phenol–formaldehyde resins and explosives). Formaldehyde is used as such in aqueous solution (formalin) as a disinfectant and preservative in many applications.

Formaldehyde occurs as a natural product in most living systems and in the environment. Common nonoccupational sources of exposure include vehicle emissions, some building materials, food, tobacco smoke and its use as a disinfectant. Levels of formaldehyde in outdoor air are generally below 0.001 mg/m^3 in remote areas and below 0.02 mg/m^3 in urban settings. The levels of formaldehyde in the indoor air of houses are typically 0.02–0.06 mg/m^3; average levels of 0.5 mg/m^3 or more have been measured in 'mobile homes' constructed with particle-board or in houses with urea–formaldehyde insulation, but the levels have declined in recent years as a result of changes in building materials.

It is estimated that several million people are exposed occupationally to formaldehyde in industrialized countries alone. The highest continuous exposures (frequently > 1 mg/m^3) have been measured in particle-board mills, during the varnishing of furniture and wooden floors, in foundries, during the finishing of textiles and in fur processing. Short-term exposures to much higher levels have been reported occasionally. Exposure to more than 1 mg/m^3 also occurs in some facilities where resins, plastics and special papers are produced. The average formaldehyde level measured in plywood mills and in embalming establishments is about 1 mg/m^3. Lower levels are encountered, for example, during the manufacture of garments, man-made mineral fibres, abrasives and rubber. Periodic occupational exposure occurs e.g. during disinfection in hospitals and in food processing plants, in some agricultural operations and during firefighting. The development of resins that release less formaldehyde and improved ventilation have resulted in decreased exposure levels in many occupational settings, such as particle-board, plywood and textile mills and foundries.

The exposures that may occur concomitantly with formaldehyde in occupational settings vary by industry, facility and period. They include other components of formaldehyde-based glues and varnishes, solvents, wood dust, wood preservatives and textile finishing agents.

5.2 Human carcinogenicity data

Excess numbers of nasopharyngeal cancers were associated with occupational exposure to formaldehyde in two of six cohort studies of industrial or professional groups, in three of four case–control studies and in meta-analyses. In one cohort study performed in 10 plants in the United States, the risk increased with category of increasing cumulative exposure. In the cohort studies that found no excess risk, no deaths were observed from nasopharyngeal cancer. In three of the case–control studies, the risk was highest in people in the highest category of exposure and among people exposed 20–25 years before death. The meta-analyses found a significantly higher risk among people estimated to have had substantial exposure than among those with low/medium or no exposure. The observed associations between exposure to formaldehyde and risk for cancer cannot reasonably be attributed to other occupational agents, including wood dust, or to tobacco smoking. Limitations of the studies include misclassification of exposure and disease and loss to follow-up, but these would tend to diminish the estimated relative risks and dilute exposure–response gradients. Taken together, the epidemiological studies suggest a causal relationship between exposure to formaldehyde and nasopharyngeal cancer, although the conclusion is tempered by the small numbers of observed and expected cases in the cohort studies.

Of the six case–control studies in which the risk for cancer of the nasal cavities and paranasal sinuses in relation to occupational exposure to formaldehyde was evaluated, three provided data on squamous-cell tumours and three on unspecified cell types. Of the three studies of squamous-cell carcinomas, two (from Denmark and the Netherlands) showed a positive association, after adjustment for exposure to wood dust, and one (from France) showed no association. Of the three studies of unspecified cell types, one (from Connecticut, United States) gave weakly positive results and two (also from the United States) reported no excess risk. The two case–control studies that considered squamous-cell tumours and gave positive results involved more exposed cases than the other case–control studies combined. In the studies of occupational cohorts overall, however, fewer cases of cancer of the nasal cavities and paranasal sinuses were observed than were expected. Because of the lack of consistency between the cohort and case–control studies, the epidemiological studies can do no more than suggest a causal role of occupational exposure to formaldehyde in squamous-cell carcinoma of the nasal cavities and paranasal sinuses.

Less information was available to evaluate the association of formaldehyde with adenocarcinoma of the nasal cavities and paranasal sinuses, and the small excess observed in one case–control study in Denmark may have been confounded by exposure to wood dust.

Neither cohort nor case–control studies showed excess risks for oropharyngeal, laryngeal or lung cancer among workers exposed to formaldehyde. The studies of industrial cohorts also showed low or no risk for lymphatic or haematopoietic cancers; however, the cohort studies of embalmers, anatomists and other professionals who use formaldehyde tended to show excess risks for cancers of the brain, although they were based on small numbers. These findings are countered by a consistent lack of excess risk for brain cancer in the studies of industrial cohorts, which generally included more direct and quantitative estimates of exposure to formaldehyde than did the cohort studies of embalmers and anatomists.

5.3 Animal carcinogenicity data

Formaldehyde was tested for carcinogenicity by inhalation in mice, rats and hamsters, by oral administration in drinking-water in rats, by skin application in mice, and by subcutaneous injection in rats. In additional studies in mice, rats and hamsters, modification of the carcinogenicity of known carcinogens was tested by administration of formaldehyde in drinking-water, by application on the skin or by inhalation.

Several studies in which formaldehyde was administered to rats by inhalation showed evidence of carcinogenicity, particularly induction of squamous-cell carcinomas of the nasal cavities, usually only at the highest exposure. Similar studies in hamsters showed no evidence of carcinogenicity. Studies in mice either showed no effect or were inadequate for evaluation. In rats administered formaldehyde in the drinking-water, increased incidences were seen of forestomach papillomas in one study and of leukaemias and gastrointestinal tract tumours in another; two other studies in which rats were treated in the drinking-water gave negative results. Studies in which formaldehyde was applied to the skin or injected subcutaneously were inadequate for evaluation.

In experiments to test the effect of formaldehyde on the carcinogenicity of known carcinogens, oral administration of formaldehyde concomitantly with *N*-nitrosodimethylamine to mice increased the incidence of tumours at various sites; skin application in addition to 7,12-dimethylbenz[*a*]anthracene reduced the latency of skin tumours. In rats, concomitant administration of formaldehyde and *N*-methyl-*N*'-nitro-*N*-nitrosoguanidine in the drinking-water increased the incidence of adenocarcinoma of the glandular stomach. Exposure of hamsters by inhalation to formaldehyde increased the multiplicity of tracheal tumours induced by subcutaneous injections of *N*-nitrosodiethylamine.

5.4 Other relevant data

The concentration of endogenous formaldehyde in human blood is about 2–3 mg/L; similar concentrations are found in the blood of monkeys and rats. Exposure of humans, monkeys or rats to formaldehyde by inhalation does not alter the concentration of formaldehyde in the blood.

Occupational exposure to formaldehyde results in damage to nasal tissues; however, these findings may have been confounded by concomitant exposures. No data were available on the induction of cell proliferation in humans. There are no conclusive data showing that formaldehyde is toxic to the immune system, to the reproductive system or to developing fetuses in humans.

More than 90% of inhaled formaldehyde gas is absorbed in the upper respiratory tract of rats and monkeys. In rats, it is absorbed in the nasal passages; in monkeys, it is also absorbed in the nasopharynx, trachea and proximal regions of the major bronchi. In mice exposed to high concentrations of formaldehyde, minute ventilation is decreased by 50% throughout exposure, resulting in a lower effective dose. This occurs only transiently in rats, as the minute ventilation is rapidly restored. Formaldehyde is rapidly oxidized to formate, which is incorporated into biological macromolecules, excreted in the urine or oxidized to carbon dioxide.

Acute or subacute exposure of rats to a concentration of 2.5 mg/m^3 appears to cause no detectable damage to the nasal epithelium and does not significantly increase rates of cell turnover. Cell turnover rates in rat nose during subchronic or chronic exposures to formaldehyde do not increase at 2.5 mg/m^3, increase marginally at concentrations of 3.7–7.4 mg/m^3 and increase substantially at concentrations of 12.3–18.4 mg/m^3. Concentration is more important than length of exposure in determining the cytotoxicity of formaldehyde.

Inhalation of formaldehyde leads to the formation of DNA–protein cross-links in the nasal respiratory mucosa of rats and monkeys. Much lower levels of DNA–protein cross-links were found in the nasopharynx, trachea and carina of some monkeys, in decreasing concentrations with passage through the respiratory tract, but none were found in the maxillary sinus. The formation of DNA–protein cross-links is a sublinear function of the formaldehyde concentration in inhaled air from 0.86 to 18.4 mg/m^3, and the yield of DNA–protein cross-links at a given inhaled concentration is approximately an order of magnitude lower in monkeys than in rats. Yields of DNA–protein cross-links are higher in the lateral meatus of the rat nose and lower in the medial and posterior meatuses. There is no detectable accumulation of DNA–protein cross-links during repeated exposure.

About 50% of formaldehyde-induced tumours in the nasal mucosa of rats have a point mutation in the *p53* tumour suppressor gene.

No adequate data were available on genetic effects of formaldehyde in humans. It is comprehensively genotoxic in a variety of experimental systems, ranging from bacteria to rodents, *in vivo*. Formaldehyde given by inhalation or gavage to rats *in vivo* induced chromosomal anomalies in lung cells, micronuclei in the gastrointestinal tract and sperm-head anomalies.

Formaldehyde induced DNA–protein cross-links, DNA single-strand breaks, chromosomal aberrations, sister chromatid exchange and gene mutation in human cells *in vitro*. It induced cell transformation, chromosomal aberrations, sister chromatid exchange, DNA strand breaks, DNA–protein cross-links and gene mutation in rodent cells *in vitro*.

Administration of formaldehyde in the diet to *Drosophila melanogaster* induced lethal and visible mutations, deficiencies, duplications, inversions, translocations and crossing-over in spermatogonia. Formaldehyde induced mutation, gene conversion, DNA strand breaks and DNA–protein cross-links in fungi and mutation and DNA damage in bacteria.

In rodents and monkeys, there is a no-observable-effect level (2.5 mg/m^3) of inhaled formaldehyde with respect to cell proliferation and tissue damage in otherwise undamaged nasal mucosa. These effects are considered to contribute to subsequent development of cancer. Although these findings provide a basis for extrapolation to humans, conclusive data demonstrating that such cellular and biochemical changes occur in humans exposed to formaldehyde are not available.

5.5 Evaluation[1]

There is *limited evidence* in humans for the carcinogenicity of formaldehyde.

There is *sufficient evidence* in experimental animals for the carcinogenicity of formaldehyde.

Overall evaluation

Formaldehyde *is probably carcinogenic to humans (Group 2A)*.

6. References

Acheson, E.D., Barnes, H.R., Gardner, M.J., Osmond, C., Pannett, B. & Taylor, C.P. (1984a) Formaldehyde in the British chemical industry. *Lancet*, **i**, 611–616

Acheson, E.D., Barnes, H.R., Gardner, M.J., Osmond, C., Pannett, B. & Taylor, C.P. (1984b) Cohort study of formaldehyde process workers. *Lancet*, **ii**, 403

Acheson, E.D., Barnes, H.R., Gardner, M.J., Osmond, C., Pannett, B. & Taylor, C.P. (1984c) Formaldehyde process workers and lung cancer. *Lancet*, **i**, 1066–1067

Adams, D.O., Hamilton, T.A., Lauer, L.D. & Dean, J.H. (1987) The effect of formaldehyde exposure upon the mononuclear phagocyte system of mice. *Toxicol. appl. Pharmacol.*, **88**, 165–174

Åhman, M., Alexandersson, R., Ekholm, U., Bergström, B., Dahlqvist, M. & Ulfvarsson, U. (1991) Impeded lung function in moulders and coremakers handling furan resin sand. *Int. Arch. occup. environ. Health*, **63**, 175–180

Albert, R.E., Sellakumar, A.R., Laskin, S., Kuschner, M., Nelson, N. & Snyder, C.A. (1982) Gaseous formaldehyde and hydrogen chloride induction of nasal cancer in the rat. *J. natl Cancer Inst.*, **68**, 597–603

Alderson, T. (1967) Induction of genetically recombinant chromosomes in the absence of induced mutation. *Nature*, **215**, 1281–1283

Alexandersson, R. & Hedenstierna, G. (1988) Respiratory hazards associated with exposure to formaldehyde and solvents in acid-curing paints. *Arch. environ. Health*, **43**, 222–227

Alexandersson, R. & Hedenstierna, G. (1989) Pulmonary function in wood workers exposed to formaldehyde: a prospective study. *Arch. environ. Health.*, **44**, 5–11

American Conference of Governmental Industrial Hygienists (1991) *Documentation of the Threshold Limit Values and Biological Exposure Indices*, 6th Ed., Vol. 2, Cincinnati, OH, pp. 664–688

---

[1] For definitions of the italicized terms, see Preamble, pp. 23–27.

American Conference of Governmental Industrial Hygienists (1993) *1993–1994 Threshold Limit Values for Chemical Substances and Physical Agents and Biological Exposure Indices*, Cincinnati, OH, p. 22

American Society for Testing and Materials (1990) *Standard Test Method for Determining Formaldehyde Levels from Wood Products Under Defined Test Conditions Using a Large Chamber* (Method E 1333-90), Philadelphia, PA

Andersen, I. & Mølhave, L. (1983) Controlled human studies with formaldehyde. In: Gibson, J.E., ed., *Formaldehyde Toxicity*, Washington DC, Hemisphere, pp. 154–165

Andersen, S.K., Jensen, O.M. & Oliva, D. (1982) Formaldehyde exposure and lung cancer in Danish doctors. *Ugeskr. Læg.*, **144**, 1571–1573 (in Danish)

Andjelkovich, D.A., Janszen, D.B., Brown, M.H., Richardson, R.B. & Miller, F.J. (1995) Mortality of iron foundry workers. IV. Analysis of a subcohort exposed to formaldehyde. *J. occup. Med.* (in press)

Anon. (1985) Facts & figures for the chemical industry. *Chem. Eng. News*, **63**, 22–86

Anon. (1989) Facts & figures for the chemical industry. *Chem. Eng. News*, **67**, 36–90

Anon. (1992) Chemical profile: formaldehyde. *Chem. Mark. Rep.*, **242**, 24,42

Anon. (1993) Facts & figures for the chemical industry. *Chem. Eng. News*, **71**, 38–83

Anon. (1994) Facts & figures for the chemical industry. *Chem. Eng. News*, **72**, 28–74

Appelman, L.M., Woutersen, R.A., Zwart, A., Falke, H.E. & Feron, V.J. (1988) One-year inhalation toxicity study of formaldehyde in male rats with a damaged or undamaged nasal mucosa. *J. appl. Toxicol.*, **8**, 85–90

Arbeidsinspectie [Labour Inspection] (1986) *De Nationale MAC-Lijst 1986* [National MAC list 1986], Voorburg, p. 13

Auerbach, C. & Moser, H. (1953) Analysis of the mutagenic action of formaldehyde on food. II. The mutagenic potentialities of the treatment. *Zeitschr. indukt. Abstamm. Verebungs.*, **85**, 547–563

Auerbach, C., Moutschen-Dahmen, M. & Moutschen, J. (1977) Genetic and cytogenetical effects of formaldehyde and related compounds. *Mutat. Res.*, **39**, 317–361

Axelson, O. & Steenland, K. (1988) Indirect methods of assessing the effects of tobacco use in occupational studies. *Am. J. ind. Hyg.*, **13**, 105–118

Ballarin, C., Sarto, F., Giacomelli, L., Battista Bartolucci, G. & Clonfero, E. (1992) Micronucleated cells in nasal mucosa of formaldehyde-exposed workers. *Mutat. Res.*, **280**, 1–7

Bardana, E.J., Jr & Montanaro, A. (1991) Formaldehyde: an analysis of its respiratory, cutaneous, and immunologic effects. *Ann. Allergy*, **66**, 441–452

Basler, A., van der Hude, H. & Scheutwinkel-Reich, M. (1985) Formaldehyde-induced sister chromatid exchanges *in vitro* and the influence of the exogenous metabolizing systems S9 mix and primary hepatocytes. *Arch. Toxicol.*, **58**, 10–13

Bauchinger, M. & Schmid, E. (1985) Cytogenetic effects in lymphocytes of formaldehyde workers of a paper factory. *Mutat. Res.*, **158**, 195–199

Beavers, J.D. (1989) Formaldehyde exposure reports (Letter to the Editor). *Am. J. ind. Med.*, **16**, 331–332

Belanger, P.L. & Kilburn, K.H. (1981) *California Society for Histotechnology, Los Angeles, CA, Health Hazard Evaluation Report* (NIOSH Report No. HETA 81-422-1387), Cincinnati, OH, US

Department of Health and Human Services, Public Health Service, Centers for Disease Control, National Institute for Occupational Safety and Health

Bender, J.R., Mullin, L.S., Graepel, G.J. & Wilson, W.E. (1983) Eye irritation response of humans to formaldehyde. *Am. ind. Hyg. Assoc. J.*, **44**, 463–465

Berglund, B. & Nordin, S. (1992) Detectability and perceived intensity for formaldehyde in smokers and non-smokers. *Chem. Senses*, **17**, 291–306

Berke, J.H. (1987) Cytologic examination of the nasal mucosa in formaldehyde-exposed workers. *J. occup. Med.*, **29**, 681–684

Bernardini, P., Carelli, G., Rimatori, V. & Contegiacomo, P. (1983) Health hazard for hospital workers from exposure to formaldehyde. *Med. Lav.*, **74**, 106–110

Bertazzi, P.A., Pesatori, A.C., Radice, L., Zocchetti, C. & Vai, T. (1986) Exposure to formaldehyde and cancer mortality in a cohort of workers producing resins. *Scand. J. Work Environ. Health*, **12**, 461–468

Bertazzi, P.A., Pesatori, A.C., Guercilena, S., Consonni, D. & Zocchetti, C. (1989) Cancer risk among workers producing formaldehyde-based resins: extension of follow-up. *Med. Lav.*, **80**, 111–122

Biagini, R.E., Moorman, W.J., Knecht, E.A., Clark, J.C. & Bernstein, I.L. (1989) Acute airway narrowing in monkeys from challenge with 2.5 ppm formaldehyde generated from formalin. *Arch. environ. Health*, **44**, 12–17

Bicking, M.K.L., Cooke, W.M., Kawahara, F.K. & Longbottom, J.E. (1988) *Method Development for the Determination of Formaldehyde in Samples of Environmental Origin* (ASTM Spec. tech. Publ. 976), Philadelphia, American Society for Testing and Materials, pp. 159–175

Binding, N. & Witting, U. (1990) Exposure to formaldehyde and glutardialdehyde in operating theatres. *Int. Arch. occup. environ. Health*, **62**, 233–238

Blade, L.M. (1983) Occupational exposure to formaldehyde—recent NIOSH involvement. In: Clary, J.J., Gibson, J.E. & Waritz, R.S., eds, *Formaldehyde—Toxicology, Epidemiology, Mechanisms*, New York, Marcel Dekker, pp. 1–23

Blair, A. & Stewart, P.A. (1989) Comments on the reanalysis of the National Cancer Institute study of workers exposed to formaldehyde. *J. occup. Med.*, **31**, 881

Blair, A. & Stewart, P.A. (1994) Misclassification of nasopharyngeal cancer—reply. *J. natl Cancer Inst.*, **86**, 1557–1558

Blair, A., Stewart, P.A., O'Berg, M., Gaffey, W., Walrath, J., Ward, J., Bales, R., Kaplan, S. & Cubit, D. (1986) Mortality among industrial workers exposed to formaldehyde. *J. natl Cancer Inst.*, **76**, 1071–1084

Blair, A., Stewart, P.A., Hoover, R.N., Fraumeni, J.F., Jr, Walrath, J., O'Berg, M. & Gaffey, W. (1987) Cancers of the nasopharynx and oropharynx and formaldehyde exposure. *J. natl Cancer Inst.*, **78**, 191–192

Blair, A., Steenland, K., Shy, C., O'Berg, M., Halperin, W. & Thomas, T. (1988) Control of smoking in occupational epidemiologic studies: methods and needs. *Am. J. ind. Med.*, **13**, 3–4

Blair, A., Saracci, R., Stewart, P.A., Hayes, R.B. & Shy, C. (1990a) Epidemiologic evidence on the relationship between formaldehyde exposure and cancer. *Scand. J. Work Environ. Health*, **16**, 381–393

Blair, A., Stewart, P.A. & Hoover, R.N. (1990b) Mortality from lung cancer among workers employed in formaldehyde industries. *Am. J. ind. Med.*, **17**, 683–699

Bogdanffy, M.S., Randall, H.W. & Morgan, K.T. (1986) Histochemical localization of aldehyde dehydrogenase in the respiratory tract of the Fischer-344 rat. *Toxicol. appl. Pharmacol.*, **82**, 560–567

Bogdanffy, M.S., Morgan, P.H., Starr, T.B. & Morgan, K.T. (1987) Binding of formaldehyde to human and rat nasal mucus and bovine serum albumin. *Toxicol. Lett.*, **38**, 145–154

Bond, G.G., Flores, G.H., Shellenberger, R.J., Cartmill, J.B., Fishbeck, W.A. & Cook, R.R. (1986) Nested case–control study of lung cancer among chemical workers. *Am. J. Epidemiol.*, **124**, 53–66

Boysen, M., Zadig, E., Digernes, V., Abeler, V. & Reith, A. (1990) Nasal mucosa in workers exposed to formaldehyde: a pilot study. *Br. J. ind. Med.*, **47**, 116–121

Bracken, M.J., Leasa, D.J. & Morgan, W.K.C. (1985) Exposure to formaldehyde: relationship to respiratory symptoms and function. *Can. J. public Health*, **76**, 312–316

Brandwein, M., Pervez, N. & Biller, H. (1987) Nasal squamous carcinoma in an undertaker—Does formaldehyde play a role? *Rhinology*, **25**, 279–284

Brinton, L.A., Blot, W.J., Becker, J.A., Winn, D.M., Browder, J.P., Farmer, J.C., Jr & Fraumeni, J.F., Jr (1984) A case–control study of cancers of the nasal cavity and paranasal sinuses. *Am. J. Epidemiol.*, **119**, 896–906

British Standards Institution (1989) *22. Determination of Extractable Formaldehyde* (BS 5669:1989), London, pp. 31–37

Broder, I., Corey, P., Brasher, P., Lipa, M. & Cole, P. (1988a) Comparison of health of occupants and characteristics of houses among control homes and homes insulated with urea formaldehyde foam. III. Health and house variables following remedial work. *Environ. Res.*, **45**, 179–203

Broder, I., Corey, P., Cole, P., Lipa, M., Mintz, S. & Nethercott, J.R. (1988b) Comparison of health of occupants and characteristics of houses among control homes and homes insulated with urea formaldehyde foam. I. Methodology. *Environ. Res.*, **45**, 141–155

Broder, I., Corey, P., Cole, P., Lipa, M., Mintz, S. & Nethercott, J.R. (1988c) Comparison of health of occupants and characteristics of houses among control homes and homes insulated with urea formaldehyde foam. II. Initial health and house variables and exposure–response relationships. *Environ. Res.*, **45**, 156–178

Broder, I., Corey, P., Brasher, P., Lipa, M. & Cole, P. (1991) Formaldehyde exposure and health status in households. *Environ. Health Perspectives*, **95**, 101–104

Buckley, L.A., Jiang, X.Z., James, R.A., Morgan, K.T. & Barrow, C.S. (1984) Respiratory tract lesions induced by sensory irritants at the RD50 concentration. *Toxicol. appl. Pharmacol.*, **74**, 417–429

Buss, J., Kuschinsky, K., Kewitz, H. & Koransky, W. (1964) Enteric resorption of formaldehyde. *Naunyn-Schmiedeberg's Arch. exp. Pathol. Pharmakol.*, **247**, 380–381 (in German)

Casanova, M. & Heck, H.d'A. (1987) Further studies of the metabolic incorporation and covalent binding of inhaled [^{3}H]- and [^{14}C]formaldehyde in Fischer-344 rats: effects of glutathione depletion. *Toxicol. appl. Pharmacol.*, **89**, 105–121

Casanova, M., Heck, H.d'A., Everitt, J.I., Harrington, W.W., Jr & Popp, J.A. (1988) Formaldehyde concentrations in the blood of rhesus monkeys after inhalation exposure. *Food chem. Toxicol.*, **26**, 715–716

Casanova, M., Deyo, D.F. & Heck, H.d'A. (1989) Covalent binding of inhaled formaldehyde to DNA in the nasal mucosa of Fischer 344 rats: analysis of formaldehyde and DNA by high-performance liquid chromatography and provisional pharmacokinetic interpretation. *Fundam. appl. Toxicol.*, **12**, 397–417

Casanova, M., Morgan, K.T., Steinhagen, W.H., Everitt, J.I., Popp, J.A. & Heck, H.d'A. (1991) Covalent binding of inhaled formaldehyde to DNA in the respiratory tract of rhesus monkeys: pharmacokinetics, rat-to-monkey interspecies scaling, and extrapolation to man. *Fundam. appl. Toxicol.*, **17**, 409–428

Casanova, M., Morgan, K.T., Gross, E.A., Moss, O.R. & Heck, H.d'A. (1995) DNA–protein cross-links and cell replication at specific sites in the nose of F344 rats exposed subchronically to formaldehyde. *Fundam. appl. Toxicol.* (in press)

Casanova-Schmitz, M. & Heck, H.d'A. (1983) Effects of formaldehyde exposure on the extractability of DNA from proteins in the rat nasal mucosa. *Toxicol. appl. Pharmacol.*, **70**, 121–132

Casanova-Schmitz, M., David, R.M. & Heck, H.d'A. (1984a) Oxidation of formaldehyde and acetaldehyde by NAD^+-dependent dehydrogenases in rat nasal mucosal homogenates. *Biochem. Pharmacol.*, **33**, 1137–1142

Casanova-Schmitz, M., Starr, T.B. & Heck, H.d'A. (1984b) Differentiation between metabolic incorporation and covalent binding in the labeling of macromolecules in the rat nasal mucosa and bone marrow by inhaled [^{14}C]- and [^{3}H]formaldehyde. *Toxicol. appl. Pharmacol.*, **76**, 26–44

Cassidy, S.L., Dix, K.M. & Jenkins, T. (1983) Evaluation of a testicular sperm head counting technique using rats exposed to dimethoxyethyl phthalate (DMEP), glycerol α-monochlorohydrin (GMCH), epichlorohydrin (ECH), formaldehyde (FA), or methyl methanesulphonate (MMS). *Arch. Toxicol.*, **53**, 71–78

Chanet, R. & von Borstel, R.C. (1979) Genetic effects of formaldehyde in yeast. III. Nuclear and cytoplasmic mutagenic effects. *Mutat. Res.*, **62**, 239–253

Chanet, R., Izard, C. & Moustacchi, E. (1975) Genetic effects of formaldehyde in yeast. I. Influence of the growth stages on killing and recombination. *Mutat. Res.*, **33**, 179–186

Chang, J.C.F. & Barrow, C.S. (1984) Sensory irritation tolerance and cross-tolerance in F-344 rats exposed to chlorine or formaldehyde gas. *Toxicol. appl. Pharmacol.*, **76**, 319–327

Chang, J.C.F., Steinhagen, W.H. & Barrow, C.S. (1981) Effect of single or repeated formaldehyde exposure on minute volume of B6C3F1 mice and F-344 rats. *Toxicol. appl. Pharmacol.*, **61**, 451–459

Chang, J.C.F., Gross, E.A., Swenberg, J.A. & Barrow, C.S. (1983) Nasal cavity deposition, histopathology and cell proliferation after single or repeated formaldehyde exposures in B6C3F1 mice and F-344 rats. *Toxicol. appl. Pharmacol.*, **68**, 161–176

Chemical Information Services Ltd (1991) *Directory of World Chemical Producers 1992/93 Edition*, Dallas, TX, pp. 296, 455, 584

China National Chemical Information Centre (1993) *World Chemical Industry Yearbook, 1993: China Chemical Industry*, English Ed., Beijing, p. 172

Coggon, D., Pannett, B. & Acheson, E.D. (1984) Use of job–exposure matrix in an occupational analysis of lung and bladder cancers on the basis of death certificates. *J. natl Cancer Inst.*, **72**, 61–65

Coldiron, V.R., Ward, J.B., Jr, Trieff, N.M., Janssen, E., Jr & Smith, J.H. (1983) Occupational exposure to formaldehyde in a medical center autopsy service. *J. occup. Med.*, **25**, 544–548

Collins, J.J., Caporossi, J.C. & Utidjian, H.M.D. (1988) Formaldehyde exposure and nasopharyngeal cancer: re-examination of the National Cancer Institute study and an update of one plant. *J. natl Cancer Inst.*, **80**, 376–377

Connor, T.H., Barrie, M.D., Theiss, J.C., Matney, T.S. & Ward, J.B., Jr (1983) Mutagenicity of formalin in the Ames assay. *Mutat. Res.*, **119**, 145-149

Connor, T.H., Theiss, J.C., Hanna, H.A., Monteith, D.K. & Matney, T.S. (1985) Genotoxicity of organic chemicals frequently found in the air of mobile homes. *Toxicol. Lett.*, **25**, 33–40

Cook, W.A. (1987) *Occupational Exposure Limits—Worldwide*, Akron, OH, American Industrial Hygiene Association, pp. 121, 141, 190

Cosma, G.N., Wilhite, A.S. & Marchok, A.C. (1988) The detection of DNA–protein cross-links in rat tracheal implants exposed *in vivo* to benzo[*a*]pyrene and formaldehyde. *Cancer Lett.*, **42**, 13–21

Cosmetic Ingredient Review Expert Panel (1984) Final report on the safety assessment of formaldehyde. *J. Am. Coll. Toxicol.*, **3**, 157–184

Coutrot, D. (1986) European formaldehyde regulations: a French view. In: Meyer, B., Kottes Andrews, B.A. & Reinhardt, R.M., eds, *Formaldehyde Release from Wood Products* (ACS Symp. Ser., 316), Washington DC, American Chemical Society, pp. 209–216

Craft, T.R., Bermudez, E. & Skopek, T.R. (1987) Formaldehyde mutagenesis and formation of DNA–protein crosslinks in human lymphoblasts *in vitro*. *Mutat. Res.*, **176**, 147–155

Cronin, E. (1991) Formaldehyde is a significant allergen in women with hand eczema. *Contact Derm.*, **25**, 276–282

Crosby, R.M., Richardson, K.K., Craft, T.R., Benforado, K.B., Liber, H.L. & Skopek, T.R. (1988) Molecular analysis of formaldehyde-induced mutations in human lymphoblasts and *E. coli*. *Environ. mol. Mutag.*, **12**, 155–166

Dalbey, W.E. (1982) Formaldehyde and tumors in hamster respiratory tract. *Toxicology*, **24**, 9–14

Dallas, C.E., Theiss, J.C., Harrist, R.B. & Fairchild, E.J. (1986) Respiratory responses in the lower respiratory tract of Sprague–Dawley rats to formaldehyde inhalation. *J. environ. Pathol. Toxicol.*, **6**, 1–12

Dallas, C.E., Scott, M.J., Ward, J.B., Jr & Theiss, J.C. (1992) Cytogenetic analysis of pulmonary lavage and bone marrow cells of rats after repeated formaldehyde inhalation. *J. appl. Toxicol.*, **12**, 199–203

Danielsson, O. & Jörnvall, H. (1992) 'Enzymogenesis': classical liver alcohol dehydrogenase origin from the glutathione-dependent formaldehyde dehydrogenase line. *Proc. natl Acad. Sci. USA*, **89**, 9247–9251

Data-Star/Dialog (1994) *CHEM-INTELL Trade and Production Statistics (PLST) Database*, Bern

Day, J.H., Lees, R.E.M., Clark, R.H. & Pattee, P.L. (1984) Respiratory response to formaldehyde and off-gas of urea formaldehyde foam insulation. *Can. med. Assoc. J.*, **131**, 1061–1065

Dean, J.H., Lauer, L.D., House, R.V., Murray, M.J., Stillman, W.S., Irons, R.D., Steinhagen, W.H., Phelps, M.C. & Adams, D.O. (1984) Studies of immune function and host resistance in B6C3F1 mice exposed to formaldehyde. *Toxicol. appl. Pharmacol.*, **72**, 519–529

Demerec, M., Bertani, G. & Flint, J. (1951) A survey of chemicals for mutagenic action on E. coli. *Am. Naturalist*, **85**, 119–136

Demkowicz-Dobrzanski, K. & Castonguay, A. (1992) Modulation by glutathione of DNA strand breaks induced by 4-(methylnitrosamino)-1-(3-pyridyl)-1-butanone and its aldehyde metabolites in rat hepatocytes. *Carcinogenesis*, **13**, 1447–1454

Deutsche Forschungsgemeinschaft (1993) *MAK- und BAT-Werte-Liste 1993* [MAK- and BAT-Values 1993] (Report No. 29), Weinheim, VCH Verlagsgesellschaft, p. 48

Dicker, E. & Cederbaum, A.I. (1984) Effect of acetaldehyde and cyanamide on the metabolism of formaldehyde by hepatocytes, mitochondria, and soluble supernatant from rat liver. *Arch. Biochem. Biophys.*, **232**, 179–188

Dicker, E. & Cederbaum, A.I. (1985) Inhibition of mitochondrial aldehyde dehydrogenase and acetaldehyde oxidation by the glutathione-depleting agents diethylmaleate and phorone. *Biochim. biophys. Acta*, **843**, 107–113

Dicker, E. & Cederbaum, A.I. (1986) Inhibition of the low-K_m mitochondrial aldehyde dehydrogenase by diethyl maleate and phorone *in vivo* and *in vitro*. *Biochem. J.*, **240**, 821–827

Dickey, F.H., Cleland, G.H. & Lotz, C. (1949) The role of organic peroxides in the induction of mutations. *Proc. natl Acad. Sci. USA*, **35**, 581–586

Dinsdale, D., Riley, R.A. & Verschoyle, R.D. (1993) Pulmonary cytochrome P450 in rats exposed to formaldehyde vapor. *Environ. Res.*, **62**, 19–27

Direktoratet for Arbeidstilsynet [Directorate of Labour Inspection] (1990) *Administrative Normer for Forurensning i Arbeidsatmosfaere* [Administrative Norms for Pollution in the Work Atmosphere], Oslo, p. 11

Doolittle, D.J., Furlong, J.W. & Butterworth, B.E. (1985) Assessment of chemically induced DNA repair in primary cultures of human bronchial epithelial cells. *Toxicol. appl. Pharmacol.*, **79**, 28–38

Dresp, J. & Bauchinger, M. (1988) Direct analysis of the clastogenic effect of formaldehyde in unstimulated human lymphocytes by means of the premature chromosome condensation technique. *Mutat. Res.*, **204**, 349–352

Dykewicz, M.S., Patterson, R., Cugell, D.W., Harris, K.E. & Wu, A.F. (1991) Serum IgE and IgG to formaldehyde–human serum albumin: lack of relation to gaseous formaldehyde exposure and symptoms. *J. Allergy clin. Immunol.*, **87**, 48–57

Edling, C., Järvholm, B., Andersson, L. & Axelson, O. (1987a) Mortality and cancer incidence among workers in an abrasive manufacturing industry. *Br. J. ind. Med.*, **44**, 57–59

Edling, C., Hellquist, H. & Ödkvist, L. (1987b) Occupational formaldehyde exposure and the nasal mucosa. *Rhinology*, **25**, 181–187

Edling, C., Hellquist, H. & Ödkvist, L. (1988) Occupational exposure to formaldehyde and histopathological changes in the nasal mucosa. *Br. J. ind. Med.*, **45**, 761–765

Elias, I. (1987) Evaluation of methods for disinfection of operating theatres in hospitals according to the concentration of formaldehyde in the air. *Zbl. Arbeitsmed.*, **37**, 389–397 (in German)

Eller, P.M., ed. (1989a) *NIOSH Manual of Analytical Methods*, 3rd Ed., Suppl. 3 (DHHS (NIOSH) Publ. No. 84-100), Washington DC, US Government Printing Office, pp. 3500-1–3500-5

Eller, P.M., ed. (1989b) *NIOSH Manual of Analytical Methods*, 3rd Ed., Suppl. 3 (DHHS (NIOSH) Publ. No. 84-100), Washington DC, US Government Printing Office, pp. 2541-1–2541-4

Elliott, L.J., Stayner, L.T., Blade, L.M., Halperin, W. & Keenlyside, R. (1987) *Formaldehyde Exposure Characterization in Garment Manufacturing Plants: A Composite Summary of Three In-depth Industrial Hygiene Surveys*, Cincinnati, OH, US Department of Health and Human Services, Public Health Service, Centers for Disease Control, National Institute for Occupational Safety and Health

Epstein, S.S. & Shafner, H. (1968) Chemical mutagens in the human environment. *Nature*, **219**, 385–387

Epstein, S.S., Arnold, E., Andrea, J., Bass, W. & Bishop, Y. (1972) Detection of chemical mutagens by the dominant lethal assay in the mouse. *Toxicol. appl. Pharmacol.*, **23**, 288–325

European Commission (1989) *Formaldehyde Emission from Wood Based Materials: Guideline for the Determination of Steady State Concentrations in Test Chambers* (EUR 12196 EN), Report No. 2, Luxembourg, European Concerted Action: Indoor Air Quality and Its Impact on Man (COST Project 613)

European Commission (1990) Proposal for a Council Directive on the approximation of the laws of the Member States relating to cosmetic products (90/C322/06). *Off. J. Eur. Commun.*, **C322**, 29-77

Fayerweather, W.E., Pell, S. & Bender, J.R. (1983) Case–control study of cancer deaths in DuPont workers with potential exposure to formaldehyde. In: Clary, J.C., Gibson, J.E. & Waritz, R.S., eds, *Formaldehyde. Toxicology, Epidemiology, and Mechanisms*, New York, Marcel Dekker, pp. 47–121

Feinman, S.E. (1988) Formaldehyde genotoxicity and teratogenicity. In: Feinman, S.E., ed., *Formaldehyde. Sensitivity and Toxicity*, Boca Raton, FL, CRC Press, pp. 167–178

Feron, V.J., Bruyntjes, J.P., Woutersen, R.A., Immel, H.R. & Appelman, L.M. (1988) Nasal tumours in rats after short-term exposure to a cytotoxic concentration of formaldehyde. *Cancer Lett.*, **39**, 101–111

Feron, V.J., Til, H.P. & Woutersen, R.A. (1990) Letter to the Editor. *Toxicol. ind. Health*, **6**, 637–639

Feron, V.J., Til, H.P., de Vrijer, F., Woutersen, R.A., Cassee, F.R. & van Bladeren, P.J. (1991) Aldehydes: occurrence, carcinogenic potential, mechanism of action and risk assessment. *Mutat. Res.*, **259**, 363–385

Finnish Institute of Occupational Health (FIOH) (1994) *Measurements of Formaldehyde, Industrial Hygiene Data Base*, Helsinki

Fleig, I., Petri, N., Stocker, W.G. & Thiess, A.M. (1982) Cytogenetic analyses of blood lymphocytes of workers exposed to formaldehyde in formaldehyde manufacturing and processing. *J. occup. Med.*, **24**, 1009–1012

Fontignie-Houbrechts, N. (1981) Genetic effects of formaldehyde in the mouse. *Mutat. Res.*, **88**, 109–114

Fornace, A.J., Jr, Lechner, J.F., Grafström, R.C. & Harris, C.C. (1982) DNA repair in human bronchial epithelial cells. *Carcinogenesis*, **3**, 1373–1377

Friedman, G.D. & Ury, H.K. (1983) Screening for possible drug carcinogenicity: second report of findings. *J. natl Cancer Inst.*, **71**, 1165–1175

Gallagher, R.P., Threlfall, W.J., Band, P.R., Spinelli, J.J. & Coldman, A.J. (1986) *Occupational Mortality in British Columbia 1950–1978*, Ottawa, Statistics Canada, Health and Welfare Canada

Gallagher, R.P., Threlfall, W.J., Band, P.R. & Spinelli, J.J. (1989) *Occupational Mortality in British Columbia 1950–1984*, Vancouver, Canadian Cancer Association of British Columbia

Gammage, R.B. & Gupta, K.C. (1984) Formaldehyde. In: Walsh, P.J., Dudney, C.S. & Copenhaver, E.D., eds, *Indoor Air Quality*, Ch. 7, Boca Raton, FL, CRC Press, pp. 109–142

Gammage, R.G. & Travis, C.C. (1989) Formaldehyde exposure and risk in mobile homes. In: Paustenbach, D.J., ed., *The Risk Assessment of Environmental and Human Health Hazards: A Textbook of Case Studies*, New York, John Wiley & Sons, pp. 601–611

Gardner, M.J., Pannett, B., Winter, P.D., & Cruddas, A.M. (1993) A cohort study of workers exposed to formaldehyde in the British chemical industry: an update. *Br. J. ind. Med.*, **50**, 827–834

Georghiou, P.E., Winsor, L., Sliwinski, J.F. & Shirtliffe, C.J. (1993) Method 11. Determination of formaldehyde in indoor air by a liquid sorbent technique. In: Seifert, B., van de Wiel, H., Dodet, B. & O'Neill, I.K., eds, *Environmental Carcinogens: Methods of Analysis and Exposure Measurement. Volume 12: Indoor Air Contaminants* (IARC Scientific Publications No. 109), Lyon, IARC, pp. 245–249

Gerberich, H.R. & Seaman, G.C. (1994) Formaldehyde. In: Kroschwitz, J.I. & Howe-Grant, M., eds, *Kirk–Othmer Encyclopedia of Chemical Technology*, 4th Ed., Vol. 11, New York, John Wiley & Sons, pp. 929–951

Gerberich, H.R., Stautzenberger, A.L. & Hopkins, W.C. (1980) Formaldehyde. In: Mark, H.F., Othmer, D.F., Overberger, C.G., Seaborg, G.T. & Grayson, N., eds, *Kirk–Othmer Encyclopedia of Chemical Technology*, 3rd Ed., Vol. 11, New York, John Wiley & Sons, pp. 231–250

Gérin, M., Siemiatycki, J., Nadon, L. Dewar, R. & Krewski, D. (1989) Cancer risks due to occupational exposure to formaldehyde: results of a multi-site case–control study in Montreal. *Int. J. Cancer*, **44**, 53–58

Gibson, J.E. (1984) Coordinated toxicology: an example study with formaldehyde. *Concepts Toxicol.*, **1**, 276–282

Gocke, E., King, M.-T., Eckhardt, K. & Wild, D. (1981) Mutagenicity of cosmetics ingredients licensed by the European Communities. *Mutat. Res.*, **90**, 91–109

Goldmacher, V.S. & Thilly, W.G. (1983) Formaldehyde is mutagenic for cultured human cells. *Mutat. Res.*, **116**, 417–422

Goldoft, M., Weiss, N., Vaughan, T. & Lee, J. (1993) Nasal melanoma. *Br. J. ind. Med.*, **50**, 767–768

Gollob, L. & Wellons, J.D. (1980) Analytical methods for formaldehyde. A review. *Forest. Prod. J.*, **30**, 27–35

Gottschling, L.M., Beaulieu, H.J. & Melvin, W.W. (1984) Monitoring of formic acid in urine of humans exposed to low levels of formaldehyde. *Am. ind. Hyg. Assoc. J.*, **45**, 19–23

Grafström, R.C (1990) In vitro studies of aldehyde effects related to human respiratory carcinogenesis. *Mutat. Res.*, **238**, 175–184

Grafström, R.C., Fornace, A., Jr & Harris, C.C. (1984) Repair of DNA damage caused by formaldehyde in human cells. *Cancer Res.*, **44**, 4323–4327

Grafström, R.C., Curren, R.D., Yang, L.L. & Harris, CC. (1985) Genotoxicity of formaldehyde in cultured human bronchial fibroblasts. *Science*, **228**, 89–91

Grafström, R.C., Willey, J.C., Sundqvist, K. & Harris, C.C. (1986) Pathobiological effects of tobacco smoke-related aldehydes in cultured human bronchial epithelial cells. In: Hoffmann, D. & Harris, C.C., eds, *Mechanisms in Tobacco Carcinogenesis* (Banbury Report 23), Cold Spring Harbor, NY, CSH Press, pp. 273–285

Grafström, R.C., Hsu, I.-C. & Harris, C.C. (1993) Mutagenicity of formaldehyde in Chinese hamster lung fibroblasts: synergy with ionizing radiation and N-nitroso-N-methylurea. *Chem.-biol. Interactions*, **86**, 41–49

Grammer, L.C., Harris, K.E., Shaughnessy, M.A., Sparks, P., Ayars, G.H., Altman, L.C. & Patterson, R. (1990) Clinical and immunologic evaluation of 37 workers exposed to gaseous formaldehyde. *J. Allergy clin. Immunol.*, **86**, 177–181

Graves, R.J., Callander, R.D. & Green, T. (1994) The role of formaldehyde and S-chloromethylglutathione in the bacterial mutagenicity of methylene chloride. *Mutat. Res.*, **320**, 235–243

Green, D.J., Sauder, L.R., Kulle, T.J. & Bascom, R. (1987) Acute response to 3.0 ppm formaldehyde in exercising healthy nonsmokers and asthmatics. *Am. Rev. respir. Dis.*, **135**, 1261–1266

Green, D.J., Bascom, R., Healey, E.M., Hebel, J.R., Sauder, L.R. & Kulle, T.J. (1989) Acute pulmonary response in healthy, nonsmoking adults to inhalation of formaldehyde and carbon. *J. Toxicol. environ. Health*, **28**, 261–275

Greenberg, G.N. & Stave, G. (1989) Formaldehyde case reports (Letter to the Editor). *Am. J. ind. Med.*, **16**, 329–330

Greenblatt, M. (1988) Formaldehyde toxicology: a review of recent research and regulatory changes. *Lab. Med.*, **19**, 425–428

Groah, W.J., Bradfield, J., Gramp, G., Rudzinski, R. & Heroux, G. (1991) Comparative response of reconstituted wood products to European and North American test methods for determining formaldehyde emissions. *Environ. Sci. Technol.*, **25**, 117–122

Gylseth, B. & Digernes, V. (1992) The European development of regulations and standards for formaldehyde in air and in wood composite boards. In: *Proceedings of the Pacific Rim Bio-based Composites Symposium, 9–13 November 1992, Rotorua, New Zealand*, Rotorua, Forest Products Research Institute, pp. 199–206

Hagberg, M., Kolmodin-Hedman, B., Lindahl, R., Nilsson, C.-A. & Nordström, Å. (1985) Irritative complaints, carboxyhemoglobin increase and minor ventilatory function changes due to exposure to chain-saw exhaust. *Eur. J. respir. Dis.*, **66**, 240–247

Hall, A., Harrington, J.M. & Aw, T.-C. (1991) Mortality study of British pathologists. *Am. J. ind. Med.*, **20**, 83–89

Halligan, A.F. (1992) The effect of formaldehyde on the future of composite products. In: *Proceedings of the Pacific Rim Bio-Based Composites Symposium, 9–13 November 1992, Rotorua, New Zealand*, Rotorua, Forest Products Research Institute, pp. 189–198

Halperin, W.E., Goodman, M., Stayner, L., Elliott, L.J., Keenlyside, R.A. & Landrigan, P.J. (1983) Nasal cancer in a worker exposed to formaldehyde. *J. Am. med. Assoc.*, **249**, 510–512

Hanrahan, L.P., Dally, K.A., Anderson, H.A., Kanarek, M.S. & Rankin, J. (1984) Formaldehyde vapor in mobile homes: a cross sectional survey of concentrations and irritant effects. *Am. J. public Health*, **74**, 1026–1027

Harrington, J.M. & Oakes, D. (1984) Mortality study of British pathologists 1974–80. *Br. J. ind. Med.*, **41**, 188–191

Harrington, J.M. & Shannon, H.S. (1975) Mortality study of pathologists and medical laboratory technicians. *Br. med. J.*, **i**, 329–332

Harving, H., Korsgaard, J., Pederson, O.F., Mølhave, L. & Dahl, R. (1990) Pulmonary function and bronchial reactivity in asthmatics during low-level formaldehyde exposure. *Lung*, **168**, 15–21

Haworth, S., Lawlor, T., Mortelmans, K., Speck, W. & Zeiger, E. (1983) Salmonella mutagenicity test results for 250 chemicals. *Environ. Mutag., Suppl 1*, 3–142

Hayes, R.B., Raatgever, J.W., De Bruyn, A. & Gérin, M. (1986) Cancer of the nasal cavity and paranasal sinuses, and formaldehyde exposure. *Int. J. Cancer*, **37**, 487–492

Hayes, R.B., Blair, A., Stewart, P.A., Herrick, R.F. & Mahar, H. (1990) Mortality of US embalmers and funeral directors. *Am. J. ind. Med.*, **18**, 641–652

Heck, H.d'A. & Casanova, M. (1987) Isotope effects and their implications for the covalent binding of inhaled [^{3}H]- and [^{14}C]formaldehyde in the rat nasal mucosa. *Toxicol. appl. Pharmacol.*, **89**, 122–134

Heck, H.d'A. & Casanova, M. (1995) Nasal dosimetry of formaldehyde: modeling site-specificity and the effects of pre-exposure. *Inhal. Toxicol.* (in press)

Heck, H.d'A. & Casanova-Schmitz, M. (1984) Biochemical toxicology of formaldehyde. *Rev. Biochem. Toxicol.*, **6**, 155–189

Heck, H.d'A., White, E.L. & Casanova-Schmitz, M. (1982) Determination of formaldehyde in biological tissues by gas chromatography/mass spectrometry. *Biomed. Mass Spectrom.*, **9**, 347–353

Heck, H.d'A., Chin, T.Y. & Schmitz, M.C. (1983) Distribution of [^{14}C]formaldehyde in rats after inhalation exposure. In: Gibson, J.E., ed., *Formaldehyde Toxicity*, Washington DC, Hemisphere, pp. 26–37

Heck, H.d'A., Casanova-Schmitz, M., Dodd, P.B., Schachter, E.N., Witek, T.J. & Tosun, T. (1985) Formaldehyde (CH_2O) concentrations in the blood of humans and Fischer-344 rats exposed to CH_2O under controlled conditions. *Am. ind. Hyg. Assoc. J.*, **46**, 1–3

Heck, H.d'A., Casanova, M., Lam, C.-W. & Swenberg, J.A. (1986) The formation of DNA–protein cross-links by aldehydes present in tobacco smoke. In: Hoffmann, D. & Harris, C.C., eds, *Mechanisms in Tobacco Carcinogenesis* (Banbury Report 23), Cold Spring Harbor, NY, CSH Press, pp. 215–230

Heck, H.d'A., Casanova, M., Steinhagen, W.H., Everitt, J.I., Morgan, K.T. & Popp, J.A. (1989) Formaldehyde toxicity: DNA–protein cross-linking studies in rats and nonhuman primates. In: Feron, V.J. & Bosland, M.C., eds, *Nasal Carcinogenesis in Rodents: Relevance to Human Risk*, Wageningen, Pudoc, pp. 159–164

Heck, H.d'A., Casanova, M. & Starr, T.B. (1990) Formaldehyde toxicity—new understanding. *CRC crit. Rev. Toxicol.*, **20**, 397–426

Heikkilä, P., Priha, E. & Savela, A. (1991) *Formaldehyde* (Exposures at Work No. 14), Helsinki, Finnish Institute of Occupational Health and Finnish Work Environment Fund (in Finnish)

Helander, I. (1977) Contact urticaria from leather containing formaldehyde. *Arch. Dermatol.*, **113**, 1443

Helrich, K., ed. (1990) *Official Methods of Analysis of the Association of Official Analytical Chemists*, 15th Ed., Vol. 2, Arlington, VA, Association of Official Analytical Chemists, pp. 1037–1038, 1149

Hemminki, K. (1984) Urinary excretion products of formaldehyde in the rat. *Chem.-biol. Interactions*, **48**, 243–248

Hemminki, K., Mutanen, P., Saloniemi, I., Niemi, M.-L. & Vainio, H. (1982) Spontaneous abortions in hospital staff engaged in sterilising instruments with chemical agents. *Br. med. J.*, **285**, 1461–1463

Hemminki, K., Kyyrönen, P. & Lindbohm, M.-L. (1985) Spontaneous abortions and malformations in the offspring of nurses exposed to anaesthetic gases, cytostatic drugs, and other potential hazards in hospitals, based on registered information of outcome. *J. Epidemiol. Community Health*, **39**, 141–147

Hernberg, S., Westerholm, P., Schultz-Larsen, K., Degerth, R., Kuosma, E., Englund, A., Engzell, U., Sand Hansen, H. & Mutanen, P. (1983) Nasal and sinonasal cancer: connection with occupational exposures in Denmark, Finland and Sweden. *Scand. J. Work Environ. Health*, **9**, 315–326

Hildesheim, A., West, S., DeVeyra, E., De Guzman, M.F., Jurado, A., Jones, C., Imai, J. & Hinuma, Y. (1992) Herbal medicine use, Epstein–Barr virus, and risk of nasopharyngeal carcinoma. *Cancer Res.*, **52**, 3048–3051

Holmquist, B. & Vallee, B.L. (1991) Human liver class III alcohol and glutathione dependent formaldehyde dehydrogenase are the same enzyme. *Biochem. biophys. Res. Commun.*, **178**, 1371–1377

Holmström, M. & Lund, V.J. (1991) Malignant melanomas of the nasal cavity after occupational exposure to formaldehyde. *Br. J. ind. Med.*, **48**, 9–11

Holmström, M. & Wilhelmsson, B. (1988) Respiratory symptoms and pathophysiological effects of occupational exposure to formaldehyde and wood dust. *Scand. J. Work Environ. Health*, **14**, 306–311

Holmström, M., Wilhelmsson, B. & Hellquist, H. (1989a) Histological changes in the nasal mucosa in rats after long-term exposure to formaldehyde and wood dust. *Acta otolaryngol.*, **108**, 274–283

Holmström, M., Wilhelmsson, B., Hellquist, H. & Rosén, G. (1989b) Histological changes in the nasal mucosa in persons occupationally exposed to formaldehyde alone and in combination with wood dust. *Acta otolaryngol.*, **107**, 120–129

Holmström, M., Rynnel-Dagöö, B. & Wilhelmsson, B. (1989c) Antibody production in rats after long-term exposure to formaldehyde. *Toxicol. appl. Pharmacol.*, **100**, 328–333

Holness, D.L. & Nethercott, J.R. (1989) Health status of funeral service workers exposed to formaldehyde. *Arch. environ. Health*, **44**, 222–228

Horton, A.W., Tye, R. & Stemmer, K.L. (1963) Experimental carcinogenesis of the lung. Inhalation of gaseous formaldehyde or an aerosol of coal tar by C3H mice. *J. natl Cancer Inst.*, **30**, 31–43

Horvath, E.P., Jr, Anderson, H., Jr, Pierce, W.E., Hanrahan, L. & Wendlick, J.D. (1988) Effects of formaldehyde on the mucous membranes and lungs. A study of an industrial population. *J. Am. med. Assoc.*, **259**, 701–707

IARC (1979) *IARC Monographs on the Evaluation of the Carcinogenic Risk of Chemicals to Humans*, Vol. 19, *Some Monomers, Plastics and Synthetic Elastomers, and Acrolein*, Lyon, pp. 314–340

IARC (1981) *IARC Monographs on the Evaluation of the Carcinogenic Risk of Chemicals to Humans*, Vol. 25, *Wood, Leather and Some Associated Industries*, Lyon

IARC (1982) *IARC Monographs on the Evaluation of the Carcinogenic Risk of Chemicals to Humans*, Vol. 29, *Some Industrial Chemicals and Dyestuffs*, Lyon, pp. 345–389

IARC (1983) *IARC Monographs on the Evaluation of the Carcinogenic Risk of Chemicals to Humans*, Vol. 32, *Polynuclear Aromatic Compounds, Part 1: Chemical, Environmental and Experimental Data*, Lyon

IARC (1984) *IARC Monographs on the Evaluation of the Carcinogenic Risk of Chemicals to Humans*, Vol. 34, *Nuclear Aromatic Compounds, Part 3, Industrial Exposures in Aluminium Production, Coal Gasification, Coke Production and Iron and Steel Founding*, Lyon

IARC (1986a) *IARC Monographs on the Evaluation of the Carcinogenic Risk of Chemicals to Humans*, Vol. 39, *Some Chemicals Used in Plastics and Elastomers*, Lyon, pp. 287–323

IARC (1986b) *IARC Monographs on the Evaluation of the Carcinogenic Risk of Chemicals to Humans*, Vol. 38, *Tobacco Smoking*, Lyon, p. 96

IARC (1987a) *IARC Monographs on the Evaluation of Carcinogenic Risks to Humans*, Suppl. 7, *Overall Evaluations of Carcinogenicity: An Updating of IARC Monographs Volumes 1–42*, Lyon, pp. 211–216

IARC (1987b) *IARC Monographs on the Evaluation of Carcinogenic Risks to Humans*, Suppl. 7, *Overall Evaluations of Carcinogenicity: An Updating of IARC Monographs Volumes 1–42*, Lyon, pp. 131–134

IARC (1987c) *IARC Monographs on the Evaluation of Carcinogenic Risks to Humans*, Vol. 42, *Silica and Silicates*, Lyon, pp. 39–143

IARC (1987d) *IARC Monographs on the Evaluation of Carcinogenic Risks to Humans*, Suppl. 7, *Overall Evaluations of Carcinogenicity: An Updating of IARC Monographs Volumes 1–42*, Lyon, pp. 106–116

IARC (1987e) *IARC Monographs on the Evaluation of Carcinogenic Risks to Humans*, Suppl. 7, *Overall Evaluations of Carcinogenicity: An Updating of IARC Monographs Volumes 1–42*, Lyon, pp. 152–154

IARC (1987f) *IARC Monographs on the Evaluation of Carcinogenic Risk of Chemicals to Humans*, Suppl. 6, *Genetic and Related Effects: An Updating of Selected IARC Monographs from Volumes 1 to 42*, Lyon, pp. 321–324

IARC (1988) *IARC Monographs on the Evaluation of Carcinogenic Risks to Humans*, Vol. 43, *Man-made Mineral Fibres and Radon*, Lyon

IARC (1989a) *IARC Monographs on the Evaluation of Carcinogenic Risks to Humans*, Vol. 47, *Some Organic Solvents, Resin Monomers and Related Compounds, Pigments and Occupational Exposures in Paint Manufacture and Painting*, Lyon, pp. 263–287

IARC (1989b) *IARC Monographs on the Evaluation of Carcinogenic Risks to Humans*, Vol. 46, *Diesel and Gasoline Engine Exhausts and Some Nitroarenes*, Lyon, pp. 41–185

IARC (1989c) *IARC Monographs on the Evaluation of Carcinogenic Risks to Humans*, Vol. 47, *Some Organic Solvents, Resin Monomers and Related Compounds, Pigments and Occupational Exposures in Paint Manufacture and Painting*, Lyon, pp. 125–156

IARC (1989d) *IARC Monographs on the Evaluation of Carcinogenic Risks to Humans*, Vol. 47, *Some Organic Solvents, Resin Monomers and Related Compounds, Pigments and Occupational Exposures in Paint Manufacture and Painting*, Lyon, pp. 79–123

IARC (1990a) *IARC Monographs on the Evaluation of Carcinogenic Risks to Humans*, Vol. 48, *Some Flame Retardants and Textile Chemicals, and Exposures in the Textile Manufacturing Industry*, Lyon, pp. 181–212

IARC (1990b) *IARC Monographs on the Evaluation of Carcinogenic Risks to Humans*, Vol. 48, *Some Flame Retardants and Textile Chemicals, and Exposures in the Textile Manufacturing Industry*, Lyon, pp. 215–280

IARC (1991) *IARC Monographs on the Evaluation of Carcinogenic Risks to Humans*, Vol. 53, *Occupational Exposures in Insecticide Application and Some Pesticides*, Lyon

IARC (1994a) *IARC Monographs on the Evaluation of Carcinogenic Risks to Humans*, Vol. 60, *Some Industrial Chemicals*, Lyon, pp. 445–474

IARC (1994b) *IARC Monographs on the Evaluation of Carcinogenic Risks to Humans*, Vol. 60, *Some Industrial Chemicals*, Lyon, pp. 73–159

ILO (1991) *Occupational Exposure Limits for Airborne Toxic Substances: Values of Selected Countries. Prepared from the ILO–CIS Data Base of Exposure Limits* (Occupational Safety and Health Series No. 37), 3rd Ed., Geneva, pp. 206–207

Imbus, H.R. & Tochilin, S.J. (1988) Acute effect upon pulmonary function of low level exposure to phenol–formaldehyde-resin-coated wood. *Am. ind. Hyg. Assoc. J.*, **49**, 434–437

Inoue, K., Nishimukai, H. & Yamasawa, K. (1979) Purification and partial characterization of aldehyde dehydrogenase from human erythrocytes. *Biochim. biophys. Acta*, **569**, 117–123

Ishidate, M., Jr, Sofuni, T. & Yoshikawa, K. (1981) Chromosomal aberration tests *in vitro* as a primary screening tool for environmental mutagens and/or carcinogens. *Gann Monogr. Cancer Res.*, **27**, 95–108

Iversen, O.H. (1986) Formaldehyde and skin carcinogenesis. *Environ. int.*, **12**, 541–544

Jaeger, R.J. & Gearhart, J.M. (1982) Respiratory and metabolic response of rats and mice to formalin vapor. *Toxicology*, **25**, 299–309

Jann, O. (1991) Present state and developments in formaldehyde regulations and testing methods in Germany. In: *Proceedings of the 25th International Particleboard/Composite Materials Symposium*, Pullman, WA, Washington State University

Japan Chemical Week, ed. (1991) *Japan Chemical Annual 1991: Japan's Chemical Industry*, Tokyo, The Chemical Daily Co. Ltd, p. 33

Japan Chemical Week, ed. (1993) *Japan Chemical Annual 1993: Japan's Chemical Industry*, Tokyo, The Chemical Daily Co. Ltd, p. 35

Jeffcoat, A.R., Chasalow, F., Feldman, D.B. & Marr, H. (1983) Disposition of [^{14}C]formaldehyde after topical exposure to rats, guinea pigs, and monkeys. In: Gibson, J.E., ed., *Formaldehyde Toxicity*, Washington DC, Hemisphere, pp. 38–50

Jensen, N.J. & Cohr, K.-H. (1983) Testing of formaldehyde in the mammalian spot test by inhalation (Abstract No. 73). *Mutat. Res.*, **113**, 266

Jensen, O.M. & Andersen, S.K. (1982) Lung cancer risk from formaldehyde. *Lancet*, **i**, 913

Jensen, K.A., Kirk, I., Kölmark, G. & Westergaard, M. (1952) Chemically induced mutations in Neurospora. *Cold Spring Harbor Symp. quant. Biol.*, **16**, 245–261

Johannsen, F.R., Levinskas, G.J. & Tegeris, A.S. (1986) Effects of formaldehyde in the rat and dog following oral exposure. *Toxicol. Lett.*, **30**, 1–6

Johnsen, R.C. & Baillie, D.L. (1988) Formaldehyde mutagenesis of the eT1 balanced region in *Caenorhabditis elegans*: dose–response curve and the analysis of mutation events. *Mutat. Res.*, **201**, 137–147

Jones, D.P., Thor, H., Andersson, B. & Orrenius, S. (1978) Detoxification reactions in isolated hepatocytes. Role of glutathione peroxidase, catalase, and formaldehyde dehydrogenase in reactions relating to *N*-demethylation by the cytochrome P-450 system. *J. biol. Chem.*, **253**, 6031–6037

Kaiser, R., Holmquist, B., Vallee, B.L. & Jörnvall, H. (1991) Human class III alcohol dehydrogenase/glutathione-dependent formaldehyde dehydrogenase. *J. Protein Chem.*, **10**, 69–73

Kaplan, W.D. (1948) Formaldehyde as a mutagen in *Drosophila*. *Science*, **108**, 43

Kauppinen, T. (1986) Occupational exposure to chemical agents in the plywood industry. *Ann. occup. Hyg.*, **30**, 19–29

Kauppinen, T. & Niemelä, R. (1985) Occupational exposure to chemical agents in the particleboard industry. *Scand. J. Work Environ. Health*, **11**, 357–363

Kauppinen, T. & Partanen, T. (1988) Use of plant- and period-specific job–exposure matrices in studies on occupational cancer. *Scand. J. Work Environ Health*, **14**, 161–167

Keller, D.A., Heck, H.d'A., Randall, H.W. & Morgan, K.T. (1990) Histochemical localization of formaldehyde dehydrogenase in the rat. *Toxicol. appl. Pharmacol.*, **106**, 311–326

Kennedy, E.R. & Hull, D. (1986) Evaluation of the Du Pont Pro-Tek® Formaldehyde Badge and the 3M Formaldehyde Monitor. *Am. ind. Hyg. Assoc. J.*, **47**, 94–105

Kennedy, E.R., Teass, A.W. & Gagnon, Y.T. (1985) Industrial hygiene sampling and analytical methods for formaldehyde. Past and present. *Adv. Chem. Ser.*, **210**, 3–12

Kerfoot, E.J. & Mooney, T.F., Jr (1975) Formaldehyde and paraformaldehyde study in funeral homes. *Am. ind. Hyg. Assoc. J.*, **36**, 533–537

Kerns, W.D., Pavkov, K.L., Donofrio, D.J., Gralla, E.J. & Swenberg, J.A. (1983a) Carcinogenicity of formaldehyde in rats and mice after long-term inhalation exposure. *Cancer Res.*, **43**, 4382–4392

Kerns, W.D., Donofrio, D.J. & Pavkov, K.L. (1983b) The chronic effects of formaldehyde inhalation in rats and mice: a preliminary report. In: Gibson, J.E., ed., *Formaldehyde Toxicity*, Washington DC, Hemisphere, pp. 111–131

Khamgaonkar, M.B. & Fulare, M.B. (1991) Pulmonary effects of formaldehyde exposure. An environmental–epidemiological study. *Indian J. Chest Dis. allied Sci.*, **33**, 9–13

Khan, A.H. (1967) The induction of crossing over in the absence of mutation. *Sind Univ. Sci. Res. J.*, **3**, 103–106

Kilburn, K.H., Warshaw, R., Boylen, C.T., Johnson, S.-J.S., Seidman, B., Sinclair, R. & Takaro, T., Jr (1985) Pulmonary and neurobehavioral effects of formaldehyde exposure. *Arch. environ. Health*, **40**, 254–260

Kimbell, J.S., Gross, E.A., Joyner, D.R., Godo, M.N. & Morgan, K.T. (1993) Application of computational fluid dynamics to regional dosimetry of inhaled chemicals in the upper respiratory tract of the rat. *Toxicol. appl. Pharmacol.*, **121**, 253–263

Kitaeva, L.V., Kitaev, E.M. & Pimenova, M.N. (1990) The cytopathic and cytogenetic effects of chronic inhalation of formaldehyde on female germ cells and bone marrow cells in rats. *Tsitologiia*, **32**, 1212–1216 (in Russian)

Kligerman, A.D., Phelps, M.C. & Erexson, G.L. (1984) Cytogenetic analysis of lymphocytes from rats following formaldehyde inhalation. *Toxicol. Lett.*, **21**, 241–246

Kölmark, G. & Westergaard, M. (1953) Further studies on chemically induced reversions at the adenine locus of *Neurospora*. *Hereditas*, **39**, 209–224

Koivusalo, M., Koivula, T. & Uotila, L. (1982) Oxidation of formaldehyde by nicotinamide nucleotide dependent dehydrogenases. In: Weiner, H. & Wermuth, B., eds, *Enzymology of Carbonyl Metabolism. Aldehyde Dehydrogenase and Aldo/Keto Reductase*, New York, Alan R. Liss, pp. 155–168

Korczynski, R.E. (1994) Formaldehyde exposure in the funeral industry. *Appl. occup. environ. Hyg.*, **9**, 575–579

Korky, J.K., Schwarz, S.R. & Lustigman, B.K. (1987) Formaldehyde concentrations in biology department teaching facilities. *Bull. environ. Contam. Toxicol.*, **38**, 907–910

Kramps, J.A., Peltenburg, L.T.C., Kerklaan, P.R.M., Spieksma, F.T.M., Valentijn, R.M. & Dijkman, J.H. (1989) Measurement of specific IgE antibodies in individuals exposed to formaldehyde. *Clin. exp. Allergy*, **19**, 509–514

Kreiger, R.A. & Garry, V.F. (1983) Formaldehyde-induced cytotoxicity and sister-chromatid exchanges in human lymphocyte cultures. *Mutat. Res.*, **120**, 51–55

Kulle, T.J., Sauder, L.R., Hebel, J.R., Green, D.J. & Chatham, M.D. (1987) Formaldehyde dose-response in healthy nonsmokers. *J. Air Pollution Control Assoc.*, **37**, 919–924

Kuykendall, J.R. & Bogdanffy, M.S. (1992) Efficiency of DNA–histone crosslinking induced by saturated and unsaturated aldehydes *in vitro*. *Mutat. Res.*, **283**, 131–136

Lam, C.-W., Casanova, M. & Heck, H.d'A. (1985) Depletion of nasal mucosal glutathione by acrolein and enhancement of formaldehyde-induced DNA–protein cross-linking by simultaneous exposure to acrolein. *Arch. Toxicol.*, **58**, 67–71

Lamont Moore, L. & Ogrodnik, E.C. (1986) Occupational exposure to formaldehyde in mortuaries. *J. environ. Health*, **49**, 32–35

Le Curieux, F., Marzin, D. & Erb, F. (1993) Comparison of three short-term assays: results on seven chemicals. Potential contribution to the control of water genotoxicity. *Mutat. Res.*, **319**, 223–236

Lee, H.K., Alarie, Y. & Karol, M.H. (1984) Induction of formaldehyde sensitivity in guinea pigs. *Toxicol. appl. Pharmacol.*, **75**, 147–155

Lehmann, W.F. & Roffael, E. (1992) International guidelines and regulations for formaldehyde emissions. In: *Proceedings of the 26th Washington State University International*

Particleboard/Composite Materials Symposium, Pullman, WA, Washington State University, pp. 124–150

Leifer, Z., Hyman, J. & Rosenkranz, H.S. (1981) Determination of genotoxic activity using DNA polymerase-deficient and -proficient E. coli. In: Stich, H.F. & San, R.H.C., eds, *Short-term Tests for Chemical Carcinogenesis*, New York, Springer, pp. 127–139

Leikauf, G.D. (1992) Formaldehyde and other aldehydes. In: Lippmann, M., ed., *Environmental Toxicants. Human Exposures and Their Health Effects*, New York, Van Nostrand Reinhold, pp. 299–330

Levine, R.J., Andjelkovich, D.A., & Shaw, L.K. (1984a) The mortality of Ontario undertakers and a review of formaldehyde-related mortality studies. *J. occup. Med.*, **26,** 740–746

Levine, R.J., DalCorso, R.D., Blunden, P.B. & Battigelli, M.C. (1984b) The effects of occupational exposure on the respiratory health of West Virginia morticians. *J. occup. Med.*, **26,** 91–98

Levy, S., Nocentini, S. & Billardon, C. (1983) Induction of cytogenetic effects in human fibroblast cultures after exposure to formaldehyde or X-rays. *Mutat. Res.*, **119**, 309–317

Liber, H.L., Benforado, K., Crosby, R.M., Simpson, D. & Skopek, T.R. (1989) Formaldehyde-induced and spontaneous alterations in human *hprt* DNA sequence and mRNA expression. *Mutat. Res.*, **226**, 31–37

Lide, D.R., ed. (1993) *CRC Handbook of Chemistry and Physics*, 74th Ed., Boca Raton, FL, CRC Press, p. 3-248

Lidén, S., Scheynius, A., Fischer, T., Johansson, S.G.O., Ruhnek-Forsbeek, M. & Stejskal, V. (1993) Absence of specific IgE antibodies in allergic contact sensitivity to formaldehyde. *Allergy*, **48**, 525–529

Liebling, T., Rosenman, K.D., Pastides, H., Griffith, R.G., & Lemeshow, S. (1984) Cancer mortality among workers exposed to formaldehyde. *Am. J. ind. Med.*, **5,** 423–428

Lindskov, R. (1982) Contact urticaria to formaldehyde. *Contact Derm.*, **8**, 333–334

Linos, A., Blair, A., Cantor, K.P., Burmeister, L., VanLier, S., Gibson, R.W., Schuman, L. & Everett, G. (1990) Leukemia and non-Hodgkin's lymphoma among embalmers and funeral directors (Letter to the Editor). *J. natl Cancer Inst.*, **82**, 66

Litvinov, N.N., Voronin, V.M. & Kazachkov, V.I. (1984) Concerning the modifying effect of aniline, lead nitrate, carbon tetrachloride and formaldehyde on chemical blastogenesis. *Vopr. Onkol.*, **30**, 56–60 (in Russian)

Liu, K.-S., Huang, F.-Y., Hayward, S.B., Wesolowski, J. & Sexton, K. (1991) Irritant effects of formaldehyde exposure in mobile homes. *Environ. Health Perspectives*, **94**, 91–94

Lodén, M. (1986) The *in vitro* permeability of human skin to benzene, ethylene glycol, formaldehyde, and n-hexane. *Acta pharmacol. toxicol.*, **58**, 382–389

Logue, J.N., Barrick, M.K. & Jessup, G.L., Jr 1986) Mortality of radiologists and pathologists in the radiation registry of physicians. *J. occup. Med.*, **28**, 91–99

Lucas, L.J. (1994) Misclassification of nasopharyngeal cancer (Letter to the Editor). *J. natl Cancer Inst.*, **86**, 1556–1557

Luce, D., Gérin, M., Leclerc, A., Morcet, J.-F., Brugère, J. & Goldberg, M. (1993) Sinonasal cancer and occupational exposure to formaldehyde and other substances. *Int. J. Cancer*, **53**, 224–231

Luker, M.A. & Van Houten, R.W. (1990) Control of formaldehyde in a garment sewing plant. *Am. ind. Hyg. Assoc. J.*, **51**, 541–544

Ma, T.-H. & Harris, M.M. (1988) Review of the genotoxicity of formaldehyde. *Mutat. Res.*, **196**, 37–59

Magaña-Schwencke, N., Ekert, B. & Moustacchi, E. (1978) Biochemical analysis of damage induced in yeast by formaldehyde. I. Induction of single-strand breaks in DNA and their repair. *Mutat. Res.*, **50**, 181–193

Magaña-Schwencke, N. & Ekert, B. (1978) Biochemical analysis of damage induced in yeast by formaldehyde. II. Induction of cross-links between DNA and protein. *Mutat. Res.*, **51**, 11–19

Magaña-Schwencke, N. & Moustacchi, E. (1980) Biochemical analysis of damage induced in yeast by formaldehyde. III. Repair of induced cross-links between DNA and proteins in the wild-type and in excision-deficient strains. *Mutat. Res.*, **70**, 29–35

Maibach, H. (1983) Formaldehyde: effects on animal and human skin. In: Gibson, J.E., ed., *Formaldehyde Toxicity*, Washington DC, Hemisphere, pp. 166–174

Main, D.M. & Hogan, T.J. (1983) Health effects of low-level exposure to formaldehyde. *J. occup. Med.*, **25**, 896–900

Malaka, T. & Kodama, A.M. (1990) Respiratory health of plywood workers occupationally exposed to formaldehyde. *Arch. environ. Health*, **45**, 288–294

Malker, H.R. & Weiner, J. (1984) *Cancer–Environment Registry: Examples of the Use of Register Epidemiology in Studies of the Work Environment* (Arbete och Hälsa 1984;9), Stockholm, Arbetarskyddsverket (in Swedish)

Malker, H.S.R., McLaughlin, J.K., Weiner, J.A., Silverman, D.T., Blot, W.J., Ericsson, J.L.E. & Fraumeni, J.F., Jr. (1990) Occupational risk factors for nasopharyngeal cancer in Sweden. *Br. J. ind. Med.*, **47**, 213–214

Malorny, G., Rietbrock, N. & Schneider, M. (1965) Oxidation of formaldehyde to formic acid in blood, a contribution to the metabolism of formaldehyde. *Naunyn–Schmiedeberg's Arch. exp. Pathol. Pharmakol.*, **250**, 419–436 (in German)

Marks, T.A., Worthy, W.C. & Staples, R.E. (1980) Influence of formaldehyde and Sonacide® (potentiated acid glutaraldehyde) on embryo and fetal development in mice. *Teratology*, **22**, 51–58

Marnett, L.J., Hurd, H.K., Hollstein, M.C., Levin, D.E., Esterbauer, H. & Ames, B.N. (1985) Naturally occurring carbonyl compounds are mutagens in Salmonella tester strain TA104. *Mutat. Res.*, **148**, 25–34

Maronpot, R.R., Miller, R.A., Clarke, W.J., Westerberg, R.B., Decker, J.R. & Moss, O.R. (1986) Toxicity of formaldehyde vapor in B6C3F1 mice exposed for 13 weeks. *Toxicology*, **41**, 253–266

Marsh, G.M. (1982) Proportional mortality patterns among chemical plant workers exposed to formaldehyde. *Br. J. ind. Med.*, **39**, 313–322

Marsh, G.M. (1983) Proportional mortality among chemical workers exposed to formaldehyde. In: Gibson, J.E., ed., *Formaldehyde Toxicity*, New York, Hemisphere, pp. 237–255

Marsh, G.M., Stone, R.A. & Henderson, V.L. (1992a) A reanalysis of the National Cancer Institute study on lung cancer mortality among industrial workers exposed to formaldehyde. *J. occup. Med.*, **34**, 42–44

Marsh, G.M., Stone, R.A. & Henderson, V.L. (1992b) Lung cancer mortality among industrial workers exposed to formaldehyde: a Poisson regression analysis of the National Cancer Institute study. *Am. ind. Hyg. Assoc. J.*, **53**, 681–691

Marsh, G.M., Stone, R.A., Esmen, N.A. & Henderson, V.L. (1994a) Mortality patterns among chemical plant workers exposed to formaldehyde and other substances (Brief communication). *J. natl Cancer Inst.*, **86**, 384–386

Marsh, G.M., Stone, R.A., Henderson, V.L. & Esmen, N.A. (1994b) Misclassification of nasopharyngeal cancer—reply. *J. natl Cancer Inst.*, **86**, 1557

Martin, W.J. (1990) A teratology study of inhaled formaldehyde in the rat. *Reprod. Toxicol.*, **4**, 237–239

Mašek, V. (1972) Aldehydes in the air of workplaces in coal coking and pitch coking plants. *Staub-Reinhalt Luft*, **32**, 335–336 (in German)

Mashford, P.M. & Jones, A.R. (1982) Formaldehyde metabolism by the rat: a re-appraisal. *Xenobiotica*, **12**, 119–124

Mashima, S. & Ikeda, Y. (1958) Selection of mutagenic agents by the *Streptomyces* reverse mutation test. *Appl. Microbiol.*, **6**, 45–49

Materna, B.L., Jones, J.R., Sutton, P.M., Rothman, N. & Harrison, R.J. (1992) Occupational exposures in California wildland fire fighting. *Am. ind. Hyg. Assoc. J.*, **53**, 69–76

McCredie, W.H. (1988) Regulations covering formaldehyde-containing resins used in pressed wood products. In: *Proceedings of the 22nd Washington State University International Particleboard/Composite Materials Symposium*, Pullman, WA, Washington State University, pp. 57–71

McCredie, W.H. (1992) Formaldehyde emissions from UF particleboard: voluntary standards vs. EPA regulations. In: *Proceedings of the 26th Washington State University International Particleboard/Composite Materials Symposium*, Pullman, WA, Washington State University, pp. 115–123

McLaughlin, J.K. (1994) Formaldehyde and cancer: a critical review. *Int. Arch. occup. environ. Health*, **66**, 295–301

Merletti, F., Boffetta, P., Ferro, G., Pisani, P. & Terracini, B. (1991) Occupation and cancer of the oral cavity or oropharynx in Turin, Italy. *Scand. J. Work Environ. Health*, **17**, 248–254

Meyer, B. (1986) Occupational and indoor air formaldehyde exposure: regulations and guidelines. In: Meyer, B., Kottes Andrews, B.A. & Reinhardt, R.M., eds, *Formaldehyde Release from Wood Products* (ACS Symposium Series No. 316), Washington DC, American Chemical Society, pp. 217–229

Migliore, L., Ventura, L., Barale, R., Loprieno, N., Castellino, S. & Pulci, R. (1989) Micronuclei and nuclear anomalies induced in the gastro-intestinal epithelium of rats treated with formaldehyde. *Mutagenesis*, **4**, 327–334

Milham, S. (1983) *Occupational Mortality in Washington State 1950–1979* (DHHS (NIOSH) Publication No. 83-116), Cincinatti, OH, National Institute for Occupational Safety and Health

Miretskaya, L.M. & Shvartsman, P.Y. (1982) Studies of chromosome aberrations in human lymphocytes under the influence of formaldehyde. 1. Formaldehyde treatment of lymphocytes *in vitro*. *Tsitologiia*, **24**, 1056–1060 (in Russian)

Monteiro-Riviere, N.A. & Popp, J.A. (1986) Ultrastructural evaluation of acute nasal toxicity in the rat respiratory epithelium in response to formaldehyde gas. *Fundam. appl. Toxicol.*, **6**, 251–262

Monticello, T.M., Morgan, K.T., Everitt, J.I. & Popp, J.A. (1989) Effects of formaldehyde gas on the respiratory tract of rhesus monkeys. Pathology and cell proliferation. *Am. J. Pathol.*, **134**, 515–527

Monticello, T.M., Miller, F.J. & Morgan, K.T. (1991) Regional increases in rat nasal epithelial cell proliferation following acute and subchronic inhalation of formaldehyde. *Toxicol. appl. Pharmacol.*, **111**, 409–421

Monticello, T.M., Gross, E.A. & Morgan, K.T. (1993) Cell proliferation and nasal carcinogenesis. *Environ. Health Perspectives*, **101** (Suppl. 5*)*, 121–124

Morgan, K.T., Jiang, X.-Z., Starr, T.B. & Kerns, W.D. (1986a) More precise localization of nasal tumors associated with chronic exposure of F-344 rats to formaldehyde gas. *Toxicol. appl. Pharmacol.*, **82**, 264–271

Morgan, K.T., Gross, E.A. & Patterson, D.L. (1986b) Distribution, progression and recovery of acute formaldehyde-induced inhibition of nasal mucociliary function in F-344 rats. *Toxicol. appl. Pharmacol.*, **86**, 448–456

Morgan, K.T., Patterson, D.L. & Gross, E.A. (1986c) Responses of the nasal mucociliary apparatus of F-344 rats to formaldehyde gas. *Toxicol. appl. Pharmacol.*, **82**, 1–13

Morgan, K.T., Kimbell, J.S., Monticello, T.M., Patra, A.L. & Fleishman, A. (1991) Studies of inspiratory airflow patterns in the nasal passages of the F344 rat and rhesus monkey using nasal molds: relevance to formaldehyde toxicity. *Toxicol. appl. Pharmacol.*, **110**, 223–240

Natarajan, A.T., Darroudi, F., Bussman, C.J.M. & van Kesteren-van Leeuwen, A.C. (1983) Evaluation of the mutagenicity of formaldehyde in mammalian cytogenetic assays *in vivo* and *in vitro*. *Mutat. Res.*, **122**, 355–360

National Particleboard Association (1983) *Formaldehyde Test Method-1—Small Scale Test Method for Determining Formaldehyde Emissions from Wood Products—Two Hour Dessicator Test—FTM 1-1983*, Gaithersburg, MD

National Particleboard Association (1992) *Voluntary Standard for Formaldehyde Emissions from Urea–Formaldehyde Bonded Particleboard Flooring Products* (NPA 10-92), Gaithersburg, MD

National Particleboard Association (1993) *American National Standard: Particleboard* (ANSI A208.1-1993), Gaithersburg, MD

National Particleboard Association (1994) *American National Standard: Medium Density Fiberboard* (ANSI A208.2-1994*)*, Gaithersburg, MD

Niemelä, R. & Vainio, H. (1981) Formaldehyde exposure in work and the general environment. Occurrence and possibilities for prevention. *Scand. J. Work Environ. Health*, **7**, 95–100

Nishioka, H. (1973) Lethal and mutagenic action of formaldehyde in Hcr+ and Hcr– strains of *Escherichia coli*. *Mutat. Res.*, **17**, 261–265

Nousiainen, P. & Lindqvist, J. (1979) *Chemical Hazards in the Textile Industry. Air Contaminants* (Tiedonanto 16), Tampere, Valtion teknillinen tutkimuskeskus (in Finnish)

Nunn, A.J., Craigen, A.A., Darbyshire, J.H., Venables, K.M. & Newman Taylor, A.J. (1990) Six year follow up of lung function in men occupationally exposed to formaldehyde. *Br. J. ind. Med.*, **47**, 747–752

Obe, G. & Beek, B. (1979) Mutagenic activity of aldehydes. *Drug Alcohol Dependence*, **4**, 91–94

O'Connor, P.M. & Fox, B.W. (1987) Comparative studies of DNA cross-linking reactions following methylene dimethanesulphonate and its hydrolytic product, formaldehyde. *Cancer Chemother. Pharmacol.*, **19**, 11–15

O'Donovan, M.R. & Mee, C.D. (1993) Formaldehyde is a bacterial mutagen in a range of Salmonella and Escherichia indicator strains. *Mutagenesis*, **8**, 577–581

Olsen, J.H. & Asnaes, S. (1986) Formaldehyde and the risk of squamous cell carcinoma of the sinonasal cavities. *Br. J. ind. Med.*, **43**, 769–774

Olsen, J.H., Plough Jensen, S., Hink, M., Faurbo, K., Breum, N.O. & Møller Jensen, O. (1984) Occupational formaldehyde exposure and increased nasal cancer risk in man. *Int. J. Cancer*, **34**, 639–644

Overman, D.O. (1985) Absence of embryotoxic effects of formaldehyde after percutaneous exposure in hamsters. *Toxicol. Lett.*, **24**, 107–110

Panfilova, Z.I., Voronina, E.N., Poslovina, A.S., Goryukhova, N.M. & Salganik, R.I. (1966) Study of the joint action of chemical mutagens and ultra-violet rays upon the appearance of back mutations in *Escherichia coli*. *Sov. Genet.*, **2**, 35–40

Partanen, T. (1993) Formaldehyde exposure and respiratory cancer—a meta-analysis of the epidemiologic evidence. *Scand. J. Work Environ. Health*, **19**, 8–15

Partanen, T., Kauppinen, T., Nurminen, M., Nickels, J., Hernberg, S., Hakulinen, T., Pukkala, E. & Savonen, E. (1985) Formaldehyde exposure and respiratory and related cancers: a case–referent study among Finnish woodworkers. *Scand. J. Work Environ. Health*, **11**, 409–415

Partanen, T., Kauppinen, T., Hernberg, S., Nickels, J., Luukkonen, R., Hakulinen, T. & Pukkala, E. (1990) Formaldehyde exposure and respiratory cancer among woodworkers—an update. *Scand. J. Work Environ. Health*, **16**, 394–400

Patterson, D.L., Gross, E.A., Bogdanffy, M.S. & Morgan, K.T. (1986) Retention of formaldehyde gas by the nasal passages of F-344 rats (Abstract 217). *Toxicologist*, **6**, 55

Patterson, R., Dykewicz, M.S., Evans, R., III, Grammer, L.C., Greenberger, P.A., Harris, K.E., Lawrence, I.D., Pruzansky, J.J., Roberts, M., Shaughnessy, M.A. & Zeiss, C.R. (1989) IgG antibody against formaldehyde human serum proteins: a comparison with other IgG antibodies against inhalant proteins and reactive chemicals. *J. Allergy clin. Immunol.*, **84**, 359–366

Peterson, G.R. & Milham, S. (1980) *Occupational Mortality in the State of California 1959–1961* (DHEW (NIOSH) Publication No. 80-104), Cincinatti, OH, National Institute for Occupational Safety and Health

Pool, B.L., Frei, E., Plesch, W.J., Romruen, K. & Wiessler, M. (1984) Formaldehyde as a possible mutagenic metabolite of *N*-nitrodimethylamine and of other agents which are suggested to yield non-alkylating species *in vitro*. *Carcinogenesis*, **5**, 809–814

Poverenny, A.M., Siomin, Y.A., Saenko, A.S. & Sinzinis, B.I. (1975) Possible mechanisms of lethal and mutagenic action of formaldehyde. *Mutat. Res.*, **27**, 123–126

Preuss, P.W., Dailey, R.L. & Lehman, E.S. (1985) Exposure to formaldehyde. In: Turoski, V., ed., *Formaldehyde. Analytical Chemistry and Toxicology* (Advances in Chemistry Series, Vol. 210), Washington DC, American Chemical Society, pp. 247–259

Priha, E., Riipinen, H. & Korhonen, K. (1986) Exposure to formaldehyde and solvents in Finnish furniture factories in 1975–1984. *Ann. occup. Hyg.*, **30**, 289–294

Priha, E., Vuorinen, R. & Schimberg, R. (1988) *Textile Finishing Agents* (Työolot 65), Helsinki, Finnish Institute of Occupational Health (in Finnish)

Pross, H.F., Day, J.H., Clark, R.H. & Lees, R.E.M. (1987) Immunologic studies of subjects with asthma exposed to formaldehyde and urea–formaldehyde foam insulation (UFFI) off products. *J. Allergy clin. Immunol.*, **79**, 797–810

Purchase, I.F.H. & Paddle, G.M. (1989) Does formaldehyde cause nasopharyngeal cancer in man? *Cancer Lett.*, **46**, 79–85

Ragan, D.L. & Boreiko, C.J. (1981) Initiation of C3H/10T1/2 cell transformation by formaldehyde. *Cancer Lett.*, **13**, 325–331

Rappaport, B.Z. & Hoffman, M.M. (1941) Urticaria due to aliphatic aldehydes. *J. Am. med. Assoc.*, **116**, 2656-2659

Ratnayake, W.E. (1968) Tests for an effect of the Y-chromosome on the mutagenic action of formaldehyde and X-rays in *Drosophila melanogaster*. *Genet. Res. Camb.*, **12**, 65-69

Ratnayake, W.E. (1970) Studies on the relationship between induced crossing-over and mutation in *Drosophila melanogaster*. *Mutat. Res.*, **9**, 71-83

Recio, L., Sisk, S., Pluta, L., Bermudez, E., Gross, E.A., Chen, Z., Morgan, K. & Walker, C. (1992) *p53* Mutations in formaldehyde-induced nasal squamous cell carcinomas in rats. *Cancer Res.*, **52**, 6113-6116

Restani, P. & Galli, C.L. (1991) Oral toxicity of formaldehyde and its derivatives. *CRC crit. Rev. Toxicol.*, **21**, 315-328

Restani, P., Restelli, A.R. & Galli, C.L. (1992) Formaldehyde and hexamethylenetetramine as food additives: chemical interactions and toxicology. *Food Addit. Contam.*, **9**, 597-605

Reuss, G., Disteldorf, W., Grundler, O. & Hilt, A. (1988) Formaldehyde. In: Gerhartz, W., Yamamoto, Y.S., Elvers, B., Rounsaville, J.F. & Schulz, G., eds, *Ullmann's Encyclopedia of Industrial Chemistry*, 5th rev. Ed., Vol. A11, New York, VCH Publishers, pp. 619-651

Reuzel, P.G.J., Wilmer, J.W.G.M., Woutersen, R.A., Zwart, A., Rombout, P.J.A. & Feron, V.J. (1990) Interactive effects of ozone and formaldehyde on the nasal respiratory lining epithelium in rats. *J. Toxicol. environ. Health*, **29**, 279-292

Riala, R.E. & Riihimäki, H.A. (1991) Solvent and formaldehyde exposure in parquet and carpet work. *Appl. occup. environ. Hyg.*, **6**, 301-308

Rietbrock, N. (1965) Formaldehyde oxidation in the rat. *Naunyn-Schmiedeberg's Arch. exp. Pathol. Pharmakol.*, **251**, 189-190 (in German)

Risby, T.H., Sehnert, S.S., Jakab, G.J. & Hemenway, D.R. (1990) Model to estimate effective doses of adsorbed pollutants on respirable particles and their subsequent release into alveolar surfactant. I. Validation of the model for the adsorption and release of formaldehyde on a respirable carbon black. *Inhal. Toxicol.*, **2**, 223-239

Ritchie, I.M. & Lehnen, R.G. (1987) Formaldehyde-related health complaints of residents living in mobile and conventional homes. *Am. J. public Health*, **77**, 323-328

Robins, J.M., Pambrun, M., Chute, C. & Blevins, D. (1988) Estimating the effect of formaldehyde exposure on lung cancer and non-malignant respiratory disease (NMRD) mortality using a new method to control for the healthy worker survivor effect. In: Hogstedt, C. & Reuterwall, C., eds, *Progress in Occupational Epidemiology*, Amsterdam, Elsevier Science, pp. 75-78

Rosén, G., Bergström, B. & Ekholm, U. (1984) Occupational exposure to formaldehyde in Sweden. *Arbete Hälsa*, **50**, 16-21 (in Swedish)

Ross, W.E. & Shipley, N. (1980) Relationship between DNA damage and survival in formaldehyde-treated mouse cells. *Mutat. Res.*, **79**, 277-283

Ross, W.E., McMillan, D.R. & Ross, C.F. (1981) Comparison of DNA damage by methylmelamines and formaldehyde. *J. natl Cancer Inst.*, **67**, 217-221

Rothenberg, S.J., Nagy, P.A., Pickrell, J.A. & Hobbs, C.H. (1989) Surface area, adsorption, and desorption studies on indoor dust samples. *Am. ind. Hyg. Assoc. J.*, **50**, 15-23

Roush, G.C., Walrath, J., Stayner, L.T., Kaplan, S.A., Flannery, J.T. & Blair, A. (1987) Nasopharyngeal cancer, sinonasal cancer, and occupations related to formaldehyde: a case-control study. *J natl Cancer Inst.*, **79**, 1221-1224

Rusch, G.M., Clary, J.J., Rinehart, W.E. & Bolte, H.F. (1983) A 26-week inhalation toxicity study with formaldehyde in the monkey, rat, and hamster. *Toxicol. appl. Pharmacol.*, **68**, 329–343

Sadtler Research Laboratories (1991) *Sadtler Standard Spectra. 1981–1991 Supplementary Index*, Philadelphia, PA

Saillenfait, A.M., Bonnet, P. & de Ceaurriz, J. (1989) The effects of maternally inhaled formaldehyde on embryonal and foetal development in rats. *Food chem. Toxicol.*, **27**, 545–548

Saladino, A.J., Willey, J.C., Lechner, J.F., Grafström, R.C., LaVeck, M. & Harris, C.C. (1985) Effects of formaldehyde, acetaldehyde, benzoyl peroxide, and hydrogen peroxide on cultured normal human bronchial epithelial cells. *Cancer Res.*, **45**, 2522–2526

Salisbury, S. (1983) *Dialysis Clinic Inc., Atlanta, GA, Health Hazard Evaluation Report* (NIOSH Report No. HETA 83-284-1536), Cincinnati, OH, US Department of Health and Human Services, Public Health Service, Centers for Disease Control, National Institute for Occupational Safety and Health

Salkie, M.L. (1991) The prevalence of atopy and hypersensitivity to formaldehyde in pathologists. *Arch. Pathol. Lab. Med.*, **115**, 614–616

Sangster, J. (1989) Octanol–water partition coefficients of simple organic compounds. *J. phys. chem. Ref. Data*, **18**, 1163

Sass-Kortsak, A.M., Holness, D.L., Pilger, C.W. & Nethercott, J.R. (1986) Wood dust and formaldehyde exposures in the cabinet-making industry. *Am. ind. Hyg. Assoc. J.*, **47**, 747–753

Sauder, L.R., Chatham, M.D., Green, D.J. & Kulle, T.J. (1986) Acute pulmonary response to formaldehyde exposure in healthy nonsmokers. *J. occup. Med.*, **28**, 420–424

Sauder, L.R., Green, D.J., Chatham, M.D. & Kulle, T.J. (1987) Acute pulmonary response of asthmatics to 3.0 ppm formaldehyde. *Toxicol. ind. Health*, **3**, 569–578

Schachter, E.N., Witek, T.J., Jr, Tosun, T. & Beck, G.J. (1986) A study of respiratory effects from exposure to 2 ppm formaldehyde in healthy subjects. *Arch. environ. Health*, **41**, 229–239

Schachter, E.N., Witek, T.J., Jr, Brody, D.J., Tosun, T., Beck, G.J. & Leaderer, B.P. (1987) A study of respiratory effects from exposure to 2.0 ppm formaldehyde in occupationally exposed workers. *Environ. Res.*, **44**, 188–205

Scheuplein, R.J. (1985) Formaldehyde: the Food and Drug Administration's perspective. *Adv. Chem. Ser.*, **210**, 237–245

Schmid, E., Göggelmann, W. & Bauchinger, M. (1986) Formaldehyde-induced cytotoxic, genotoxic and mutagenic response in human lymphocytes and *Salmonella typhimurium*. *Mutagenesis*, **1**, 427–431

Schreider, J.P. (1986) Comparative anatomy and function of the nasal passages. In: Barrow, C.S., ed., *Toxicology of the Nasal Passages*, Washington DC, Hemisphere, pp. 1–25

Sellakumar, A.R., Snyder, C.A., Solomon, J.J. & Albert, R.E. (1985) Carcinogenicity of formaldehyde and hydrogen chloride in rats. *Toxicol. appl. Pharmacol.*, **81**, 401–406

de Serres, F.J., Brockman, H.E. & Hung, C.Y. (1988) Effect of the homokaryotic state of the *uvs*-2 allele in *Neurospora crassa* on formaldehyde-induced killing and *ad*-3 mutation. *Mutat. Res.*, **199**, 235–242

Sexton, K., Petreas, M.X. & Liu, K.-S. (1989) Formaldehyde exposures inside mobile homes. *Environ. Sci. Technol.*, **23**, 985–988

Siboulet, R., Grinfeld, S., Deparis, P. & Jaylet, A. (1984) Micronuclei in red blood cells of the newt *Pleurodeles waltl* Michah: induction with X-rays and chemicals. *Mutat. Res.*, **125**, 275–281

Siemiatycki, J., Wacholder, S., Dewar, R., Wald, L., Bégin, D., Richardson, L., Rosenman, K. & Gérin, M. (1988) Smoking and degree of occupational exposure: are internal analyses in cohort studies likely to be confounded by smoking status? *Am. J. ind. Med.*, **13**, 59–69

Sittig, M. (1985) *Handbook of Toxic and Hazardous Chemicals and Carcinogens*, 2nd Ed., Park Ridge, NJ, Noyes Publications, pp. 462–463

Skisak, C.M. (1983) Formaldehyde vapor exposures in anatomy laboratories. *Am. ind. Hyg. Assoc. J.*, **44**, 948–950

Smith, R. (1993) Environmental economics and the new paradigm. *Chem. Ind. Newsl.*, Nov.–Dec., 8

Smith, E.L., Hill, R.L., Lehman, I.R., Lefkowitz, R.J., Handler, P. & White, A. (1983) *Principles of Biochemistry: Mammalian Biochemistry*, New York, McGraw-Hill, pp. 3–4, 142

Snyder, R.D. & Van Houten, B. (1986) Genotoxicity of formaldehyde and an evaluation of its effects on the DNA repair process in human diploid fibroblasts. *Mutat. Res.*, **165**, 21–30

Sobels, F.H. & van Steenis, H. (1957) Chemical induction of crossing-over in *Drosophila* males. *Nature*, **179**, 29–31

Soffritti, M., Maltoni, C., Maffei, F. & Biagi, R. (1989) Formaldehyde: an experimental multipotential carcinogen. *Toxicol. ind. Health*, **5**, 699–730

Šrám, R.J. (1970) The effect of storage on the frequency of dominant lethals in *Drosophila melanogaster*. *Mol. gen. Genet.*, **106**, 286–288

Stayner, L., Smith, A.B., Reeve, G., Blade, L., Elliott, L., Keenlyside, R. & Halperin, W. (1985) Proportionate mortality study of workers in the garment industry exposed to formaldehyde. *Am. J. ind. Med.*, **7**, 229–240

Stayner, L.T., Elliott, L., Blade, L., Keenlyside, R. & Halperin, W. (1988) A retrospective cohort mortality study of workers exposed to formaldehyde in the garment industry. *Am. J. ind. Med.*, **13**, 667–681

Steinhagen, W.H. & Barrow, C.S. (1984) Sensory irritation structure–activity study of inhaled aldehydes in B6C3F1 mice and Swiss–Webster mice. *Toxicol. appl. Pharmacol.*, **72**, 495–503

Sterling, T.D. & Weinkam, J.J. (1976) Smoking characteristics by type of employment. *J. occup. Med.*, **18**, 743–754

Sterling, T.D. & Weinkam, J.J. (1988) Reanalysis of lung cancer mortality in a National Cancer Institute study on mortality among industrial workers exposed to formaldehyde. *J. occup. Med.*, **30**, 895–901

Sterling, T.D. & Weinkam, J.J. (1989a) Reanalysis of lung cancer mortality in a National Cancer Institute study on 'Mortality among industrial workers exposed to formaldehyde'. *Exp. Pathol.*, **37**, 128–132

Sterling, T.D. & Weinkam, J.J. (1989b) Reanalysis of lung cancer mortality in a National Cancer Institute study of 'Mortality among industrial workers exposed to formaldehyde': additional discussion. *J. occup. Med.*, **31**, 881–884

Sterling, T.D. & Weinkam, J.J. (1994) Mortality from respiratory cancers (including lung cancer) among workers employed in formaldehyde industries. *Am. J. ind. Med.*, **25**, 593–602

Stewart, P.A., Blair, A., Cubit, D.A., Bales, R.E., Kaplan, S.A., Ward, J., Gaffey, W., O'Berg, M.T. & Walrath, J. (1986) Estimating historical exposures to formaldehyde in a retrospective mortality study. *Appl. ind. Hyg.*, **1**, 34–41

Stewart, P.A., Cubit, D.A., Blair, A. & Spirtas, R. (1987a) Performance of two formaldehyde passive dosimeters. *Appl. ind. Hyg.*, **2**, 61–65

Stewart, P.A., Cubit, D.A. & Blair, A. (1987b) Formaldehyde levels in seven industries. *Appl. ind. Hyg.*, **2**, 231–236

Stewart, P.A., Schairer, C. & Blair, A. (1990) Comparison of jobs, exposures, and mortality risks for short-term and long-term workers. *J. occup. Med.*, **32**, 703–708

Stewart, P.A., Herrick, R.F., Feigley, C.E., Utterback, D.F., Hornung, R., Mahar, H., Hayes, R., Douthit, D.E. & Blair, A. (1992) Study design for assessing exposures of embalmers for a case–control study. Part I. Monitoring results. *Appl. occup. environ. Hyg.*, **7**, 532–540

Stroup, N.E., Blair, A. & Erikson, G.E. (1986) Brain cancer and other causes of death in anatomists. *J. natl Cancer Inst.*, **77**, 1217–1224

Stumm-Tegethoff, B.F.A. (1969) Formaldehyde-induced mutations in *Drosophila melanogaster* in dependence of the presence of acids. *Theoret. appl. Genet.*, **39**, 330–334

Sundin, E.B. (1985) The formaldehyde situation in Europe. In: *Proceedings of the 19th Washington State University International Particleboard/Composite Materials Symposium*, Pullman, WA, Washington State University, pp. 255–275

Suruda, A., Schulte, P., Boeniger, M., Hayes, R.B., Livingston, G.K., Steenland, K., Stewart, P., Herrick, R., Douthit, D. & Fingerhut, M.A. (1993) Cytogenetic effects of formaldehyde exposure in students of mortuary science. *Cancer Epidemiol. Biomarkers Prev.*, **2**, 453–460

Swenberg, J.A., Gross, E.A., Randall, H.W. & Barrow, C.S. (1983) The effect of formaldehyde exposure on cytotoxicity and cell proliferation. In: Clary, J.J., Gibson, J.E. & Waritz, R.S., eds, *Formaldehyde: Toxicity, Epidemiology, Mechanisms*, New York, Marcel Decker, pp. 225–236

Swiecichowski, A.L., Long, K.J., Miller, M.L. & Leikauf, G.D. (1993) Formaldehyde-induced airway hyperreactivity *in vivo* and *ex vivo* in guinea pigs. *Environ. Res.*, **61**, 185–199

Takahashi, K., Morita, T. & Kawazoe, Y. (1985) Mutagenic characteristics of formaldehyde on bacterial systems. *Mutat. Res.*, **156**, 153–161

Takahashi, M., Hasegawa, R., Furukawa, F., Toyoda, K., Sato, H. & Hayashi, Y. (1986) Effects of ethanol, potassium metabisulfite, formaldehyde and hydrogen peroxide on gastric carcinogenesis in rats after initiation with *N*-methyl-*N'*-nitro-*N*-nitrosoguanidine. *Jpn. J. Cancer Res.*, **77**, 118–124

Tamburro, C.H. & Waddell, W.J. (1987) Re: Cancers of the nasopharynx and oropharynx and formaldehyde exposure. *J. natl Cancer Inst.*, **79**, 605

Taskinen, H., Kyyrönen, P., Hemminki, K., Hoikkala, M., Lajunen, K. & Lindbohm, M.-L. (1994) Laboratory work and pregnancy outcome. *J. occup. Med.*, **36**, 311–319

Temcharoen, P. & Thilly, W.G. (1983) Toxic and mutagenic effects of formaldehyde in *Salmonella typhimurium*. *Mutat. Res.*, **119**, 89–93

Thomson, E.J., Shackleton, S. & Harrington, J.M. (1984) Chromosome aberrations and sister-chromatid exchange frequencies in pathology staff occupationally exposed to formaldehyde. *Mutat. Res.*, **141**, 89–93

Thrasher, J.D., Wojdani, A., Cheung, G. & Heuser, G. (1987) Evidence for formaldehyde antibodies and altered cellular immunity in subjects exposed to formaldehyde in mobile homes. *Arch. environ. Health*, **42**, 347–350

Thrasher, J.D., Broughton, A. & Micevich, P. (1988) Antibodies and immune profiles of individuals occupationally exposed to formaldehyde: six case reports. *Am. J. ind. Med.*, **14**, 479–488

Thrasher, J.D., Broughton, A. & Madison, R. (1990) Immune activation and autoantibodies in humans with long-term inhalation exposure to formaldehyde. *Arch. environ. Health*, **45**, 217–223

Til, H.P., Woutersen, R.A., Feron, V.J. & Clary, J.J. (1988) Evaluation of the oral toxicity of acetaldehyde and formaldehyde in a 4-week drinking-water study in rats. *Food chem. Toxicol.*, **26**, 447–452

Til, H.P., Woutersen, R.A., Feron, V.J., Hollanders, V.H.M. & Falke, H.E. (1989) Two-year drinking-water study of formaldehyde in rats. *Food. chem. Toxicol.*, **27**, 77–87

Tobe, M., Naito, K. & Kurokawa, Y. (1989) Chronic toxicity study on formaldehyde administered orally to rats. *Toxicology*, **56**, 79–86

Triebig, G., Schaller, K.-H., Berger, B., Müller, J. & Valentin, H. (1989) Formaldehyde exposure at various workplaces. *Sci. total Environ.*, **79**, 191–195

Työministeriö [Ministry of Labour] (1993) *Limit Values 1993*, Helsinki, p. 12 (in Finnish)

Uba, G., Pachorek, D., Bernstein, J., Garabrant, D.H., Balmes, J.R., Wright, W.E. & Amar, R.B. (1989) Prospective study of respiratory effects of formaldehyde among healthy and asthmatic medical students. *Am. J. ind. Med.*, **15**, 91–101

United Kingdom Health and Safety Executive (1992) *EH40/92 Occupational Exposure Limits 1992*, London, Her Majesty's Sationery Office, p. 11

United States Department of Housing and Urban Development (1994) Housing and urban development. *US Code fed. Regul.*, **Title 24**, Part 3280.308, pp. 233–234

United States Environmental Protection Agency (1988a) *Compendium of Methods for the Determination of Toxic Organic Compounds in Ambient Air* (EPA Report No. EPA-600/4-89-017; US NTIS PB90-116989), Research Triangle Park, NC, Office of Research and Development, pp. TO5-1–TO5-22

United States Environmental Protection Agency (1988b) *Compendium of Methods for the Determination of Toxic Organic Compounds in Ambient Air* (EPA Report No. EPA-600/4-89-017; US NTIS PB90-116989), Research Triangle Park, NC, Office of Research and Development, pp. TO11-1–TO11-38

United States Environmental Protection Agency (1993) Protection of environment. *US Code fed. Regul.*, **Title 40**, Parts 180.1001, 180.1024, 180.1032, 185.4650, pp. 435–458, 461, 463, 494

United States Food and Drug Administration (1994a) Food and drugs. *US Code fed. Regul.*, **Title 21**, Parts 173.340, 175.105, 176.170, 176.180, 176.200, 176.210, 177.2800, 178.3120, pp. 121–122, 127–142, 168–200, 308–310, 337–338

United States Food and Drug Administration (1994b) Food and drugs. *US Code fed. Regul.*, **Title 21**, Parts 529.1030, 573.460, pp. 345, 513

United States National Institute for Occupational Safety and Health (1976) *Criteria for a Recommended Standard... Occupational Exposure to Formaldehyde* (DHEW (NIOSH) Publ No. 77-126), Washington DC, US Government Printing Office

United States National Institute of Occupational Safety and Health (1990) *National Occupational Exposure Survey 1981–83*, Cincinnati, OH

United States National Institute of Occupational Safety and Health (1992) *NIOSH Recommendations for Occupational Safety and Health. Compendium of Policy Documents and Statements* (DHHS (NIOSH) Publ. No. 92-100), Cincinnati, OH, Division of Standards Development and Technology Transfer, p. 84

United States National Research Council (1980) *Formaldehyde: An Assessment of Its Health Effects*, Washington DC, National Academy Press

United States National Research Council (1981) Health effects of formaldehyde. In: *Formaldehyde and Other Aldehydes*, Washington DC, National Academy Press, pp. 175–220, 306–340

United States Occupational Safety and Health Administration (1990) *OSHA Analytical Methods Manual*, 2nd Ed., Part 1, Vol. 2 (Methods 29–54), Salt Lake City, UT, pp. 52-1–52-38

United States Occupational Safety and Health Administration (1991) *OSHA Analytical Methods Manual*, 2nd Ed., Part 2, Vol. 2 (Methods ID-160 to ID-210), Salt Lake City, UT, Method ID-205

United States Occupational Safety and Health Administration (1993) Formaldehyde. *US Code fed. Regul.*, **Title 29**, Part 1910.1048, pp. 410–442

Uotila, L. & Koivusalo, M. (1974a) Formaldehyde dehydrogenase from human liver. Purification, properties, and evidence for the formation of glutathione thiol esters by the enzyme. *J. biol. Chem.*, **249**, 7653–7663

Uotila, L. & Koivusalo, M. (1974b) Purification and properties of S-formylglutathione hydrolase from human liver. *J. biol. Chem.*, **249**, 7664–7672

Uotila, L. & Koivusalo, M. (1987) Multiple forms of formaldehyde dehydrogenase from human red blood cells. *Hum. Hered.*, **37**, 102–106

Vargová, M., Janota, S., Karelová, J., Barančokova, M. & Šulcová, M. (1992) Analysis of the health risk of occupational exposure to formaldehyde using biological markers. *Analysis*, **20**, 451–454

Vaughan, T.L., Strader, C., Davis, S. & Daling, J.R. (1986a) Formaldehyde and cancers of the pharynx, sinus and nasal cavity: I. Occupational exposures. *Int. J. Cancer*, **38**, 677–683

Vaughan, T.L., Strader, C., Davis, S. & Daling, J.R. (1986b) Formaldehyde and cancers of the pharynx, sinus and nasal cavity: II. Residential exposures. *Int. J. Cancer*, **38**, 685–688

Vaught, C. (1991) *Locating and Estimating Air Emissions From Sources of Formaldehyde (Revised)* (Report No. EPA-450/4-91-012; US NTIS PB91-181842), Research Triangle Park, NC, Environmental Protection Agency

Vinzents, P. & Laursen, B. (1993) A national cross-sectional study of the working environment in the Danish wood and furniture industry—air pollution and noise. *Ann. occup. Hyg.*, **37**, 25–34

Walrath, J. & Fraumeni, J.F., Jr (1983) Mortality patterns among embalmers. *Int. J. Cancer*, **31**, 407–411

Walrath, J. & Fraumeni, J.F., Jr (1984) Cancer and other causes of death among embalmers. *Cancer Res.*, **44**, 4638–4641

Walrath, J., Rogot, E., Murray, J. & Blair, A. (1985) *Mortality Patterns among US Veterans by Occupation and Smoking Status* (NIH Publ. No. 85-2756), Bethesda, MD, Department of Health and Human Services

Watanabe, F., Matsunaga, T., Soejima, T. & Iwata, Y. (1954) Study on the carcinogenicity of aldehyde. First report. Experimentally produced rat sarcomas by repeated injections of aqueous solution of formaldehyde. *Gann*, **45**, 451–452 (in Japanese)

Ward, J.B., Jr, Hokanson, J.A., Smith, E.R., Chang, L.W., Pereira, M.A., Whorton, E.B., Jr & Legator, M.S. (1984) Sperm count, morphology and fluorescent body frequency in autopsy service workers exposed to formaldehyde. *Mutat. Res.*, **130**, 417–424

Waydhas, C., Weigl, K. & Sies, H. (1978) The disposition of formaldehyde and formate arising from drug N-demethylations dependent on cytochrome P-450 in hepatocytes and in perfused rat liver. *Eur. J. Biochem.*, **89**, 143–150

Weast, R.C. & Astle, M.J., eds (1985) *CRC Handbook of Data on Organic Compounds*, Vol. I, Boca Raton, FL, CRC Press, p. 641

Weber-Tschopp, A., Fischer, T. & Grandjean, E. (1977) Irritating effects of formaldehyde (HCHO) in man. *Int. Arch. occup. environ. Health*, **39**, 207–218 (in German)

West, S., Hildesheim, A. & Dosemeci, M. (1993) Non-viral risk factors for nasopharyngeal carcinoma in the Philippines: results from a case–control study. *Int. J. Cancer*, **55**, 722–727

WHO (1989) *Formaldehyde* (Environmental Health Criteria 89), Geneva, International Programme on Chemical Safety

WHO (1991) *Formaldehyde Health and Safety Guide* (Health and Safety Guide No. 57), Geneva, International Programme on Chemical Safety

Wilkins, R.J., & MacLeod, H.D. (1976) Formaldehyde induced DNA–protein crosslinks in *Escherichia coli*. *Mutat. Res.*, **36**, 11–16

Williams, T.M., Levine, R.J. & Blunden, P.B. (1984) Exposure of embalmers to formaldehyde and other chemicals. *Am. ind. Hyg. Assoc. J.*, **45**, 172–176

Wilmer, J.W.G.M., Woutersen, R.A., Appelman, L.M., Leeman, W.R. & Feron, V.J. (1987) Subacute (4-week) inhalation toxicity study of formaldehyde in male rats: 8-hour intermittent *versus* 8-hour continuous exposures. *J. appl. Toxicol.*, **7**, 15–16

Wilmer, J.W.G.M., Woutersen, R.A., Appelman, L.M., Leeman, W.R. & Feron, V.J. (1989) Subchronic (13-week) inhalation toxicity study of formaldehyde in male rats: 8-hour intermittent versus 8-hour continuous exposures. *Toxicol. Lett.*, **47**, 287–293

Witek, T.J., Jr, Schachter, E.N., Tosun, T., Beck, G.J. & Leaderer, B.P. (1987) An evaluation of respiratory effects following exposure to 2.0 ppm formaldehyde in asthmatics: lung function, symptoms, and airway reactivity. *Arch. environ. Health*, **42**, 230–237

Wong, O. (1983) An epidemiologic mortality study of a cohort of chemical workers potentially exposed to formaldehyde, with a discussion on SMR and PMR. In: Gibson, J.E., ed., *Formaldehyde Toxicity*, New York, Hemisphere, pp. 256–272

Wortley, P., Vaughan, T.L., Davis, S., Morgan, M.S. & Thomas, D.B. (1992) A case–control study of occupational risk factors for laryngeal cancer. *Br. J. ind. Med.*, **49**, 837–844

Woutersen, R.A., Appelman, L.M., Wilmer, J.W.G.M., Falke, H.E. & Feron, V.J. (1987) Subchronic (13-week) inhalation toxicity study of formaldehyde in rats. *J. appl. Toxicol.*, **7**, 43–49

Woutersen, R.A., van Garderen-Hoetmer, A., Bruijntjes, J.P., Zwart, A. & Feron, V.J. (1989) Nasal tumours in rats after severe injury to the nasal mucosa and prolonged exposure to 10 ppm formaldehyde. *J. appl. Toxicol.*, **9**, 39–46

Yager, J.W., Cohn, K.L., Spear, R.C., Fisher, J.M. & Morse, L.M. (1986) Sister-chromatid exchanges in lymphocytes of anatomy students exposed to formaldehyde-embalming solution. *Mutat. Res.*, **174**, 135–139

Zijlstra, J.A. (1989) Liquid holding increases mutation induction by formaldehyde and some other cross-linking agents in *Escherichia coli* K12. *Mutat. Res.*, **210**, 255–261

Zimmermann, F.K. & Mohr, A. (1992) Formaldehyde, glyoxal, urethane, methyl carbamate, 2,3-butenedione, 2,3-hexanedione, ethyl acrylate, dibromoacetonitrile and 2-hydroxypropionitrile induce chromosome loss in *Saccharomyces cerevisiae*. *Mutat. Res.*, **270**, 151–166

Zwart, A., Woutersen, R.A., Wilmer, J.W.G.M., Spit, B.J. & Feron, V.J. (1988) Cytotoxic and adaptive effects in rat nasal epithelium after 3-day and 13-week exposure to low concentrations of formaldehyde vapour. *Toxicology*, **51**, 87–99

APPENDIX 1

SUMMARY TABLE OF
GENETIC AND RELATED EFFECTS

APPENDIX 1

Summary table of genetic and related effects of formaldehyde

Non-mammalian systems											Mammalian systems																					
Prokaryotes		Lower eukaryotes				Plants				Insects					Animal cells (*In vitro*)								Human cells						Animal (*In vivo*)					Humans			
D	G	D	R	G	A	D	G	C	R	G	C	A	D	G	S	M	C	A	T	I	D	G	S	M	C	A	T	I	D	G	S	M	C	DL	A	D	S	M	C	A
+	+	+	+	+			+¹		+	+	+	+	+	+¹	+	+		+		+¹	+	+	+	+		+			+	⁻	+¹	?	?	?	?	?	?¹	?

A, aneuploidy; C, chromosomal aberrations; D, DNA damage; DL, dominant lethal mutation; G, gene mutation; I, inhibition of intercellular communication; M, micronuclei; R, mitotic recombination and gene conversion; S, sister chromatid exchange; T, cell transformation

In completing the table, the following symbols indicate the consensus of the Working Group with regard to the results for each end-point:

+ considered to be positive for the specific end-point and level of biological complexity
+¹ considered to be positive, but only one valid study was available to the Working Group
− considered to be negative
−¹ considered to be negative, but only one valid study was available to the Working Group
? considered to be equivocal or inconclusive (e.g. there were contradictory results from different laboratories; there were confounding exposures; the results were equivocal)

APPENDIX 2

**ACTIVITY PROFILE FOR
GENETIC AND RELATED EFFECTS**

APPENDIX 2

ACTIVITY PROFILE FOR GENETIC AND RELATED EFFECTS

Methods

The x-axis of the activity profile (Waters *et al.*, 1987, 1988) represents the bioassays in phylogenetic sequence by end-point, and the values on the y-axis represent the logarithmically transformed lowest effective doses (LED) and highest ineffective doses (HID) tested. The term 'dose', as used in this report, does not take into consideration length of treatment or exposure and may therefore be considered synonymous with concentration. In practice, the concentrations used in all the in-vitro tests were converted to µg/ml, and those for in-vivo tests were expressed as mg/kg bw. Because dose units are plotted on a log scale, differences in the relative molecular masses of compounds do not, in most cases, greatly influence comparisons of their activity profiles. Conventions for dose conversions are given below.

Profile-line height (the magnitude of each bar) is a function of the LED or HID, which is associated with the characteristics of each individual test system—such as population size, cell-cycle kinetics and metabolic competence. Thus, the detection limit of each test system is different, and, across a given activity profile, responses will vary substantially. No attempt is made to adjust or relate responses in one test system to those of another.

Line heights are derived as follows: for negative test results, the highest dose tested without appreciable toxicity is defined as the HID. If there was evidence of extreme toxicity, the next highest dose is used. A single dose tested with a negative result is considered to be equivalent to the HID. Similarly, for positive results, the LED is recorded. If the original data were analysed statistically by the author, the dose recorded is that at which the response was significant ($p < 0.05$). If the available data were not analysed statistically, the dose required to produce an effect is estimated as follows: when a dose-related positive response is observed with two or more doses, the lower of the doses is taken as the LED; a single dose resulting in a positive response is considered to be equivalent to the LED.

In order to accommodate both the wide range of doses encountered and positive and negative responses on a continuous scale, doses are transformed logarithmically, so that effective (LED) and ineffective (HID) doses are represented by positive and negative numbers, respectively. The response, or logarithmic dose unit (LDUij), for a given test system i and chemical j is represented by the expressions

$LDU_{ij} = -\log_{10}$ (dose), for HID values; LDU ≤ 0
and (1)
$LDU_{ij} = -\log_{10}$ (dose $\times 10^{-5}$), for LED values; LDU ≥ 0.

These simple relationships define a dose range of 0 to -5 logarithmic units for ineffective doses (1–100 000 µg/ml or mg/kg bw) and 0 to +8 logarithmic units for effective doses (100 000–0.001 µg/ml or mg/kg bw). A scale illustrating the LDU values is shown in Figure 1. Negative responses at doses less than 1 µg/ml (mg/kg bw) are set equal to 1. Effectively, an LED value $\geq$ 100 000 or an HID value $\leq$ 1 produces an LDU = 0; no quantitative information is gained from such extreme values. The dotted lines at the levels of log dose units 1 and -1 define a 'zone of uncertainty' in which positive results are reported at such high doses (between 10 000 and 100 000 mg/ml or mg/kg bw) or negative results are reported at such low doses (1 to 10 mg/ml or mg/kg bw) as to call into question the adequacy of the test.

Fig. 1. Scale of log dose units used on the y-axis of activity profiles

Positive (µg/ml or mg/kg bw)		Log dose units	
0.001		8	----
0.01		7	--
0.1		6	--
1.0		5	--
10		4	--
100		3	--
1000		2	--
10 000		1	--
100 000	1	0	----
	10	-1	--
	100	-2	--
	1000	-3	--
	10 000	-4	--
	100 000	-5	----

Negative
(µg/ml or mg/kg bw)

In practice, an activity profile is computer generated. A data entry programme is used to store abstracted data from published reports. A sequential file (in ASCII) is created for each compound, and a record within that file consists of the name and Chemical Abstracts Service number of the compound, a three-letter code for the test system (see below), the qualitative test result (with and without an exogenous metabolic system), dose (LED or HID), citation number and additional source information. An abbreviated citation for each publication is stored in a segment of a record accessing both the test data file and the citation file. During processing of the data file, an average of the logarithmic values of the data subset is calculated, and the length of the profile line represents this average value. All dose values are plotted for each profile line, regardless of whether results are positive or negative. Results obtained in the absence of an exogenous metabolic system are indicated by a bar (–), and results obtained in the presence of an exogenous metabolic system are indicated by an upward-directed arrow (↑). When all results for a given assay are either positive or negative, the mean of the LDU values is plotted as a solid line; when conflicting data are reported for the same assay (i.e. both positive and negative results), the majority data are shown by a solid line and the minority data by a dashed line (drawn to the extreme conflicting response). In the few cases in which the numbers of positive and negative results are equal, the solid line is drawn in the positive direction and the maximal negative response is indicated with a dashed line. Profile lines are identified by three-letter code words representing the commonly used tests. Code words for most of the test systems in current use in genetic toxicology were defined for the US Environmental Protection Agency's GENE-TOX Program (Waters, 1979; Waters & Auletta, 1981). For *IARC Monographs* Supplement 6, Volume 44 and subsequent volumes, including this publication, codes were redefined in a manner that should facilitate inclusion of additional tests. Naming conventions are described below.

Data listings are presented in the text and include end-point and test codes, a short test code definition, results, either with (M) or without (NM) an exogenous activation system, the associated LED or HID value and a short citation. Test codes are organized phylogenetically and by end-point from left to right across each activity profile and from top to bottom of the corresponding data listing. End-points are defined as follows: A, aneuploidy; C, chromosomal aberrations; D, DNA damage; F, assays of body fluids; G, gene mutation; H, host-mediated assays; I, inhibition of intercellular communication; M, micronuclei; P, sperm morphology; R, mitotic recombination or gene conversion; S, sister chromatid exchange; and T, cell transformation.

Dose conversions for activity profiles

Doses are converted to µg/ml for in-vitro tests and to mg/kg bw per day for in-vivo experiments.

1. In-vitro test systems

    (a) Weight/volume converts directly to µg/ml.

    (b) Molar (M) concentration × molecular weight = mg/ml = 10^3 mg/ml; mM concentration × molecular weight = µg/ml.

    (c) Soluble solids expressed as % concentration are assumed to be in units of mass per volume (i.e. 1% = 0.01 g/ml = 10 000 µg/ml; also, 1 ppm = 1 µg/ml).

    (d) Liquids and gases expressed as % concentration are assumed to be given in units of volume per volume. Liquids are converted to weight per volume using the density (D) of the solution (D = g/ml). Gases are converted from volume to mass using the ideal gas law, PV = nRT. For exposure at 20–37 °C at standard atmospheric pressure, 1% (v/v) = 0.4 µg/ml × molecular weight of the gas. Also, 1 ppm (v/v) = 4×10^5 µg/ml × molecular weight.

(e) In microbial plate tests, it is usual for the doses to be reported as weight/plate, whereas concentrations are required to enter data on the activity profile chart. While remaining cognisant of the errors involved in the process, it is assumed that a 2-ml volume of top agar is delivered to each plate and that the test substance remains in solution within it; concentrations are derived from the reported weight/plate values by dividing by this arbitrary volume. For spot tests, a 1-ml volume is used in the calculation.

(f) Conversion of particulate concentrations given in µg/cm^2 is based on the area (A) of the dish and the volume of medium per dish; i.e. for a 100-mm dish: $A = \pi R^2 = \pi \times (5\text{ cm})^2 = 78.5\text{ cm}^2$. If the volume of medium is 10 ml, then 78.5 cm^2 = 10 ml and 1 cm^2 = 0.13 ml.

2. In-vitro systems using in-vivo activation

For the body fluid-urine (BF-) test, the concentration used is the dose (in mg/kg bw) of the compound administered to test animals or patients.

3. In-vivo test systems

(a) Doses are converted to mg/kg bw per day of exposure, assuming 100% absorption. Standard values are used for each sex and species of rodent, including body weight and average intake per day, as reported by Gold *et al.* (1984). For example, in a test using male mice fed 50 ppm of the agent in the diet, the standard food intake per day is 12% of body weight, and the conversion is dose = 50 ppm × 12% = 6 mg/kg bw per day.

Standard values used for humans are: weight—males, 70 kg; females, 55 kg; surface area, 1.7 m^2; inhalation rate, 20 L/min for light work, 30 L/min for mild exercise.

(b) When reported, the dose at the target site is used. For example, doses given in studies of lymphocytes of humans exposed *in vivo* are the measured blood concentrations in µg/ml.

Codes for test systems

For specific nonmammalian test systems, the first two letters of the three-symbol code word define the test organism (e.g. SA- for *Salmonella typhimurium*, EC- for *Escherichia coli*). If the species is not known, the convention used is -S-. The third symbol may be used to define the tester strain (e.g. SA8 for *S. typhimurium* TA1538, ECW for *E. coli* WP2*uvr*A). When strain designation is not indicated, the third letter is used to define the specific genetic end-point under investigation (e.g. --D for differential toxicity, --F for forward mutation, --G for gene conversion or genetic crossing-over, --N for aneuploidy, --R for reverse mutation, --U for unscheduled DNA synthesis). The third letter may also be used to define the general end-point under investigation when a more complete definition is not possible or relevant (e.g. --M for mutation, --C for chromosomal aberration). For mammalian test systems, the first letter of the three-letter code word defines the genetic end-point under investigation: A-- for aneuploidy, B-- for binding, C-- for chromosomal aberration, D-- for DNA strand breaks, G-- for gene mutation, I-- for inhibition of intercellular communication, M-- for micronucleus formation, R-- for DNA repair, S-- for sister chromatid exchange, T-- for cell transformation and U-- for unscheduled DNA synthesis.

For animal (i.e. non-human) test systems *in vitro*, when the cell type is not specified, the code letters -IA are used. For such assays *in vivo*, when the animal species is not specified, the code letters -VA are used. Commonly used animal species are identified by the third letter (e.g. --C for Chinese hamster, --M for mouse, --R for rat, --S for Syrian hamster).

For test systems using human cells *in vitro*, when the cell type is not specified, the code letters -IH are used. For assays on humans *in vivo*, when the cell type is not specified, the code letters -VH are used. Otherwise, the second letter specifies the cell type under investigation (e.g. -BH for bone marrow, -LH for lymphocytes).

Some other specific coding conventions used for mammalian systems are as follows: BF- for body fluids, HM- for host-mediated, --L for leukocytes or lymphocytes *in vitro* (-AL, animals; -HL, humans), -L- for leukocytes *in vivo* (-LA, animals; -LH, humans), --T for transformed cells.

Note that these are examples of major conventions used to define the assay code words. The alphabetized listing of codes must be examined to confirm a specific code word. As might be expected from the limitation to three symbols, some codes do not fit the naming conventions precisely. In a few cases, test systems are defined by first-letter code words, for example: MST, mouse spot test; SLP, mouse specific locus mutation, postspermatogonia; SLO, mouse specific locus mutation, other stages; DLM, dominant lethal mutation in mice; DLR, dominant lethal mutation in rats; MHT, mouse heritable translocation.

The genetic activity profiles and listings were prepared in collaboration with Environmental Health Research and Testing Inc. (EHRT) under contract to the United States Environmental Protection Agency; EHRT also determined the doses used. The references cited in each genetic activity profile listing can be found in the list of references in the appropriate monograph.

References

Garrett, N.E., Stack, H.F., Gross, M.R. & Waters, M.D. (1984) An analysis of the spectra of genetic activity produced by known or suspected human carcinogens. *Mutat. Res.*, **134**, 89–111

Gold, L.S., Sawyer, C.B., Magaw, R., Backman, G.M., de Veciana, M., Levinson, R., Hooper, N.K., Havender, W.R., Bernstein, L., Peto, R., Pike, M.C. & Ames, B.N. (1984) A carcinogenic potency database of the standardized results of animal bioassays. *Environ. Health Perspect.*, **58**, 9–319

Waters, M.D. (1979) *The GENE-TOX program*. In: Hsie, A.W., O'Neill, J.P. & McElheny, V.K., eds, *Mammalian Cell Mutagenesis: The Maturation of Test Systems* (Banbury Report 2), Cold Spring Harbor, NY, CSH Press, pp. 449–467

Waters, M.D. & Auletta, A. (1981) The GENE-TOX program: genetic activity evaluation. *J. chem. Inf. comput. Sci.*, **21**, 35–38

Waters, M.D., Stack, H.F., Brady, A.L., Lohman, P.H.M., Haroun, L. & Vainio, H. (1987) Appendix 1: Activity profiles for genetic and related tests. In: *IARC Monographs on the Evaluation of the Carcinogenic Risk of Chemicals to Humans*, Suppl. 6, *Genetic and Related Effects: An Updating of Selected* IARC Monographs *from Volumes 1 to 42*, Lyon, IARC, pp. 687–696

Waters, M.D., Stack, H.F., Brady, A.L., Lohman, P.H.M., Haroun, L. & Vainio, H. (1988) Use of computerized data listings and activity profiles of genetic and related effects in the review of 195 compounds. *Mutat. Res.*, **205**, 295–312

APPENDIX 2

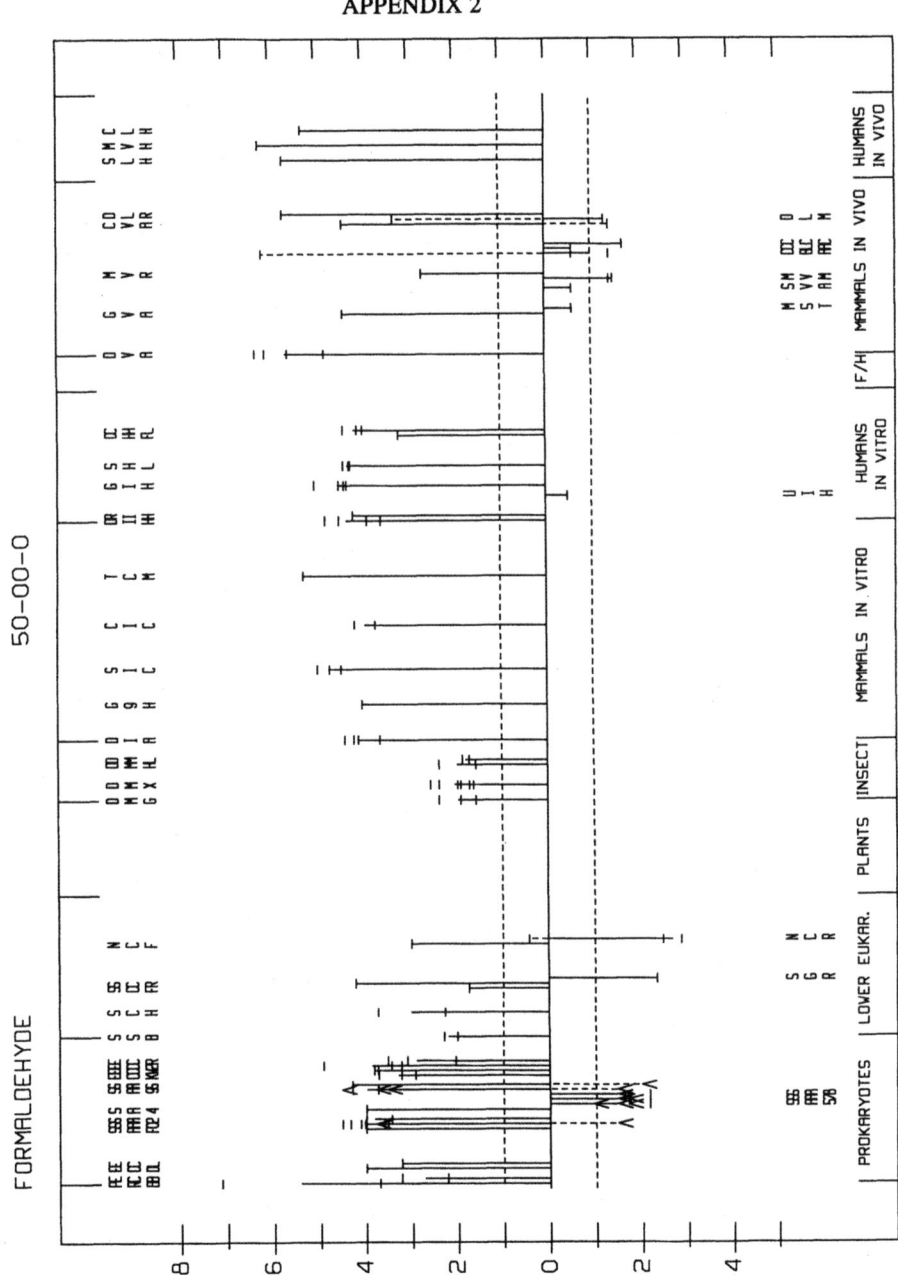

CUMULATIVE CROSS INDEX TO *IARC MONOGRAPHS ON THE EVALUATION OF CARCINOGENIC RISKS TO HUMANS*

The volume, page and year of publication are given. References to corrigenda are given in parentheses.

A

A-α-C	*40*, 245 (1986); *Suppl. 7*, 56 (1987)
Acetaldehyde	*36*, 101 (1985) (*corr. 42*, 263); *Suppl. 7*, 77 (1987)
Acetaldehyde formylmethylhydrazone (*see* Gyromitrin)	
Acetamide	*7*, 197 (1974); *Suppl. 7*, 389 (1987)
Acetaminophen (*see* Paracetamol)	
Acridine orange	*16*, 145 (1978); *Suppl. 7*, 56 (1987)
Acriflavinium chloride	*13*, 31 (1977); *Suppl. 7*, 56 (1987)
Acrolein	*19*, 479 (1979); *36*, 133 (1985); *Suppl. 7*, 78 (1987)
Acrylamide	*39*, 41 (1986); *Suppl. 7*, 56 (1987); *60*, 389 (1994)
Acrylic acid	*19*, 47 (1979); *Suppl. 7*, 56 (1987)
Acrylic fibres	*19*, 86 (1979); *Suppl. 7*, 56 (1987)
Acrylonitrile	*19*, 73 (1979); *Suppl. 7*, 79 (1987)
Acrylonitrile-butadiene-styrene copolymers	*19*, 91 (1979); *Suppl. 7*, 56 (1987)
Actinolite (*see* Asbestos)	
Actinomycins	*10*, 29 (1976) (*corr. 42*, 255); *Suppl. 7*, 80 (1987)
Adriamycin	*10*, 43 (1976); *Suppl. 7*, 82 (1987)
AF-2	*31*, 47 (1983); *Suppl. 7*, 56 (1987)
Aflatoxins	*1*, 145 (1972) (*corr. 42*, 251); *10*, 51 (1976); *Suppl. 7*, 83 (1987); *56*, 245 (1993)
Aflatoxin B_1 (*see* Aflatoxins)	
Aflatoxin B_2 (*see* Aflatoxins)	
Aflatoxin G_1 (*see* Aflatoxins)	
Aflatoxin G_2 (*see* Aflatoxins)	
Aflatoxin M_1 (*see* Aflatoxins)	
Agaritine	*31*, 63 (1983); *Suppl. 7*, 56 (1987)
Alcohol drinking	*44* (1988)
Aldicarb	*53*, 93 (1991)
Aldrin	*5*, 25 (1974); *Suppl. 7*, 88 (1987)
Allyl chloride	*36*, 39 (1985); *Suppl. 7*, 56 (1987)
Allyl isothiocyanate	*36*, 55 (1985); *Suppl. 7*, 56 (1987)

Allyl isovalerate	*36*, 69 (1985); *Suppl. 7*, 56 (1987)
Aluminium production	*34*, 37 (1984); *Suppl. 7*, 89 (1987)
Amaranth	*8*, 41 (1975); *Suppl. 7*, 56 (1987)
5-Aminoacenaphthene	*16*, 243 (1978); *Suppl. 7*, 56 (1987)
2-Aminoanthraquinone	*27*, 191 (1982); *Suppl. 7*, 56 (1987)
para-Aminoazobenzene	*8*, 53 (1975); *Suppl. 7*, 390 (1987)
ortho-Aminoazotoluene	*8*, 61 (1975) (*corr. 42*, 254); *Suppl. 7*, 56 (1987)
para-Aminobenzoic acid	*16*, 249 (1978); *Suppl. 7*, 56 (1987)
4-Aminobiphenyl	*1*, 74 (1972) (*corr. 42*, 251); *Suppl. 7*, 91 (1987)
2-Amino-3,4-dimethylimidazo[4,5-*f*]quinoline (*see* MeIQ)	
2-Amino-3,8-dimethylimidazo[4,5-*f*]quinoxaline (*see* MeIQx)	
3-Amino-1,4-dimethyl-5*H*-pyrido[4,3-*b*]indole (*see* Trp-P-1)	
2-Aminodipyrido[1,2-*a*:3′,2′-*d*]imidazole (*see* Glu-P-2)	
1-Amino-2-methylanthraquinone	*27*, 199 (1982); *Suppl. 7*, 57 (1987)
2-Amino-3-methylimidazo[4,5-*f*]quinoline (*see* IQ)	
2-Amino-6-methyldipyrido[1,2-*a*:3′,2′-*d*]imidazole (*see* Glu-P-1)	
2-Amino-1-methyl-6-phenylimidazo[4,5-*b*]pyridine (*see* PhIP)	
2-Amino-3-methyl-9*H*-pyrido[2,3-*b*]indole (*see* MeA-α-C)	
3-Amino-1-methyl-5*H*-pyrido[4,3-*b*]indole (*see* Trp-P-2)	
2-Amino-5-(5-nitro-2-furyl)-1,3,4-thiadiazole	*7*, 143 (1974); *Suppl. 7*, 57 (1987)
4-Amino-2-nitrophenol	*16*, 43 (1978); *Suppl. 7*, 57 (1987)
2-Amino-4-nitrophenol	*57*, 167 (1993)
2-Amino-5-nitrophenol	*57*, 177 (1993)
2-Amino-5-nitrothiazole	*31*, 71 (1983); *Suppl. 7*, 57 (1987)
2-Amino-9*H*-pyrido[2,3-*b*]indole (*see* A-α-C)	
11-Aminoundecanoic acid	*39*, 239 (1986); *Suppl. 7*, 57 (1987)
Amitrole	*7*, 31 (1974); *41*, 293 (1986) (*corr. 52*, 513; *Suppl. 7*, 92 (1987)
Ammonium potassium selenide (*see* Selenium and selenium compounds)	
Amorphous silica (*see also* Silica)	*42*, 39 (1987); *Suppl. 7*, 341 (1987)
Amosite (*see* Asbestos)	
Ampicillin	*50*, 153 (1990)
Anabolic steroids (*see* Androgenic (anabolic) steroids)	
Anaesthetics, volatile	*11*, 285 (1976); *Suppl. 7*, 93 (1987)
Analgesic mixtures containing phenacetin (*see also* Phenacetin)	*Suppl. 7*, 310 (1987)
Androgenic (anabolic) steroids	*Suppl. 7*, 96 (1987)
Angelicin and some synthetic derivatives (*see also* Angelicins)	*40*, 291 (1986)
Angelicin plus ultraviolet radiation (*see also* Angelicin and some synthetic derivatives)	*Suppl. 7*, 57 (1987)
Angelicins	*Suppl. 7*, 57 (1987)
Aniline	*4*, 27 (1974) (*corr. 42*, 252); *27*, 39 (1982); *Suppl. 7*, 99 (1987)
ortho-Anisidine	*27*, 63 (1982); *Suppl. 7*, 57 (1987)
para-Anisidine	*27*, 65 (1982); *Suppl. 7*, 57 (1987)
Anthanthrene	*32*, 95 (1983); *Suppl. 7*, 57 (1987)
Anthophyllite (*see* Asbestos)	
Anthracene	*32*, 105 (1983); *Suppl. 7*, 57 (1987)
Anthranilic acid	*16*, 265 (1978); *Suppl. 7*, 57 (1987)
Antimony trioxide	*47*, 291 (1989)

Antimony trisulfide	47, 291 (1989)
ANTU (see 1-Naphthylthiourea)	
Apholate	9, 31 (1975); Suppl. 7, 57 (1987)
Aramite®	5, 39 (1974); Suppl. 7, 57 (1987)
Areca nut (see Betel quid)	
Arsanilic acid (see Arsenic and arsenic compounds)	
Arsenic and arsenic compounds	1, 41 (1972); 2, 48 (1973); 23, 39 (1980); Suppl. 7, 100 (1987)
Arsenic pentoxide (see Arsenic and arsenic compounds)	
Arsenic sulfide (see Arsenic and arsenic compounds)	
Arsenic trioxide (see Arsenic and arsenic compounds)	
Arsine (see Arsenic and arsenic compounds)	
Asbestos	2, 17 (1973) (corr. 42, 252); 14 (1977) (corr. 42, 256); Suppl. 7, 106 (1987) (corr. 45, 283)
Atrazine	53, 441 (1991)
Attapulgite	42, 159 (1987); Suppl. 7, 117 (1987)
Auramine (technical-grade)	1, 69 (1972) (corr. 42, 251); Suppl. 7,118 (1987)
Auramine, manufacture of (see also Auramine, technical-grade)	Suppl. 7, 118 (1987)
Aurothioglucose	13, 39 (1977); Suppl. 7, 57 (1987)
Azacitidine	26, 37 (1981); Suppl. 7, 57 (1987); 50, 47 (1990)
5-Azacytidine (see Azacitidine)	
Azaserine	10, 73 (1976) (corr. 42, 255); Suppl. 7, 57 (1987)
Azathioprine	26, 47 (1981); Suppl. 7, 119 (1987)
Aziridine	9, 37 (1975); Suppl. 7, 58 (1987)
2-(1-Aziridinyl)ethanol	9, 47 (1975); Suppl. 7, 58 (1987)
Aziridyl benzoquinone	9, 51 (1975); Suppl. 7, 58 (1987)
Azobenzene	8, 75 (1975); Suppl. 7, 58 (1987)

B

Barium chromate (see Chromium and chromium compounds)	
Basic chromic sulfate (see Chromium and chromium compounds)	
BCNU (see Bischloroethyl nitrosourea)	
Benz[a]acridine	32, 123 (1983); Suppl. 7, 58 (1987)
Benz[c]acridine	3, 241 (1973); 32, 129 (1983); Suppl. 7, 58 (1987)
Benzal chloride (see also α-Chlorinated toluenes)	29, 65 (1982); Suppl. 7, 148 (1987)
Benz[a]anthracene	3, 45 (1973); 32, 135 (1983); Suppl. 7, 58 (1987)
Benzene	7, 203 (1974) (corr. 42, 254); 29, 93, 391 (1982); Suppl. 7, 120 (1987)
Benzidine	1, 80 (1972); 29, 149, 391 (1982); Suppl. 7, 123 (1987)
Benzidine-based dyes	Suppl. 7, 125 (1987)

Benzo[*b*]fluoranthene	*3*, 69 (1973); *32*, 147 (1983); *Suppl. 7*, 58 (1987)
Benzo[*j*]fluoranthene	*3*, 82 (1973); *32*, 155 (1983); *Suppl. 7*, 58 (1987)
Benzo[*k*]fluoranthene	*32*, 163 (1983); *Suppl. 7*, 58 (1987)
Benzo[*ghi*]fluoranthene	*32*, 171 (1983); *Suppl. 7*, 58 (1987)
Benzo[*a*]fluorene	*32*, 177 (1983); *Suppl. 7*, 58 (1987)
Benzo[*b*]fluorene	*32*, 183 (1983); *Suppl. 7*, 58 (1987)
Benzo[*c*]fluorene	*32*, 189 (1983); *Suppl. 7*, 58 (1987)
Benzo[*ghi*]perylene	*32*, 195 (1983); *Suppl. 7*, 58 (1987)
Benzo[*c*]phenanthrene	*32*, 205 (1983); *Suppl. 7*, 58 (1987)
Benzo[*a*]pyrene	*3*, 91 (1973); *32*, 211 (1983); *Suppl. 7*, 58 (1987)
Benzo[*e*]pyrene	*3*, 137 (1973); *32*, 225 (1983); *Suppl. 7*, 58 (1987)
para-Benzoquinone dioxime	*29*, 185 (1982); *Suppl. 7*, 58 (1987)
Benzotrichloride (*see also* α-Chlorinated toluenes)	*29*, 73 (1982); *Suppl. 7*, 148 (1987)
Benzoyl chloride	*29*, 83 (1982) (*corr. 42*, 261); *Suppl. 7*, 126 (1987)
Benzoyl peroxide	*36*, 267 (1985); *Suppl. 7*, 58 (1987)
Benzyl acetate	*40*, 109 (1986); *Suppl. 7*, 58 (1987)
Benzyl chloride (*see also* α-Chlorinated toluenes)	*11*, 217 (1976) (*corr. 42*, 256); *29*, 49 (1982); *Suppl. 7*, 148 (1987)
Benzyl violet 4B	*16*, 153 (1978); *Suppl. 7*, 58 (1987)
Bertrandite (*see* Beryllium and beryllium compounds)	
Beryllium and beryllium compounds	*1*, 17 (1972); *23*, 143 (1980) (*corr. 42*, 260); *Suppl. 7*, 127 (1987); *58*, 41 (1993)
Beryllium acetate (*see* Beryllium and beryllium compounds)	
Beryllium acetate, basic (*see* Beryllium and beryllium compounds)	
Beryllium-aluminium alloy (*see* Beryllium and beryllium compounds)	
Beryllium carbonate (*see* Beryllium and beryllium compounds)	
Beryllium chloride (*see* Beryllium and beryllium compounds)	
Beryllium-copper alloy (*see* Beryllium and beryllium compounds)	
Beryllium-copper-cobalt alloy (*see* Beryllium and beryllium compounds)	
Beryllium fluoride (*see* Beryllium and beryllium compounds)	
Beryllium hydroxide (*see* Beryllium and beryllium compounds)	
Beryllium-nickel alloy (*see* Beryllium and beryllium compounds)	
Beryllium oxide (*see* Beryllium and beryllium compounds)	
Beryllium phosphate (*see* Beryllium and beryllium compounds)	
Beryllium silicate (*see* Beryllium and beryllium compounds)	
Beryllium sulfate (*see* Beryllium and beryllium compounds)	
Beryl ore (*see* Beryllium and beryllium compounds)	
Betel quid	*37*, 141 (1985); *Suppl. 7*, 128 (1987)
Betel-quid chewing (*see* Betel quid)	
BHA (*see* Butylated hydroxyanisole)	
BHT (*see* Butylated hydroxytoluene)	
Bis(1-aziridinyl)morpholinophosphine sulfide	*9*, 55 (1975); *Suppl. 7*, 58 (1987)
Bis(2-chloroethyl)ether	*9*, 117 (1975); *Suppl. 7*, 58 (1987)

N,N-Bis(2-chloroethyl)-2-naphthylamine	*4*, 119 (1974) (*corr. 42*, 253); *Suppl. 7*, 130 (1987)
Bischloroethyl nitrosourea (see also Chloroethyl nitrosoureas)	*26*, 79 (1981); *Suppl. 7*, 150 (1987)
1,2-Bis(chloromethoxy)ethane	*15*, 31 (1977); *Suppl. 7*, 58 (1987)
1,4-Bis(chloromethoxymethyl)benzene	*15*, 37 (1977); *Suppl. 7*, 58 (1987)
Bis(chloromethyl)ether	*4*, 231 (1974) (*corr. 42*, 253); *Suppl. 7*, 131 (1987)
Bis(2-chloro-1-methylethyl)ether	*41*, 149 (1986); *Suppl. 7*, 59 (1987)
Bis(2,3-epoxycyclopentyl)ether	*47*, 231 (1989)
Bisphenol A diglycidyl ether (*see* Glycidyl ethers)	
Bisulfites (see Sulfur dioxide and some sulfites, bisulfites and metabisulfites)	
Bitumens	*35*, 39 (1985); *Suppl. 7*, 133 (1987)
Bleomycins	*26*, 97 (1981); *Suppl. 7*, 134 (1987)
Blue VRS	*16*, 163 (1978); *Suppl. 7*, 59 (1987)
Boot and shoe manufacture and repair	*25*, 249 (1981); *Suppl. 7*, 232 (1987)
Bracken fern	*40*, 47 (1986); *Suppl. 7*, 135 (1987)
Brilliant Blue FCF, disodium salt	*16*, 171 (1978) (*corr. 42*, 257); *Suppl. 7*, 59 (1987)
Bromochloroacetonitrile (*see* Halogenated acetonitriles)	
Bromodichloromethane	*52*, 179 (1991)
Bromoethane	*52*, 299 (1991)
Bromoform	*52*, 213 (1991)
1,3-Butadiene	*39*, 155 (1986) (*corr. 42*, 264); *Suppl. 7*, 136 (1987); *54*, 237 (1992)
1,4-Butanediol dimethanesulfonate	*4*, 247 (1974); *Suppl. 7*, 137 (1987)
n-Butyl acrylate	*39*, 67 (1986); *Suppl. 7*, 59 (1987)
Butylated hydroxyanisole	*40*, 123 (1986); *Suppl. 7*, 59 (1987)
Butylated hydroxytoluene	*40*, 161 (1986); *Suppl. 7*, 59 (1987)
Butyl benzyl phthalate	*29*, 193 (1982) (*corr. 42*, 261); *Suppl. 7*, 59 (1987)
β-Butyrolactone	*11*, 225 (1976); *Suppl. 7*, 59 (1987)
γ-Butyrolactone	*11*, 231 (1976); *Suppl. 7*, 59 (1987)

C

Cabinet-making (*see* Furniture and cabinet#making)	
Cadmium acetate (*see* Cadmium and cadmium compounds)	
Cadmium and cadmium compounds	*2*, 74 (1973); *11*, 39 (1976) (*corr. 42*, 255); *Suppl. 7*, 139 (1987); *58*, 119 (1993)
Cadmium chloride (*see* Cadmium and cadmium compounds)	
Cadmium oxide (*see* Cadmium and cadmium compounds)	
Cadmium sulfate (*see* Cadmium and cadmium compounds)	
Cadmium sulfide (*see* Cadmium and cadmium compounds)	
Caffeic acid	*56*, 115 (1993)
Caffeine	*51*, 291 (1991)
Calcium arsenate (*see* Arsenic and arsenic compounds)	
Calcium chromate (see Chromium and chromium compounds)	
Calcium cyclamate (*see* Cyclamates)	
Calcium saccharin (*see* Saccharin)	

Cantharidin	*10*, 79 (1976); *Suppl. 7*, 59 (1987)
Caprolactam	*19*, 115 (1979) (*corr. 42*, 258); *39*, 247 (1986) (*corr. 42*, 264); *Suppl. 7*, 390 (1987)
Captafol	*53*, 353 (1991)
Captan	*30*, 295 (1983); *Suppl. 7*, 59 (1987)
Carbaryl	*12*, 37 (1976); *Suppl. 7*, 59 (1987)
Carbazole	*32*, 239 (1983); *Suppl. 7*, 59 (1987)
3-Carbethoxypsoralen	*40*, 317 (1986); *Suppl. 7*, 59 (1987)
Carbon blacks	*3*, 22 (1973); *33*, 35 (1984); *Suppl. 7*, 142 (1987)
Carbon tetrachloride	*1*, 53 (1972); *20*, 371 (1979); *Suppl. 7*, 143 (1987)
Carmoisine	*8*, 83 (1975); *Suppl. 7*, 59 (1987)
Carpentry and joinery	*25*, 139 (1981); *Suppl. 7*, 378 (1987)
Carrageenan	*10*, 181 (1976) (*corr. 42*, 255); *31*, 79 (1983); *Suppl. 7*, 59 (1987)
Catechol	*15*, 155 (1977); *Suppl. 7*, 59 (1987)
CCNU (*see* 1-(2-Chloroethyl)-3-cyclohexyl-1-nitrosourea)	
Ceramic fibres (see Man#made mineral fibres)	
Chemotherapy, combined, including alkylating agents (*see* MOPP and other combined chemotherapy including alkylating agents)	
Chlorambucil	*9*, 125 (1975); *26*, 115 (1981); *Suppl. 7*, 144 (1987)
Chloramphenicol	*10*, 85 (1976); *Suppl. 7*, 145 (1987); *50*, 169 (1990)
Chlordane (*see also* Chlordane/Heptachlor)	*20*, 45 (1979) (*corr. 42*, 258)
Chlordane/Heptachlor	*Suppl. 7*, 146 (1987); *53*, 115 (1991)
Chlordecone	*20*, 67 (1979); *Suppl. 7*, 59 (1987)
Chlordimeform	*30*, 61 (1983); *Suppl. 7*, 59 (1987)
Chlorendic acid	*48*, 45 (1990)
Chlorinated dibenzodioxins (other than TCDD)	*15*, 41 (1977); *Suppl. 7*, 59 (1987)
Chlorinated drinking-water	*52*, 45 (1991)
Chlorinated paraffins	*48*, 55 (1990)
α-Chlorinated toluenes	*Suppl. 7*, 148 (1987)
Chlormadinone acetate (*see also* Progestins; Combined oral contraceptives)	*6*, 149 (1974); *21*, 365 (1979)
Chlornaphazine (*see N,N*-Bis(2-chloroethyl)-2-naphthylamine)	
Chloroacetonitrile (*see* Halogenated acetonitriles)	
para-Chloroaniline	*57*, 305 (1993)
Chlorobenzilate	*5*, 75 (1974); *30*, 73 (1983); *Suppl. 7*, 60 (1987)
Chlorodibromomethane	*52*, 243 (1991)
Chlorodifluoromethane	*41*, 237 (1986) (*corr. 51*, 483); *Suppl. 7*, 149 (1987)
Chloroethane	*52*, 315 (1991)
1-(2-Chloroethyl)-3-cyclohexyl-1-nitrosourea (*see also* Chloroethyl nitrosoureas)	*26*, 137 (1981) (*corr. 42*, 260); *Suppl. 7*, 150 (1987)
1-(2-Chloroethyl)-3-(4-methylcyclohexyl)-1-nitrosourea (*see also* Chloroethyl nitrosoureas)	*Suppl. 7*, 150 (1987)
Chloroethyl nitrosoureas	*Suppl. 7*, 150 (1987)

Chlorofluoromethane	*41*, 229 (1986); *Suppl. 7*, 60 (1987)
Chloroform	*1*, 61 (1972); *20*, 401 (1979)
	Suppl. 7, 152 (1987)
Chloromethyl methyl ether (technical-grade) (*see also* Bis(chloromethyl)ether)	*4*, 239 (1974); *Suppl. 7*, 131 (1987)
(4-Chloro-2-methylphenoxy)acetic acid (*see* MCPA)	
Chlorophenols	*Suppl. 7*, 154 (1987)
Chlorophenols (occupational exposures to)	*41*, 319 (1986)
Chlorophenoxy herbicides	*Suppl. 7*, 156 (1987)
Chlorophenoxy herbicides (occupational exposures to)	*41*, 357 (1986)
4-Chloro-*ortho*-phenylenediamine	*27*, 81 (1982); *Suppl. 7*, 60 (1987)
4-Chloro-*meta*-phenylenediamine	*27*, 82 (1982); *Suppl. 7*, 60 (1987)
Chloroprene	*19*, 131 (1979); *Suppl. 7*, 160 (1987)
Chloropropham	*12*, 55 (1976); *Suppl. 7*, 60 (1987)
Chloroquine	*13*, 47 (1977); *Suppl. 7*, 60 (1987)
Chlorothalonil	*30*, 319 (1983); *Suppl. 7*, 60 (1987)
para-Chloro-*ortho*-toluidine and its strong acid salts (*see also* Chlordimeform)	*16*, 277 (1978); *30*, 65 (1983); *Suppl. 7*, 60 (1987); *48*, 123 (1990)
Chlorotrianisene (*see also* Nonsteroidal oestrogens)	*21*, 139 (1979)
2-Chloro-1,1,1-trifluoroethane	*41*, 253 (1986); *Suppl. 7*, 60 (1987)
Chlorozotocin	*50*, 65 (1990)
Cholesterol	*10*, 99 (1976); *31*, 95 (1983);
	Suppl. 7, 161 (1987)
Chromic acetate (*see* Chromium and chromium compounds)	
Chromic chloride (*see* Chromium and chromium compounds)	
Chromic oxide (*see* Chromium and chromium compounds)	
Chromic phosphate (*see* Chromium and chromium compounds)	
Chromite ore (*see* Chromium and chromium compounds)	
Chromium and chromium compounds	*2*, 100 (1973); *23*, 205 (1980); *Suppl. 7*, 165 (1987); *49*, 49 (1990) (*corr. 51*, 483)
Chromium carbonyl (*see* Chromium and chromium compounds)	
Chromium potassium sulfate (*see* Chromium and chromium compounds)	
Chromium sulfate (*see* Chromium and chromium compounds)	
Chromium trioxide (*see* Chromium and chromium compounds)	
Chrysazin (*see* Dantron)	
Chrysene	*3*, 159 (1973); *32*, 247 (1983); *Suppl. 7*, 60 (1987)
Chrysoidine	*8*, 91 (1975); *Suppl. 7*, 169 (1987)
Chrysotile (*see* Asbestos)	
CI Acid Orange 3	*57*, 121 (1993)
CI Acid Red 114	*57*, 247 (1993)
CI Basic Red 9	*57*, 215 (1993)
Ciclosporin	*50*, 77 (1990)
CI Direct Blue 15	*57*, 235 (1993)
CI Disperse Yellow 3 (*see* Disperse Yellow 3)	
Cimetidine	*50*, 235 (1990)
Cinnamyl anthranilate	*16*, 287 (1978); *31*, 133 (1983); *Suppl. 7*, 60 (1987)
CI Pigment Red 3	*57*, 259 (1993)
CI Pigment Red 53:1 (*see* D&C Red No. 9)	

Cisplatin	*26*, 151 (1981); *Suppl. 7*, 170 (1987)
Citrinin	*40*, 67 (1986); *Suppl. 7*, 60 (1987)
Citrus Red No. 2	*8*, 101 (1975) (*corr. 42*, 254)
	Suppl. 7, 60 (1987)
Clofibrate	*24*, 39 (1980); *Suppl. 7*, 171 (1987)
Clomiphene citrate	*21*, 551 (1979); *Suppl. 7*, 172 (1987)
Clonorchis sinensis (infection with)	*61*, 121 (1994)
Coal gasification	*34*, 65 (1984); *Suppl. 7*, 173 (1987)
Coal-tar pitches (*see also* Coal#tars)	*35*, 83 (1985); *Suppl. 7*, 174 (1987)
Coal-tars	*35*, 83 (1985); *Suppl. 7*, 175 (1987)
Cobalt[III] acetate (*see* Cobalt and cobalt compounds)	
Cobalt-aluminium-chromium spinel (*see* Cobalt and cobalt compounds)	
Cobalt and cobalt compounds	*52*, 363 (1991)
Cobalt[II] chloride (*see* Cobalt and cobalt compounds)	
Cobalt-chromium alloy (*see* Chromium and chromium compounds)	
Cobalt-chromium-molybdenum alloys (*see* Cobalt and cobalt compounds)	
Cobalt metal powder (*see* Cobalt and cobalt compounds)	
Cobalt naphthenate (*see* Cobalt and cobalt compounds)	
Cobalt[II] oxide (*see* Cobalt and cobalt compounds)	
Cobalt[II,III] oxide (*see* Cobalt and cobalt compounds)	
Cobalt[II] sulfide (*see* Cobalt and cobalt compounds)	
Coffee	*51*, 41 (1991) (*corr. 52*, 513)
Coke production	*34*, 101 (1984); *Suppl. 7*, 176 (1987)
Combined oral contraceptives (*see also* Oestrogens, progestins and combinations)	*Suppl. 7*, 297 (1987)
Conjugated oestrogens (*see also* Steroidal oestrogens)	*21*, 147 (1979)
Contraceptives, oral (*see* Combined oral contraceptives; Sequential oral contraceptives)	
Copper 8-hydroxyquinoline	*15*, 103 (1977); *Suppl. 7*, 61 (1987)
Coronene	*32*, 263 (1983); *Suppl. 7*, 61 (1987)
Coumarin	*10*, 113 (1976); *Suppl. 7*, 61 (1987)
Creosotes (*see also* Coal-tars)	*35*, 83 (1985); *Suppl. 7*, 177 (1987)
meta-Cresidine	*27*, 91 (1982); *Suppl. 7*, 61 (1987)
para-Cresidine	*27*, 92 (1982); *Suppl. 7*, 61 (1987)
Crocidolite (*see* Asbestos)	
Crude oil	*45*, 119 (1989)
Crystalline silica (*see also* Silica)	*42*, 39 (1987); *Suppl. 7*, 341 (1987)
Cycasin	*1*, 157 (1972) (*corr. 42*, 251); *10*, 121 (1976); *Suppl. 7*, 61 (1987)
Cyclamates	*22*, 55 (1980); *Suppl. 7*, 178 (1987)
Cyclamic acid (*see* Cyclamates)	
Cyclochlorotine	*10*, 139 (1976); *Suppl. 7*, 61 (1987)
Cyclohexanone	*47*, 157 (1989)
Cyclohexylamine (*see* Cyclamates)	
Cyclopenta[*cd*]pyrene	*32*, 269 (1983); *Suppl. 7*, 61 (1987)
Cyclopropane (*see* Anaesthetics, volatile)	
Cyclophosphamide	*9*, 135 (1975); *26*, 165 (1981); *Suppl. 7*, 182 (1987)

D

2,4-D (*see also* Chlorophenoxy herbicides; Chlorophenoxy herbicides, occupational exposures to)	*15*, 111 (1977)
Dacarbazine	*26*, 203 (1981); *Suppl. 7*, 184 (1987)
Dantron	*50*, 265 (1990) (*corr. 59*, 257)
D&C Red No. 9	*8*, 107 (1975); *Suppl. 7*, 61 (1987); *57*, 203 (1993)
Dapsone	*24*, 59 (1980); *Suppl. 7*, 185 (1987)
Daunomycin	*10*, 145 (1976); *Suppl. 7*, 61 (1987)
DDD (*see* DDT)	
DDE (*see* DDT)	
DDT	*5*, 83 (1974) (*corr. 42*, 253); *Suppl. 7*, 186 (1987); *53*, 179 (1991)
Decabromodiphenyl oxide	*48*, 73 (1990)
Deltamethrin	*53*, 251 (1991)
Deoxynivalenol (*see* Toxins derived from *Fusarium graminearum, F. culmorum* and *F. crookwellense*)	
Diacetylaminoazotoluene	*8*, 113 (1975); *Suppl. 7*, 61 (1987)
N,N'-Diacetylbenzidine	*16*, 293 (1978); *Suppl. 7*, 61 (1987)
Diallate	*12*, 69 (1976); *30*, 235 (1983); *Suppl. 7*, 61 (1987)
2,4-Diaminoanisole	*16*, 51 (1978); *27*, 103 (1982); *Suppl. 7*, 61 (1987)
4,4'-Diaminodiphenyl ether	*16*, 301 (1978); *29*, 203 (1982); *Suppl. 7*, 61 (1987)
1,2-Diamino-4-nitrobenzene	*16*, 63 (1978); *Suppl. 7*, 61 (1987)
1,4-Diamino-2-nitrobenzene	*16*, 73 (1978); *Suppl. 7*, 61 (1987); *57*, 185 (1993)
2,6-Diamino-3-(phenylazo)pyridine (*see* Phenazopyridine hydrochloride)	
2,4-Diaminotoluene (*see also* Toluene diisocyanates)	*16*, 83 (1978); *Suppl. 7*, 61 (1987)
2,5-Diaminotoluene (*see also* Toluene diisocyanates)	*16*, 97 (1978); *Suppl. 7*, 61 (1987)
ortho-Dianisidine (*see* 3,3'-Dimethoxybenzidine)	
Diazepam	*13*, 57 (1977); *Suppl. 7*, 189 (1987)
Diazomethane	*7*, 223 (1974); *Suppl. 7*, 61 (1987)
Dibenz[*a,h*]acridine	*3*, 247 (1973); *32*, 277 (1983); *Suppl. 7*, 61 (1987)
Dibenz[*a,j*]acridine	*3*, 254 (1973); *32*, 283 (1983); *Suppl. 7*, 61 (1987)
Dibenz[*a,c*]anthracene	*32*, 289 (1983) (*corr. 42*, 262); *Suppl. 7*, 61 (1987)
Dibenz[*a,h*]anthracene	*3*, 178 (1973) (*corr. 43*, 261); *32*, 299 (1983); *Suppl. 7*, 61 (1987)
Dibenz[*a,j*]anthracene	*32*, 309 (1983); *Suppl. 7*, 61 (1987)
7*H*-Dibenzo[*c,g*]carbazole	*3*, 260 (1973); *32*, 315 (1983); *Suppl. 7*, 61 (1987)
Dibenzodioxins, chlorinated (other than TCDD) [*see* Chlorinated dibenzodioxins (other than TCDD)]	
Dibenzo[*a,e*]fluoranthene	*32*, 321 (1983); *Suppl. 7*, 61 (1987)
Dibenzo[*h,rst*]pentaphene	*3*, 197 (1973); *Suppl. 7*, 62 (1987)

Dibenzo[*a,e*]pyrene	3, 201 (1973); 32, 327 (1983); Suppl. 7, 62 (1987)
Dibenzo[*a,h*]pyrene	3, 207 (1973); 32, 331 (1983); Suppl. 7, 62 (1987)
Dibenzo[*a,i*]pyrene	3, 215 (1973); 32, 337 (1983); Suppl. 7, 62 (1987)
Dibenzo[*a,l*]pyrene	3, 224 (1973); 32, 343 (1983); Suppl. 7, 62 (1987)
Dibromoacetonitrile (*see* Halogenated acetonitriles)	
1,2-Dibromo-3-chloropropane	15, 139 (1977); 20, 83 (1979); Suppl. 7, 191 (1987)
Dichloroacetonitrile (*see* Halogenated acetonitriles)	
Dichloroacetylene	39, 369 (1986); Suppl. 7, 62 (1987)
ortho-Dichlorobenzene	7, 231 (1974); 29, 213 (1982); Suppl. 7, 192 (1987)
para-Dichlorobenzene	7, 231 (1974); 29, 215 (1982); Suppl. 7, 192 (1987)
3,3'-Dichlorobenzidine	4, 49 (1974); 29, 239 (1982); Suppl. 7, 193 (1987)
trans-1,4-Dichlorobutene	15, 149 (1977); Suppl. 7, 62 (1987)
3,3'-Dichloro-4,4'-diaminodiphenyl ether	16, 309 (1978); Suppl. 7, 62 (1987)
1,2-Dichloroethane	20, 429 (1979); Suppl. 7, 62 (1987)
Dichloromethane	20, 449 (1979); 41, 43 (1986); Suppl. 7, 194 (1987)
2,4-Dichlorophenol (*see* Chlorophenols; Chlorophenols, occupational exposures to)	
(2,4-Dichlorophenoxy)acetic acid (*see* 2,4-D)	
2,6-Dichloro-*para*-phenylenediamine	39, 325 (1986); Suppl. 7, 62 (1987)
1,2-Dichloropropane	41, 131 (1986); Suppl. 7, 62 (1987)
1,3-Dichloropropene (technical-grade)	41, 113 (1986); Suppl. 7, 195 (1987)
Dichlorvos	20, 97 (1979); Suppl. 7, 62 (1987); 53, 267 (1991)
Dicofol	30, 87 (1983); Suppl. 7, 62 (1987)
Dicyclohexylamine (*see* Cyclamates)	
Dieldrin	5, 125 (1974); Suppl. 7, 196 (1987)
Dienoestrol (*see also* Nonsteroidal oestrogens)	21, 161 (1979)
Diepoxybutane	11, 115 (1976) (*corr.* 42, 255); Suppl. 7, 62 (1987)
Diesel and gasoline engine exhausts	46, 41 (1989)
Diesel fuels	45, 219 (1989) (*corr.* 47, 505)
Diethyl ether (*see* Anaesthetics, volatile)	
Di(2-ethylhexyl)adipate	29, 257 (1982); Suppl. 7, 62 (1987)
Di(2-ethylhexyl)phthalate	29, 269 (1982) (*corr.* 42, 261); Suppl. 7, 62 (1987)
1,2-Diethylhydrazine	4, 153 (1974); Suppl. 7, 62 (1987)
Diethylstilboestrol	6, 55 (1974); 21, 173 (1979) (*corr.* 42, 259); Suppl. 7, 273 (1987)
Diethylstilboestrol dipropionate (*see* Diethylstilboestrol)	
Diethyl sulfate	4, 277 (1974); Suppl. 7, 198 (1987); 54, 213 (1992)

Diglycidyl resorcinol ether	*11*, 125 (1976); *36*, 181 (1985); Suppl. 7, 62 (1987)
Dihydrosafrole	*1*, 170 (1972); *10*, 233 (1976); Suppl. 7, 62 (1987)
1,8-Dihydroxyanthraquinone (*see* Dantron)	
Dihydroxybenzenes (*see* Catechol; Hydroquinone; Resorcinol)	
Dihydroxymethylfuratrizine	*24*, 77 (1980); Suppl. 7, 62 (1987)
Diisopropyl sulfate	*54*, 229 (1992)
Dimethisterone (*see also* Progestins; Sequential oral contraceptives	*6*, 167 (1974); *21*, 377 (1979))
Dimethoxane	*15*, 177 (1977); Suppl. 7, 62 (1987)
3,3'-Dimethoxybenzidine	*4*, 41 (1974); Suppl. 7, 198 (1987)
3,3'-Dimethoxybenzidine-4,4'-diisocyanate	*39*, 279 (1986); Suppl. 7, 62 (1987)
para-Dimethylaminoazobenzene	*8*, 125 (1975); Suppl. 7, 62 (1987)
para-Dimethylaminoazobenzenediazo sodium sulfonate	*8*, 147 (1975); Suppl. 7, 62 (1987)
trans-2-[(Dimethylamino)methylimino]-5-[2-(5-nitro-2-furyl)-vinyl]-1,3,4-oxadiazole	*7*, 147 (1974) (*corr. 42*, 253); Suppl. 7, 62 (1987)
4,4'-Dimethylangelicin plus ultraviolet radiation (*see also* Angelicin and some synthetic derivatives)	Suppl. 7, 57 (1987)
4,5'-Dimethylangelicin plus ultraviolet radiation (*see also* Angelicin and some synthetic derivatives)	Suppl. 7, 57 (1987)
2,6-Dimethylaniline	*57*, 323 (1993)
N,N-Dimethylaniline	*57*, 337 (1993)
Dimethylarsinic acid (*see* Arsenic and arsenic compounds)	
3,3'-Dimethylbenzidine	*1*, 87 (1972); Suppl. 7, 62 (1987)
Dimethylcarbamoyl chloride	*12*, 77 (1976); Suppl. 7, 199 (1987)
Dimethylformamide	*47*, 171 (1989)
1,1-Dimethylhydrazine	*4*, 137 (1974); Suppl. 7, 62 (1987)
1,2-Dimethylhydrazine	*4*, 145 (1974) (*corr. 42*, 253); Suppl. 7, 62 (1987)
Dimethyl hydrogen phosphite	*48*, 85 (1990)
1,4-Dimethylphenanthrene	*32*, 349 (1983); Suppl. 7, 62 (1987)
Dimethyl sulfate	*4*, 271 (1974); Suppl. 7, 200 (1987)
3,7-Dinitrofluoranthene	*46*, 189 (1989)
3,9-Dinitrofluoranthene	*46*, 195 (1989)
1,3-Dinitropyrene	*46*, 201 (1989)
1,6-Dinitropyrene	*46*, 215 (1989)
1,8-Dinitropyrene	*33*, 171 (1984); Suppl. 7, 63 (1987); *46*, 231 (1989)
Dinitrosopentamethylenetetramine	*11*, 241 (1976); Suppl. 7, 63 (1987)
1,4-Dioxane	*11*, 247 (1976); Suppl. 7, 201 (1987)
2,4'-Diphenyldiamine	*16*, 313 (1978); Suppl. 7, 63 (1987)
Direct Black 38 (*see also* Benzidine-based dyes)	*29*, 295 (1982) (*corr. 42*, 261)
Direct Blue 6 (*see also* Benzidine-based dyes)	*29*, 311 (1982)
Direct Brown 95 (*see also* Benzidine-based dyes)	*29*, 321 (1982)
Disperse Blue 1	*48*, 139 (1990)
Disperse Yellow 3	*8*, 97 (1975); Suppl. 7, 60 (1987); *48*, 149 (1990)
Disulfiram	*12*, 85 (1976); Suppl. 7, 63 (1987)
Dithranol	*13*, 75 (1977); Suppl. 7, 63 (1987)
Divinyl ether (*see* Anaesthetics, volatile)	
Dulcin	*12*, 97 (1976); Suppl. 7, 63 (1987)

E

Endrin	5, 157 (1974); *Suppl. 7*, 63 (1987)
Enflurane (*see* Anaesthetics, volatile)	
Eosin	15, 183 (1977); *Suppl. 7*, 63 (1987)
Epichlorohydrin	11, 131 (1976) (*corr. 42*, 256); *Suppl. 7*, 202 (1987)
1,2-Epoxybutane	47, 217 (1989)
1-Epoxyethyl-3,4-epoxycyclohexane (*see* 4-Vinylcyclohexene diepoxide)	
3,4-Epoxy-6-methylcyclohexylmethyl-3,4-epoxy-6-methyl-cyclohexane carboxylate	11, 147 (1976); *Suppl. 7*, 63 (1987)
cis-9,10-Epoxystearic acid	11, 153 (1976); *Suppl. 7*, 63 (1987)
Erionite	42, 225 (1987); *Suppl. 7*, 203 (1987)
Ethinyloestradiol (*see also* Steroidal oestrogens)	6, 77 (1974); 21, 233 (1979)
Ethionamide	13, 83 (1977); *Suppl. 7*, 63 (1987)
Ethyl acrylate	19, 57 (1979); 39, 81 (1986); *Suppl. 7*, 63 (1987)
Ethylene	19, 157 (1979); *Suppl. 7*, 63 (1987); 60, 45 (1994)
Ethylene dibromide	15, 195 (1977); *Suppl. 7*, 204 (1987)
Ethylene oxide	11, 157 (1976); 36, 189 (1985) (*corr. 42*, 263); *Suppl. 7*, 205 (1987); 60, 73 (1994)
Ethylene sulfide	11, 257 (1976); *Suppl. 7*, 63 (1987)
Ethylene thiourea	7, 45 (1974); *Suppl. 7*, 207 (1987)
2-Ethylhexyl acrylate	60, 475 (1994)
Ethyl methanesulfonate	7, 245 (1974); *Suppl. 7*, 63 (1987)
N-Ethyl-*N*-nitrosourea	1, 135 (1972); 17, 191 (1978); *Suppl. 7*, 63 (1987)
Ethyl selenac (*see also* Selenium and selenium compounds)	12, 107 (1976); *Suppl. 7*, 63 (1987)
Ethyl tellurac	12, 115 (1976); *Suppl. 7*, 63 (1987)
Ethynodiol diacetate (*see also* Progestins; Combined oral contraceptives)	6, 173 (1974); 21, 387 (1979)
Eugenol	36, 75 (1985); *Suppl. 7*, 63 (1987)
Evans blue	8, 151 (1975); *Suppl. 7*, 63 (1987)

F

Fast Green FCF	16, 187 (1978); *Suppl. 7*, 63 (1987)
Fenvalerate	53, 309 (1991)
Ferbam	12, 121 (1976) (*corr. 42*, 256); *Suppl. 7*, 63 (1987)
Ferric oxide	1, 29 (1972); *Suppl. 7*, 216 (1987)
Ferrochromium (*see* Chromium and chromium compounds)	
Fluometuron	30, 245 (1983); *Suppl. 7*, 63 (1987)
Fluoranthene	32, 355 (1983); *Suppl. 7*, 63 (1987)
Fluorene	32, 365 (1983); *Suppl. 7*, 63 (1987)
Fluorescent lighting (exposure to) (*see* Ultraviolet radiation)	
Fluorides (inorganic, used in drinking-water)	27, 237 (1982); *Suppl. 7*, 208 (1987)
5-Fluorouracil	26, 217 (1981); *Suppl. 7*, 210 (1987)

Fluorspar (*see* Fluorides)	
Fluosilicic acid (*see* Fluorides)	
Fluroxene (*see* Anaesthetics, volatile)	
Formaldehyde	*29*, 345 (1982); *Suppl. 7*, 211 (1987); *62*, 217 (1995)
2-(2-Formylhydrazino)-4-(5-nitro-2-furyl)thiazole	*7*, 151 (1974) (*corr. 42*, 253); *Suppl. 7*, 63 (**1987**)
Frusemide (*see* Furosemide)	
Fuel oils (heating oils)	*45*, 239 (1989) (*corr. 47*, 505)
Fumonisin B_1 (*see* Toxins derived from Fusarium moniliforme)	
Fumonisin B_2 (*see* Toxins derived from Fusarium moniliforme)	
Furazolidone	*31*, 141 (1983); *Suppl. 7*, 63 (1987)
Furniture and cabinet-making	*25*, 99 (1981); *Suppl. 7*, 380 (1987)
Furosemide	*50*, 277 (1990)
2-(2-Furyl)-3-(5-nitro-2-furyl)acrylamide (*see* AF-2)	
Fusarenon-X (*see* Toxins derived from *Fusarium graminearum*, *F. culmorum* and *F. crookwellense*)	
Fusarenone-X (*see* Toxins derived from *Fusarium graminearum*, *F. culmorum* and *F. crookwellense*)	
Fusarin C (*see* Toxins derived from *Fusarium moniliforme*)	

G

Gasoline	45, 159 (1989) (corr. 47, 505)
Gasoline engine exhaust (see Diesel and gasoline engine exhausts)	
Glass fibres (see Man-made mineral fibres)	
Glass manufacturing industry, occupational exposures in	58, 347 (1993)
Glasswool (*see* Man-made mineral fibres)	
Glass filaments (*see* Man-made mineral fibres)	
Glu-P-1	*40*, 223 (1986); *Suppl. 7*, 64 (1987)
Glu-P-2	*40*, 235 (1986); *Suppl. 7*, 64 (1987)
L-Glutamic acid, 5-[2-(4-hydroxymethyl)phenylhydrazide] (*see* Agaritine)	
Glycidaldehyde	*11*, 175 (1976); *Suppl. 7*, 64 (1987)
Glycidyl ethers	*47*, 237 (1989)
Glycidyl oleate	*11*, 183 (1976); *Suppl. 7*, 64 (1987)
Glycidyl stearate	*11*, 187 (1976); *Suppl. 7*, 64 (1987)
Griseofulvin	*10*, 153 (1976); *Suppl. 7*, 391 (1987)
Guinea Green B	*16*, 199 (1978); *Suppl. 7*, 64 (1987)
Gyromitrin	*31*, 163 (1983); *Suppl. 7*, 391 (1987)

H

Haematite	*1*, 29 (1972); *Suppl. 7*, 216 (1987)
Haematite and ferric oxide	*Suppl. 7*, 216 (1987)
Haematite mining, underground, with exposure to radon	*1*, 29 (1972); *Suppl. 7*, 216 (1987)
Hairdressers and barbers (occupational exposure as)	57, 43 (1993)
Hair dyes, epidemiology of	*16*, 29 (1978); *27*, 307 (1982);
Halogenated acetonitriles	52, 269 (1991)

Halothane (*see* Anaesthetics, volatile)
HC Blue No. 1   *57*, 129 (1993)
HC Blue No. 2   *57*, 143 (1993)
α-HCH (*see* Hexachlorocyclohexanes)
β-HCH (*see* Hexachlorocyclohexanes)
γ-HCH (*see* Hexachlorocyclohexanes)
HC Red No. 3   *57*, 153 (1993)
HC Yellow No. 4   *57*, 159 (1993)
Heating oils (*see* Fuel oils)
Helicobacter pylori (infection with)   *61*, 177 (1994)
Hepatitis B virus   *59*, 45 (1994)
Hepatitis C virus   *59*, 165 (1994)
Hepatitis D virus   *59*, 223 (1994)
Heptachlor (*see also* Chlordane/Heptachlor)   *5*, 173 (1974); *20*, 129 (1979)
Hexachlorobenzene   *20*, 155 (1979); *Suppl. 7*, 219 (1987)
Hexachlorobutadiene   *20*, 179 (1979); *Suppl. 7*, 64 (1987)
Hexachlorocyclohexanes   *5*, 47 (1974); *20*, 195 (1979)
                                                           (*corr. 42*, 258); *Suppl. 7*, 220 (1987)
Hexachlorocyclohexane, technical-grade (*see* Hexachlorocyclohexanes)
Hexachloroethane   *20*, 467 (1979); *Suppl. 7*, 64 (1987)
Hexachlorophene   *20*, 241 (1979); *Suppl. 7*, 64 (1987)
Hexamethylphosphoramide   *15*, 211 (1977); *Suppl. 7*, 64 (1987)
Hexoestrol (*see* Nonsteroidal oestrogens)
Hycanthone mesylate   *13*, 91 (1977); *Suppl. 7*, 64 (1987)
Hydralazine   *24*, 85 (1980); *Suppl. 7*, 222 (1987)
Hydrazine   *4*, 127 (1974); *Suppl. 7*, 223 (1987)
Hydrochloric acid   *54*, 189 (1992)
Hydrochlorothiazide   *50*, 293 (1990)
Hydrogen peroxide   *36*, 285 (1985); *Suppl. 7*, 64 (1987)
Hydroquinone   *15*, 155 (1977); *Suppl. 7*, 64 (1987)
4-Hydroxyazobenzene   *8*, 157 (1975); *Suppl. 7*, 64 (1987)
17α-Hydroxyprogesterone caproate (*see also* Progestins)   *21*, 399 (1979) (*corr. 42*, 259)
8-Hydroxyquinoline   *13*, 101 (1977); *Suppl. 7*, 64 (1987)
8-Hydroxysenkirkine   *10*, 265 (1976); *Suppl. 7*, 64 (1987)
Hypochlorite salts   *52*, 159 (1991)

I

Indeno[1,2,3-*cd*]pyrene   *3*, 229 (1973); *32*, 373 (1983);
                                                           Suppl. 7, 64 (1987)

Inorganic acids (see Sulfuric acid and other strong inorganic acids,
    occupational exposures to mists and vapours from)
Insecticides, occupational exposures in spraying and application of   *53*, 45 (1991)
IQ   *40*, 261 (1986); *Suppl. 7*, 64 (1987);
                                                           56, 165 (1993)
Iron and steel founding   *34*, 133 (1984); *Suppl. 7*, 224 (1987)
Iron-dextran complex   *2*, 161 (1973); *Suppl. 7*, 226 (1987)
Iron-dextrin complex   *2*, 161 (1973) (*corr. 42*, 252);
                                                           Suppl. 7, 64 (1987)
Iron oxide (*see* Ferric oxide)

Iron oxide, saccharated (*see* Saccharated iron oxide)	
Iron sorbitol-citric acid complex	*2*, 161 (1973); *Suppl. 7*, 64 (1987)
Isatidine	*10*, 269 (1976); *Suppl. 7*, 65 (1987)
Isoflurane (*see* Anaesthetics, volatile)	
Isoniazid (*see* Isonicotinic acid hydrazide)	
Isonicotinic acid hydrazide	*4*, 159 (1974); *Suppl. 7*, 227 (1987)
Isophosphamide	*26*, 237 (1981); *Suppl. 7*, 65 (1987)
Isoprene	*60*, 215 (1994)
Isopropanol	*5*, 223 (1977); *Suppl. 7*, 229 (1987)
Isopropanol manufacture (strong-acid process)	*Suppl. 7*, 229 (1987)
(*see* also Isopropyl alcohol; Sulfuric acid and other strong inorganic acids, occupational exposures to mists and vapours from)	
Isopropyl oils	*15*, 223 (1977); *Suppl. 7*, 229 (1987)
Isosafrole	*1*, 169 (1972); *10*, 232 (1976); *Suppl. 7*, 65 (1987)

J

Jacobine	*10*, 275 (1976); *Suppl. 7*, 65 (1987)
Jet fuel	*45*, 203 (1989)
Joinery (*see* Carpentry and joinery	

K

Kaempferol	*31*, 171 (1983); *Suppl. 7*, 65 (1987)
Kepone (*see* Chlordecone)	

L

Lasiocarpine	*10*, 281 (1976); *Suppl. 7*, 65 (1987)
Lauroyl peroxide	*36*, 315 (1985); *Suppl. 7*, 65 (1987)
Lead acetate (*see* Lead and lead compounds)	
Lead and lead compounds	*1*, 40 (1972) (*corr. 42*, 251); *2*, 52, 150 (1973); *12*, 131 (1976); *23*, 40, 208, 209, 325 (1980); *Suppl. 7*, 230 (1987)
Lead arsenate (*see* Arsenic and arsenic compounds)	
Lead carbonate (*see* Lead and lead compounds)	
Lead chloride (*see* Lead and lead compounds)	
Lead chromate (*see* Chromium and chromium compounds)	
Lead chromate oxide (*see* Chromium and chromium compounds)	
Lead naphthenate (*see* Lead and lead compounds)	
Lead nitrate (*see* Lead and lead compounds)	
Lead oxide (*see* Lead and lead compounds)	
Lead phosphate (*see* Lead and lead compounds)	
Lead subacetate (*see* Lead and lead compounds)	
Lead tetroxide (*see* Lead and lead compounds)	

Leather goods manufacture	25, 279 (1981); Suppl. 7, 235 (1987)
Leather industries	25, 199 (1981); Suppl. 7, 232 (1987)
Leather tanning and processing	25, 201 (1981); Suppl. 7, 236 (1987)
Ledate (see also Lead and lead compounds)	12, 131 (1976)
Light Green SF	16, 209 (1978); Suppl. 7, 65 (1987)
d-Limonene	56, 135 (1993)
Lindane (see Hexachlorocyclohexanes)	
Liver flukes (see Clonorchis sinensis, Opisthorchis felineus and Opisthorchis viverrini)	
The lumber and sawmill industries (including logging)	25, 49 (1981); Suppl. 7, 383 (1987)
Luteoskyrin	10, 163 (1976); Suppl. 7, 65 (1987)
Lynoestrenol (see also Progestins; Combined oral contraceptives)	21, 407 (1979)

M

Magenta	4, 57 (1974) (corr. 42, 252); Suppl. 7, 238 (1987); 57, 215 (1993)
Magenta, manufacture of (see also Magenta)	Suppl. 7, 238 (1987)
Malathion	30, 103 (1983); Suppl. 7, 65 (1987)
Maleic hydrazide	4, 173 (1974) (corr. 42, 253); Suppl. 7, 65 (1987)
Malonaldehyde	36, 163 (1985); Suppl. 7, 65 (1987)
Maneb	12, 137 (1976); Suppl. 7, 65 (1987)
Man-made mineral fibres	43, 39 (1988)
Mannomustine	9, 157 (1975); Suppl. 7, 65 (1987)
Mate	51, 273 (1991)
MCPA (see also Chlorophenoxy herbicides; Chlorophenoxy herbicides, occupational exposures to)	30, 255 (1983)
MeA-α-C	40, 253 (1986); Suppl. 7, 65 (1987)
Medphalan	9, 168 (1975); Suppl. 7, 65 (1987)
Medroxyprogesterone acetate	6, 157 (1974); 21, 417 (1979) (corr. 42, 259); Suppl. 7, 289 (1987)
Megestrol acetate (see also Progestins; Combined oral contraceptives)	
MeIQ	40, 275 (1986); Suppl. 7, 65 (1987); 56, 197 (1993)
MeIQx	40, 283 (1986); Suppl. 7, 65 (1987); 56, 211 (1993)
Melamine	39, 333 (1986); Suppl. 7, 65 (1987)
Melphalan	9, 167 (1975); Suppl. 7, 239 (1987)
6-Mercaptopurine	26, 249 (1981); Suppl. 7, 240 (1987)
Mercuric chloride (see Mercury and mercury compounds)	
Mercury and mercury compounds	58, 239 (1993)
Merphalan	9, 169 (1975); Suppl. 7, 65 (1987)
Mestranol (see also Steroidal oestrogens)	6, 87 (1974); 21, 257 (1979) (corr. 42, 259)
Metabisulfites (see Sulfur dioxide and some sulfites, bisulfites and metabisulfites)	
Metallic mercury (see Mercury and mercury compounds)	
Methanearsonic acid, disodium salt (see Arsenic and arsenic compounds)	

Methanearsonic acid, monosodium salt (*see* Arsenic and arsenic compounds	
Methotrexate	26, 267 (1981); *Suppl. 7*, 241 (1987)
Methoxsalen (*see* 8-Methoxypsoralen)	
Methoxychlor	5, 193 (1974); 20, 259 (1979); *Suppl. 7*, 66 (1987)
Methoxyflurane (*see* Anaesthetics, volatile)	
5-Methoxypsoralen	40, 327 (1986); *Suppl. 7*, 242 (1987)
8-Methoxypsoralen (*see also* 8-Methoxypsoralen plus ultraviolet radiation)	24, 101 (1980)
8-Methoxypsoralen plus ultraviolet radiation	*Suppl. 7*, 243 (1987)
Methyl acrylate	19, 52 (1979); 39, 99 (1986); *Suppl. 7*, 66 (1987)
5-Methylangelicin plus ultraviolet radiation (*see also* Angelicin and some synthetic derivatives)	*Suppl. 7*, 57 (1987)
2-Methylaziridine	9, 61 (1975); *Suppl. 7*, 66 (1987)
Methylazoxymethanol acetate	1, 164 (1972); 10, 131 (1976); *Suppl. 7*, 66 (1987)
Methyl bromide	41, 187 (1986) (*corr. 45*, 283); *Suppl. 7*, 245 (1987)
Methyl carbamate	12, 151 (1976); *Suppl. 7*, 66 (1987)
Methyl-CCNU [*see* 1-(2-Chloroethyl)-3-(4-methylcyclohexyl)-1-nitrosourea]	
Methyl chloride	41, 161 (1986); *Suppl. 7*, 246 (1987)
1-, 2-, 3-, 4-, 5- and 6-Methylchrysenes	32, 379 (1983); *Suppl. 7*, 66 (1987)
N-Methyl-N,4-dinitrosoaniline	1, 141 (1972); *Suppl. 7*, 66 (1987)
4,4'-Methylene bis(2-chloroaniline)	4, 65 (1974) (*corr. 42*, 252); *Suppl. 7*, 246 (1987); 57, 271 (1993)
4,4'-Methylene bis(N,N-dimethyl)benzenamine	27, 119 (1982); *Suppl. 7*, 66 (1987)
4,4'-Methylene bis(2-methylaniline)	4, 73 (1974); *Suppl. 7*, 248 (1987)
4,4'-Methylenedianiline	4, 79 (1974) (*corr. 42*, 252); 39, 347 (1986); *Suppl. 7*, 66 (1987)
4,4'-Methylenediphenyl diisocyanate	19, 314 (1979); *Suppl. 7*, 66 (1987)
2-Methylfluoranthene	32, 399 (1983); *Suppl. 7*, 66 (1987)
3-Methylfluoranthene	32, 399 (1983); *Suppl. 7*, 66 (1987)
Methylglyoxal	51, 443 (1991)
Methyl iodide	15, 245 (1977); 41, 213 (1986); *Suppl. 7*, 66 (1987)
Methylmercury chloride (*see* Mercury and mercury compounds)	
Methylmercury compounds (*see* Mercury and mercury compounds)	
Methyl methacrylate	19, 187 (1979); *Suppl. 7*, 66 (1987); 60, 445 (1994)
Methyl methanesulfonate	7, 253 (1974); *Suppl. 7*, 66 (1987)
2-Methyl-1-nitroanthraquinone	27, 205 (1982); *Suppl. 7*, 66 (1987)
N-Methyl-N'-nitro-N-nitrosoguanidine	4, 183 (1974); *Suppl. 7*, 248 (1987)
3-Methylnitrosaminopropionaldehyde [*see* 3-(N-Nitrosomethylamino)-propionaldehyde]	
3-Methylnitrosaminopropionitrile [*see* 3-(N-Nitrosomethylamino)-propionitrile]	
4-(Methylnitrosamino)-4-(3-pyridyl)-1-butanal-[*see* 4-(N-Nitrosomethylamino)-4-(3-pyridyl)-1-butanal]	

4-(Methylnitrosamino)-1-(3-pyridyl)-1-butanone [see 4-(-Nitrosomethyl-
    amino)-1-(3-pyridyl)-1-butanone]
N-Methyl-N-nitrosourea   *1*, 125 (1972); *17*, 227 (1978);
                                                                    Suppl. *7*, 66 (1987)
N-Methyl-N-nitrosourethane   *4*, 211 (1974); Suppl. *7*, 66 (1987)
N-Methylolacrylamide   *60*, 435 (1994)
Methyl parathion   *30*, 131 (1983); Suppl. *7*, 392 (1987)
1-Methylphenanthrene   *32*, 405 (1983); Suppl. *7*, 66 (1987)
7-Methylpyrido[3,4-*c*]psoralen   *40*, 349 (1986); Suppl. *7*, 71 (1987)
Methyl red   *8*, 161 (1975); Suppl. *7*, 66 (1987)
Methyl selenac (*see also* Selenium and selenium compounds)   *12*, 161 (1976); Suppl. *7*, 66 (1987)
Methylthiouracil   *7*, 53 (1974); Suppl. *7*, 66 (1987)
Metronidazole   *13*, 113 (1977); Suppl. *7*, 250 (1987)
Mineral oils   *3*, 30 (1973); *33*, 87 (1984)
                                                                    (*corr. 42*, 262); Suppl. *7*, 252 (1987)
Mirex   *5*, 203 (1974); *20*, 283 (1979)
                                                                    (*corr. 42*, 258); Suppl. *7*, 66 (1987)
Mitomycin C   *10*, 171 (1976); Suppl. *7*, 67 (1987)
MNNG [see N-Methyl-N'-nitro-N-nitrosoguanidine]
MOCA [see 4,4'-Methylene bis(2-chloroaniline)]
Modacrylic fibres   *19*, 86 (1979); Suppl. *7*, 67 (1987)
Monocrotaline   *10*, 291 (1976); Suppl. *7*, 67 (1987)
Monuron   *12*, 167 (1976); Suppl. *7*, 67 (1987);
                                                                    53, 467 (1991)
MOPP and other combined chemotherapy including   Suppl. *7*, 254 (1987)
    alkylating agents
Morpholine   *47*, 199 (1989)
5-(Morpholinomethyl)-3-[(5-nitrofurfurylidene)amino]-2-   *7*, 161 (1974); Suppl. *7*, 67 (1987)
    oxazolidinone
Mustard gas   *9*, 181 (1975) (*corr. 42*, 254);
    Suppl. *7*, 259 (1987)
Myleran (*see* 1,4-Butanediol dimethanesulfonate)

N

Nafenopin   *24*, 125 (1980); Suppl. *7*, 67 (1987)
1,5-Naphthalenediamine   *27*, 127 (1982); Suppl. *7*, 67 (1987)
1,5-Naphthalene diisocyanate   *19*, 311 (1979); Suppl. *7*, 67 (1987)
1-Naphthylamine   *4*, 87 (1974) (*corr. 42*, 253);
                                                                    Suppl. *7*, 260 (1987)
2-Naphthylamine   *4*, 97 (1974); Suppl. *7*, 261 (1987)
1-Naphthylthiourea   *30*, 347 (1983); Suppl. *7*, 263 (1987)
Nickel acetate (*see* Nickel and nickel compounds)
Nickel ammonium sulfate (*see* Nickel and nickel compounds)
Nickel and nickel compounds   *2*, 126 (1973) (*corr. 42*, 252); *11*, 75
                                                                    (1976); Suppl. *7*, 264 (1987)
                                                                    (*corr. 45*, 283); *49*, 257 (1990)
Nickel carbonate (*see* Nickel and nickel compounds)
Nickel carbonyl (*see* Nickel and nickel compounds)
Nickel chloride (*see* Nickel and nickel compounds)

Nickel-gallium alloy (see Nickel and nickel compounds)
Nickel hydroxide (see Nickel and nickel compounds)
Nickelocene (see Nickel and nickel compounds)
Nickel oxide (see Nickel and nickel compounds)
Nickel subsulfide (see Nickel and nickel compounds)
Nickel sulfate (see Nickel and nickel compounds)

Niridazole	*13*, 123 (1977); *Suppl. 7*, 67 (1987)
Nithiazide	*31*, 179 (1983); *Suppl. 7*, 67 (1987)
Nitrilotriacetic acid and its salts	*48*, 181 (1990)
5-Nitroacenaphthene	*16*, 319 (1978); *Suppl. 7*, 67 (1987)
5-Nitro-*ortho*-anisidine	*27*, 133 (1982); *Suppl. 7*, 67 (1987)
9-Nitroanthracene	*33*, 179 (1984); *Suppl. 7*, 67 (1987)
7-Nitrobenz[*a*]anthracene	*46*, 247 (1989)
6-Nitrobenzo[*a*]pyrene	*33*, 187 (1984); *Suppl. 7*, 67 (1987); *46*, 255 (1989)
4-Nitrobiphenyl	*4*, 113 (1974); *Suppl. 7*, 67 (1987)
6-Nitrochrysene	*33*, 195 (1984); *Suppl. 7*, 67 (1987); *46*, 267 (1989)
Nitrofen (technical-grade)	*30*, 271 (1983); *Suppl. 7*, 67 (1987)
3-Nitrofluoranthene	*33*, 201 (1984); *Suppl. 7*, 67 (1987)
2-Nitrofluorene	*46*, 277 (1989)
Nitrofural	*7*, 171 (1974); *Suppl. 7*, 67 (1987); *50*, 195 (1990)
5-Nitro-2-furaldehyde semicarbazone (see Nitrofural)	
Nitrofurantoin	*50*, 211 (1990)
Nitrofurazone (see Nitrofural)	
1-[(5-Nitrofurfurylidene)amino]-2-imidazolidinone	*7*, 181 (1974); *Suppl. 7*, 67 (1987)
N-[4-(5-Nitro-2-furyl)-2-thiazolyl]acetamide	*1*, 181 (1972); *7*, 185 (1974); *Suppl. 7*, 67 (1987)
Nitrogen mustard	*9*, 193 (1975); *Suppl. 7*, 269 (1987)
Nitrogen mustard *N*-oxide	*9*, 209 (1975); *Suppl. 7*, 67 (1987)
1-Nitronaphthalene	*46*, 291 (1989)
2-Nitronaphthalene	*46*, 303 (1989)
3-Nitroperylene	*46*, 313 (1989)
2-Nitro-*para*-phenylenediamine (see 1,4-Diamino-2-nitrobenzene)	
2-Nitropropane	*29*, 331 (1982); *Suppl. 7*, 67 (1987)
1-Nitropyrene	*33*, 209 (1984); *Suppl. 7*, 67 (1987); *46*, 321 (1989)
2-Nitropyrene	*46*, 359 (1989)
4-Nitropyrene	*46*, 367 (1989)
N-Nitrosatable drugs	*24*, 297 (1980) (*corr. 42*, 260)
N-Nitrosatable pesticides	*30*, 359 (1983)
N'-Nitrosoanabasine	*37*, 225 (1985); *Suppl. 7*, 67 (1987)
N'-Nitrosoanatabine	*37*, 233 (1985); *Suppl. 7*, 67 (1987)
N-Nitrosodi-*n*-butylamine	*4*, 197 (1974); *17*, 51 (1978); *Suppl. 7*, 67 (1987)
N-Nitrosodiethanolamine	*17*, 77 (1978); *Suppl. 7*, 67 (1987)
N-Nitrosodiethylamine	*1*, 107 (1972) (*corr. 42*, 251); *17*, 83 (1978) (*corr. 42*, 257); *Suppl. 7*, 67 (1987)

N-Nitrosodimethylamine	*1*, 95 (1972); *17*, 125 (1978) (*corr. 42*, 257); *Suppl. 7*, 67 (1987)
N-Nitrosodiphenylamine	*27*, 213 (1982); *Suppl. 7*, 67 (1987)
para-Nitrosodiphenylamine	*27*, 227 (1982) (*corr. 42*, 261); *Suppl. 7*, 68 (1987)
N-Nitrosodi-*n*-propylamine	*17*, 177 (1978); *Suppl. 7*, 68 (1987)
N-Nitroso-*N*-ethylurea (*see N*-Ethyl-*N*-nitrosourea)	
N-Nitrosofolic acid	*17*, 217 (1978); *Suppl. 7*, 68 (1987)
N-Nitrosoguvacine	*37*, 263 (1985); *Suppl. 7*, 68 (1987)
N-Nitrosoguvacoline	*37*, 263 (1985); *Suppl. 7*, 68 (1987)
N-Nitrosohydroxyproline	*17*, 304 (1978); *Suppl. 7*, 68 (1987)
3-(*N*-Nitrosomethylamino)propionaldehyde	*37*, 263 (1985); *Suppl. 7*, 68 (1987)
3-(*N*-Nitrosomethylamino)propionitrile	*37*, 263 (1985); *Suppl. 7*, 68 (1987)
4-(*N*-Nitrosomethylamino)-4-(3-pyridyl)-1-butanal	*37*, 205 (1985); *Suppl. 7*, 68 (1987)
4-(*N*-Nitrosomethylamino)-1-(3-pyridyl)-1-butanone	*37*, 209 (1985); *Suppl. 7*, 68 (1987)
N-Nitrosomethylethylamine	*17*, 221 (1978); *Suppl. 7*, 68 (1987)
N-Nitroso-*N*-methylurea (*see N*-Methyl-*N*-nitrosourea)	
N-Nitroso-*N*-methylurethane (*see N*-Methyl-*N*-nitrosourethane)	
N-Nitrosomethylvinylamine	*17*, 257 (1978); *Suppl. 7*, 68 (1987)
N-Nitrosomorpholine	*17*, 263 (1978); *Suppl. 7*, 68 (1987)
N'-Nitrosonornicotine	*17*, 281 (1978); *37*, 241 (1985); *Suppl. 7*, 68 (1987)
N-Nitrosopiperidine	*17*, 287 (1978); *Suppl. 7*, 68 (1987)
N-Nitrosoproline	*17*, 303 (1978); *Suppl. 7*, 68 (1987)
N-Nitrosopyrrolidine	*17*, 313 (1978); *Suppl. 7*, 68 (1987)
N-Nitrososarcosine	*17*, 327 (1978); *Suppl. 7*, 68 (1987)
Nitrosoureas, chloroethyl (*see* Chloroethyl nitrosoureas)	
5-Nitro-*ortho*-toluidine	*48*, 169 (1990)
Nitrous oxide (*see* Anaesthetics, volatile)	
Nitrovin	*31*, 185 (1983); *Suppl. 7*, 68 (1987)
Nivalenol (*see* Toxins derived from *Fusarium graminearum*, *F. culmorum* and *F. crookwellense*)	
NNA [*see* 4-(*N*-Nitrosomethylamino)-4-(3-pyridyl)-1-butanal]	
NNK [*see* 4-(*N*-Nitrosomethylamino)-1-(3-pyridyl)-1-butanone]	
Nonsteroidal oestrogens (*see also* Oestrogens, progestins and combinations)	*Suppl. 7*, 272 (1987)
Norethisterone (*see also* Progestins; Combined oral contraceptives)	*6*, 179 (1974); *21*, 461 (1979)
Norethynodrel (*see also* Progestins; Combined oral contraceptives	*6*, 191 (1974); *21*, 461 (1979) (*corr. 42*, 259)
Norgestrel (*see also* Progestins, Combined oral contraceptives)	*6*, 201 (1974); *21*, 479 (1979)
Nylon 6	*19*, 120 (1979); *Suppl. 7*, 68 (1987)

O

Ochratoxin A	*10*, 191 (1976); *31*, 191 (1983) (*corr. 42*, 262); *Suppl. 7*, 271 (1987); *56*, 489 (1993)
Oestradiol-17β (*see also* Steroidal oestrogens)	*6*, 99 (1974); *21*, 279 (1979)
Oestradiol 3-benzoate (*see* Oestradiol-17β)	
Oestradiol dipropionate (*see* Oestradiol-17β)	

Oestradiol mustard	9, 217 (1975)
Oestradiol-17β-valerate (see Oestradiol-17β)	
Oestriol (see also Steroidal oestrogens)	6, 117 (1974); 21, 327 (1979)
Oestrogen-progestin combinations (see Oestrogens, progestins and combinations)	
Oestrogen-progestin replacement therapy (see also Oestrogens, progestins and combinations)	Suppl. 7, 308 (1987)
Oestrogen replacement therapy (see also Oestrogens, progestins and combinations)	Suppl. 7, 280 (1987)
Oestrogens (see Oestrogens, progestins and combinations)	
Oestrogens, conjugated (see Conjugated oestrogens)	
Oestrogens, nonsteroidal (see Nonsteroidal oestrogens)	
Oestrogens, progestins and combinations	6 (1974); 21 (1979); Suppl. 7, 272 (1987)
Oestrogens, steroidal (see Steroidal oestrogens)	
Oestrone (see also Steroidal oestrogens)	6, 123 (1974); 21, 343 (1979) (corr. 42, 259)
Oestrone benzoate (see Oestrone)	
Oil Orange SS	8, 165 (1975); Suppl. 7, 69 (1987)
Opisthorchis felineus (infection with)	61, 121 (1994)
Opisthorchis viverrini (infection with)	61, 121 (1994)
Oral contraceptives, combined (see Combined oral contraceptives)	
Oral contraceptives, investigational (see Combined oral contraceptives)	
Oral contraceptives, sequential (see Sequential oral contraceptives)	
Orange I	8, 173 (1975); Suppl. 7, 69 (1987)
Orange G	8, 181 (1975); Suppl. 7, 69 (1987)
Organolead compounds (see also Lead and lead compounds)	Suppl. 7, 230 (1987)
Oxazepam	13, 58 (1977); Suppl. 7, 69 (1987)
Oxymetholone [see also Androgenic (anabolic) steroids]	13, 131 (1977)
Oxyphenbutazone	13, 185 (1977); Suppl. 7, 69 (1987)

P

Paint manufacture and painting (occupational exposures in)	47, 329 (1989)
Panfuran S (see also Dihydroxymethylfuratrizine)	24, 77 (1980); Suppl. 7, 69 (1987)
Paper manufacture (see Pulp and paper manufacture)	
Paracetamol	50, 307 (1990)
Parasorbic acid	10, 199 (1976) (corr. 42, 255); Suppl. 7, 69 (1987)
Parathion	30, 153 (1983); Suppl. 7, 69 (1987)
Patulin	10, 205 (1976); 40, 83 (1986); Suppl. 7, 69 (1987)
Penicillic acid	10, 211 (1976); Suppl. 7, 69 (1987)
Pentachloroethane	41, 99 (1986); Suppl. 7, 69 (1987)
Pentachloronitrobenzene (see Quintozene)	
Pentachlorophenol (see also Chlorophenols; Chlorophenols, occupational exposures to)	20, 303 (1979); 53, 371 (1991)
Permethrin	53, 329 (1991)
Perylene	32, 411 (1983); Suppl. 7, 69 (1987)
Petasitenine	31, 207 (1983); Suppl. 7, 69 (1987)

Petasites japonicus (*see* Pyrrolizidine alkaloids)
Petroleum refining (occupational exposures in)   *45*, 39 (1989)
Some petroleum solvents   *47*, 43 (1989)
Phenacetin   *13*, 141 (1977); *24*, 135 (1980);
    Suppl. *7*, 310 (1987)
Phenanthrene   *32*, 419 (1983); Suppl. *7*, 69 (1987)
Phenazopyridine hydrochloride   *8*, 117 (1975); *24*, 163 (1980)
    (corr. *42*, 260); Suppl. *7*, 312 (1987)
Phenelzine sulfate   *24*, 175 (1980); Suppl. *7*, 312 (1987)
Phenicarbazide   *12*, 177 (1976); Suppl. *7*, 70 (1987)
Phenobarbital   *13*, 157 (1977); Suppl. *7*, 313 (1987)
Phenol   *47*, 263 (1989) (corr. *50*, 385)
Phenoxyacetic acid herbicides (*see* Chlorophenoxy herbicides)
Phenoxybenzamine hydrochloride   *9*, 223 (1975); *24*, 185 (1980);
    Suppl. *7*, 70 (1987)
Phenylbutazone   *13*, 183 (1977); Suppl. *7*, 316 (1987)
meta-Phenylenediamine   *16*, 111 (1978); Suppl. *7*, 70 (1987)
para-Phenylenediamine   *16*, 125 (1978); Suppl. *7*, 70 (1987)
Phenyl glycidyl ether (*see* Glycidyl ethers)
N-Phenyl-2-naphthylamine   *16*, 325 (1978) (corr. *42*, 257);
    Suppl. *7*, 318 (1987)
ortho-Phenylphenol   *30*, 329 (1983); Suppl. *7*, 70 (1987)
Phenytoin   *13*, 201 (1977); Suppl. *7*, 319 (1987)
PhIP   *56*, 229 (1993)
Pickled vegetables   *56*, 83 (1993)
Picloram   *53*, 481 (1991)
Piperazine oestrone sulfate (*see* Conjugated oestrogens)
Piperonyl butoxide   *30*, 183 (1983); Suppl. *7*, 70 (1987)
Pitches, coal-tar (*see* Coal-tar pitches)
Polyacrylic acid   *19*, 62 (1979); Suppl. *7*, 70 (1987)
Polybrominated biphenyls   *18*, 107 (1978); *41*, 261 (1986);
    Suppl. *7*, 321 (1987)
Polychlorinated biphenyls   *7*, 261 (1974); *18*, 43 (1978)
    (corr. *42*, 258); Suppl. *7*, 322 (1987)
Polychlorinated camphenes (*see* Toxaphene)
Polychloroprene   *19*, 141 (1979); Suppl. *7*, 70 (1987)
Polyethylene   *19*, 164 (1979); Suppl. *7*, 70 (1987)
Polymethylene polyphenyl isocyanate   *19*, 314 (1979); Suppl. *7*, 70 (1987)
Polymethyl methacrylate   *19*, 195 (1979); Suppl. *7*, 70 (1987)
Polyoestradiol phosphate (*see* Oestradiol-17β)
Polypropylene   *19*, 218 (1979); Suppl. *7*, 70 (1987)
Polystyrene   *19*, 245 (1979); Suppl. *7*, 70 (1987)
Polytetrafluoroethylene   *19*, 288 (1979); Suppl. *7*, 70 (1987)
Polyurethane foams   *19*, 320 (1979); Suppl. *7*, 70 (1987)
Polyvinyl acetate   *19*, 346 (1979); Suppl. *7*, 70 (1987)
Polyvinyl alcohol   *19*, 351 (1979); Suppl. *7*, 70 (1987)
Polyvinyl chloride   *7*, 306 (1974); *19*, 402 (1979);
    Suppl. *7*, 70 (1987)
Polyvinyl pyrrolidone   *19*, 463 (1979); Suppl. *7*, 70 (1987)
Ponceau MX   *8*, 189 (1975); Suppl. *7*, 70 (1987)
Ponceau 3R   *8*, 199 (1975); Suppl. *7*, 70 (1987)

Ponceau SX	8, 207 (1975); *Suppl. 7*, 70 (1987)
Potassium arsenate (*see* Arsenic and arsenic compounds)	
Potassium arsenite (*see* Arsenic and arsenic compounds)	
Potassium bis(2-hydroxyethyl)dithiocarbamate	12, 183 (1976); *Suppl. 7*, 70 (1987)
Potassium bromate	40, 207 (1986); *Suppl. 7*, 70 (1987)
Potassium chromate (*see* Chromium and chromium compounds)	
Potassium dichromate (*see* Chromium and chromium compounds)	
Prednimustine	50, 115 (1990)
Prednisone	26, 293 (1981); *Suppl. 7*, 326 (1987)
Procarbazine hydrochloride	26, 311 (1981); *Suppl. 7*, 327 (1987)
Proflavine salts	24, 195 (1980); *Suppl. 7*, 70 (1987)
Progesterone (*see also* Progestins; Combined oral contraceptives)	6, 135 (1974); 21, 491 (1979) (*corr. 42*, 259)
Progestins (*see also* Oestrogens, progestins and combinations)	*Suppl. 7*, 289 (1987)
Pronetalol hydrochloride	13, 227 (1977) (*corr. 42*, 256); *Suppl. 7*, 70 (1987)
1,3-Propane sultone	4, 253 (1974) (*corr. 42*, 253); *Suppl. 7*, 70 (1987)
Propham	12, 189 (1976); *Suppl. 7*, 70 (1987)
β-Propiolactone	4, 259 (1974) (*corr. 42*, 253); *Suppl. 7*, 70 (1987)
n-Propyl carbamate	12, 201 (1976); *Suppl. 7*, 70 (1987)
Propylene	19, 213 (1979); *Suppl. 7*, 71 (1987); 60, 161 (1994)
Propylene oxide	11, 191 (1976); 36, 227 (1985) (*corr. 42*, 263); *Suppl. 7*, 328 (1987); 60, 181 (1994)
Propylthiouracil	7, 67 (1974); *Suppl. 7*, 329 (1987)
Ptaquiloside (*see also* Bracken fern)	40, 55 (1986); *Suppl. 7*, 71 (1987)
Pulp and paper manufacture	25, 157 (1981); *Suppl. 7*, 385 (1987)
Pyrene	32, 431 (1983); *Suppl. 7*, 71 (1987)
Pyrido[3,4-*c*]psoralen	40, 349 (1986); *Suppl. 7*, 71 (1987)
Pyrimethamine	13, 233 (1977); *Suppl. 7*, 71 (1987)
Pyrrolizidine alkaloids (*see* Hydroxysenkirkine; Isatidine; Jacobine; Lasiocarpine; Monocrotaline; Retrorsine; Riddelliine; Seneciphylline; Senkirkine)	

Q

Quercetin (*see also* Bracken fern)	31, 213 (1983); *Suppl. 7*, 71 (1987)
para-Quinone	15, 255 (1977); *Suppl. 7*, 71 (1987)
Quintozene	5, 211 (1974); *Suppl. 7*, 71 (1987)

R

Radon	43, 173 (1988) (*corr. 45*, 283)
Reserpine	10, 217 (1976); 24, 211 (1980) (*corr. 42*, 260); *Suppl. 7*, 330 (1987)
Resorcinol	15, 155 (1977); *Suppl. 7*, 71 (1987)

Retrorsine	*10*, 303 (1976); *Suppl. 7*, 71 (1987)
Rhodamine B	*16*, 221 (1978); *Suppl. 7*, 71 (1987)
Rhodamine 6G	*16*, 233 (1978); *Suppl. 7*, 71 (1987)
Riddelliine	*10*, 313 (1976); *Suppl. 7*, 71 (1987)
Rifampicin	*24*, 243 (1980); *Suppl. 7*, 71 (1987)
Rockwool (*see* Man-made mineral fibres)	
The rubber industry	28 (1982) (*corr. 42*, 261); *Suppl. 7*, 332 (1987)
Rugulosin	*40*, 99 (1986); *Suppl. 7*, 71 (1987)

S

Saccharated iron oxide	*2*, 161 (1973); *Suppl. 7*, 71 (1987)
Saccharin	*22*, 111 (1980) (*corr. 42*, 259); *Suppl. 7*, 334 (1987)
Safrole	*1*, 169 (1972); *10*, 231 (1976); *Suppl. 7*, 71 (1987)
Salted fish	*56*, 41 (1993)
The sawmill industry (including logging) [*see* The lumber and sawmill industry (including logging)]	
Scarlet Red	*8*, 217 (1975); *Suppl. 7*, 71 (1987)
Schistosoma haematobium (infection with)	*61*, 45 (1994)
Schistosoma japonicum (infection with)	*61*, 45 (1994)
Schistosoma mansoni (infection with)	*61*, 45 (1994)
Selenium and selenium compounds	*9*, 245 (1975) (*corr. 42*, 255); *Suppl. 7*, 71 (1987)
Selenium dioxide (*see* Selenium and selenium compounds)	
Selenium oxide (*see* Selenium and selenium compounds)	
Semicarbazide hydrochloride	*12*, 209 (1976) (*corr. 42*, 256); *Suppl. 7*, 71 (1987)
Senecio jacobaea L. (*see* Pyrrolizidine alkaloids)	
Senecio longilobus (*see* Pyrrolizidine alkaloids)	
Seneciphylline	*10*, 319, 335 (1976); *Suppl. 7*, 71 (1987)
Senkirkine	*10*, 327 (1976); *31*, 231 (1983); *Suppl. 7*, 71 (1987)
Sepiolite	*42*, 175 (1987); *Suppl. 7*, 71 (1987)
Sequential oral contraceptives (*see also* Oestrogens, progestins and combinations)	*Suppl. 7*, 296 (1987)
Shale-oils	*35*, 161 (1985); *Suppl. 7*, 339 (1987)
Shikimic acid (*see also* Bracken fern)	*40*, 55 (1986); *Suppl. 7*, 71 (1987)
Shoe manufacture and repair (*see* Boot and shoe manufacture and repair)	
Silica (*see also* Amorphous silica; Crystalline silica)	*42*, 39 (1987)
Simazine	*53*, 495 (1991)
Slagwool (*see* Man-made mineral fibres)	
Sodium arsenate (*see* Arsenic and arsenic compounds)	
Sodium arsenite (*see* Arsenic and arsenic compounds)	
Sodium cacodylate (*see* Arsenic and arsenic compounds)	
Sodium chlorite	*52*, 145 (1991)

Sodium chromate (*see* Chromium and chromium compounds)	
Sodium cyclamate (*see* Cyclamates)	
Sodium dichromate (*see* Chromium and chromium compounds)	
Sodium diethyldithiocarbamate	*12*, 217 (1976); *Suppl. 7*, 71 (1987)
Sodium equilin sulfate (*see* Conjugated oestrogens)	
Sodium fluoride (*see* Fluorides)	
Sodium monofluorophosphate (*see* Fluorides)	
Sodium oestrone sulfate (*see* Conjugated oestrogens)	
Sodium *ortho*-phenylphenate (*see also* ortho-Phenylphenol)	*30*, 329 (1983); *Suppl. 7*, 392 (1987)
Sodium saccharin (*see* Saccharin)	
Sodium selenate (*see* Selenium and selenium compounds)	
Sodium selenite (*see* Selenium and selenium compounds)	
Sodium silicofluoride (*see* Fluorides)	
Solar radiation	*55* (1992)
Soots	*3*, 22 (1973); *35*, 219 (1985); *Suppl. 7*, 343 (1987)
Spironolactone	*24*, 259 (1980); *Suppl. 7*, 344 (1987)
Stannous fluoride (*see* Fluorides)	
Steel founding (*see* Iron and steel founding)	
Sterigmatocystin	*1*, 175 (1972); *10*, 245 (1976); *Suppl. 7*, 72 (1987)
Steroidal oestrogens (*see also* Oestrogens, progestins and combinations)	*Suppl. 7*, 280 (1987)
Streptozotocin	*4*, 221 (1974); *17*, 337 (1978); *Suppl. 7*, 72 (1987)
Strobaner (*see* Terpene polychlorinates)	
Strontium chromate (*see* Chromium and chromium compounds)	
Styrene	*19*, 231 (1979) (*corr. 42*, 258); *Suppl. 7*, 345 (1987); *60*, 233 (1994)
Styrene-acrylonitrile-copolymers	*19*, 97 (1979); *Suppl. 7*, 72 (1987)
Styrene-butadiene copolymers	*19*, 252 (1979); *Suppl. 7*, 72 (1987)
Styrene-7,8-oxide	*11*, 201 (1976); *19*, 275 (1979); *36*, 245 (1985); *Suppl. 7*, 72 (1987); *60*, 321 (1994)
Succinic anhydride	*15*, 265 (1977); *Suppl. 7*, 72 (1987)
Sudan I	*8*, 225 (1975); *Suppl. 7*, 72 (1987)
Sudan II	*8*, 233 (1975); *Suppl. 7*, 72 (1987)
Sudan III	*8*, 241 (1975); *Suppl. 7*, 72 (1987)
Sudan Brown RR	*8*, 249 (1975); *Suppl. 7*, 72 (1987)
Sudan Red 7B	*8*, 253 (1975); *Suppl. 7*, 72 (1987)
Sulfafurazole	*24*, 275 (1980); *Suppl. 7*, 347 (1987)
Sulfallate	*30*, 283 (1983); *Suppl. 7*, 72 (1987)
Sulfamethoxazole	*24*, 285 (1980); *Suppl. 7*, 348 (1987)
Sulfites (*see* Sulfur dioxide and some sulfites, bisulfites and metabisulfites)	
Sulfur dioxide and some sulfites, bisulfites and metabisulfites	*54*, 131 (1992)
Sulfur mustard (*see* Mustard gas)	
Sulfuric acid and other strong inorganic acids, occupational exposures to mists and vapours from	*54*, 41 (1992)
Sulfur trioxide	*54*, 121 (1992)
Sulphisoxazole (*see* Sulfafurazole)	
Sunset Yellow FCF	*8*, 257 (1975); *Suppl. 7*, 72 (1987)

Symphytine   *31*, 239 (1983); *Suppl. 7*, 72 (1987)

T

2,4,5-T (*see also* Chlorophenoxy herbicides; Chlorophenoxy herbicides, occupational exposures to)   *15*, 273 (1977)
Talc   *42*, 185 (1987); Suppl. 7, 349 (1987)
Tannic acid   *10*, 253 (1976) (*corr. 42*, 255); *Suppl. 7*, 72 (1987)
Tannins (*see also* Tannic acid)   *10*, 254 (1976); *Suppl. 7*, 72 (1987)
TCDD (*see* 2,3,7,8-Tetrachlorodibenzo-*para*-dioxin)
TDE (*see* DDT)
Tea   *51*, 207 (1991)
Terpene polychlorinates   *5*, 219 (1974); *Suppl. 7*, 72 (1987)
Testosterone (*see also* Androgenic (anabolic) steroids)   *6*, 209 (1974); *21*, 519 (1979)
Testosterone oenanthate (*see* Testosterone)
Testosterone propionate (*see* Testosterone)
2,2′,5,5′-Tetrachlorobenzidine   *27*, 141 (1982); *Suppl. 7*, 72 (1987)
2,3,7,8-Tetrachlorodibenzo-*para*-dioxin   *15*, 41 (1977); *Suppl. 7*, 350 (1987)
1,1,1,2-Tetrachloroethane   *41*, 87 (1986); *Suppl. 7*, 72 (1987)
1,1,2,2-Tetrachloroethane   *20*, 477 (1979); *Suppl. 7*, 354 (1987)
Tetrachloroethylene   *20*, 491 (1979); *Suppl. 7*, 355 (1987)
2,3,4,6-Tetrachlorophenol (*see* Chlorophenols; Chlorophenols, occupational exposures to)
Tetrachlorvinphos   *30*, 197 (1983); *Suppl. 7*, 72 (1987)
Tetraethyllead (*see* Lead and lead compounds)
Tetrafluoroethylene   *19*, 285 (1979); *Suppl. 7*, 72 (1987)
Tetrakis(hydroxymethyl) phosphonium salts   *48*, 95 (1990)
Tetramethyllead (*see* Lead and lead compounds)
Textile manufacturing industry, exposures in   *48*, 215 (1990) (*corr. 51*, 483)
Theobromine   *51*, 421 (1991)
Theophylline   *51*, 391 (1991)
Thioacetamide   *7*, 77 (1974); *Suppl. 7*, 72 (1987)
4,4′-Thiodianiline   *16*, 343 (1978); *27*, 147 (1982); *Suppl. 7*, 72 (1987)
Thiotepa   *9*, 85 (1975); *Suppl. 7*, 368 (1987); *50*, 123 (1990)
Thiouracil   *7*, 85 (1974); *Suppl. 7*, 72 (1987)
Thiourea   *7*, 95 (1974); *Suppl. 7*, 72 (1987)
Thiram   *12*, 225 (1976); *Suppl. 7*, 72 (1987); *53*, 403 (1991)
Titanium dioxide   *47*, 307 (1989)
Tobacco habits other than smoking (*see* Tobacco products, smokeless)
Tobacco products, smokeless   *37* (1985) (*corr. 42*, 263; *52*, 513); *Suppl. 7*, 357 (1987)
Tobacco smoke   *38* (1986) (*corr. 42*, 263); *Suppl. 7*, 357 (1987)
Tobacco smoking (*see* Tobacco smoke)
ortho-Tolidine (*see* 3,3′-Dimethylbenzidine)
2,4-Toluene diisocyanate (*see* also Toluene diisocyanates)   *19*, 303 (1979); *39*, 287 (1986)

CUMULATIVE INDEX

2,6-Toluene diisocyanate (*see* also Toluene diisocyanates)	*19*, 303 (1979); *39*, 289 (1986)
Toluene	*47*, 79 (1989)
Toluene diisocyanates	*39*, 287 (1986) (*corr. 42*, 264); *Suppl. 7*, 72 (1987)
Toluenes, α-chlorinated (*see* α-Chlorinated toluenes)	
ortho-Toluenesulfonamide (*see* Saccharin)	
ortho-Toluidine	*16*, 349 (1978); *27*, 155 (1982); *Suppl. 7*, 362 (1987)
Toxaphene	*20*, 327 (1979); *Suppl. 7*, 72 (1987)
T-2 Toxin (*see* Toxins derived from *Fusarium sporotrichioides*)	
Toxins derived from *Fusarium graminearum*, *F. culmorum* and *F. crookwellense*	*11*, 169 (1976); *31*, 153, 279 (1983); *Suppl. 7*, 64, 74 (1987); *56*, 397 (1993)
Toxins derived from *Fusarium moniliforme*	*56*, 445 (1993)
Toxins derived from *Fusarium sporotrichioides*	*31*, 265 (1983); *Suppl. 7*, 73 (1987); *56*, 467 (1993)
Tremolite (*see* Asbestos)	
Treosulfan	*26*, 341 (1981); *Suppl. 7*, 363 (1987)
Triaziquone [*see* Tris(aziridinyl)-*para*-benzoquinone]	
Trichlorfon	*30*, 207 (1983); *Suppl. 7*, 73 (1987)
Trichlormethine	*9*, 229 (1975); *Suppl. 7*, 73 (1987); *50*, 143 (1990)
Trichloroacetonitrile (*see* Halogenated acetonitriles)	
1,1,1-Trichloroethane	*20*, 515 (1979); *Suppl. 7*, 73 (1987)
1,1,2-Trichloroethane	*20*, 533 (1979); *Suppl. 7*, 73 (1987); *52*, 337 (1991)
Trichloroethylene	*11*, 263 (1976); *20*, 545 (1979); *Suppl. 7*, 364 (1987)
2,4,5-Trichlorophenol (*see also* Chlorophenols; Chlorophenols occupational exposures to)	*20*, 349 (1979)
2,4,6-Trichlorophenol (*see also* Chlorophenols; Chlorophenols, occupational exposures to)	*20*, 349 (1979)
(2,4,5-Trichlorophenoxy)acetic acid (*see* 2,4,5-T)	
Trichlorotriethylamine-hydrochloride (*see* Trichlormethine)	
T$_2$-Trichothecene (*see* Toxins derived from *Fusarium sporotrichioides*)	
Triethylene glycol diglycidyl ether	*11*, 209 (1976); *Suppl. 7*, 73 (1987)
Trifluralin	*53*, 515 (1991)
4,4′,6-Trimethylangelicin plus ultraviolet radiation (*see also* Angelicin and some synthetic derivatives)	*Suppl. 7*, 57 (1987)
2,4,5-Trimethylaniline	*27*, 177 (1982); *Suppl. 7*, 73 (1987)
2,4,6-Trimethylaniline	*27*, 178 (1982); *Suppl. 7*, 73 (1987)
4,5′,8-Trimethylpsoralen	*40*, 357 (1986); *Suppl. 7*, 366 (1987)
Trimustine hydrochloride (*see* Trichlormethine)	
Triphenylene	*32*, 447 (1983); *Suppl. 7*, 73 (1987)
Tris(aziridinyl)-*para*-benzoquinone	*9*, 67 (1975); *Suppl. 7*, 367 (1987)
Tris(1-aziridinyl)phosphine-oxide	*9*, 75 (1975); *Suppl. 7*, 73 (1987)
Tris(1-aziridinyl)phosphine-sulphide (*see* Thiotepa)	
2,4,6-Tris(1-aziridinyl)-*s*-triazine	*9*, 95 (1975); *Suppl. 7*, 73 (1987)
Tris(2-chloroethyl) phosphate	*48*, 109 (1990)
1,2,3-Tris(chloromethoxy)propane	*15*, 301 (1977); *Suppl. 7*, 73 (1987)
Tris(2,3-dibromopropyl)phosphate	*20*, 575 (1979); *Suppl. 7*, 369 (1987)

Tris(2-methyl-1-aziridinyl)phosphine#oxide	9, 107 (1975); *Suppl. 7*, 73 (1987)
Trp-P-1	31, 247 (1983); *Suppl. 7*, 73 (1987)
Trp-P-2	31, 255 (1983); *Suppl. 7*, 73 (1987)
Trypan blue	8, 267 (1975); *Suppl. 7*, 73 (1987)
Tussilago farfara L. (*see* Pyrrolizidine alkaloids)	

U

Ultraviolet radiation	40, 379 (1986); 55 (1992)
Underground haematite mining with exposure to radon	1, 29 (1972); *Suppl. 7*, 216 (1987)
Uracil mustard	9, 235 (1975); *Suppl. 7*, 370 (1987)
Urethane	7, 111 (1974); *Suppl. 7*, 73 (1987)

V

Vat Yellow 4	48, 161 (1990)
Vinblastine sulfate	26, 349 (1981) (*corr. 42*, 261); *Suppl. 7*, 371 (1987)
Vincristine sulfate	26, 365 (1981); *Suppl. 7*, 372 (1987)
Vinyl acetate	19, 341 (1979); 39, 113 (1986); *Suppl. 7*, 73 (1987)
Vinyl bromide	19, 367 (1979); 39, 133 (1986); *Suppl. 7*, 73 (1987)
Vinyl chloride	7, 291 (1974); 19, 377 (1979) (*corr. 42*, 258); *Suppl. 7*, 373 (1987)
Vinyl chloride-vinyl acetate copolymers	7, 311 (1976); 19, 412 (1979) (*corr. 42*, 258); *Suppl. 7*, 73 (1987)
4-Vinylcyclohexene	11, 277 (1976); 39, 181 (1986) *Suppl. 7*, 73 (1987); 60, 347 (1994)
4-Vinylcyclohexene diepoxide	11, 141 (1976); *Suppl. 7*, 63 (1987); 60, 361 (1994)
Vinyl fluoride	39, 147 (1986); *Suppl. 7*, 73 (1987)
Vinylidene chloride	19, 439 (1979); 39, 195 (1986); *Suppl. 7*, 376 (1987)
Vinylidene chloride-vinyl chloride copolymers	19, 448 (1979) (*corr. 42*, 258); *Suppl. 7*, 73 (1987)
Vinylidene fluoride	39, 227 (1986); *Suppl. 7*, 73 (1987)
N-Vinyl-2-pyrrolidone	19, 461 (1979); *Suppl. 7*, 73 (1987)
Vinyl toluene	60, 373 (1994)

W

Welding	49, 447 (1990) (*corr. 52*, 513)
Wollastonite	42, 145 (1987); *Suppl. 7*, 377 (1987)
Wood dust	62, 35 (1995)

Wood industries   *25* (1981); *Suppl. 7*, 378 (1987)

X

Xylene   *47*, 125 (1989)
2,4-Xylidine   *16*, 367 (1978); *Suppl. 7*, 74 (1987)
2,5-Xylidine   *16*, 377 (1978); *Suppl. 7*, 74 (1987)
2,6-Xylidine (*see* 2,6-Dimethylaniline)

Y

Yellow AB   *8*, 279 (1975); *Suppl. 7*, 74 (1987)
Yellow OB   *8*, 287 (1975); *Suppl. 7*, 74 (1987)

Z

Zearalenone (*see* Toxins derived from *Fusarium graminearum*,
   F. culmorum and *F. crookwellense*)
Zectran   *12*, 237 (1976); *Suppl. 7*, 74 (1987)
Zinc beryllium silicate (*see* Beryllium and beryllium compounds)
Zinc chromate (*see* Chromium and chromium compounds)
Zinc chromate hydroxide (*see* Chromium and chromium compounds)
Zinc potassium chromate (*see* Chromium and chromium
   compounds)
Zinc yellow (*see* Chromium and chromium compounds)
Zineb   *12*, 245 (1976); *Suppl. 7*, 74 (1987)
Ziram   *12*, 259 (1976); *Suppl. 7*, 74 (1987);
                                                    53, 423 (1991)

PUBLICATIONS OF THE INTERNATIONAL AGENCY FOR RESEARCH ON CANCER

Scientific Publications Series

(Available from Oxford University Press through local bookshops)

No. 1 **Liver Cancer**
1971; 176 pages (*out of print*)

No. 2 **Oncogenesis and Herpesviruses**
Edited by P.M. Biggs, G. de-Thé and L.N. Payne
1972; 515 pages (*out of print*)

No. 3 **N-Nitroso Compounds: Analysis and Formation**
Edited by P. Bogovski, R. Preussman and E.A. Walker
1972; 140 pages (*out of print*)

No. 4 **Transplacental Carcinogenesis**
Edited by L. Tomatis and U. Mohr
1973; 181 pages (*out of print*)

No. 5/6 **Pathology of Tumours in Laboratory Animals, Volume 1, Tumours of the Rat**
Edited by V.S. Turusov
1973/1976; 533 pages (*out of print*)

No. 7 **Host Environment Interactions in the Etiology of Cancer in Man**
Edited by R. Doll and I. Vodopija
1973; 464 pages (*out of print*)

No. 8 **Biological Effects of Asbestos**
Edited by P. Bogovski, J.C. Gilson, V. Timbrell and J.C. Wagner
1973; 346 pages (*out of print*)

No. 9 **N-Nitroso Compounds in the Environment**
Edited by P. Bogovski and E.A. Walker
1974; 243 pages (*out of print*)

No. 10 **Chemical Carcinogenesis Essays**
Edited by R. Montesano and L. Tomatis
1974; 230 pages (*out of print*)

No. 11 **Oncogenesis and Herpesviruses II**
Edited by G. de-Thé, M.A. Epstein and H. zur Hausen
1975; Part I: 511 pages
Part II: 403 pages (*out of print*)

No. 12 **Screening Tests in Chemical Carcinogenesis**
Edited by R. Montesano, H. Bartsch and L. Tomatis
1976; 666 pages (*out of print*)

No. 13 **Environmental Pollution and Carcinogenic Risks**
Edited by C. Rosenfeld and W. Davis
1975; 441 pages (*out of print*)

No. 14 **Environmental N-Nitroso Compounds. Analysis and Formation**
Edited by E.A. Walker, P. Bogovski and L. Griciute
1976; 512 pages (*out of print*)

No. 15 **Cancer Incidence in Five Continents, Volume III**
Edited by J.A.H. Waterhouse, C. Muir, P. Correa and J. Powell
1976; 584 pages (*out of print*)

No. 16 **Air Pollution and Cancer in Man**
Edited by U. Mohr, D. Schmähl and L. Tomatis
1977; 328 pages (*out of print*)

No. 17 **Directory of On-going Research in Cancer Epidemiology 1977**
Edited by C.S. Muir and G. Wagner
1977; 599 pages (*out of print*)

No. 18 **Environmental Carcinogens. Selected Methods of Analysis. Volume 1: Analysis of Volatile Nitrosamines in Food**
Editor-in-Chief: H. Egan
1978; 212 pages (*out of print*)

No. 19 **Environmental Aspects of N-Nitroso Compounds**
Edited by E.A. Walker, M. Castegnaro, L. Griciute and R.E. Lyle
1978; 561 pages (*out of print*)

No. 20 **Nasopharyngeal Carcinoma: Etiology and Control**
Edited by G. de-Thé and Y. Ito
1978; 606 pages (*out of print*)

No. 21 **Cancer Registration and its Techniques**
Edited by R. MacLennan, C. Muir, R. Steinitz and A. Winkler
1978; 235 pages (*out of print*)

No. 22 **Environmental Carcinogens. Selected Methods of Analysis. Volume 2: Methods for the Measurement of Vinyl Chloride in Poly(vinyl chloride), Air, Water and Foodstuffs**
Editor-in-Chief: H. Egan
1978; 142 pages (*out of print*)

No. 23 **Pathology of Tumours in Laboratory Animals. Volume II: Tumours of the Mouse**
Editor-in-Chief: V.S. Turusov
1979; 669 pages (*out of print*)

No. 24 **Oncogenesis and Herpesviruses III**
Edited by G. de-Thé, W. Henle and F. Rapp
1978; Part I: 580 pages, Part II: 512 pages (*out of print*)

Prices, valid for February 1995, are subject to change without notice. Limited supplies of some books marked 'out of print' are available directly from IARC.

List of IARC Publications

No. 25 Carcinogenic Risk. Strategies for Intervention
Edited by W. Davis and C. Rosenfeld
1979; 280 pages (*out of print*)

No. 26 Directory of On-going Research in Cancer Epidemiology 1978
Edited by C.S. Muir and G. Wagner
1978; 550 pages (*out of print*)

No. 27 Molecular and Cellular Aspects of Carcinogen Screening Tests
Edited by R. Montesano, H. Bartsch and L. Tomatis
1980; 372 pages £30.00

No. 28 Directory of On-going Research in Cancer Epidemiology 1979
Edited by C.S. Muir and G. Wagner
1979; 672 pages (*out of print*)

No. 29 Environmental Carcinogens. Selected Methods of Analysis. Volume 3: Analysis of Polycyclic Aromatic Hydrocarbons in Environmental Samples
Editor-in-Chief: H. Egan
1979; 240 pages (*out of print*)

No. 30 Biological Effects of Mineral Fibres
Editor-in-Chief: J.C. Wagner
1980; Volume 1: 494 pages Volume 2: 513 pages (*out of print*)

No. 31 N-Nitroso Compounds: Analysis, Formation and Occurrence
Edited by E.A. Walker, L. Griciute, M. Castegnaro and M. Börzsönyi
1980; 835 pages (*out of print*)

No. 32 Statistical Methods in Cancer Research. Volume 1. The Analysis of Case-control Studies
By N.E. Breslow and N.E. Day
1980; 338 pages £18.00

No. 33 Handling Chemical Carcinogens in the Laboratory
Edited by R. Montesano *et al.*
1979; 32 pages (*out of print*)

No. 34 Pathology of Tumours in Laboratory Animals. Volume III. Tumours of the Hamster
Editor-in-Chief: V.S. Turusov
1982; 461 pages (*out of print*)

No. 35 Directory of On-going Research in Cancer Epidemiology 1980
Edited by C.S. Muir and G. Wagner
1980; 660 pages (*out of print*)

No. 36 Cancer Mortality by Occupation and Social Class 1851-1971
Edited by W.P.D. Logan
1982; 253 pages (*out of print*)

No. 37 Laboratory Decontamination and Destruction of Aflatoxins B_1, B_2, G_1, G_2 in Laboratory Wastes
Edited by M. Castegnaro *et al.*
1980; 56 pages (*out of print*)

No. 38 Directory of On-going Research in Cancer Epidemiology 1981
Edited by C.S. Muir and G. Wagner
1981; 696 pages (*out of print*)

No. 39 Host Factors in Human Carcinogenesis
Edited by H. Bartsch and B. Armstrong
1982; 583 pages (*out of print*)

No. 40 Environmental Carcinogens. Selected Methods of Analysis. Volume 4: Some Aromatic Amines and Azo Dyes in the General and Industrial Environment
Edited by L. Fishbein, M. Castegnaro, I.K. O'Neill and H. Bartsch
1981; 347 pages (*out of print*)

No. 41 N-Nitroso Compounds: Occurrence and Biological Effects
Edited by H. Bartsch, I.K. O'Neill, M. Castegnaro and M. Okada
1982; 755 pages (*out of print*)

No. 42 Cancer Incidence in Five Continents, Volume IV
Edited by J. Waterhouse, C. Muir, K. Shanmugaratnam and J. Powell
1982; 811 pages (*out of print*)

No. 43 Laboratory Decontamination and Destruction of Carcinogens in Laboratory Wastes: Some N-Nitrosamines
Edited by M. Castegnaro *et al.*
1982; 73 pages £7.50

No. 44 Environmental Carcinogens. Selected Methods of Analysis. Volume 5: Some Mycotoxins
Edited by L. Stoloff, M. Castegnaro, P. Scott, I.K. O'Neill and H. Bartsch
1983; 455 pages (*out of print*)

No. 45 Environmental Carcinogens. Selected Methods of Analysis. Volume 6: N-Nitroso Compounds
Edited by R. Preussmann, I.K. O'Neill, G. Eisenbrand, B. Spiegelhalder and H. Bartsch
1983; 508 pages (*out of print*)

No. 46 Directory of On-going Research in Cancer Epidemiology 1982
Edited by C.S. Muir and G. Wagner
1982; 722 pages (*out of print*)

No. 47 Cancer Incidence in Singapore 1968-1977
Edited by K. Shanmugaratnam, H.P. Lee and N.E. Day
1983; 171 pages (*out of print*)

No. 48 Cancer Incidence in the USSR (2nd Revised Edition)
Edited by N.P. Napalkov, G.F. Tserkovny, V.M. Merabishvili, D.M. Parkin, M. Smans and C.S. Muir
1983; 75 pages (*out of print*)

No. 49 Laboratory Decontamination and Destruction of Carcinogens in Laboratory Wastes: Some Polycyclic Aromatic Hydrocarbons
Edited by M. Castegnaro *et al.*
1983; 87 pages (*out of print*)

No. 50 Directory of On-going Research in Cancer Epidemiology 1983
Edited by C.S. Muir and G. Wagner
1983; 731 pages (*out of print*)

No. 51 Modulators of Experimental Carcinogenesis
Edited by V. Turusov and R. Montesano
1983; 307 pages (*out of print*)

List of IARC Publications

No. 52 Second Cancers in Relation to Radiation Treatment for Cervical Cancer: Results of a Cancer Registry Collaboration
Edited by N.E. Day and J.C. Boice, Jr
1984; 207 pages (*out of print*)

No. 53 Nickel in the Human Environment
Editor-in-Chief: F.W. Sunderman, Jr
1984; 529 pages (*out of print*)

No. 54 Laboratory Decontamination and Destruction of Carcinogens in Laboratory Wastes: Some Hydrazines
Edited by M. Castegnaro et al.
1983; 87 pages (*out of print*)

No. 55 Laboratory Decontamination and Destruction of Carcinogens in Laboratory Wastes: Some N-Nitrosamides
Edited by M. Castegnaro et al.
1984; 66 pages (*out of print*)

No. 56 Models, Mechanisms and Etiology of Tumour Promotion
Edited by M. Börzsönyi, N.E. Day, K. Lapis and H. Yamasaki
1984; 532 pages (*out of print*)

No. 57 N-Nitroso Compounds: Occurrence, Biological Effects and Relevance to Human Cancer
Edited by I.K. O'Neill, R.C. von Borstel, C.T. Miller, J. Long and H. Bartsch
1984; 1013 pages (*out of print*)

No. 58 Age-related Factors in Carcinogenesis
Edited by A. Likhachev, V. Anisimov and R. Montesano
1985; 288 pages (*out of print*)

No. 59 Monitoring Human Exposure to Carcinogenic and Mutagenic Agents
Edited by A. Berlin, M. Draper, K. Hemminki and H. Vainio
1984; 457 pages (*out of print*)

No. 60 Burkitt's Lymphoma: A Human Cancer Model
Edited by G. Lenoir, G. O'Conor and C.L.M. Olweny
1985; 484 pages (*out of print*)

No. 61 Laboratory Decontamination and Destruction of Carcinogens in Laboratory Wastes: Some Haloethers
Edited by M. Castegnaro et al.
1985; 55 pages (*out of print*)

No. 62 Directory of On-going Research in Cancer Epidemiology 1984
Edited by C.S. Muir and G. Wagner
1984; 717 pages (*out of print*)

No. 63 Virus-associated Cancers in Africa
Edited by A.O. Williams, G.T. O'Conor, G.B. de-Thé and C.A. Johnson
1984; 773 pages (*out of print*)

No. 64 Laboratory Decontamination and Destruction of Carcinogens in Laboratory Wastes: Some Aromatic Amines and 4-Nitrobiphenyl
Edited by M. Castegnaro et al.
1985; 84 pages (*out of print*)

No. 65 Interpretation of Negative Epidemiological Evidence for Carcinogenicity
Edited by N.J. Wald and R. Doll
1985; 232 pages (*out of print*)

No. 66 The Role of the Registry in Cancer Control
Edited by D.M. Parkin, G. Wagner and C.S. Muir
1985; 152 pages £10.00

No. 67 Transformation Assay of Established Cell Lines: Mechanisms and Application
Edited by T. Kakunaga and H. Yamasaki
1985; 225 pages (*out of print*)

No. 68 Environmental Carcinogens. Selected Methods of Analysis. Volume 7. Some Volatile Halogenated Hydrocarbons
Edited by L. Fishbein and I.K. O'Neill
1985; 479 pages (*out of print*)

No. 69 Directory of On-going Research in Cancer Epidemiology 1985
Edited by C.S. Muir and G. Wagner
1985; 745 pages (*out of print*)

No. 70 The Role of Cyclic Nucleic Acid Adducts in Carcinogenesis and Mutagenesis
Edited by B. Singer and H. Bartsch
1986; 467 pages (*out of print*)

No. 71 Environmental Carcinogens. Selected Methods of Analysis. Volume 8: Some Metals: As, Be, Cd, Cr, Ni, Pb, Se, Zn
Edited by I.K. O'Neill, P. Schuller and L. Fishbein
1986; 485 pages (*out of print*)

No. 72 Atlas of Cancer in Scotland, 1975–1980. Incidence and Epidemiological Perspective
Edited by I. Kemp, P. Boyle, M. Smans and C.S. Muir
1985; 285 pages (*out of print*)

No. 73 Laboratory Decontamination and Destruction of Carcinogens in Laboratory Wastes: Some Antineoplastic Agents
Edited by M. Castegnaro et al.
1985; 163 pages £13.50

No. 74 Tobacco: A Major International Health Hazard
Edited by D. Zaridze and R. Peto
1986; 324 pages £24.00

No. 75 Cancer Occurrence in Developing Countries
Edited by D.M. Parkin
1986; 339 pages £24.00

No. 76 Screening for Cancer of the Uterine Cervix
Edited by M. Hakama, A.B. Miller and N.E. Day
1986; 315 pages £31.50

No. 77 Hexachlorobenzene: Proceedings of an International Symposium
Edited by C.R. Morris and J.R.P. Cabral
1986; 668 pages (*out of print*)

No. 78 Carcinogenicity of Alkylating Cytostatic Drugs
Edited by D. Schmähl and J.M. Kaldor
1986; 337 pages (*out of print*)

No. 79 Statistical Methods in Cancer Research. Volume III: The Design and Analysis of Long-term Animal Experiments
By J.J. Gart, D. Krewski, P.N. Lee, R.E. Tarone and J. Wahrendorf
1986; 213 pages £23.50

List of IARC Publications

No. 80 Directory of On-going Research in Cancer Epidemiology 1986
Edited by C.S. Muir and G. Wagner
1986; 805 pages (*out of print*)

No. 81 Environmental Carcinogens: Methods of Analysis and Exposure Measurement. Volume 9: Passive Smoking
Edited by I.K. O'Neill, K.D. Brunnemann, B. Dodet and D. Hoffmann
1987; 383 pages £37.00

No. 82 Statistical Methods in Cancer Research. Volume II: The Design and Analysis of Cohort Studies
By N.E. Breslow and N.E. Day
1987; 404 pages £25.00

No. 83 Long-term and Short-term Assays for Carcinogens: A Critical Appraisal
Edited by R. Montesano, H. Bartsch, H. Vainio, J. Wilbourn and H. Yamasaki
1986; 575 pages £37.00

No. 84 The Relevance of N-Nitroso Compounds to Human Cancer: Exposure and Mechanisms
Edited by H. Bartsch, I.K. O'Neill and R. Schulte-Hermann
1987; 671 pages (*out of print*)

No. 85 Environmental Carcinogens: Methods of Analysis and Exposure Measurement. Volume 10: Benzene and Alkylated Benzenes
Edited by L. Fishbein and I.K. O'Neill
1988; 327 pages £42.00

No. 86 Directory of On-going Research in Cancer Epidemiology 1987
Edited by D.M. Parkin and J. Wahrendorf
1987; 676 pages (*out of print*)

No. 87 International Incidence of Childhood Cancer
Edited by D.M. Parkin, C.A. Stiller, C.A. Bieber, G.J. Draper, B. Terracini and J.L. Young
1988; 401 pages £35.00

No. 88 Cancer Incidence in Five Continents Volume V
Edited by C. Muir, J. Waterhouse, T. Mack, J. Powell and S. Whelan
1987; 1004 pages £58.00

No. 89 Method for Detecting DNA Damaging Agents in Humans: Applications in Cancer Epidemiology and Prevention
Edited by H. Bartsch, K. Hemminki and I.K. O'Neill
1988; 518 pages £50.00

No. 90 Non-occupational Exposure to Mineral Fibres
Edited by J. Bignon, J. Peto and R. Saracci
1989; 500 pages £52.50

No. 91 Trends in Cancer Incidence in Singapore 1968–1982
Edited by H.P. Lee, N.E. Day and K. Shanmugaratnam
1988; 160 pages (*out of print*)

No. 92 Cell Differentiation, Genes and Cancer
Edited by T. Kakunaga, T. Sugimura, L. Tomatis and H. Yamasaki
1988; 204 pages £29.00

No. 93 Directory of On-going Research in Cancer Epidemiology 1988
Edited by M. Coleman and J. Wahrendorf
1988; 662 pages (*out of print*)

No. 94 Human Papillomavirus and Cervical Cancer
Edited by N. Muñoz, F.X. Bosch and O.M. Jensen
1989; 154 pages £22.50

No. 95 Cancer Registration: Principles and Methods
Edited by O.M. Jensen, D.M. Parkin, R. MacLennan, C.S. Muir and R. Skeet
1991; 288 pages £28.00

No. 96 Perinatal and Multigeneration Carcinogenesis
Edited by N.P. Napalkov, J.M. Rice, L. Tomatis and H. Yamasaki
1989; 436 pages £52.50

No. 97 Occupational Exposure to Silica and Cancer Risk
Edited by L. Simonato, A.C. Fletcher, R. Saracci and T. Thomas
1990; 124 pages £24.00

No. 98 Cancer Incidence in Jewish Migrants to Israel, 1961–1981
Edited by R. Steinitz, D.M. Parkin, J.L. Young, C.A. Bieber and L. Katz
1989; 320 pages £37.00

No. 99 Pathology of Tumours in Laboratory Animals, Second Edition, Volume 1, Tumours of the Rat
Edited by V.S. Turusov and U. Mohr
740 pages £90.00

No. 100 Cancer: Causes, Occurrence and Control
Editor-in-Chief L. Tomatis
1990; 352 pages £25.50

No. 101 Directory of On-going Research in Cancer Epidemiology 1989/90
Edited by M. Coleman and J. Wahrendorf
1989; 818 pages £42.00

No. 102 Patterns of Cancer in Five Continents
Edited by S.L. Whelan, D.M. Parkin & E. Masuyer
1990; 162 pages £26.50

No. 103 Evaluating Effectiveness of Primary Prevention of Cancer
Edited by M. Hakama, V. Beral, J.W. Cullen and D.M. Parkin
1990; 250 pages £34.00

No. 104 Complex Mixtures and Cancer Risk
Edited by H. Vainio, M. Sorsa and A.J. McMichael
1990; 442 pages £40.00

No. 105 Relevance to Human Cancer of N-Nitroso Compounds, Tobacco Smoke and Mycotoxins
Edited by I.K. O'Neill, J. Chen and H. Bartsch
1991; 614 pages £74.00

No. 106 Atlas of Cancer Incidence in the Former German Democratic Republic
Edited by W.H. Mehnert, M. Smans, C.S. Muir, M. Möhner & D. Schön
1992; 384 pages £52.50

List of IARC Publications

No. 107 Atlas of Cancer Mortality in the European Economic Community
Edited by M. Smans, C.S. Muir and P. Boyle
1992; 280 pages £35.00

No. 108 Environmental Carcinogens: Methods of Analysis and Exposure Measurement. Volume 11: Polychlorinated Dioxins and Dibenzofurans
Edited by C. Rappe, H.R. Buser, B. Dodet and I.K. O'Neill
1991; 426 pages £47.50

No. 109 Environmental Carcinogens: Methods of Analysis and Exposure Measurement. Volume 12: Indoor Air Contaminants
Edited by B. Seifert, H. van de Wiel, B. Dodet and I.K. O'Neill
1993; 384 pages £45.00

No. 110 Directory of On-going Research in Cancer Epidemiology 1991
Edited by M. Coleman and J. Wahrendorf
1991; 753 pages £40.00

No. 111 Pathology of Tumours in Laboratory Animals, Second Edition, Volume 2, Tumours of the Mouse
Edited by V.S. Turusov and U. Mohr
1993; 776 pages; £90.00

No. 112 Autopsy in Epidemiology and Medical Research
Edited by E. Riboli and M. Delendi
1991; 288 pages £26.50

No. 113 Laboratory Decontamination and Destruction of Carcinogens in Laboratory Wastes: Some Mycotoxins
Edited by M. Castegnaro, J. Barek, J.-M. Frémy, M. Lafontaine, M. Miraglia, E.B. Sansone and G.M. Telling
1991; 64 pages £12.00

No. 114 Laboratory Decontamination and Destruction of Carcinogens in Laboratory Wastes: Some Polycyclic Heterocyclic Hydrocarbons
Edited by M. Castegnaro, J. Barek J. Jacob, U. Kirso, M. Lafontaine, E.B. Sansone, G.M. Telling and T. Vu Duc
1991; 50 pages £8.00

No. 115 Mycotoxins, Endemic Nephropathy and Urinary Tract Tumours
Edited by M. Castegnaro, R. Plestina, G. Dirheimer, I.N. Chernozemsky and H Bartsch
1991; 340 pages £47.50

No. 116 Mechanisms of Carcinogenesis in Risk Identification
Edited by H. Vainio, P.N. Magee, D.B. McGregor & A.J. McMichael
1992; 616 pages £69.00

No. 117 Directory of On-going Research in Cancer Epidemiology 1992
Edited by M. Coleman, J. Wahrendorf & E. Démaret
1992; 773 pages £44.50

No. 118 Cadmium in the Human Environment: Toxicity and Carcinogenicity
Edited by G.F. Nordberg, R.F.M. Herber & L. Alessio
1992; 470 pages £60.00

No. 119 The Epidemiology of Cervical Cancer and Human Papillomavirus
Edited by N. Muñoz, F.X. Bosch, K.V. Shah & A. Meheus
1992; 288 pages £29.50

No. 120 Cancer Incidence in Five Continents, Volume VI
Edited by D.M. Parkin, C.S. Muir, S.L. Whelan, Y.T. Gao, J. Ferlay & J. Powell
1992; 1080 pages £120.00

No. 121 Trends in Cancer Incidence and Mortality
M.P. Coleman, J. Estève, P. Damiecki, A. Arslan and H. Renard
1993; 806 pages, £120.00

No. 122 International Classification of Rodent Tumours. Part 1. The Rat
Editor-in-Chief: U. Mohr
1992/95; 10 fascicles of 60–100 pages, £120.00

No. 123 Cancer in Italian Migrant Populations
Edited by M. Geddes, D.M. Parkin, M. Khlat, D. Balzi and E. Buiatti
1993; 292 pages, £40.00

No. 124 Postlabelling Methods for Detection of DNA Adducts
Edited by D.H. Phillips, M. Castegnaro and H. Bartsch
1993; 392 pages, £46.00

No. 125 DNA Adducts: Identification and Biological Significance
Edited by K. Hemminki, A. Dipple, D. Shuker, F.F. Kadlubar, D. Segerbäck and H. Bartsch
1994; 480 pages; £52.00

No. 127 Butadiene and Styrene: Assessment of Health Hazards
Edited by M. Sorsa, K. Peltonen, H. Vainio and K. Hemminki
1993; 412 pages; £54.00

No. 128 Statistical Methods in Cancer Research. Volume IV. Descriptive Epidemiology
By J. Estève, E. Benhamou & L. Raymond
1994; 302 pages; £25.00

No. 129 Occupational Cancer in Developing Countries
Edited by N. Pearce, E. Matos, H. Vainio, P. Boffetta & M. Kogevinas
1994; 192 pages £20.00

No. 130 Directory of On-going Research in Cancer Epidemiology 1994
Edited by R. Sankaranarayanan, J. Wahrendorf and E. Démaret
1994; 792 pages, £46.00

No. 132 Survival of Cancer Patients in Europe. The EUROCARE Study
Edited by F. Berrino, M. Sant, A. Verdecchia, R. Capocaccia, T. Hakulinen and J. Estève
1994; 463 pages; £45.00

List of IARC Publications

IARC MONOGRAPHS ON THE EVALUATION OF CARCINOGENIC RISKS TO HUMANS

(Available from booksellers through the network of WHO Sales Agents)

Volume 1 Some Inorganic Substances, Chlorinated Hydrocarbons, Aromatic Amines, *N*-Nitroso Compounds, and Natural Products
1972; 184 pages (*out of print*)

Volume 2 Some Inorganic and Organometallic Compounds
1973; 181 pages (*out of print*)

Volume 3 Certain Polycyclic Aromatic Hydrocarbons and Heterocyclic Compounds
1973; 271 pages (*out of print*)

Volume 4 Some Aromatic Amines, Hydrazine and Related Substances, *N*-Nitroso Compounds and Miscellaneous Alkylating Agents
1974; 286 pages Sw. fr. 18.–

Volume 5 Some Organochlorine Pesticides
1974; 241 pages (*out of print*)

Volume 6 Sex Hormones
1974; 243 pages (*out of print*)

Volume 7 Some Anti-Thyroid and Related Substances, Nitrofurans and Industrial Chemicals
1974; 326 pages (*out of print*)

Volume 8 Some Aromatic Azo Compounds
1975; 357 pages Sw. fr. 44.–

Volume 9 Some Aziridines, *N*-, *S*- and *O*-Mustards and Selenium
1975; 268 pages Sw.fr. 33.–

Volume 10 Some Naturally Occurring Substances
1976; 353 pages (*out of print*)

Volume 11 Cadmium, Nickel, Some Epoxides, Miscellaneous Industrial Chemicals and General Considerations on Volatile Anaesthetics
1976; 306 pages (*out of print*)

Volume 12 Some Carbamates, Thiocarbamates and Carbazides
1976; 282 pages Sw. fr. 41.-

Volume 13 Some Miscellaneous Pharmaceutical Substances
1977; 255 pages Sw. fr. 36.–

Volume 14 Asbestos
1977; 106 pages (*out of print*)

Volume 15 Some Fumigants, The Herbicides 2,4-D and 2,4,5-T, Chlorinated Dibenzodioxins and Miscellaneous Industrial Chemicals
1977; 354 pages (*out of print*)

Volume 16 Some Aromatic Amines and Related Nitro Compounds - Hair Dyes, Colouring Agents and Miscellaneous Industrial Chemicals
1978; 400 pages Sw. fr. 60.–

Volume 17 Some *N*-Nitroso Compounds
1978; 365 pages Sw. fr. 60.–

Volume 18 Polychlorinated Biphenyls and Polybrominated Biphenyls
1978; 140 pages Sw. fr. 24.–

Volume 19 Some Monomers, Plastics and Synthetic Elastomers, and Acrolein
1979; 513 pages (*out of print*)

Volume 20 Some Halogenated Hydrocarbons
1979; 609 pages (*out of print*)

Volume 21 Sex Hormones (II)
1979; 583 pages Sw. fr. 72.–

Volume 22 Some Non-Nutritive Sweetening Agents
1980; 208 pages Sw. fr. 30.–

Volume 23 Some Metals and Metallic Compounds
1980; 438 pages (*out of print*)

Volume 24 Some Pharmaceutical Drugs
1980; 337 pages Sw. fr. 48.–

Volume 25 Wood, Leather and Some Associated Industries
1981; 412 pages Sw. fr. 72.–

Volume 26 Some Antineoplastic and Immunosuppressive Agents
1981; 411 pages Sw. fr. 75.–

Volume 27 Some Aromatic Amines, Anthraquinones and Nitroso Compounds, and Inorganic Fluorides Used in Drinking Water and Dental Preparations
1982; 341 pages Sw. fr. 48.–

Volume 28 The Rubber Industry
1982; 486 pages Sw. fr. 84.–

Volume 29 Some Industrial Chemicals and Dyestuffs
1982; 416 pages Sw. fr. 72.–

Volume 30 Miscellaneous Pesticides
1983; 424 pages Sw. fr. 72.–

Volume 31 Some Food Additives, Feed Additives and Naturally Occurring Substances
1983; 314 pages Sw. fr. 66.–

Volume 32 Polynuclear Aromatic Compounds, Part 1: Chemical, Environmental and Experimental Data
1983; 477 pages Sw. fr. 88.–

Volume 33 Polynuclear Aromatic Compounds, Part 2: Carbon Blacks, Mineral Oils and Some Nitroarenes
1984; 245 pages (*out of print*)

Volume 34 Polynuclear Aromatic Compounds, Part 3: Industrial Exposures in Aluminium Production, Coal Gasification, Coke Production, and Iron and Steel Founding
1984; 219 pages Sw. fr. 53.–

Volume 35 Polynuclear Aromatic Compounds, Part 4: Bitumens, Coal-tars and Derived Products, Shale-oils and Soots
1985; 271 pages Sw. fr. 77.–

List of IARC Publications

Volume 36 Allyl Compounds, Aldehydes, Epoxides and Peroxides
1985; 369 pages Sw. fr. 77.–

Volume 37 Tobacco Habits Other than Smoking: Betel-quid and Areca-nut Chewing; and some Related Nitrosamines
1985; 291 pages Sw. fr. 77.–

Volume 38 Tobacco Smoking
1986; 421 pages Sw. fr. 83.–

Volume 39 Some Chemicals Used in Plastics and Elastomers
1986; 403 pages Sw. fr. 83.–

Volume 40 Some Naturally Occurring and Synthetic Food Components, Furocoumarins and Ultraviolet Radiation
1986; 444 pages Sw. fr. 83.–

Volume 41 Some Halogenated Hydrocarbons and Pesticide Exposures
1986; 434 pages Sw. fr. 83.–

Volume 42 Silica and Some Silicates
1987; 289 pages Sw. fr. 72.–

Volume 43 Man-Made Mineral Fibres and Radon
1988; 300 pages Sw. fr. 72.–

Volume 44 Alcohol Drinking
1988; 416 pages Sw. fr. 83.–

Volume 45 Occupational Exposures in Petroleum Refining; Crude Oil and Major Petroleum Fuels
1989; 322 pages Sw. fr. 72.–

Volume 46 Diesel and Gasoline Engine Exhausts and Some Nitroarenes
1989; 458 pages Sw. fr. 83.–

Volume 47 Some Organic Solvents, Resin Monomers and Related Compounds, Pigments and Occupational Exposures in Paint Manufacture and Painting
1989; 535 pages Sw. fr. 94.–

Volume 48 Some Flame Retardants and Textile Chemicals, and Exposures in the Textile Manufacturing Industry
1990; 345 pages Sw. fr. 72.–

Volume 49 Chromium, Nickel and Welding
1990; 677 pages Sw. fr. 105.–

Volume 50 Pharmaceutical Drugs
1990; 415 pages Sw. fr. 93.–

Volume 51 Coffee, Tea, Mate, Methylxanthines and Methylglyoxal
1991; 513 pages Sw. fr. 88.–

Volume 52 Chlorinated Drinking-water; Chlorination By-products; Some Other Halogenated Compounds; Cobalt and Cobalt Compounds
1991; 544 pages Sw. fr. 88.–

Volume 53 Occupational Exposures in Insecticide Application and some Pesticides
1991; 612 pages Sw. fr. 105.–

Volume 54 Occupational Exposures to Mists and Vapours from Strong Inorganic Acids; and Other Industrial Chemicals
1992; 336 pages Sw. fr. 72.–

Volume 55 Solar and Ultraviolet Radiation
1992; 316 pages Sw. fr. 65.–

Volume 56 Some Naturally Occurring Substances: Food Items and Constituents, Heterocyclic Aromatic Amines and Mycotoxins
1993; 600 pages Sw. fr. 95.–

Volume 57 Occupational Exposures of Hairdressers and Barbers and Personal Use of Hair Colourants; Some Hair Dyes, Cosmetic Colourants, Industrial Dyestuffs and Aromatic Amines
1993; 428 pages Sw. fr. 75.–

Volume 58 Beryllium, Cadmium, Mercury and Exposures in the Glass Manufacturing Industry
1993; 426 pages Sw. fr. 75.–

Volume 59 Hepatitis Viruses
1994; 286 pages Sw. fr. 65.–

Volume 60 Some Industrial Chemicals
1994; 560 pages Sw. fr. 90.–

Volume 61 Schistosomes, Liver Flukes and Helicobacter pylori
1994; 270 pages Sw. fr. 70.–

Volume 62 Wood Dust and Formaldehyde
1995; 406 pages Sw. fr. 80.–

Supplement No. 1
Chemicals and Industrial Processes Associated with Cancer in Humans (IARC Monographs, Volumes 1 to 20)
1979; 71 pages (*out of print*)

Supplement No. 2
Long-term and Short-term Screening Assays for Carcinogens: A Critical Appraisal
1980; 426 pages Sw. fr. 40.–

Supplement No. 3
Cross Index of Synonyms and Trade Names in Volumes 1 to 26
1982; 199 pages (*out of print*)

Supplement No. 4
Chemicals, Industrial Processes and Industries Associated with Cancer in Humans (IARC Monographs, Volumes 1 to 29)
1982; 292 pages (*out of print*)

Supplement No. 5
Cross Index of Synonyms and Trade Names in Volumes 1 to 36
1985; 259 pages (*out of print*)

Supplement No. 6
Genetic and Related Effects: An Updating of Selected IARC Monographs from Volumes 1 to 42
1987; 729 pages Sw. fr. 80.–

Supplement No. 7
Overall Evaluations of Carcinogenicity: An Updating of IARC Monographs Volumes 1-42
1987; 440 pages Sw. fr. 65.–

Supplement No. 8
Cross Index of Synonyms and Trade Names in Volumes 1 to 46
1990; 346 pages Sw. fr. 60.–

List of IARC Publications

IARC TECHNICAL REPORTS*

No. 1 Cancer in Costa Rica
Edited by R. Sierra,
R. Barrantes, G. Muñoz Leiva, D.M. Parkin, C.A. Bieber and
N. Muñoz Calero
1988; 124 pages Sw. fr. 30.-

No. 2 SEARCH: A Computer Package to Assist the Statistical Analysis of Case-control Studies
Edited by G.J. Macfarlane,
P. Boyle and P. Maisonneuve
1991; 80 pages (*out of print*)

No. 3 Cancer Registration in the European Economic Community
Edited by M.P. Coleman and
E. Démaret
1988; 188 pages Sw. fr. 30.-

No. 4 Diet, Hormones and Cancer: Methodological Issues for Prospective Studies
Edited by E. Riboli and
R. Saracci
1988; 156 pages Sw. fr. 30.-

No. 5 Cancer in the Philippines
Edited by A.V. Laudico,
D. Esteban and D.M. Parkin
1989; 186 pages Sw. fr. 30.-

No. 6 La genèse du Centre International de Recherche sur le Cancer
Par R. Sohier et A.G.B. Sutherland
1990; 104 pages Sw. fr. 30.-

No. 7 Epidémiologie du cancer dans les pays de langue latine
1990; 310 pages Sw. fr. 30.-

No. 8 Comparative Study of Antismoking Legislation in Countries of the European Economic Community
Edited by A. Sasco, P. Dalla Vorgia and P. Van der Elst
1992; 82 pages Sw. fr. 30.-

No. 9 Epidemiologie du cancer dans les pays de langue latine
991 346 pages Sw. fr. 30.-

No. 11 Nitroso Compounds: Biological Mechanisms, Exposures and Cancer Etiology
Edited by I.K. O'Neill & H. Bartsch
1992; 149 pages Sw. fr. 30.-

No. 12 Epidémiologie du cancer dans les pays de langue latine
1992; 375 pages Sw. fr. 30.-

No. 13 Health, Solar UV Radiation and Environmental Change
By A. Kricker, B.K. Armstrong, M.E. Jones and R.C. Burton
1993; 216 pages Sw.fr. 30.–

No. 14 Epidémiologie du cancer dans les pays de langue latine
1993; 385 pages Sw. fr. 30.-

No. 15 Cancer in the African Population of Bulawayo, Zimbabwe, 1963–1977: Incidence, Time Trends and Risk Factors
By M.E.G. Skinner, D.M. Parkin, A.P. Vizcaino and A. Ndhlovu
1993; 123 pages Sw. fr. 30.-

No. 16 Cancer in Thailand, 1988–1991
By V. Vatanasapt, N. Martin, H. Sriplung, K. Vindavijak, S. Sontipong, S. Sriamporn, D.M. Parkin and J. Ferlay
1993; 164 pages Sw. fr. 30.-

No. 18 Intervention Trials for Cancer Prevention
By E. Buiatti
1994; 52 pages Sw. fr. 30.-

No. 19 Comparability and Quality Control in Cancer Registration
By D.M. Parkin, V.W. Chen, J. Ferlay, J. Galceran, H.H. Storm and S.L. Whelan
1994; 110 pages plus diskette
Sw. fr. 40.-

No. 20 Epidémiologie du cancer dans les pays de langue latine
1994; 346 pages Sw. fr. 30.-

No. 21 ICD Conversion Programs for Cancer
By J. Ferlay
1994; 24 pages plus diskette
Sw. fr. 30.-

No. 22 Cancer Incidence by Occupation and Industry in Tianjin, China, 1981–1987
By Q.S. Wang, P. Boffetta, M. Kogevinas and D.M. Parkin
1994; 96 pages Sw. fr. 30.–

DIRECTORY OF AGENTS BEING TESTED FOR CARCINOGENICITY (Until Vol. 13 Information Bulletin on the Survey of Chemicals Being Tested for Carcinogenicity)*

No. 8 Edited by M.-J. Ghess,
H. Bartsch and L. Tomatis
1979; 604 pages Sw. fr. 40.-

No. 9 Edited by M.-J. Ghess,
J.D. Wilbourn, H. Bartsch and
L. Tomatis
1981; 294 pages Sw. fr. 41.-

No. 10 Edited by M.-J. Ghess,
J.D. Wilbourn and H. Bartsch
1982; 362 pages Sw. fr. 42.-

No. 11 Edited by M.-J. Ghess,
J.D. Wilbourn, H. Vainio and
H. Bartsch
1984; 362 pages Sw. fr. 50.-

No. 12 Edited by M.-J. Ghess,
J.D. Wilbourn, A. Tossavainen and H. Vainio
1986; 385 pages Sw. fr. 50.-

No. 13 Edited by M.-J. Ghess,
J.D. Wilbourn and A. Aitio 1988; 404 pages Sw. fr. 43.-

No. 14 Edited by M.-J. Ghess,
J.D. Wilbourn and H. Vainio
1990; 370 pages Sw. fr. 45.-

No. 15 Edited by M.-J. Ghess, J.D. Wilbourn and H. Vainio
1992; 318 pages Sw. fr. 45.-

No. 16 Edited by M.-J. Ghess, J.D. Wilbourn and H. Vainio
1994; 294 pages Sw. fr. 50.-

NON-SERIAL PUBLICATIONS

Alcool et Cancer†
By A. Tuyns (in French only)
1978; 42 pages Fr. fr. 35.-

Cancer Morbidity and Causes of Death Among Danish Brewery Workers†
By O.M. Jensen
1980; 143 pages Fr. fr. 75.-

Directory of Computer Systems Used in Cancer Registries†
By H.R. Menck and D.M. Parkin
1986; 236 pages Fr. fr. 50.-

Facts and Figures of Cancer in the European Community*
Edited by J. Estève, A. Kricker, J. Ferlay and D.M. Parkin
1993; 52 pages Sw. fr. 10.-

* Available from booksellers through the network of WHO Sales agents.   † Available directly from IARC.

IARC Monographs are distributed
by the
World Health Organization,
Distribution and Sales Service,
1211 Geneva 27, Switzerland
and are available from booksellers
through the network of WHO Sales Agents.

A list of these Agents may be obtained
by writing to the above address.

www.ingramcontent.com/pod-product-compliance
Ingram Content Group UK Ltd.
Pitfield, Milton Keynes, MK11 3LW, UK
UKHW051258180426
11947UKWH00020B/1779